THE INTRIGUING LIFE OF MASSIVE GALAXIES

IAU SYMPOSIUM No. 295

COVER ILLUSTRATION: CONFERENCE POSTER

The cover picture is the official conference poster of the IAUS295. It shows a reproduction of the Great Wall in China, meant to symbolise the path of life of galaxies. The background image is a picture of the elliptical galaxy M87 (Courtesy NASA/ESA and the Hubble Heritage Team). The movie cut outs show galaxy mergers that form key events in the evolution of galaxies.

The picture illustrates the aim of the symposium to discuss the lives of massive galaxies in the real time direction from the highest redshifts to the local Universe both from an observational and theoretical perspective.

The life of a massive galaxy is like walking along the Great Wall. It is an exhausting up and down. You get squeezed and squashed in the beginning just like in a massive galaxy's early life, which is full of mergers and violent star formation. But suddenly, after only a few billion years this phase is over, and the massive galaxy begins to evolve passively without further major perturbations. Just like on the Great Wall, when the dense crowd of tourists thins out the more you walk along.

IAU SYMPOSIUM PROCEEDINGS SERIES

Chief Editor

THIERRY MONTMERLE, IAU General Secretary
Institut d'Astrophysique de Paris,
98bis, Bd Arago, 75014 Paris, France
montmerle@iap.fr

Editor

PIERO BENVENUTI, IAU Assistant General Secretary
University of Padua, Dept of Physics and Astronomy,
Vicolo dell'Osservatorio, 3, 35122 Padova, Italy
piero.benvenuti@unipd.it

INTERNATIONAL ASTRONOMICAL UNION

UNION ASTRONOMIQUE INTERNATIONALE

THE INTRIGUING LIFE OF MASSIVE GALAXIES

PROCEEDINGS OF THE 295th SYMPOSIUM
OF THE INTERNATIONAL ASTRONOMICAL
UNION HELD IN BEIJING, CHINA
AUGUST 27–31, 2012

Edited by

DANIEL THOMAS
University of Portsmouth, UK

ANNA PASQUALI
University of Heidelberg, GERMANY

and

IGNACIO FERRERAS
University College London, UK

CAMBRIDGE UNIVERSITY PRESS

CAMBRIDGE UNIVERSITY PRESS
The Edinburgh Building, Cambridge CB2 2RU, United Kingdom
32 Avenue of the Americas, New York, NY 10013 2473, USA
10 Stamford Road, Oakleigh, Melbourne 3166, Australia

First published 2013

Printed in the UK by Bell & Bain

Typeset in System LaTeX 2_ε

A catalogue record for this book is available from the British Library

Library of Congress Cataloguing in Publication data

ISBN 9781107033849 hardback
ISSN 1743-9213

Table of Contents

The first galaxies in the very early universe

The first few billion years

Evolution of massive galaxies in the second half

Massive galaxies today

Future prospects and final discussion

Preface

Massive galaxies live an exciting and eventful life. Most of them are "dead" by today and their morphology is typical of early-type galaxies. But they might have looked very different in the past. While their predecessors in the very early Universe might well have been small, they must have soon become massive, vigorously star forming objects. Just shortly after this violent phase, possibly triggered by galaxy mergers, star formation in massive galaxies got quenched, followed by a long phase of passive evolution. Following their genealogical/merger tree, they get quenched and rejuvenated, they get strangulated, they starve, they cannibalise their smaller neighbours and merge with their peers. Massive galaxies are responsible for most of the chemical enrichment in the Universe, and many eventually end their lives clustered together. The most amazing fact about massive galaxies is that they constitute, today, a surprisingly homogeneous class of objects. Today's massive galaxies are almost featureless with elliptical morphology, they have little or no gas, and show no signs of significant star formation activity, while their cores are often characterised by surprisingly complex kinematics.

We know little about the significance of transitions in the formation and evolution of massive galaxies. Clearly, star formation activity needs to be quenched somewhere along the evolutionary path of a galaxy and its progenitors. But what is the physical mechanism of this quenching? What are the relative roles of feedback from supernovae and super-massive black holes? Where does the environment and galaxy mergers come in? Does cold accretion at high redshift solve the problem by boosting the formation of stars in massive galaxies? Each of these scenarios have their successes and pitfalls in shaping massive galaxies along their travels through cosmic time. Do we need a combination of these? Or remains the true mechanism yet to be discovered?

The observational key lies in detailed studies of both the fossil record in the population of today's massive galaxies and galaxy properties as a function of redshift. However, the further we get back in time with our observations, the more we have to worry about the link to the present galaxy population, a problem known as the progenitor bias. In a cold dark matter dominated Universe we expect galaxies to evolve along a merger tree throughout their life, that fans out as soon as we go back in time. We coin terms for galaxy populations mostly by the selection criteria we use to observe them at any given epoch. We still do not know how well EROs, DRGs, or heavily star forming galaxies such as SCUBA sources qualify as the progenitors of today's massive galaxies. The evolution of which objects do we actually trace when we climb up the redshift ladder? What galaxies do we need to pick if we want to trace the life of a massive galaxy? When did today's massive galaxies assemble? The major challenge for observations of galaxies at high redshift is exactly to establish this link to the local galaxy population that we ultimately want to understand.

If we want to understand the intriguing life of massive galaxies, we need to map their entire evolution over cosmic time, and this requires very different observational and theoretical approaches. However, the links between the various research groups that study different evolutionary stages of massive galaxies at different redshifts are loose. After the overwhelming progress in observational and theoretical studies of galaxy evolution over the past decade, the time was ripe to bring these communities together, both theorists and observers. Recent and near-future advances in telescope technology and computer power for large-scale simulations, as well as the launch of massive galaxy surveys will lead to a further leap in our understanding of galaxy formation. Revolutionary new observatories such as the James Webb Space Telescope or the next generation of ground-based Extremely Large Telescopes are within reach, and it was exciting to discuss both theoretical predictions and expected advances in observational astronomy.

The IAUS 295 has brought together observers and theorists to discuss recent progress in the field and to plan ahead for future challenges. The symposium covered the life of massive galaxies from the formation of the first galaxies in the early Universe, through their evolution with redshift to massive galaxies in the local Universe touching upon all kinds of issues relates to the life of massive galaxies including gas accretion and star formation, feedback and quenching, black hole growth, mass assembly, galaxy mergers and interactions, chemical enrichment and stellar populations, dark matter, galaxy environment, galaxy haloes, and satellite accretion both from a theoretical and observational perspective.

We believe that the excellent presentations and lively discussions at the symposium have shed light on the intriguing life of massive galaxies. We hope that these proceedings will provide a useful summary of the many topics discussed at the meeting. We are very grateful to all participants for their contributions, in particular to those who have contributed to this book.

Daniel Thomas, Anna Pasquali, and Ignacio Ferreras, co-chairs SOC,
Portsmouth (UK), Heidelberg (Germany), London (UK), April 18, 2013

THE ORGANIZING COMMITTEE

Scientific

Roger Davies (UK)
Avishai Dekel (Israel)
Richard Ellis (USA)
Ignacio Ferreras (co-chair, UK)
Yipeng Jing (China Nanjing)
Xu Kong (China Nanjing)
Shude Mao (China Nanjing)
Anna Pasquali (co-chair, Germany)

Eric Peng (China Nanjing)
Alvio Renzini (Italy)
Rachel Somerville (USA)
Ian Smail (UK)
Linda Tacconi (Germany)
Daniel Thomas (co-chair, UK)
Christy Tremonti (USA)
XianZhong Zheng (China Nanjing)

Local

Jun Yan (Co-Chair)
Gang Zhao (Co-Chair)
Yue Chen
Chenzhou Cui
Frankie Gao
Qili Hei
Hairong Jing
Haining Li
Yanchun Liang
Nancy Liu
Yujuan Liu
Ye Lu

Michael Podt
Raymond Qian
Yi Wang
Haoyi Wan
Xiaochun Sun
Ang Xu
Suijian Xue
Deting Yang
Xiaoshan Yun
Isabella Zhang
Bing Zhao
Jin Zhu

Acknowledgements

The symposium is sponsored and supported by the IAU Division VIII (Galaxies & the Universe) and by the IAU Commission No. 28 (Galaxies).

Funding by the
Royal Astronomical Society,
International Astronomical Union,
and
Science, Technology and Facilities Council (UK)
is gratefully acknowledged.

An attentive audience at the IAUS 295.

Oleg Gnedin and Simon Lilly.

Bianca Poggianti.

Simon Driver and Oleg Gnedin.

John Kormendy during his plenary talk on black holes.

John Kormendy and Daniel Thomas.

Participants

Paula **Aguirre**, PUC/UNAB, Chile — aguirre.paula@gmail.com
Philippe **Amram**, Laboratoire d'Astrophysique de Marseille, France — Philippe.Amram@oamp.fr
Jacob **Arnold**, UCSC, United States — jacob@ucolick.org
Zhongrui **Bai**, NAOC, China — zhrbai@gmail.com
Guillermo **Barro**, University of California Santa Cruz, United States — gbarro@ucolick.org
Carlton **Baugh**, Durham University, United Kingdom — c.m.baugh@durham.ac.uk
Richard **Beare**, Monash Centre for Astrophysics, Australia — beares@beares.net
Rachel **Bezanson**, Yale University, United States — rachel.bezanson@yale.edu
Paolo **Bonfini**, University of Crete, Greece — paolo@physics.uoc.gr
Rychard **Bouwens**, Leiden Observatory, Netherlands — bouwens@strw.leidenuniv.nl
Rebecca **Bowler**, Institute for Astronomy, University of Edinburgh, United Kingdom — raab@roe.ac.uk
Gabriel **Brammer**, European Southern Observatory, Chile — gbrammer@gmail.com
Volker **Bromm**, University of Texas at Austin, United States — vbromm@astro.as.utexas.edu
Sarah **Brough**, AAO, Australia — sb@aao.gov.au
Michael **Brown**, Monash University, Australia — michael.brown@monash.edu
Victoria **Bruce**, Institute for Astronomy, University of Edinburgh, United Kingdom — vab@roe.ac.uk
Gustavo **Bruzual**, CRyA, Morelia, Mexico Mexico — g.bruzual@crya.unam.mx
Fernando **Buitrago**, University of Edinburgh, United Kingdom — fb@roe.ac.uk
Martin **Bureau**, University of Oxford, United Kingdom — bureau@astro.ox.ac.uk
Claire **Burke**, Liverpool John Moores University, United Kingdom — cb@astro.livjm.ac.uk
Diego **Capozzi**, ICG, University of Portsmouth, United Kingdom — diego.capozzi@port.ac.uk
Michele **Cappellari**, University of Oxford, United Kingdom — cappellari@astro.ox.ac.uk
Evelyn **Caris**, Swinburne, University Australia — ecaris@astro.swin.edu.au
Marcio **Catelan**, Pontificia Universidad Catolica de Chile, Chile — mcatelan@astro.puc.cl
Chin-Wei **Chen**, Academia Sinica Taiwan, China — chinwei.chen@asiaa.sinica.edu.tw
Ke-Jung **Chen**, Kavli Institute for Theoretical Physics, UC Santa Barbara, USA — chen1399@umn.edu
Yanmei **Chen**, NJU, China — chenym@nju.edu.cn
Ana **Chies-Santos**, University of Nottingham, United Kingdom — Ana.Chies_Santos@nottingham.ac.uk
Igor **Chilingarian**, CfA / SAI MSU, United States — igor.chilingarian@cfa.harvard.edu
Hyejeon **Cho**, Yonsei University, South Korea — hyejeon.cho@gmail.com
Andrew **Cooper**, Max-Planck Institute For Astrophysics, Germany — acooper@mpa-garching.mpg.de
Enrico Maria **Corsini**, Dipartimento di Fisica e Astronomia,
Università di Padova, Italy — enricomaria.corsini@unipd.it
Jorge **Cuadra**, PUC, Chile — jcuadra@astro.puc.cl
Emanuele **Daddi**, CEA Saclay, France — edaddi@cea.fr
Elena **Dalla Bontà**, Department of Physics and Astronomy,
University of Padua, Italy — elena.dallabonta@unipd.it
Claudio **Dalla Vecchia**, MPE, Germany — caius@mpe.mpg.de
Ivana **Damjanov**, Harvard-Smithsonian Center for Astrophysics, United States — idamjanov@cfa.harvard.edu
Roger **Davies**, University of Oxford United Kingdom — rld@astro.ox.ac.uk
Timothy **Davis**, European Southern Observatory, Germany — tdavis@eso.org
Sperello **di Serego Alighieri**, INAF, Osservatorio Astrofisico di Arcetri, Italy — sperello@arcetri.astro.it
Simon **Driver**, UWA/St Andrews, Australia — simon.driver@icrar.org
Cuihua **Du**, College of Physical Sciences, Graduate university of China — ducuihua@gucas.ac.cn
Pierre-Alain **Duc**, AIM Paris-Saclay, France — paduc@cea.fr
Allan **Ernest**, Charles Sturt University, Australia — aernest@csu.edu.au
Renato **Falomo**, INAF- Osservatorio Astronomico di Padova, Italy — renato.falomo@oapd.inaf.it
Lulu **Fan**, University of Science and Technology of China, China — llfan@ustc.edu.cn
Mirian **Fernandez Lorenzo**, IAA-CSIC, Spain — mirian@iaa.es
Anna **Ferre-Mateu**, Instituto de Astrofisica de Canarias, Spain — aferre@iac.es
Ignacio **Ferreras**, University College London, United Kingdom — ipf@mssl.ucl.ac.uk
Duncan **Forbes**, Swinburne University, Australia — dforbes@swin.edu.au
William **Forman**, SAO, United States — wrf@cfa.harvard.edu
Sebastien **Foucaud**, National Taiwan Normal University Taiwan, China — foucaud@ntnu.edu.tw
Carlos **Frenk**, Institute for Computational Cosmology,
Durham University, United Kingdom — c.s.frenk@durham.ac.uk
Hai **Fu**, University of California, Irvine United States — haif@uci.edu
Mattia **Fumagalli**, Leiden Observatory, Netherlands — fumagalli@strw.leidenuniv.nl
Jared **Gabor**, CEA Saclay, France — jared.gabor@cea.fr
Dimitri **Gadotti**, ESO, Germany — dgadotti@eso.org
Jesus **Gallego**, Dpto. Astrofisica y CC de la Atmosfera
Universidad Complutense de Madrid, Spain — j.gallego@fis.ucm.es
Ortwin **Gerhard**, gerhard@mpe.mpg.de MPE, Germany — gerhard@mpe.mpg.de
Karl **Glazebrook**, Swinburne, Australia — karl@astro.swin.edu.au
Oleg **Gnedin**, University of Michigan, United States — ognedin@umich.edu
Leith **Godfrey**, ICRAR/Curtin University, Australia — l.godfrey@curtin.edu.au
Oleksiy **Golubov**, ARI, Heidelberg University, Germany — golubov@ari.uni-heidelberg.de
Thiago **Goncalves**, Universidade Federal do Rio de Janeiro, Brazil — tsg@astro.ufrj.br
Rosa **Gonzalez Delgado**, IAA (CSIC), Spain — rosa@iaa.es
Alister **Graham**, Swinburne University of Technology, Australia — agraham@astro.swin.edu.au
Sebastian **Haan**, CSIRO ATNF, Australia — sebastian.haan@csiro.au
Yunkun **Han**, Yunnan Observatory, China — hanyk@ynao.ac.cn
Will **Hartley**, University of Nottingham, United Kingdom — will.hartley@nottingham.ac.uk
Mike **Hudson**, Univ. of Waterloo, Canada — mjhudson@uwaterloo.ca
Thomas **Hughes**, KIAA/PKU, China — tmhughes@pku.edu.cn
Bernd **Husemann**, Leibniz-Institute for Astrophysics Potsdam, Germany — bhusemann@aip.de
Woong-Seob **Jeong**, KASI, South Korea — jeongws@kasi.re.kr
Linhua **Jiang**, Arizona State University, United States — linhua.jiang@asu.edu
Yipeng **Jing**, Shanghai Astronomical Observatory, China — ypjing@shao.ac.cn
Jonas **Johansson**, Max-Planck Institute for Astrophysics, Garching, Germany — jonasj@mpa-garching.mpg.de
Peter H. **Johansson**, University of Helsinki, Finland — peter.johansson@helsinki.fi
Christine **Jones**, Harvard-Smithsonian CfA, United States — cjones@cfa.harvard.edu
Marios **Karouzos**, CEOU-Seoul National University, South Korea — mkarouzos@astro.snu.ac.kr
Sugata **Kaviraj**, Imperial College London, United Kingdom — s.kaviraj@imperial.ac.uk
Ryan **Keenan**, ASIAA, Taiwan — rykeenan@gmail.com
Simon **Kemp**, Instituto de Astronomia, Universidad de Guadalajara, Mexico — snk@astro.iam.udg.mx
Jae-Woo **Kim**, Seoul National University, South Korea — kjw0704@googlemail.com
Anatoly **Klypin**, NMSU, United States — aklypin@nmsu.edu
Tadayuki **Kodama**, Subaru Telescope, NAOJ, Japan — t.kodama@nao.ac.jp
Yutaka **Komiya**, National Astronomical Observatory of Japan, Japan — yutaka.komiya@nao.ac.jp
Xu **Kong**, Center for astrophysics,
University of Science and Technology of China, Anhui, China — xkong@ustc.edu.cn

John **Kormendy**, University of Texas at Austin, United States	kormendy@astro.as.utexas.edu
Ralf **Kotulla**, University of Wisconsin-Milwaukee, United States	kotulla@uwm.edu
Renee **Kraan-Korteweg**, University of Cape Town, South Africa	kraan@ast.uct.ac.za
Davor **Krajnovic**, ESO, Germany	dkrajnov@eso.org
Michal **Krizek**, Institute of Matmematics, Academy of Sciences, Prague, Czech Republic	krizek@cesnet.cz
Ivo **Labbe**, Leiden Observatory, Netherlands	ivo@strw.leidenuniv.nl
Cedric **Lacey**, ICC, Durham University, United Kingdom	cedric.lacey@durham.ac.uk
Myung Gyoon **Lee**, Seoul National University, South Korea	mglee@astro.snu.ac.kr
Joel **Leja**, Yale University, United States	joel.leja@yale.edu
Cheng **Li**, Shanghai Astronomical Observatory, China	leech@shao.ac.cn
Yanchun **Liang**, NAOC, China	ycliang@bao.ac.cn
Simon **Lilly**, ETH Zurich, Switzerland	simon.lilly@phys.ethz.ch
YenTing **Lin**, Institute of Astronomy and Astrophysics, Academia Sinica Taiwan, China	ytl@asiaa.sinica.edu.tw
Lukas **Lindroos**, Chalmers University of Technology, Sweden	lindroos@chalmers.se
Gaochao **Liu**, NAOC, China	
Xin **Liu**, Harvard College Observatory, United States	xinliu@cfa.harvard.edu
Colin **Lonsdale**, MIT Haystack Observatory, United States	cjl@haystack.mit.edu
Ilani **Loubser**, North-West University, South Africa	ilani.loubser@nwu.ac.za
He **Ma**, National Astronomical Observatory, China	joker.mahe@gmail.com
Christina **Magoulas**, University of Melbourne, Australia	c.magoulas@pgrad.unimelb.edu.au
Claudia **Maraston**, Institute of Cosmology and Gravitation, University of Portsmouth, United Kingdom	claudia.maraston@port.ac.uk
Esther **Marmol-Queralto**, Instituto de Astrof?sica de Canarias, Spain	emq@iac.es
Davide **Martizzi**, Institute for Theoretical Physics, University of Zurich, Switzerland	martdav@physik.uzh.ch
Richard **McDermid**, Gemini Observatory, United States	rmcdermid@gemini.edu
Karin **Menendez-Delmestre**, Valongo Observatory, Federal University of Rio de Janeiro, Brazil	kmd@astro.ufrj.br
Attila **Meszaros**, Charles University, Czech Republic	meszaros@cesnet.cz
Areg **Mickaelian**, Byurakan Astrophysical Observatory (BAO), Armenia	aregmick@yahoo.com
Mireia **Montes**, IAC, Spain	mmontes@iac.es
Christopher **Moody**, UC Santa Cruz, United States	cemoody@ucsc.edu
Mark **Mozena**, University of California-Santa Cruz, United States	mmozena@ucolick.org
Thorsten **Naab**, Max-Planck-Institute for Astrophysics, Germany	naab@mpa-garching.mpg.de
Taira **Oogi**, Hokkaido University, Japan	oogi@astro1.sci.hokudai.ac.jp
Alvaro **Orsi**, Pontificia Universidad Catolica de Chile, Chile	aaorsi@astro.puc.cl
Ludwig **Oser**, Max-Planck-Institute for Astrophysics, Germany	oser@mpa-garching.mpg.de
Jamie **Ownsworth**, University of Nottingham, United Kingdom	ppxjo1@nottingham.ac.uk
Nelson **Padilla**, Universidad Cat?lica de Chile, Chile	npadilla@astro.puc.cl
Anna **Pasquali**, ARI-Heidelberg, Germany	pasquali@ari.uni-heidelberg.de
Xiyan **Peng**, Graduate University of the Chinese Academy of Sciences, China	pengxiyan09@mails.gucas.ac.cn
Yingjie **Peng**, Institute of Astronomy, ETH Zurich, Switzerland	peng@phys.ethz.ch
Pablo G. **Perez-Gonzalez**, UCM, Spain	pgperez@fis.ucm.es
Bianca **Poggianti**, INAF-Astronomical Observatory of Padova, Italy	bianca.poggianti@oapd.inaf.it
Lauren **Porter**, University of California, Santa Cruz, United States	laporter@ucsc.edu
Michael **Pracy**, Sydney University, Australia	laporter@ucsc.edu
Leila **Powell**, MPE, Germany	lpowell@mpe.mpg.de
Ando **Ratsimbazafy**, University of the Western Cape, South Africa	raljha.a@gmail.com
Rhea-Silvia **Remus**, University Observatory Munich / MPE, Germany	rhea@usm.lmu.de
Elena **Ricciardelli**, Universitat de Valencia, Spain	elena.ricciardelli@uv.es
Brigitte **Rocca-Volmerange**, Institut d'Astrophysique de Paris, France	rocca@iap.fr
Margarita **Rosado**, UNAM, Mexico	margarit@astro.unam.mx
Anna **Saburova**, Sternberg Astronomical Institute, Moscow State University, Russian Fed	saburovaann@gmail.com
Jose Ramon **Sanchez-Gallego**, Instituto de Astrof?sica de Canarias, Spain	jrsg@iac.es
Utane **Sawangwit**, National Astronomical Research Institute of Thailand, Thailand	utane@narit.or.th
Andreas **Schulze**, Kavli Institute for Astronomy and Astrophysics, China	aschulze@pku.edu.cn
Xu **Shao**, NAOC, China	xshao@nao.cas.cn
Maryam **Shirazi**, Leiden Observatory, Netherlands	shirazi@strw.leidenuniv.nl
Yiping **Shu**, University of Utah, United States	yiping.shu@utah.edu
Chiara **Spiniello**, Kapteyn Astronomical Institute, Groningen, Netherlands	spiniello@astro.rug.nl
Daniel **Stark**, University of Arizona, United States	dpstark@email.arizona.edu
Oliver **Steele**, Institute of Cosmology and Gravitation, University of Portsmouth, United Kingdom	oliver.steele@port.ac.uk
Veronica **Strazzullo**, CEA-Saclay, France	veronica.strazzullo@cea.fr
Daniel **Szomoru**, Leiden Observatory, Netherlands	szomoru@strw.leidenuniv.nl
Qinghua **Tan**, CEA Saclay, France	qinghua.tan@cea.fr
Thomas **Targett**, Institute for Astronomy, University of Edinburgh, United Kingdom	tat@roe.ac.uk
James **Taylor**, University of Waterloo, Canada	taylor@uwaterloo.ca
Daniel **Thomas**, Institute of Cosmology and Gravitation, University of Portsmouth, United Kingdom	daniel.thomas@port.ac.uk
Sune **Toft**, Dark Cosmology Centre, Denmark	sune@dark-cosmology.dk
Chiara **Tonini**, Centre for Astrophysics and Supercomputing, Swinburne University, Australia	ctonini@astro.swin.edu.au
Tommaso **Treu**, UCSB, United States	tt@physics.ucsb.edu
Ignacio **Trujillo**, IAC, Spain	trujillo@iac.es
Jesse **van de Sande**, Leiden Observatory, Netherlands	jesse.vd.sande@gmail.com
Sjoert **van Velzen**, Radboud University Nijmegen	s.vanvelzen@astro.ru.nl
Jozsef **Varga**, Department of Physics of Complex Systems, Eotvos Lorand University, Hungary	varga@caesar.elte.hu
Daniel **Viljoen**, North-West University, South Africa	20569513@nwu.ac.za
Jakob **Walcher**, Leibniz Institute for Astrophysics (AIP), Germany	jwalcher@aip.de
Li Wei **Wang**, China	wlw001@163.com
Zhonglue **Wen**, NAO, CAS, China	zhonglue@nao.cas.cn
Gillian **Wilson**, University of California Riverside, United States	gillianw@ucr.edu
Stijn **Wuyts**, MPE, Germany	swuyts@mpe.mpg.de
Renbin **Yan**, New York University, United States	yanrenbin@gmail.com
Haibo **Yuan**, Kavli Institute for Astronomy and Astrophysics, Peking University, China	yuanhb4861@pku.edu.cn
Yu **Zhang**, Yunan Observatory, China	zhy@ynao.ac.cn
Zhitai **Zhang**, NAOC, China	ztzhang@nao.cas.cn
XianZhong **Zheng**, Purple Mountain Observatory, CAS, China	xzzheng@pmo.ac.cn

The first galaxies in the very early universe

The intriguing life of massive galaxies
Proceedings IAU Symposium No. 295, 2012
D. Thomas, A. Pasquali & I. Ferreras, eds.

© International Astronomical Union 2013
doi:10.1017/S1743921313004146

Simulating the First Galaxies

Volker Bromm

Department of Astronomy, University of Texas,
2511 Speedway, RLM 13.116, Austin, TX 78712, U.S.A.
email: vbromm@astro.as.utexas.edu

Abstract. The formation of the first galaxies marks the end of the cosmic dark ages, initiating the prolonged process of reionization and enriching the pristine intergalactic medium with the first heavy chemical elements. It is now possible to simulate this process with ever greater detail of physical realism, while still considering the proper cosmological context. The simulations have taught us that the feedback from Population III stars is vital in shaping the properties of early galaxies. We are close in pushing ab initio simulations to the point where future instruments, such as the *James Webb Space Telescope*, can directly test theoretical predictions.

Keywords. cosmology: theory, early universe, galaxies: formation, stars: formation

1. Introduction

The first sources of light fundamentally transformed the early universe, from the simple initial state of the cosmic dark ages into one of proliferating complexity (Barkana & Loeb 2001; Bromm *et al.* 2009; Loeb 2010). This process began with the formation of the first stars, the so-called Population III (Pop III), at redshifts $z \sim 20 - 30$. These stars are predicted to form in dark matter minihalos, comprising total masses of $\sim 10^6 M_\odot$. Current models suggest that Pop III stars were typically massive, or even very massive, with $M_* \sim 10 - 100 M_\odot$; these models also predict that the first stars formed in small groups, including binaries or higher-order multiples. These developments are further discussed below.

Once the first stars had formed, feedback processes began to modify the surrounding intergalactic medium (IGM). It is useful to classify them into 3 categories (Ciardi & Ferrara 2005): radiative, mechanical, and chemical. The first feedback consists of the hydrogen-ionizing photons emitted by Pop III stars, as well as the less energetic, molecule-dissociating radiation in the Lyman-Werner (LW) bands. When the first stars die, after their short life of a few million years, they will explode as a supernova (SN), or directly collapse into massive black holes. In the SN case, mechanical and chemical feedback come into play. The SN blastwave exerts a direct, possibly very disruptive, effect on its host system, whereas the chemical feedback acts in a more indirect way, as follows: The first stars, forming out of metal-free, primordial gas, are predicted to be characterized by a top-heavy initial mass function (IMF). Once the gas had been enriched to a threshold level, termed "critical metallicity" ($Z_{\rm crit}$), the mode of star formation would revert to a more normal IMF, which is dominated by lower mass stars (Frebel *et al.* 2007). Chemical feedback refers to this transition in star formation mode, implying that less massive stars have a less disruptive impact on their surroundings. Pre-galactic metal enrichment, the transport and mixing of the first heavy elements into the pristine IGM, thus governs the Pop III – Pop II transition (Tornatore *et al.* 2007; Maio *et al.* 2010). Although crucially important, this early enrichment episode is still poorly understood (reviewed in Karlsson *et al.* 2012).

To gauge the strength of the feedback exerted by Pop III stars, the key is to consider that their formation sites, the minihalos mentioned above, have shallow gravitational potential wells. The corresponding virial temperatures, $T_{\mathrm{vir}} \sim 10^3$ K, indicate that such halos cannot confine photoheated gas. As Pop III stars were typically massive, they would quickly exert a strong negative feedback on their host systems. Numerical simulations indicate that in extreme cases this feedback completely destroys the host, in the sense of heating and evacuating all remaining gas. There would therefore be no opportunity for a second burst of star formation in a minihalo. In the case of a less top-heavy IMF, the initial negative feedback would still be strong, but the recovery timescale for enabling second-generation star formation could be significantly reduced. Since all (most?) Pop III stars are massive enough to quickly die, there would be no long-lived system of low-mass stars left behind. The Pop III forming minihalos, therefore, are *not* galaxies, if a bona-fide galaxy is meant to imply a long-lived stellar system, embedded in a dark matter halo. The question: *What is a galaxy, and, more specifically, what is a first galaxy?*, however, is a matter of ongoing debate (Bromm *et al.* 2009; Forbes & Kroupa 2011; Willman & Strader 2012). As we have seen, this question is intimately tied up with the feedback from the first stars, which in turn is governed by the Pop III IMF (top-heavy or more normal).

Theorists are currently exploring the hypothesis that "atomic cooling halos" are viable hosts for the true first galaxies (Oh & Haiman 2002). These halos have deeper potential wells, compared to minihalos, characterized by virial temperatures of $T_{\mathrm{vir}} \simeq 10^4$ K, enabling the primordial gas to cool via efficient line emission from atomic hydrogen. It is useful to keep in mind that observers and theorists often employ different definitions for 'first galaxy'. As a theorist, you wish to identify the first, i.e., lowest-mass, dark matter halos that satisfy the conditions for a galaxy. Observers, on the other hand, usually aim at detecting truly metal-free, primordial systems. Recent simulation results, however, suggest that such metal-free galaxies do not exist. The reason being that rapid SN enrichment from Pop III stars, formed in the galaxy's minihalo progenitors, provided a bedrock of heavy elements (Greif *et al.* 2010; Bromm & Yoshida 2011; Wise *et al.* 2012). Any second generation stars would then already belong to Population II (Pop II).

The first star and galaxy field is just entering a dynamic phase of rapid discovery. This development is primarily driven by new technology, on the theory side by ever more powerful supercomputers, reaching peta-scale machines, and on the observational side by next-generation telescopes and facilities. Among them are the *James Webb Space Telescope (JWST)*, planned for launch in ~ 2018 (Gardner *et al.* 2006), and the suite of extremely large, ground-based telescopes, such as the GMT, TMT, and E-ELT. Complementary to them are ongoing and future meter-wavelength radio arrays, designed to detect the redshifted 21cm radiation from the neutral hydrogen in the early universe (Furlanetto *et al.* 2006). A further intriguing window into the epoch of the first stars is provided by high-redshift gamma-ray bursts (GRBs). These are extremely bright, relativistic explosions, triggered when a rapidly rotating massive star is collapsing into a black hole (Bloom 2011). The first stars are promising GRB progenitors, thus possibly enabling what has been termed "GRB cosmology" (Lamb & Reichart 2000; Bromm & Loeb 2006). Of great promise is to use such high-z bursts as bright background sources to probe the ionization and metal enrichment state of the early IGM (Wang *et al.* 2012).

There is a second approach to study the ancient past, nicely complementary to the *in situ* observation of high-redshift sources. This alternative channel, often termed "Near-Field Cosmology" (Freeman & Bland-Hawthorn 2002), is provided by local fossils that have survived since early cosmic times. Among them are extremely metal-poor stars found in the halo of the Milky Way. The idea here is to scrutinize their chemical abundance patterns and derive constraints on the properties of the first SNe, and, indirectly, of the Pop III progenitor stars, such as their mass and rate of rotation (Beers & Christlieb 2005; Frebel 2010; Tumlinson 2010). Another class of relic objects is made up of the newly discovered extremely faint dwarf galaxies in the Local Group. These ultra-faint dwarf (UFD) galaxies consist of only a few hundred stars, and reside in very low-mass dark matter halos. Their chemical and structural history is therefore much simpler than what is encountered in massive, mature galaxies, and it should be much more straightforward to make the connection with the primordial building blocks (Salvadori & Ferrara 2009; Bovill & Ricotti 2011a, Bovill & Ricotti 2011b; Frebel & Bromm 2012).

In the following, we will briefly review some of the key physical ingedients involved in simulating the formation of primordial galaxies (for further details, see Bromm & Yoshida 2011).

2. Feedback from the first stars

The assembly process of the first galaxies sensitively depends on the feedback exerted by Pop III stars formed in the progenitor minihalos. Different assumptions on the primordial IMF will thus lead to divergent first galaxy formation histories and resulting properties (Pawlik *et al.* 2011, Pawlik *et al.* 2013). Here, we will comment on some issues that are vigorously debated, without attempting a complete exposition of the vast subject of feedback (for further discussion, see Ciardi & Ferrara 2005). In this context, it is important to keep in mind two related points. First, even a single Pop III star can exert an impact on a cosmological scale, possibly affecting the entire Lagrangian volume that is destined to collapse into the first galaxy later on. Second, the number of possible progenitor stars is quite limited (Wise & Abel 2007; Greif *et al.* 2008). The merger tree of an atomic cooling halo contains of order 10 minihalos. If one then assumes that only a small multiple of Pop III stars forms per minihalo, as suggested by recent simulations, one has the same order of progenitor Pop III stars. The Pop III feedback events will thus sample a possibly broad primordial IMF, thus likely introducing a range in first star properties, such as stellar mass, average metallicity, and luminosity.

2.1. *Population III Mass Scale*

The longstanding consensus view has been that the conditions in the early universe favored the formation of predominantly massive stars, such that the Pop III IMF was top-heavy (Abel *et al.* 2002; Bromm *et al.* 2002; Bromm & Larson 2004; Glover 2005). This expectation rests on the much less efficient cooling in pure H/He gas, where the only viable cooling agent is molecular hydrogen. The primordial gas can then cool to only about $\sim 200\,\mathrm{K}$, compared to the $10\,\mathrm{K}$ reached in dust-cooled molecular clouds in the present-day Milky Way. The correspondingly enhanced thermal pressure is reflected in a Jeans mass that is larger by one to two orders of magnitude in the Pop III case. If cooling due to hydrogen deuteride, HD, were to become important, the primordial gas could reach somewhat lower temperatures, possibly down to the temperature of the cosmic microwave background (Johnson & Bromm 2006). The Jeans mass, and therefore the scale of fragmentation, may then also have been somewhat lower, giving rise to what has been termed 'Pop III.2' stars (McKee & Tan 2008). Another element of this 'standard model' of primordial star formation has been that the first stars formed typically in isolation, one per minihalo.

Recently, beginning with work done in 2009, this traditional paradigm has been refined in important ways (Turk *et al.* 2009; Stacy *et al.* 2010; Clark *et al.* 2011b; Greif *et al.* 2011, Greif *et al.* 2012). Supercomputing power, as well as algorithmic advances, now enable us to follow the protostellar collapse to densities where the initial hydrostatic core forms in the center of the cloud, at $n \sim 10^{22}\,\mathrm{cm^{-3}}$ (Yoshida *et al.* 2006, Yoshida *et al.* 2008), and, crucially, beyond this stage into the main accretion phase. These simulations have demonstrated that accretion is mediated through a near-Keplerian disk. The hot conditions in the surrounding cloud result in extremely large rates of infall onto the disk ($\dot{M} \propto T^{3/2}$); this rapid mass loading drives the disk inevitably towards gravitational instability, such that a small multiple of Pop III protostars emerges, often dominated by a binary system (see Fig. 1). It is not yet possible to push such *ab initio* simulations all the way to the completion of the protostellar assembly process; the final mass of Pop III stars and their final IMF are thus still subject to considerable uncertainty.

However, first attempts to carry out the radiation-hydrodynamical calculations required to treat the late accretion phase, where protostellar feedback tends to limit further infall, have confirmed the basic prediction: the first stars were typically massive, with masses of a few $\sim 10 M_\odot$, although rarely very massive ($> 100 M_\odot$), as previously thought (McKee & Tan 2008; Hosokawa *et al.* 2011; Stacy *et al.* 2012). More specifically, the radiative feedback operates through the agency of an ultra-compact H II region, expanding perpendicularly to the disk, once the first protostar has reached a mass, $M_* \gtrsim 10 M_\odot$, and surface temperature that is sufficiently hot to produce ionizing photons. This hourglass-shaped ionized region then acts back on the dense, neutral gas in the disk, slowly photo-evaporating it (McKee & Tan 2008). With the reservoir for further growth gone, the Pop III star has reached its final mass. This process will depend on the rotation state of the accreting protostar (Stacy *et al.* 2013), because different rotation speeds lead to different protostellar radii and therefore effective temperatures. There is a similar dependence of radius on the rate of accretion (Smith *et al.* 2012). The upper mass limit, set via radiative feedback, will thus exhibit a possibly wide range.

2.2. *Supernova Feedback*

Supernova feedback impacts first galaxy formation in multiple ways (Salvaterra *et al.* 2011). The related energy input tends to partially disrupt the minihalo host, delaying any second-generation star formation. Within the previous paradigm of extremely massive Pop III stars, attention has been focused on pair-instability supernovae (PISNe), with explosion energies that can be larger than for conventional core-collapse events by up to two orders of magnitude (Hummel *et al.* 2012; Pan *et al.* 2012). In this case, the negative feedback is severe, and next-generation star formation is delayed by a considerable fraction of the local Hubble time (Wise & Abel 2008; Greif *et al.* 2010). Indeed, the severity of such energetic SN feedback is largely responsible for shifting the mass scale of first galaxy, and therefore second-generation star formation, hosts to $\sim 10^8 M_\odot$ (atomic cooling halos). If, on the other hand, one considers Pop III progenitor masses that are less extreme, in line with the most recent findings, any feedback effects will be less disruptive, and recovery timescales to enable next-generation star formation could be less, typically a few million years (see Fig. 2). As a corrolary, one then also gets less massive host systems for the first galaxies, possibly below the threshold for atomic hydrogen cooling (Ricotti *et al.* 2002a, Ricotti *et al.* 2002b). One should keep in mind, however, that extremely energetic explosions may be triggered even in the death of progenitor stars of more modest mass, provided that they are rapidly rotating. Recent simulations have indeed found hints for such high-spin conditions (Stacy *et al.* 2011, Stacy *et al.* 2013), possibly giving rise to hypernova explosions which would have a very similar effect on first galaxy assembly as a PISN (Greif *et al.* 2010).

Next, the Pop III SNe disperse heavy-elements into the surrounding medium, thus fundamentally changing the conditions for second-generation star formation. Once

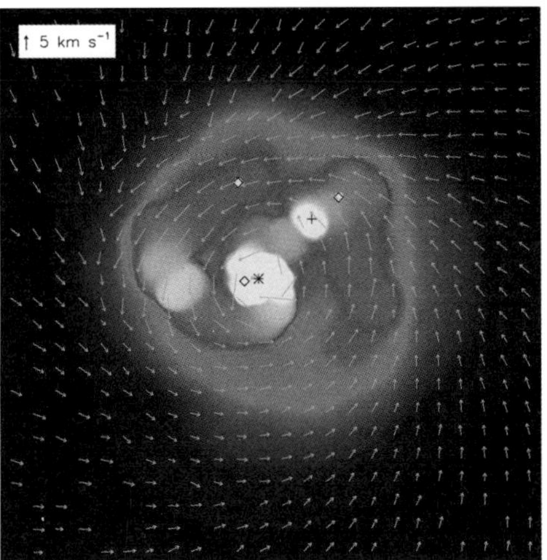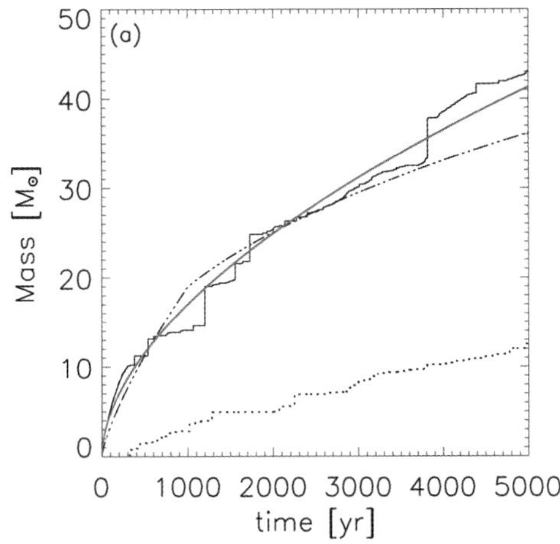

Figure 1. Simulating Pop III star formation (from Stacy *et al.* 2010). *Left panel:* Multiple protostars embedded in an accretion disk. The colors symbolize the underlying density field (yellow marks highest density) within the central 5000 AU. Here, protostars are represented by sink particles, such that the asterisk marks the location of the most massive sink, the cross that of the second most massive one, and diamonds represent the other, smaller sinks. Shown is the situation 5000 yr after initial sink formation. At this point, an ordered, nearly Keplerian velocity structure has been established within the disk. *Right panel:* Sink mass vs. time. The solid line shows the mass of the first sink particle, fitted by a power law according to $M \propto t^{0.55}$ (red line). The dash-dotted line depicts the growth law found in an earlier, lower-resolution simulation (Bromm & Loeb 2004). The dotted line traces the mass growth of the second largest sink. As can be seen, sinks grow to masses $> 10 M_\odot$ within a few 1,000 yr.

metallicity levels exceed the critical value, Z_{crit}, predicted by theory, stars will form with a more normal IMF, giving rise to the first low-mass (Pop II) stars. The detailed physics of this Pop III to II transition is complex; one issue being whether cooling from fine-structure lines dominates (Bromm & Loeb 2003a), or that from dust grains (Schneider *et al.* 2006). As even a single Pop III SN can already enrich the gas in the center of a first galaxy to levels of $Z > 10^{-3} Z_\odot$, larger than *any* of the cooling thresholds (dust or other), one gets the robust prediction: The majority of the first galaxies were already metal-enriched, and thus hosted long-lived Pop II stars (Johnson *et al.* 2008; Wise & Abel 2008; Greif *et al.* 2010; Wise *et al.* 2012). The presence of heavy-element coolants may also be able to significantly boost star formation efficiencies. Simulations tell us that these are rather low in primordial gas: $\eta_* = M_*/M_{\mathrm{gas}} \sim 10^{-3}$, where M_{gas} is the total baryonic (=gas) mass in a given host system (here a minihalo). The same inefficiency may also apply to primordial star formation in more massive host systems (Safranek-Shrader *et al.* 2012). A metal-induced boost in η_* is thus essential to form massive stellar (Pop II) clusters. It is an interesting problem for 'stellar archaeologists' to (spectroscopically) identify those fossil Pop II stars in our Milky Way and Local Group (see the references given above), providing us with a powerful probe of the conditions in the first galaxies.

2.3. *Black Hole Feedback*

For a range of Pop III progenitor masses, the star is predicted to collapse, directly or in a delayed fashion, into a black hole (BH). This opens up the possibility for a qualitatively different class of feedback effects. Accretion onto the BH, either from the surrounding diffuse medium or a less-massive companion in a binary (HMXB) system, will

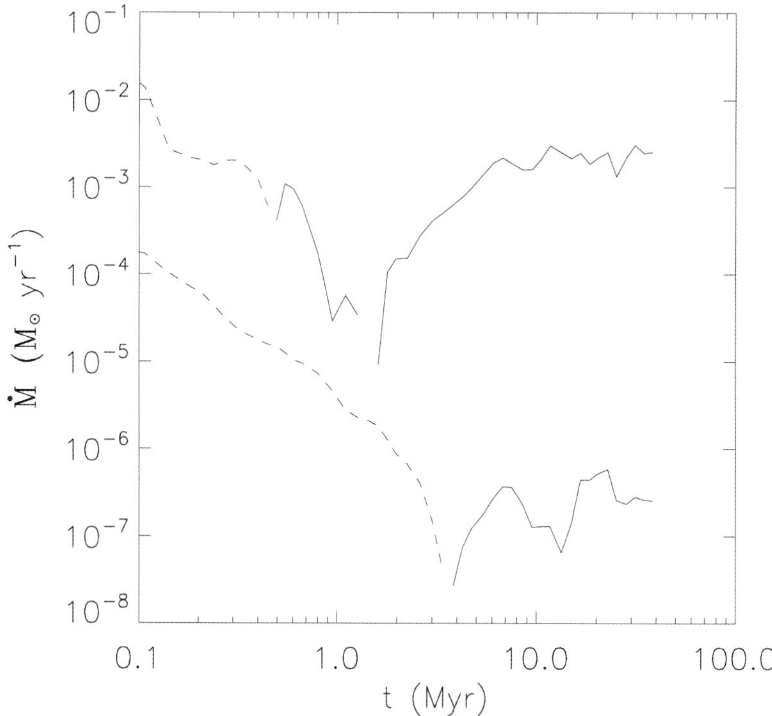

Figure 2. Pop III Supernova feedback (from Ritter *et al.* 2012). Here, the simulation traces the case of a less energetic explosion ($E_{\mathrm{SN}} = 10^{51}$ ergs), corresponding to a progenitor star of $\sim 40 M_\odot$. Metal (lower curve) and total baryon (upper curve) net mass flow through a sphere of radius 20 pc, centered on the gravitational potential minimum. Dashed lines indicate outflows and solid lines inflows. Net outflow reverses into an inflow earlier in the baryons because of the presence of cold filaments delivering metal-free gas from the cosmic web into the halo center.

lead to non-thermal X-ray emission. Accretion is typically weak in the former case, where Bondi-Hoyle accretion occurs, and the resulting emission luminosities are strongly sub-Eddington (Milosavljević *et al.* 2009a, Milosavljević *et al.* 2009b). Even then, however, the corresponding photoionization heating has a strong negative effect locally, but virtually none on larger scales (see Fig. 3). In the case of a HMXB source, on the other hand, where X-ray fluxes are typically very high for a limited time, there is a global impact on star formation (Mirabel *et al.* 2011); now, the strong X-ray flux can partially ionize the IGM out to large distances, including the gas in neighboring minihalos. The boost in the free-electron fraction in turn catalyzes the increased formation of H_2 molecules, such that cooling, and therefore Pop III star formation, is enhanced as well. Consequently, the global feedback here is positive, whereas locally the effect is again negative (Jeon *et al.* 2012).

A related problem is whether Pop III BH seeds may be able to grow into supermassive BHs that can power the high-redshift quasars already found at $z \sim 6 - 7$. The answer is to the affirmative, provided that growth can proceed at near-Eddington rates (Loeb 2010). However, strong negative feedback effectively limits growth to small fractions of the Eddington rate, because the photoionization-heating around the BH progenitor star evaporates the gas from the center, such that the holes find themselves in a virtual vacuum for a substantial part of the local Hubble time (Johnson & Bromm 2007). The correspondingly low rates of accretion severely challenge any theory that tries to explain the high-z quasar hosts with accretion onto (Pop III) stellar seeds. Alternative, more

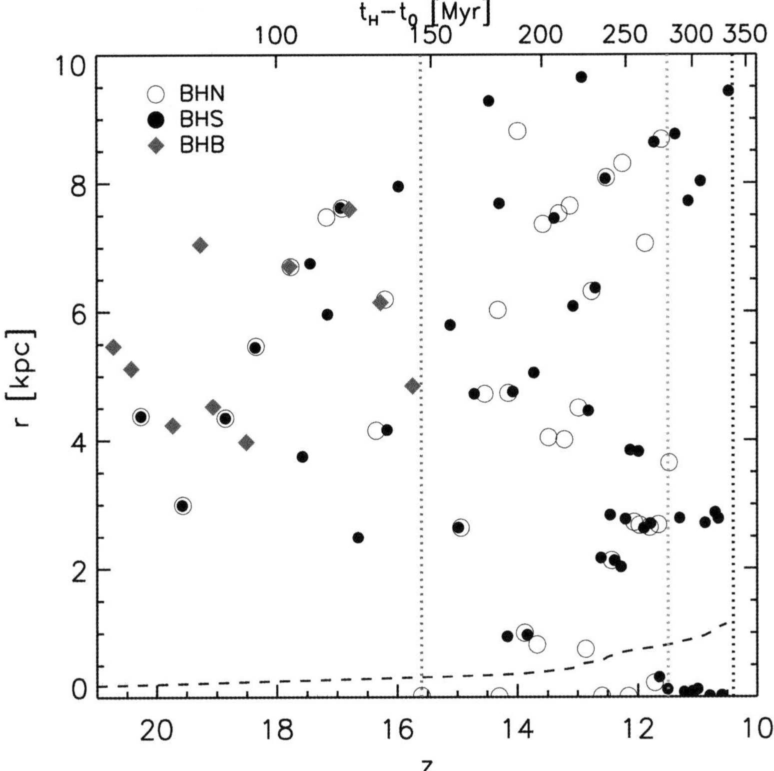

Figure 3. Black hole feedback during first galaxy assembly (from Jeon *et al.* 2012). Distances between newly formed Pop III stars and the BH at the center of the emerging first galaxy as a function of redshift. The symbols refer to three related simulations, where the only difference is what is assumed for the BH feedback. Specifically, BHN refers to no BH feedback present, BHS to feedback from a single BH, and BHB to that from a binary (HMXB) source. In the BHB run, positive feedback is evident far away from the source, where gas collapses into distant minihalos, facilitated via H_2 cooling promoted by the strong X-ray emission; locally, star formation is suppressed. The virial radius of the halo hosting the active BH or the HMXB is shown for reference (*dashed line*).

exotic, scenarios have therefore been explored, such as the direct collapse of a primordial gas cloud under strong LW radiation. Molecules may then never form, and consequently no star formation would occur; the cloud may then be able to directly collapse into a (super-) massive BH seed in the absence of any stellar feedback (Bromm & Loeb 2003b; Volonteri 2012).

3. Turbulence in the first galaxies

With the emergence of the first galaxies, we witness the onset of supersonic turbulence, which is expected to have important consequences for star formation (Wise & Abel 2007; Greif *et al.* 2010; Prieto *et al.* 2011). Indeed, the Reynolds number in the center of the first galaxies is very large, $Re \sim 10^9$, indicating a highly-turbulent situation, and the Mach number, $Ma \sim V/c_s \sim v_{\rm vir}/c_s \sim 10$, indicates supersonic flows. In the last estimate, we have used the virial velocity inside an atomic cooling halo, $v_{\rm vir} \sim 10\,{\rm km\,s}^{-1}$, and the sound-speed of H_2-cooled gas ($c_s \sim 1\,{\rm km\,s}^{-1}$). In the minihalos, due to their smaller

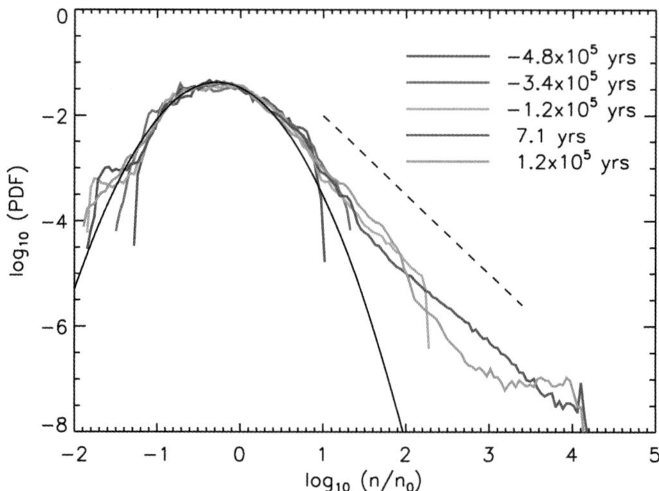

Figure 4. Density structure in turbulent primordial gas (from Safranek-Shrader *et al.* 2012). The presence of supersonic turbulence is manifested in the characteristic log-normal probability distribution. At late times, the effect of self-gravity imprints a power-law extension towards the highest densities. It is possible that the turbulently structured gas will give rise to a high-mass slope in the stellar IMF similar to the present-day Salpeter one.

virial velocities, which are of order the sound speed, any low-level turbulence is mostly subsonic (Clark *et al.* 2011a).

Supersonic turbulence generates density fluctuations in the central gas cloud. Statistically, these can be described with a log-normal probability density function (PDF):

$$f(x)dx = \frac{1}{\sqrt{2\pi\sigma_x^2}} \exp\left[-\frac{(x-\mu_x)^2}{2\sigma_x^2}\right] dx \ , \tag{3.1}$$

where $x \equiv \ln(\rho/\bar{\rho})$, and μ_x and σ_x^2 are the mean and dispersion of the distribution, respectively. The latter two are connected: $\mu_x = -\sigma_x^2/2$ (McKee & Ostriker 2007). Numerical simulations have shown that the dispersion of the density PDF is connected to the Mach number of the flow: $\sigma_x^2 \simeq \ln(1 + 0.25Ma^2)$. Inside the first galaxies, one finds values close to $\sigma_x \simeq 1$ (Safranek-Shrader *et al.* 2012). Similar to the well-studied case of isothermal, supersonic turbulence, the central gas in the first galaxies exhibits the imprint of self-gravity (Fig. 4): a power-law tail toward the highest densities, on top of the log-normal PDF at lower densities, which is generated by purely hydrodynamical effects (Safranek-Shrader *et al.* 2012).

4. Future prospects

The coming decade will likely see a flurry of discoveries in the pre-reionization universe, getting us closer to answering some of the questions of the ages: What are our cosmic origins and how did it all begin? Technology, involving both next-generation observational facilities and peta-scale supercomputing, will likely play a prominent role in this endeavor. Thus probing the first galaxies may provide us with an ideal, simplified laboratory for the otherwise exceedingly complex problem of galaxy formation and evolution in general. Simulation-based predictions will be crucial in guiding the design and observing strategies of future telescopes, in effect providing signposts in the dark, at the very frontier of our knowledge of the early universe. It is clear that serendipity will be involved, but with a bit of luck, dark-age cosmology should soon come into its own.

Acknowledgments

I would like to thank the organizers for putting together a stimulating program, providing an exciting new perspective on massive galaxy formation. Support from NSF grant AST-1009928 and NASA ATFP grant NNX09AJ33G is gratefully acknowledged.

References

Abel, T., Bryan, G. L., & Norman, M. L. 2002, *Science*, 295, 93
Barkana, R. & Loeb, A. 2001, *Phys. Rep.*, 349, 125
Beers, T. C. & Christlieb, N. 2005, *ARAA*, 43, 531
Bloom, J. S. 2011, *What are Gamma-ray Bursts?* (Princeton: Princeton Univ. Press)
Bovill, M. & Ricotti, M. 2011a, *ApJ*, 741, 17
Bovill, M. & Ricotti, M. 2011b, *ApJ*, 741, 18
Bromm, V., Coppi, P. S., & Larson, R. B. 2002, *ApJ*, 564, 23
Bromm, V. & Larson, R. B. 2004, *ARAA*, 42, 79
Bromm, V. & Loeb, A. 2003a, *Nature*, 425, 812
Bromm, V. & Loeb, A. 2003b, *ApJ*, 596, 34
Bromm, V. & Loeb, A. 2004, *New Astron.*, 9, 353
Bromm, V. & Loeb, A. 2006, *ApJ*, 642, 382
Bromm, V., Yoshida, N., Hernquist, L., & McKee, C. F. 2009, *Nature*, 459, 49
Bromm, V. & Yoshida, N. 2011, *ARAA*, 49, 373
Ciardi, B. & Ferrara, A. 2005, *Space Science Rev.*, 116, 625
Clark, P. C., Glover, S. C. O., Klessen, R. S., & Bromm, V. 2011a, *ApJ*, 727, 110
Clark, P. C., Glover, S. C. O., Smith, R. J., Greif, T. H., Klessen, R. S., & Bromm, V. 2011b, *Science*, 331, 1040
Forbes, D. A. & Kroupa, P. 2011, *PASA*, 28, 77
Frebel, A. 2010, *Astronomische Nachrichten*, 331, 474
Frebel, A. & Bromm, V. 2012, *ApJ*, 759, 115
Frebel, A., Johnson, J. L., & Bromm, V. 2007, *MNRAS* (Letters), 380, L40
Freeman, K. & Bland-Hawthorn, J. 2002, *ARAA*, 40, 487
Furlanetto, S. R., Oh, S. P., & Briggs, F. H. 2006, *Phys. Rep.*, 433, 181
Gardner, J. P., et al. 2006, *Space Science Rev.*, 123, 485
Glover, S. C. O. 2005, *Space Science Rev.*, 117, 445
Greif, T. H., Johnson, J. L., Klessen, R. S., & Bromm, V. 2008, *MNRAS*, 387, 1021
Greif, T. H., Glover, S. C. O., Bromm, V., & Klessen, R. S. 2010, *ApJ*, 716, 510
Greif, T. H., Springel, V., White, S. D. M., Glover, S. C. O., Clark, P. C., Smith, R. J., Klessen, R. S., & Bromm, V. 2011, *ApJ*, 737, 75
Greif, T. H., Bromm, V., Clark, P. C., Glover, S. C. O., Smith, R. J., Klessen, R. S., Yoshida, N., & Springel, V. 2012, *MNRAS*, 424, 399
Hosokawa, T., Omukai, K., Yoshida, N., & Yorke, H. W. 2011, *Science*, 334, 1250
Hummel, J., Pawlik, A., Milosavljevic, M., & Bromm, V. 2012, *ApJ*, 755, 72
Jeon, M., Pawlik, A. H., Greif, T. H., Glover, S. C. O., Bromm, V., Milosavljevic, M., & Klessen, R. S. 2012, *ApJ*, 754, 34
Johnson, J. L. & Bromm, V. 2006, *MNRAS*, 366, 247
Johnson, J. L. & Bromm, V. 2007, *MNRAS*, 374, 1557
Johnson, J. L., Greif, T. H., & Bromm, V. 2008, *MNRAS*, 388, 26
Karlsson, T., Bromm, V., & Bland-Hawthorn, J. 2012, *Rev. Mod. Phys.*, in press (arXiv:1101.4024)
Lamb, D. Q. & Reichart, D. E. 2000, *ApJ*, 536, 1
Loeb, A. 2010, *How did the first stars and galaxies form?* (Princeton: Princeton Univ. Press)

Maio, U., Ciardi, B., Dolag, K., Tornatore, L., & Khochfar, S. 2010, *MNRAS*, 407, 1003

McKee, C. F. & Ostriker, E. C. 2007, *ARAA*, 45 565

McKee, C. F. & Tan, J. C. 2008, *ApJ*, 681, 771

Milosavljević, M., Bromm, V., Couch, S. M., & Oh, S. P. 2009a, *ApJ*, 698, 766

Milosavljević, M., Couch, S. M., & Bromm, V. 2009b, *ApJ* (Letters), 696, L146

Mirabel, I. F., Dijkstra, M., Laurent, P., Loeb, A., & Pritchard, J. R. 2011, *A&A*, 528, 149

Oh, S. P. & Haiman, Z. 2002, *ApJ*, 569, 558

Pan, T., Kasen, D., & Loeb, A. 2012, *MNRAS*, 422, 2701

Pawlik, A., Milosavljević, M., & Bromm, V. 2011, *ApJ*, 731, 54

Pawlik, A., Milosavljević, M., & Bromm, V. 2013, *ApJ*, 767, 59

Prieto, J., Padoan, P., Jimenez, R., & Infante, L. 2011, *ApJ* (Letters), 731, L38

Ricotti, M., Gnedin, N. Y., & Shull, M. J. 2002a, *ApJ*, 575, 33

Ricotti, M., Gnedin, N. Y., & Shull, M. J. 2002b, *ApJ*, 575, 49

Ritter, J. S., Safranek-Shrader, C., Gnat, O., Milosavljević, M., & Bromm, V. 2012, *ApJ*, 761, 56

Safranek-Shrader, C., Agarwal, M., Federrath, C., Dubey, A., Milosavljević, M., & Bromm, V. 2012, *MNRAS*, 426, 1159

Salvadori, S. & Ferrara, A. 2009, *MNRAS* (Letters), 395, L6

Salvaterra, R., Ferrara, A., & Dayal, P. 2011, *MNRAS*, 414, 847

Schneider, R., Omukai, K., Inoue, A. K., & Ferrara, A. 2006, *MNRAS*, 369, 1437

Smith, R. J., Hosokawa, T., Omukai, K., Glover, S. C. O., & Klessen, R. S. 2012 *MNRAS*, 424, 457

Stacy, A., Greif, T. H., & Bromm, V. 2010, *MNRAS*, 403, 45

Stacy, A., Bromm, V., & Loeb, A. 2011, *MNRAS*, 413, 543

Stacy, A., Greif, T. H., & Bromm, V. 2012, *MNRAS*, 422, 290

Stacy, A., Greif, T. H., Klessen, R. S., Bromm, V., & Loeb, A. 2013, *MNRAS*, 431, 1470

Tornatore, L., Ferrara, A., & Schneider, R. 2007, *MNRAS*, 382, 945.

Tumlinson, J. 2010, *ApJ*, 708, 1398

Turk, M. J., Abel, T., & O'Shea, B. W. 2009, *Science*, 325, 601

Volonteri, M. 2012, *Science*, 337, 544

Wang, F. Y., Bromm, V., Greif, T. H., Stacy, A., Dai, Z. G., Loeb, A., & Cheng, K. S. 2012, *ApJ*, 760, 27

Willman, B. & Strader, J. 2012, *AJ*, 144, 76

Wise, J. H. & Abel, T. 2007, *ApJ*, 665, 899

Wise, J. H. & Abel, T. 2008, *ApJ*, 685, 40

Wise, J. H., Turk, M. J., Norman, M. L., & Abel, T. 2012, *ApJ*, 745, 50

Yoshida, N., Omukai, K., Hernquist, L., & Abel, T. 2006, *ApJ*, 652, 6

Yoshida, N., Omukai, K., & Hernquist, L. 2008, *Science*, 321, 669

The intriguing life of massive galaxies
Proceedings IAU Symposium No. 295, 2012
D. Thomas, A. Pasquali & I. Ferreras, eds.

© International Astronomical Union 2013
doi:10.1017/S1743921313004158

Enhancing and inhibiting star formation: high-resolution simulation studies of the impact of cold accretion, mergers and feedback on individual massive galaxies

Leila C. Powell[1], Frederic Bournaud[2], Damien Chapon[2], Julien Devriendt[3], Volker Gaibler[4], Sadegh Khochfar[1], Adrianne Slyz[3] and Romain Teyssier[2]

[1]Max Planck Institute for extraterrestrial Physics, PO Box 1312, Giessenbachstr., 85741 Garching, Germany
email: lpowell@mpe.mpg.de
[2]Service d'astrophysique, CEA, Orme des Merisiers, Gif-sur-Yvette Cedex, France
[3]Oxford astrophysics, Denys Wilkinson Building, Keble Road, OX1 3RH
[4]Universität Heidelberg, Zentrum für Astronomie, Institut für Theoretische Astrophysik, Albert-Ueberle-Str. 2, 69120 Heidelberg, Germany

Abstract. The quest for a better understanding of the evolution of massive galaxies can be broadly summarised with 2 questions: how did they build up their large (stellar) masses and what eventually quenched their star formation (SF)? To tackle these questions, we use high-resolution RAMSES simulations (Teyssier 2002) to study several aspects of the detailed interplay between accretion (mergers and cold flows), SF and feedback in individual galaxies. We examine SF in major mergers; a process crucial to stellar mass assembly. We explore whether the merger-induced, clustered SF is as important a mechanism in average mergers, as it is in extreme systems like the Antennae. We find that interaction-induced turbulence drives up the velocity dispersion, and that there is a correlated rise in SFR in all our simulated mergers as the density pdf evolves to have an excess of very dense gas. Next, we introduce a new study into whether mechanical jet feedback can impact upon the ability of hot gas haloes to provide a supply of fuel for SF during mergers and in their remnants. Finally, we briefly review our recent study, in which we examine the effect of supernova (SN) feedback on galaxies accreting via the previously overlooked cold-mode, by resimulating a stream-fed galaxy at $z \sim 9$. A far-reaching galactic wind results yet it cannot suppress the cold, filamentary accretion or eject significant mass in order to reduce the SFR, suggesting that SN feedback may not be as effective as is often assumed.

Keywords. methods: numerical, galaxies: evolution, galaxies: star clusters, galaxies: interactions, galaxies: high-redshift, galaxies: starburst, galaxies: jets, stars: supernovae: general

---

1. Clumpy star formation induced by mergers

It has been widely demonstrated with previous generations of merger simulations, that merging galaxies undergo a nuclear starburst. In this picture, gas is driven to the centre by tidal torques, where it is compressed and turns into stars at a very high rate (Barnes & Hernquist 1991). This was compatible with observations of ULIRGs which exhibit centrally concentrated star formation. However, there is now significant observational evidence of the existence of merging galaxies whose star formation is extended and clustered (e.g. the Antennae), as well as attempts by simulators to model these specific systems (Teyssier *et al.* 2010). We undertake a more general study of clustered star formation by performing 5 idealised simulations of equal-mass mergers (with ~ 5pc

13

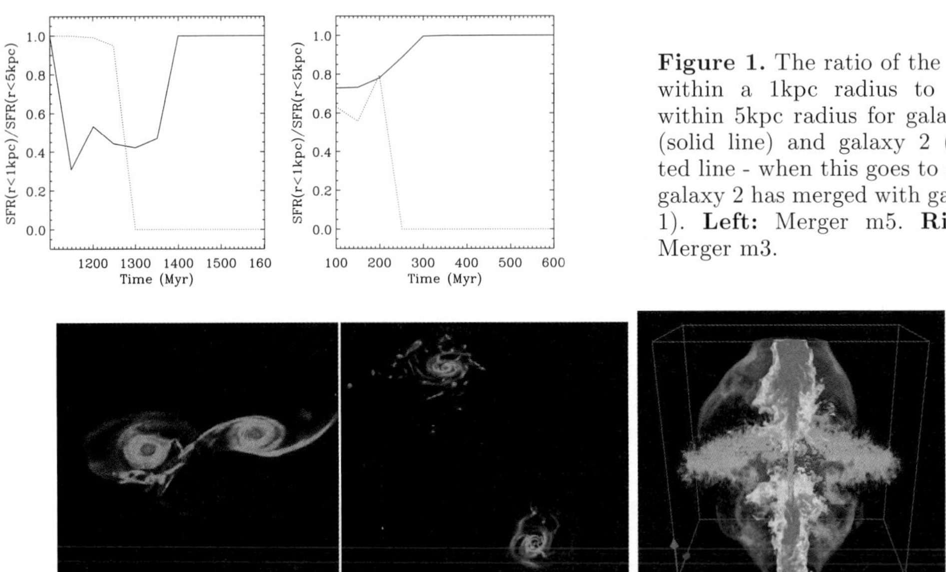

Figure 1. The ratio of the SFR within a 1kpc radius to that within 5kpc radius for galaxy 1 (solid line) and galaxy 2 (dotted line - when this goes to zero, galaxy 2 has merged with galaxy 1). **Left:** Merger m5. **Right:** Merger m3.

Figure 2. Left(2 panels): Gas density map to illustrate the increased clumpiness of the gas as merger m5 progresses. There is a 100Myr time difference between the left and right panels - initially both discs are relatively smooth, but one becomes noticeably more clumpy. **Right:** Image of jet interacting with cold gas disc and a hot ambient medium. See Gaibler *et al.* (2012) for details of the jet modeling.

resolution) with orbital parameters that can be considered 'average' for a LCDM cosmology. The full results of this study will be presented in Powell *et al.* (2012, in prep).

1.1. *Extended, clustered star formation*

In Fig. 1 we show the ration of the SFR within 1 kpc to that within 5 kpc for galaxy 1 (solid line) and galaxy 2 (dotted line) in mergers m5 (left-hand panel) and m3 (right-hand panel). When this quantity is 1, all SF is within the central 1 kpc and so it could be considered to be a 'nuclear' starburst. When the quantity is lower it indicates some of the SF is occurring beyond 1 kpc and so could be considered 'extended'. Here we see two extreme examples. Merger m3 has very concentrated SF, whereas merger m5 has a period of more extended star formation where the ratio reaches values of ~ 0.5, in between two phases where the SF is centrally concentrated. The appearance of new/larger clumps in merger m5 is illustrated in the density maps in Fig. 2. The SF of all the mergers tends towards being centrally concentrated as the galaxies approach coalescence. Although it seems that stars formed in the nucleus dominate the SFR at the very peak of the starburst, extended/clumpy star formation still occurs to a greater or lesser extent in most of the mergers in our sample. This process could lead to significant phases of enhanced SF before the main starburst, alter the stellar distribution in massive ellipticals and even have implications for the origin of globular clusters.

1.2. *How does the merger affect the gas properties?*

In order to try to understand the mechanisms responsible for the merger-induced clumpy SF we measure the evolution of some of the gas properties. Fig. 3 (left-hand panel) shows the time evolution of the 1D velocity dispersion (solid line) measured in 100pc cells and averaged over a cube of side-length 30kpc, centred on one of the galaxies in merger m4. The dotted line shows the global SFR at the same times. It is interesting that the increase

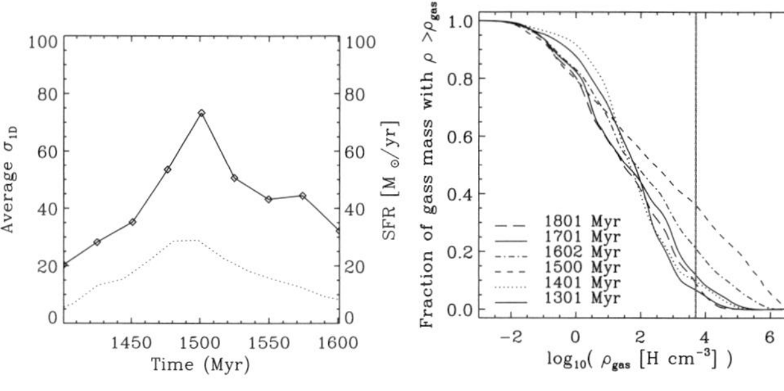

Figure 3. Left: The average 1D velocity dispersion within a 15kpc radius (solid line) and the global SFR (dotted line) for one of the mergers (m4). **Right** The cumulative gas density pdf within a radius of 15kpc, also for merger m4. The vertical line shows the threshold for star formation.

in velocity dispersion is correlated with the increase in SFR and we note that this result is seen for all the mergers in our sample. There are two main possible sources of turbulence that could drive up the velocity dispersion: supernova feedback (which is included in the simulations) or the interaction itself. We rerun merger m4 for the duration of the starburst with supernova feedback switched off and find that the velocity dispersion curve remains the same, suggesting the interaction is driving increased turbulence.

Fig. 3 (right-hand panel) shows the time evolution of the gas density pdf for merger m4. It is clear that there is significant evolution of the pdf as the merger progresses (and as the velocity dispersion goes up, as just demonstrated). A large excess of very dense gas is created and this increases the amount of gas eligible for SF (the vertical line marks the SF threshold in the simulations). This trend is observed for all mergers.

1.3. *Revisiting the Kennicutt-Schmidt Relation*

There is ongoing debate about whether quiescent and starbursting galaxies form one or more sequences on the Kennicutt-Schmidt (KS) plot. Daddi *et al.* (2010) recently showed it was possible to fit two separate sequences and Teyssier *et al.* (2010) suggested that resolving clumpy star formation could be the key to reproducing/explaining this with simulations. Fig. 4 shows examples of the evolution of a merging galaxy on the KS plot. We do not see strong evidence for a clear bimodality, but rather some scatter in between the two relations. We note that merger m5 (left-hand panel) occurs after a long time has elapsed, which is why it starts below the disc sequence (the pre-merger disc has evolved away from the disc sequence).

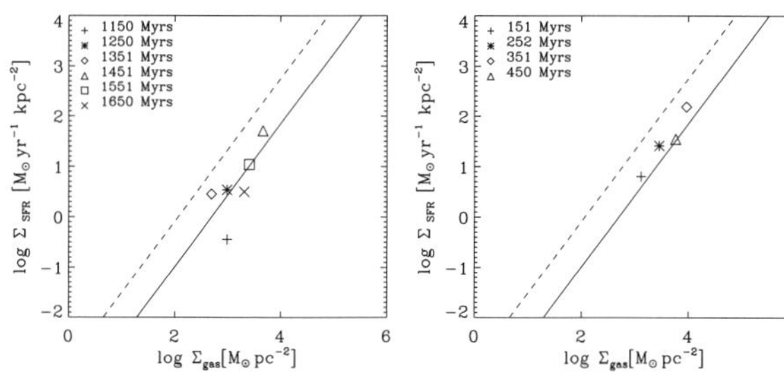

Figure 4. The time evolution of one galaxy from a merger on the Kennicutt-Schmidt plot. Fits from Daddi *et al.* (2010) for quiescent (solid line) and starbursting (dotted line) galaxies are overplotted. **Left:** Merger m5 a 'slow' merger. **Right** Merger m3, a 'rapid' merger.

2. Star formation fueled by hot gas haloes in mergers

Idealised merger simulations have been typically performed without hot gas haloes (including those discussed above). Recently, Moster *et al.* (2011) included this component and found it affects the remnant properties as it acted as a reservoir of fuel for later SF. However, the impact of strong feedback, such as that from an active galactic nuclei has not yet been taken into account. Jets could potentially keep the hot halo hot, preventing it from being a fuel supply. We are currently testing simulations which combine idealised merger simulations including hot gas haloes with the jet module presented in Gaibler *et al.* (2012) (see Fig. 2, right-hand panel), in order to address this issue.

3. Star formation fueled by cold flows

In addition to mergers, cold flows are another important source of fuel for SF. This is particularly true for low mass galaxies (which can't support stable virial shocks, Birnboim & Dekel 2003) and at high redshift (where the filaments are narrower and denser). In the hierarchical picture, however, this also has a knock-on effect for galaxies at lower redshifts and higher masses (Keres *et al.* 2005). We also observe supernova-driven winds at high-redshift and we expect feedback from these winds to be particularly effective at reducing, or shutting down star formation in low-mass galaxies (Dekel & Silk 1986). Essentially then, we expect both cold flows and galactic winds to be effective simultaneously.

We perform ultra-high resolution (~ 0.5pc) cosmological resimulations of the $z \sim 9$ progenitor of a Milky Way-like galaxy to follow how its evolution is influenced by the combined effects of cold gas accretion via filaments and a supernova-driven galactic wind. This resolution allows us to directly model the Sedov blastwave phase of supernovae. Due to the expansion and overlapping of 'bubbles' formed by the blastwaves, a high-speed wind is driven to several times the galaxy's virial radius. By comparison with a simulation without supernovae, we establish that the wind has had no effect on the accretion rate via the filaments and does not eject significant mass, so it is not able to reduce the SFR. This suggests SN feedback is not as effective (in this regime) as previously thought. We refer the reader to Powell *et al.* (2011) for the full details and results.

4. Conclusions

We have highlighted aspects of galaxy evolution in which refinements to the classical pictures are now required because of more detailed information gleaned from observations and/or simulations: the addition of clumpy star formation to the nuclear starburst model; the inclusion in merger studies of not just hot gas haloes, but the feedback that may affect them; and the impact that filaments (often neglected) can have on the predicted efficiency of supernova feedback. We have demonstrated that hydrodynamical simulations are an invaluable tool to investigate all these refinements, but note that having high resolution is absolutely crucial, or these affects are simply not captured.

References

Barnes, J. E. & Hernquist, L. E. 1991, *ApJ*, 370, L65

Birnboim, Y. & Dekel, A. 2003, *MNRAS*, 345, 349

Daddi, E., Elbaz, D., Walter, F., Bournaud, F., Salmi, F., Carilli, C., Dannerbauer, H., Dickinson, M., Monaco, P., & Riechers, D. 2010, *ApJ* (Letters), 714, L118

Dekel, A. & Silk, J. 1986, *ApJ*, 303, 39

Kereŝ, D., Katz, N., Weinberg, D. H., & Davé, R. 2003, *MNRAS*, 363, 2

Gaibler, V., Khochfar, S., Krause, M., & Silk, J. 2012, *MNRAS*, 425, 438

Moster, B. P., Macciò, A. V., Somerville, R. S., Naab, T., & Cox, T. J. 2011, *MNRAS*, 415, 3750

Powell, L. C., Slyz, A., & Devriendt, J. 2011, *MNRAS*, 414, 3671

Teyssier, R. 2002, *A&A*, 385, 337

Teyssier, R., Chapon, D., & Bournaud, F. 2010, *ApJ* (Letters), 720, L149

The intriguing life of massive galaxies
Proceedings IAU Symposium No. 295, 2012
D. Thomas, A. Pasquali & I. Ferreras, eds.

© International Astronomical Union 2013
doi:10.1017/S174392131300416X

The First Billion Years simulation project. Galactic outflows and metal enrichment

Claudio Dalla Vecchia,[1,a] Sadegh Khochfar[1,b] and Joop Schaye[2]

[1] Max Planck Institute for Extraterrestrial Physics,
Giessenbach-Straße, D-85748, Garching, Germany
emails: *a*) caius@mpe.mpg.de; *b*) sadeghk@mpe.mpg.de
[2] Leiden Observatory, University of Leiden,
Niels Bohrweg 2, NL-2333CA, Leiden, The Netherlands
email: schaye@strw.leidenuniv.nl

Abstract. The *First Billion Years* is one of the largest and most comprehensive cosmological, hydrodynamical simulation of the early Universe to date. It studies the formation of the first proto-galaxies down to redshift 6, and provides the path of the evolution from proto-galaxies to the observed galaxy population at redshift 4. It also provides predictions for future observations.

Keywords. galaxies: evolution, galaxies: formation, galaxies: high-redshift, galaxies: ISM, methods: numerical

1. Introduction

In the last decade, deep observations of the Universe have been providing high quality data in the redshift range $4 < z \lesssim 6$. In the near future, the *James Webb Space Telescope* will be probing even higher redshifts with observations of proto-galaxies formed when the Universe was less that one billion years old. Providing a detailed, theoretical framework of the formation of these proto-galaxies will help to predict, as well as to interpret, what will be observed. The *First Billion Years* simulation project (*FiBY*) has been conducted with this aim.

FiBY is one of the largest and most comprehensive cosmological, hydrodynamical simulation of the early Universe to date. A volume of $(8 \text{ Mpc})^3$ containing 2×1368^3 dark matter and gas particles is evolved down to redshift $z = 6$. The gas mass resolution is 1250 $M_\odot$, whereas the spatial resolution is a few tens of physical parsecs. Such high resolution is necessary to resolve the collapse of the first structures—the mini-haloes that host the first generation of stars (population-III stars, hereafter pop-III).

The collapse of gas of primordial composition into mini-haloes is driven by molecular hydrogen cooling. This gas is the fuel for the formation of pop-III stars. These stars are responsible for the early enrichment of the inter-galactic medium (IGM) through very energetic pair-instability supernova (PISN) explosions, creating the conditions for the formation of the metal-enriched, second generation stars (population-II stars, hereafter pop-II). Therefore, molecular formation and molecular cooling, pop-III and pop-II star formation and stellar feedback are the main ingredients for building the first galaxies.

A set of simulations with increasing volume and decreasing resolution have been performed to provide a larger range of galaxy masses, to study the effect of numerical resolution, and to test variations of the physics and parameters in the model. When combined, these simulations extend the low mass end of the simulated galaxy mass function down to 10^5 $M_\odot$. This corresponds to resolving galaxies with masses two orders of magnitude smaller than in simulations that are currently state-of-the-art (e.g. Finlator *et al.* 2011).

Table 1. Summary of simulation parameters: simulation name; co-moving box length, L; number of particles, N_{tot}; gas particle mass, m_b; dark matter particle mass, m_{DM}; co-moving gravitational softening, ϵ.

Simulation	L (h^{-1} Mpc)	N_{tot}	m_b (h^{-1} M$_\odot$)	m_{DM} (h^{-1} M$_\odot$)	ϵ (h^{-1} kpc)
FiBY08XL	5.68	2×1368^3	8.87×10^2	4.36×10^3	0.17
FiBY04M	2.84	2×684^3	8.87×10^2	4.36×10^3	0.17
FiBY08M	5.68	2×684^3	7.10×10^3	3.48×10^4	0.33
FiBY16M	11.36	2×684^3	5.68×10^4	2.79×10^5	0.66
FiBY32M	22.72	2×684^3	4.54×10^5	2.23×10^6	1.33

The list of *FiBY* simulations is in Table 1. A brief description of the code employed in the project is in the next section.

2. Numerical tool

The code used for the project is an extension of that employed in the *OWLS* project (Schaye *et al.* 2010). The *OWLS* code is a modified version of GADGET (Springel 2005) with a novel treatment of star formation (Schaye & Dalla Vecchia 2008), stellar evolution (Wiersma *et al.* 2009b), metal line cooling (Wiersma *et al.* 2009a) and black hole feedback (Booth & Schaye 2009). It was updated by introducing molecular hydrogen formation and cooling, gas shielding from the UV background (e.g. Nagamine *et al.* 2010), pop-III star formation and feedback, and stellar dust yields. Stellar feedback has been improved by adopting the model of Dalla Vecchia & Schaye (2012), where thermal energy from SN explosions is injected in the surrounding gas. As shown by these authors, the effect

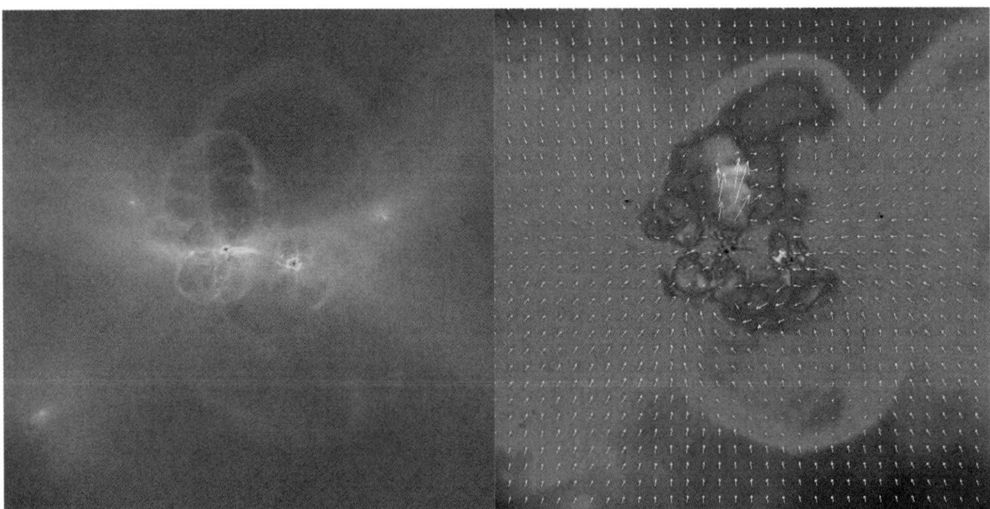

Figure 1. Projected density (left panel) and mass-weighted temperature (right panel) maps of a box of 900 co-moving kpc centred on the most massive halo in *FiBY04M*. The colour scale is logarithmic. Two interacting galaxies are in a filament which is horizontal in the picture. Several shock fronts surrounding underdense bubbles are clearly visible in the density map and are the result of several stellar feedback episodes. The temperature of the gas inside the bubbles (right panel) is larger than the background temperature and can be as high as 10^{5-6} K. The over-plotted velocity field shows the accretion along the filament as well as the galactic outflow.

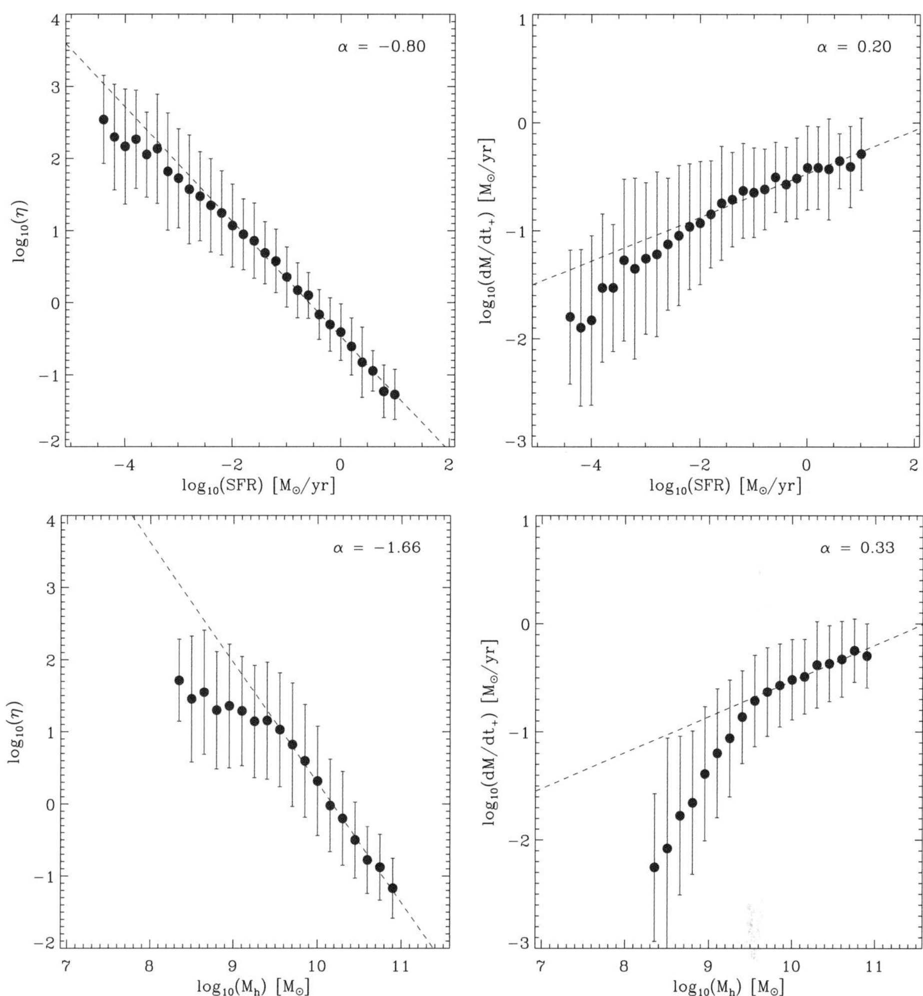

Figure 2. Outflow mass loading (left panels), η, and rate (right panels) as function of star formation rate (top row) and halo mass (bottom row). The dashed lines are fits of the bin averages for halo masses $M_{\rm h} > 10^{9.5}$ M$_\odot$. The slope of the fitting relations is reported in the top-right corner of each plot. The break in the relations is evident when plotting the values against halo mass.

of feedback depends on resolution. The resolution criterion they propose is matched in all but the largest volume simulations. The effect of feedback in creating large galactic outflows is depicted in Fig. 1. Hot and metal-rich bubbles extend well into the IGM contributing to the enrichment of a large volume. The metal volume-filling factor at $z = 6$ is $\sim 10^{-3}$ (Johnson *et al.* 2013).

3. Preliminary results

Fig. 2 shows outflow properties as function of star formation rate (SFR) and halo mass. The mass loading (the ratio between mass outflow rate and SFR), η, is plotted in the left panels, whereas the mass outflow rate is plotted in the right panels. All gas particles at the virial radius that are moving with (positive) radial velocity larger than the halo escape velocity are considered to be outflowing. There is a clear break in the relations at halo masses $M_\mathrm{h} \simeq 10^{9.5}$ $M_\odot$. This is not present in the galaxies mass and luminosity functions which match the observed data at $z \sim 6$. Therefore, the stellar mass of haloes does not seem to play a role.

What plays a role in defining the break is the stellar feedback. Feedback is more effective in removing gas from small haloes. Haloes in a low density environment cannot accrete gas efficiently, therefore, after feedback has removed a large part of their gas content, their stellar mass does not grow further. This becomes clear when plotting the ratio of stellar mass to gas mass, $f_* = M_*/M_\mathrm{g}$, where, for $M_\mathrm{h} < 10^{9.5}$ $M_\odot$, the ratio flattens. The flattening of f_* is in the numerically resolved mass range.

Okamoto *et al.* (2008) showed that the same characteristic halo mass (the so called filtering mass) is imprinted by re-ionisation processes, and galaxies in halos smaller than $10^{9.5}$ $M_\odot$ cannot grow efficiently. However, in *FiBY*, the flattening of f_* is observed well before re-ionisation happens in the simulations. The conclusion is that stellar feedback may be the main driver defining the filtering mass at any redshift before re-ionisation.

References

Booth, C. M. & Schaye, J. 2009, *MNRAS*, 398, 53
Dalla Vecchia, C. & Schaye, J. 2012, *MNRAS*, 426, 140
Finlator, K., Oppenheimer, B. D., & Davé, R. 2011, *MNRAS* 410, 1703
Johnson, J. L., Dalla Vecchia, C., & Khochfar, S. 2013, *MNRAS*, 428, 1857
Nagamine, K., Choi, J.-H., & Yajima, H. 2010, *ApJ*, 725, L219
Okamoto, T., Gao, L., & Theuns, T. 2008, *MNRAS*, 390, 920
Schaye, J., Dalla Vecchia, C., Booth, C. M., *et al.* 2010, *MNRAS*, 402, 1536
Schaye, J. & Dalla Vecchia, C. 2008, *MNRAS*, 383, 1210
Springel, V. 2005, *MNRAS*, 364, 1105
Wiersma, R. P. C., Schaye, J., & Smith, B. D. 2009, *MNRAS*, 393, 99
Wiersma, R. P. C., Schaye, J., Theuns, T., Dalla Vecchia, C., & Tornatore, L. 2009, *MNRAS*, 399, 574

The intriguing life of massive galaxies
Proceedings IAU Symposium No. 295, 2012
D. Thomas, A. Pasquali & I. Ferreras, eds.

© International Astronomical Union 2013
doi:10.1017/S1743921313004171

Impact of the First Stars to the First Galaxy Formation

Ke-Jung Chen[1], Myoungwon Jeon[2], Thomas Greif[3], Volker Bromm[2], and Alexander Heger[4]

[1]Minnesota Institute for Astrophysics, University of Minnesota, Minneapolis, MN 55455, USA

[2]Department of Astronomy, University of Texas, Austin, TX 78712, USA

[3]Institute for Theory and Computation, Harvard-Smithsonian Center for Astrophysics, Cambridge, MA 02138, USA

[4]Monash Centre for Astrophysics, Monash University, Victoria 3800, Australia

Abstract. We present the results from our cosmological simulations of the first stages of galaxy formation. We use `Gadget-2` (Springel 2005), modified to include detailed cooling, chemistry, and radiative transfer of primordial gas to study the impact of the first stars on galaxy formation. In contrast to previous work, we apply a realistic treatment of stellar feedback by using updated stellar models for the first stars. In this proceeding, we briefly summarize how stellar feedback from the first stars affects the primordial IGM inside the first galaxies.

Keywords. First stars, early Universe

Summary

In our simulations, the first-ever star with a mass of $60\,\mathrm{M}_\odot$ forms at a redshift of $z \sim 28$ inside a $5 \times 10^5\,\mathrm{M}_\odot$ dark matter halo. Once this star evolves to the main sequence, when the stable hydrogen burning at the core occurs, its surface temperature quickly rises to $\mathrm{T} \sim 10^5\,\mathrm{K}$ and begins to emit a large amount of photons that ionize hydrogen and helium. The gas inside the host halo is strongly photo-heated to temperatures, of $\mathrm{T} \sim 2 \times 10^4\,\mathrm{K}$, allowing the gas to escape the gravitational well of the host halo, forming an outflow. When the star dies, the I-front ultimately creates an extensive H II region of size $\sim 4\,\mathrm{kpc}$. It implies that radiative feedback of the first stars can significantly ionize the gas of the IGM and changes the collapse mass scale for the first galaxies (e.g., Bromm & Yoshida 2011). Besides radiative feedback, the first stars synthesize the first heavy chemical elements beyond hydrogen and helium during their evolution. These metals are eventually dispersed in the IGM when the stars die as supernovae. In our simulation, we assume the first-ever $60\,\mathrm{M}_\odot$ star dies as a core collapse supernova with explosion energy of $3 \times 10^{51}\,\mathrm{erg}$. Such a supernova explosion can efficiently spread the metals over a region of size about 1 kpc in a few million years and enrich the metallicity of pristine gas inside IGM to $10^{-3} - 10^{-6}\,\mathrm{Z}_\odot$. If the chemical enrichment is over a critical metallicity ($\sim 10^{-3}\,\mathrm{Z}_\odot$), it results in the formation of Pop II stars. Overall, the stellar feedback of the first stars can significantly affect the later star formation history and the assembly of the first galaxies.

References

Bromm, V. & Yoshida, N. 2011, *ARAA*, 49, 373
Springel, V. 2005, *MNRAS*, 364, 1105

The intriguing life of massive galaxies
Proceedings IAU Symposium No. 295, 2012
D. Thomas, A. Pasquali & I. Ferreras, eds.

© International Astronomical Union 2013
doi:10.1017/S1743921313004183

Discovery of bright z ~ 7 galaxies in the UltraVISTA survey

Rebecca A. A. Bowler, James S. Dunlop and Ross J. McLure

SUPA, Institute for Astronomy, University of Edinburgh, Edinburgh, EH9 3HJ
email: raab@roe.ac.uk

Abstract. We have exploited the new, deep, near-infrared Y, J, H, K_s UltraVISTA imaging of the COSMOS field, in tandem with deep optical and mid-infrared imaging, to conduct a new search for luminous galaxies at redshifts $z \simeq 7$. We have utilised this unique multi-wavelength dataset to select galaxy candidates at redshifts $z > 6.5$ by searching first for $Y + J$-detected objects which are undetected in the CFHT and HST optical data. This sample was then refined using a photometric redshift fitting code, enabling the rejection of lower-redshift galaxy contaminants and cool galactic M, L, T dwarf stars. The final result of this process is a small sample of (at most) ten credible galaxy candidates at $z > 6.5$ (from over 200,000 galaxies detected in the year-one UltraVISTA data). The new $z \simeq 7$ galaxies reported here are the first credible $z \simeq 7$ Lyman-break galaxies discovered in the COSMOS field and, as the most UV-luminous discovered to date at these redshifts, are prime targets for deep follow-up spectroscopy. We explore their physical properties, and briefly consider the implications of their inferred number density for the form of the galaxy luminosity function at $z \simeq 7$.

Keywords. galaxies: evolution, galaxies: formation, galaxies: high-redshift

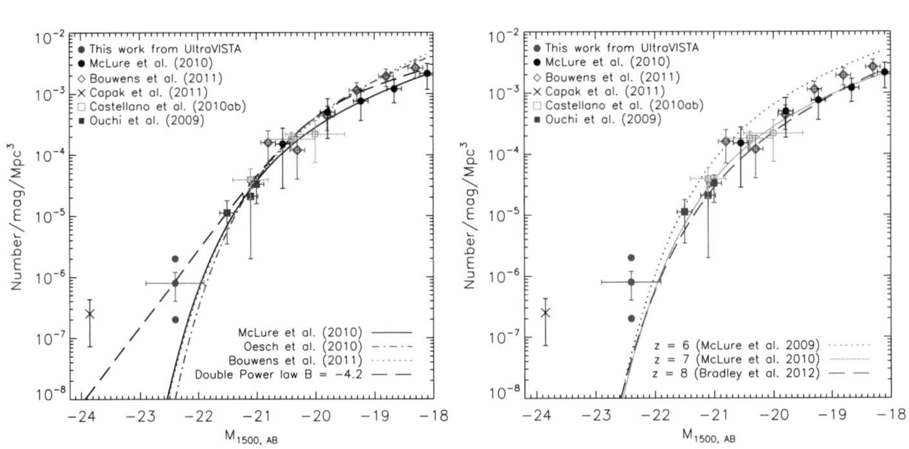

Figure 1. The above plots show a compilation of the current Schechter function determinations and data points for the $z = 7$ galaxy UV ($\simeq 1500$ Å) luminosity function. The estimated data–point from our new UltraVISTA study is shown in red, with Poissonian errors on the number density, with the upper, central and lower point given by ten, four or only one candidate in our sample being confirmed as a $z > 6.5$ galaxy. A double power-law curve is also included for comparison. In the right-hand plot we show the same data, but this time compared with the $z = 6, 7, 8$ luminosity functions.

References

Bowler, R. A. A., *et al.* 2012, *MNRAS*, 426, 2772

The intriguing life of massive galaxies
Proceedings IAU Symposium No. 295, 2012
D. Thomas, A. Pasquali & I. Ferreras, eds.

© International Astronomical Union 2013
doi:10.1017/S1743921313004195

Looking for molecular gas in a massive lyman break galaxy at z = 4.05

Qinghua Tan[1,2], Emanuele Daddi[1], Mark Sargent[1], Jackie Hodge[3], and Yu Gao[2]

[1] CEA, Laboratoire AIM, Irfu/SAp, F-91191 Gif-sur-Yvette, France
email: edaddi@cea.fr

[2] Purple Mountain Observatory, CAS, 2 West Beijing Road, 210008 Nanjing, China

[3] Max-Planck Institute for Astronomy, Königstuhl 17, 69117 Heidelberg, Germany

Abstract. We present a search for CO emission in a massive lyman break galaxy at z ~ 4.05.

Keywords. galaxies: high-redshift, galaxies: star formation, radio lines: galaxies

We have observed the CO(4-3) and CO (6-5) lines with the Plateau de Bure Interferometer. The observations of each individual configuration show a tentative detection at the $\sim 3\sigma$ level of CO emission at the position of the ACS/HST source. The signal is improved to S/N ~ 5 when combining CO (4-3) and CO (6-5) observations (Fig. 1). We have run extensive simulations to estimate that the chance probability of such a signal in our combined datacubes is $\sim 2 \times 10^{-4}$. Assuming that both detections are real, we infer a molecular gas mass of $\sim 1.4 \times 10^{11}$ M$\odot$ by adopting a conversion factor of $\alpha_{CO} \sim$ 7.0, which is based on the α_{CO} - metallicity relation (Magdis *et al.* 2011; Sargent *et al.* 2012b). The location of this galaxy in the L_{IR} - L'_{CO} plane suggests little variation from the trend defined by normal star-forming galaxies over $0 < z < 2.2$, possible evidence against a too strong evolution of the conversion factor to higher redshifts. The molecular gas ratio ($\sim 68\%$) is found to be comparable to the ratios observed at $z = 2$ (Magdis *et al.* 2012a), providing additional support for the existence of a plateau in the redshift evolution of the specific SFR of normal galaxies at $z > 3$. However, we need more CO observations to make a definitive detection and thus further confirm these conclusions.

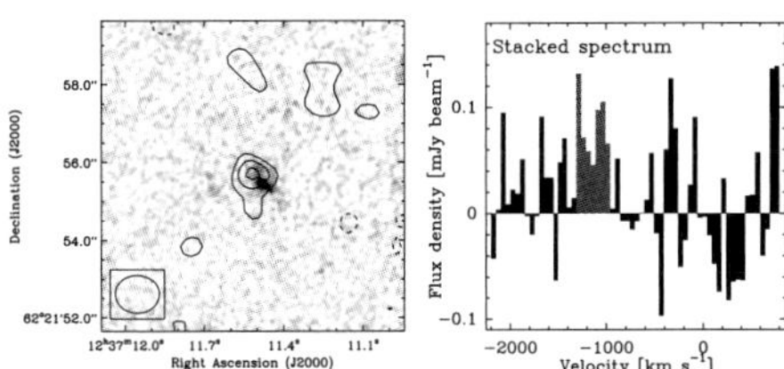

Figure 1. Left: Countours of stacked CO (4-3) and CO (6-5) overlaid on HST+WFC3 F140W image. Countour levels start at $\pm 2\sigma$ and are in steps of 1σ, with positive(negative) countours shown as solid (dashed) lines. **Right**: Combined CO spectrum adopting average line ratio from GN20 and M82 total SLED models. The red color indicates the maximum emission region.

References

Magdis, G., Daddi, E., Elbaz, D., *et al.* 2011, *ApJ*, 740, 15
Magdis, G., Daddi, E., Sargent, M., *et al.* 2012a, *ApJ*, 758, 9
Sargent, M., *et al.* 2012b, in preparation

The first few billion years

The intriguing life of massive galaxies
Proceedings IAU Symposium No. 295, 2012
D. Thomas, A. Pasquali & I. Ferreras, eds.

© International Astronomical Union 2013
doi:10.1017/S1743921313004201

The size and mass evolution of the massive galaxies over cosmic time

Ignacio Trujillo[1,2]

[1]Instituto de Astrofísica de Canarias, c/ Vía Láctea s/n, E-38205, La Laguna, Tenerife, Spain

[2]Departamento de Astrofísica, Universidad de La Laguna, E-38205, La Laguna, Tenerife, Spain
email: trujillo@iac.es

Abstract. Once understood as the paradigm of passively evolving objects, the discovery that massive galaxies experienced an enormous structural evolution in the last ten billion years has opened an active line of research. The most significant pending question in this field is the following: which mechanism has made galaxies to grow largely in size without altering their stellar populations properties dramatically? The most viable explanation is that massive galaxies have undergone a significant number of minor mergers which have deposited most of their material in the outer regions of the massive galaxies. This scenario, although appealing, is still far from be observationally proved since the number of satellite galaxies surrounding the massive objects appears insufficient at all redshifts. The presence also of a population of nearby massive compact galaxies with mixture stellar properties is another piece of the puzzle that still does not nicely fit within a comprehensive scheme. I will review these and other intriguing properties of the massive galaxies in this contribution.

Keywords. galaxies: elliptical and lenticular, cD, galaxies: evolution, galaxies: formation, galaxies: fundamental parameters, galaxies: high-redshift, galaxies: structure

1. Introduction

The discovery that massive galaxies were much more compact in the past (Daddi *et al.* 2005; Trujillo *et al.* 2006) revolutionized our traditional picture of how these objects have developed with cosmic time. A monolithic-like scenario, where the bulk of the stellar population as well as the structure of these galaxies are formed in a single dissipative event followed by a passive evolution, is not longer supported by the observations.

As any shift in scientific paradigm, there has been an enormous debate about the reality of this huge structural evolution. Most of the critics against this discovery focused on the reliability of the size estimations and the accuracy of the stellar mass determinations of these z>1 objects (e.g. Mancini *et al.* 2010; Muzzin *et al.* 2009). Today, ultra-deep observations of these galaxies (e.g. Carrasco *et al.* 2010; Cassata *et al.* 2010) as well as the first dynamical estimations of their masses (e.g. Cenarro & Trujillo 2009; Cappellari *et al.* 2009) have inclined the vast majority of the community to accept as real the size evolution of the massive galaxies. But not only the size of the massive galaxies have dramatically changed with cosmic time, also the morphological content among the family of massive galaxies has drastically varied as redshift decreases (see Fig. 1). In fact, present-day massive galaxies are composed mostly by objects with spheroidal-like appearance. At high-z, the most common morphology of the massive galaxies resembled disk-like structures (e.g. van der Wel *et al.* 2011; Buitrago *et al.* 2013).

The stellar mass-size relation of massive galaxies seem to be at place (although with a different "zeropoint" position than in the present-day universe) since at least z∼3 (e.g. Trujillo *et al.* 2007; Buitrago *et al.* 2008) and the scatter along this relation has not significantly changed since then (see Fig. 2). However, the number of galaxies that populate these relations have grown with time as the number density of massive galaxies have continuously increasing since that epoch (e.g. Pérez-González *et al.* 2008). That means that the new massive galaxies that are incorporated in the stellar mass-size relation are located in such sense that do not alter dramatically this relation. In order to maintain the scatter of this relation relatively constant with time, the newcomers should evolve later in size similarly as the older galaxies that already populated the stellar mass-size relation.

On what follows I will summarize the different scenarios that have been proposed to explain the significant structural change of the massive galaxies as well as the observational evidence favoring the different mechanisms. We adopt a cosmology with $\Omega_m = 0.3$, $\Omega_\Lambda = 0.7$ and $H_0 = 70$ km s^{-1} Mpc^{-1}.

2. What is the physical mechanism behind the size evolution?

If we accept the reality of the structural evolution of the massive galaxies, the next question to solve is how these objects have reached their present configuration. We can summarize the different proposed scenarios in three categories. It is worth stressing that the following mechanisms can take all place simultaneously and certainly they should all have a role in the evolution of the massive galaxies. Consequently, when we use the word rejected or supported by the observations we will be referring to the role of such mechanism as the main driver of the size evolution.

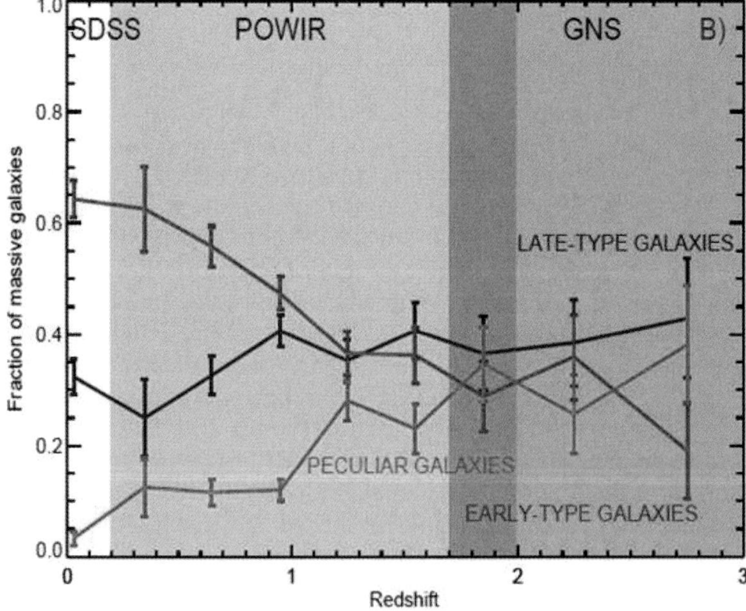

Figure 1. Fraction of massive ($M_\star \sim 10^{11} M_{sun}$) galaxies as function of redshift segregating the objects according to their visual morphological classification. Blue color represents late type (S) objects and red early type (E+S0) galaxies, while peculiar (ongoing mergers and irregulars) galaxies are tagged in green. Different color backgrounds indicate the redshift range expanded for each survey used: SDSS, POWIR/DEEP2 and GNS. Error bars are estimated following a binomial distribution. Figure taken from Buitrago *et al.* (2013).

• Major mergers. This was the earliest theoretical suggestion (e.g. Naab *et al.* 2007; Nipoti *et al.* 2010) and it was also the first hypothesis rejected by the observations (e.g. Bundy *et al.* 2009; Wild *et al.* 2009; de Ravel *et al.* 2009; Bluck *et al.* 2009; López-San Juan *et al.* 2010). Simply, there is not enough number of major mergers that can account by the huge size evolution observed (a factor of 4 since z∼2; Trujillo et al. 2007) plus the relatively modest evolution in stellar mass (a factor of 2 since z∼2; van Dokkum et al. 2010). The predicted size evolution as a function of the increase in mass goes as: $\Delta r_e \propto \Delta M$ in major mergers (e.g. Ciotti & van Albada 2001; Boylan-Kolchin *et al.* 2006) which is insufficient to produce the observed size evolution.

• Puffing up. Fan *et al.* (2008; 2010) as well as Damjanov *et al.* (2009) proposed a scenario where the size evolution is connected with the massive expulsion of gas by the effect of an AGN (Fan *et al.*) or stellar winds (Damjanov *et al.*). According to this mechanism, the removal of gas changes the gravitational potential of the galaxy making the object to puff up to its new (larger) configuration. This evolution is fast ($\lesssim 1$ Gyr; Ragone-Figueroa & Granato 2011) and the model predicts a dichotomy of massive objects at all redshifts: young ones (<1 Gyr) with small sizes and high velocity dispersions (∼400 km/s) and old ones (>1 Gyr) with present-day sizes and moderate velocity dispersion (∼200 km/s). This is not observed in nature: massive compact galaxies at high-z are "old" at those epochs and there is not an age segregation in the stellar mass-size relation since, at least, z=1 for objects with spheroid-like morphologies (Trujillo *et al.* 2011). Summarizing, this scenario is also not favored observationally.

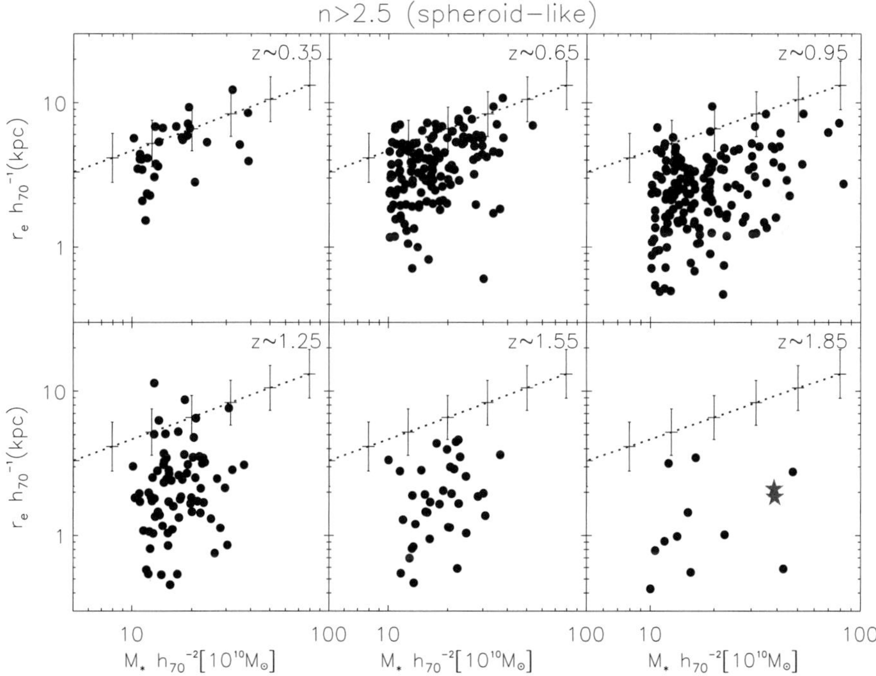

Figure 2. Stellar mass-size distribution of our high-concentrated (spheroid-like) galaxies. Over–plotted on the observed distribution of points are the mean and dispersion of the distribution of the Sérsic half-light radius of the SDSS early-type (n>2.5; Shen *et al.* 2003) galaxies as a function of the stellar mass. For clarity, individual error bars are not shown. The mean size relative error is <30 per cent. Uncertainties in the stellar mass are ∼0.2 dex. Solid black points are from the ACS sample of Trujillo *et al.* (2007), red stars from Carrasco *et al.* (2010) Gemini high resolution imaging.

- Minor mergers. This model (Khochfar & Burkert 2006; Maller *et al.* 2006; Hopkins *et al.* 2009b; Naab *et al.* 2009; Sommer-Larsen & Toft 2010; Oser *et al.* 2010) proposes that most of the size evolution of the massive galaxies has taken place due to the continuous accretion of minor bodies. The stars of these merged satellites are mainly located in the periphery of the main body, making this mechanism an excellent vehicle for the size evolution. The predicted increase in size as a function of the increase in mass goes as: $\Delta r_e \propto \Delta M^2$ (e.g. Naab *et al.* 2009). This evolutionary path predicts the following observables: a continuous increase in size of the global population of massive galaxies, a size growth not related with the age of the main galaxy, a mild velocity dispersion evolution of the massive galaxy with time (Hopkins *et al.* 2009b).

3. Observational evidence favoring the minor merging scenario

There are many observational evidences favoring the minor merging hypothesis as the main channel of massive galaxies growth. We can summarize them in three groups:

- The size evolution of the spheroid-like massive galaxies is not related with the age of their stellar population. Since $z \sim 1$, spheroid-like massive galaxies, at a given fixed stellar mass, still need to grow by a factor ~ 2 to reach their present configuration. This significant size evolution is observed, at all redshifts, to be independent of the stellar age of the massive galaxies (Trujillo *et al.* 2011). This observation points out to a size growth mechanism that does not know about the age of the main galaxy. An external accretion of stars (where the infalling satellites do not have previous knowledge about the age of the central galaxy) fits well within this scheme.

- There is a progressive and steady formation of the outer galaxy envelopes. The central stellar mass density of the massive galaxies at high-z do not dramatically differ from the central stellar mass density of the nearby massive galaxies (Bezanson *et al.* 2009; Hopkins *et al.* 2009a). The majority of the evolution of the stellar mass density profile of the massive galaxies has taken place at their extended wings. Massive galaxies have steadily increased their number of stars at farther distances (van Dokkum *et al.* 2010). This progressive build-up is very suggestive of a continuous accretion of new stars with cosmic time in the periphery of these galaxies.

- At a fixed stellar mass, the velocity dispersion of the massive galaxies has mildly declined since $z \sim 2$. Cenarro & Trujillo (2009) compiled from the literature the velocity dispersions of many massive ($M_\star \sim 10^{11} M_{sun}$) galaxies since $z \sim 2$. This compilation took data from van der Wel et al (2005; 2008) at $0.5 < z < 1$ and di Serego Alighieri *et al.* (2005) at $z \sim 1$. This data was complemented with the measurement of the velocity dispersions of massive galaxies in the SDSS (for having a local reference) and with the first estimation of the velocity dispersion of massive galaxies at $z > 1.5$ (using the published stacked spectrum of Cimatti *et al.* 2008). All this data together (see Fig. 3) clearly indicated that the evolution of the velocity dispersion of the massive galaxies, at a fixed stellar mass, has only moderately declined with cosmic time. This result has been later confirmed by many new estimations of the velocity dispersion of massive galaxies at $1.5 < z < 2$ (e.g. Cappellari *et al.* 2009; Onodera *et al.* 2010; van de Sande *et al.* 2011; Newman *et al.* 2010, Toft *et al.* 2012). This mild evolution of the velocity dispersion is in good agreement with the idea that most of the structural evolution of the massive galaxies has taken place in their outer regions. This again fits well with a scenario of accretion of new stars that is smooth and mostly locate stars in the periphery of these objects. The fact that the central stellar mass density of the massive galaxies has only changed mildly since $z \sim 2$ also agrees with the fact that the central velocity dispersion of these objects have not changed significantly (see a much elaborated discussion of this point in Trujillo *et al.* 2012).

4. Some puzzling observations

So far, both the cosmological simulations as well as the observational evidence favor the minor merging scenario as the main driver of size and mass evolution of the massive galaxies. However, there are two observational evidences that are not easy to understand, at least with the present theoretical development, within the minor merging hypothesis. These two puzzling observations are: the scarcity of massive compact galaxies in the local universe and the factor of 2 less satellites surrounding the massive galaxies at all redshifts compared with the model predictions.

4.1. *Nearby massive compact galaxies: relics of the early universe?*

After the discovery that massive galaxies at high-z were compact, there was an observational effort to try finding massive ($M_\star \sim 10^{11} M_{sun}$) and compact ($r_e \sim 1$ kpc) objects in the nearby Universe (see Fig. 4). According to the theoretical predictions (Hopkins *et al.* 2009b) around 10% of the massive compact galaxies since z$\sim$2 should have survived intact due to the stochastic nature of the merging channel. Taking into account that the

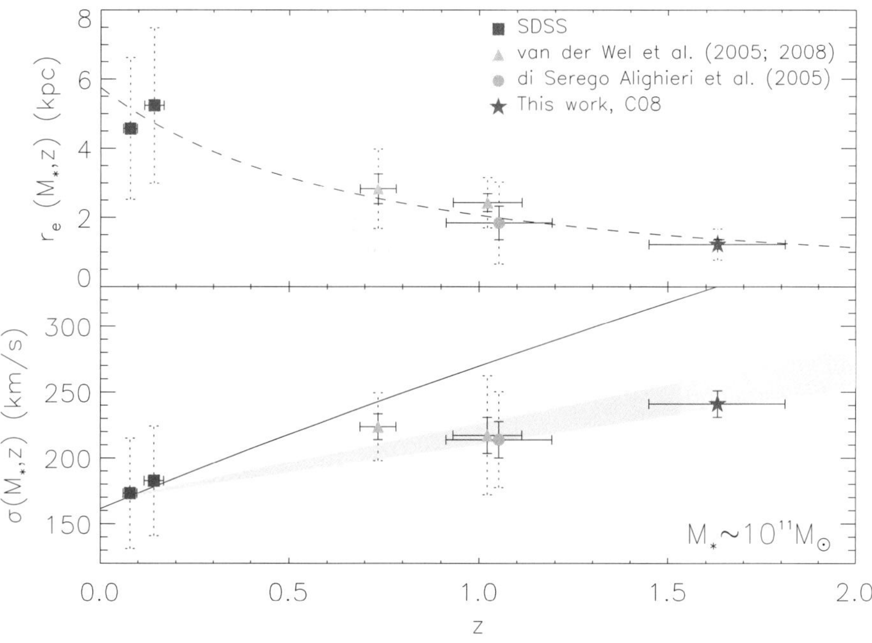

Figure 3. Top panel: size evolution of $M_\star \sim 10^{11} M_{sun}$ spheroid-like galaxies as a function of redshift. Different symbols show the median values of the effective radii for the different galaxy sets considered in this work (see Section 3), as indicated in the labels. Dashed error bars, if available, show the dispersion of the sample, whereas the solid error bars indicate the uncertainty of the median value. The dashed line represents the observed evolution of sizes $r_e(z) \propto (1 + z)^{-1.48}$ found in Buitrago *et al.* (2008) for galaxies of similar stellar mass. Bottom panel: velocity dispersion evolution of the spheroid-like galaxies as a function of redshift, with symbols as given above. Assuming the Buitrago *et al.*'s (2008) size evolution, the solid line represents the prediction from the "puffing-up" scenario (Fan *et al.* 2008), whereas the gray area illustrates the velocity dispersion evolution within the merger scenario of Hopkins *et al.* (2009b) for $1 < \gamma < 2$. Figure from Cenarro & Trujillo (2009).

extensive dynamical modelling effort of Cappellari *et al.* (2012a) discovered a systematic trend in the IMF of ETGs. They found a mass normalization for the IMF varying between values consistent with a Kroupa/Chabrier IMF to heavier than a Salpeter IMF (assuming the population models are correct) as a function of either the mass-to-light ratio (M/L) or the galaxy stellar velocity dispersion σ. This IMF trend implies potential systematic errors of up to a factor 2–3 in the galaxies stellar masses derived using stellar population models.

3. Uncertainties of galaxy sizes

A second observable that is often used to constrain galaxy formation scenarios, in combination with galaxy mass, is the galaxy size, commonly parametrized by the projected half-light radius R_e. The evolution of galaxy size with its mass depends sensitively on the galaxy assembly mechanism (see Ciotti 2009 for a comprehensive review). R_e increases nearly proportionally to mass in the case of a galaxy assembly process via gas-poor major (1:1) mergers (Hernquist *et al.* 1993), due to energy conservation, while R_e increases more rapidly during gas-poor minor (1:3 or less) merging (Nipoti *et al.* 2003; Boylan-Kolchin *et al.* 2006). R_e *decreases* during gas-rich mergers or cold accretion, when gas accumulates towards the centre (e.g. Mihos & Hernquist 1994).

Massive (stellar mass $M_\star > 10^{11}$ M$_\odot$) passive galaxies at $z \sim 2$ were found to have much smaller R_e than their local counterparts of the same mass (Daddi *et al.* 2005; di Serego Alighieri *et al.* 2005; Trujillo *et al.* 2006; Longhetti *et al.* 2007; Toft *et al.* 2007; Cimatti *et al.* 2008; van Dokkum *et al.* 2008). This suggest a scenario in which the progenitors of today's massive ETGs assemble most of their mass via minor dry mergers (e.g. Naab *et al.* 2009; Bezanson *et al.* 2009; Hopkins *et al.* 2009), to account for the rapid size increase.

Unfortunately the half-light radius R_e is an observationally ill-defined quantity. By definition measuring R_e requires an extrapolation of the galaxy surface brightness profile to infinite radii, to estimate the galaxy *total* stellar light. This seemed a sensible approach a few decades ago, when the technique was used to quantify the sizes of elliptical galaxies, which were thought to be all well described by homologous $R^{1/4}$ de Vaucouleurs (1948) surface-brightness profiles. But we now know that ETGs are best represented by Sersic (1968) profiles, with concentration that systematically increases with galaxy mass (Caon *et al.* 1993). Moreover the ATLAS^{3D} volume-limited ($M_\star > 6 \times 10^9$ M$_\odot$) survey of nearby ETGs (Cappellari *et al.* 2011a) has revealed that the majority of the systems are more closely related to spiral galaxies than to genuine spheroidal ellipticals (Cappellari *et al.* 2011b). In fact as much as 2/3 of the galaxies classified as ellipticals are dominated by stellar disks, which are often missed by the photometry, but are clearly revealed by their integral-field stellar kinematics (Krajnović *et al.* 2011; Emsellem *et al.* 2011).

The complex and multi-component nature of ETGs makes it impossible to know how the outer unobservable surface brightness profile should be extrapolated. Different authors use photometry of different quality and make different choices for the profile extrapolation, as well as for its extraction. For these reason, even from very good quality photometry of nearby galaxies, R_e is often not known with an accuracy better than about a factor 2×, with errors dominated by systematics effects. This is illustrated in the left panel of Fig. 1 (from Chen *et al.* 2010), which provides a comparison between independent determinations of R_e for well-studied galaxies in the Virgo cluster.

Uncertainties and systematic biases in R_e are likely to be more severe when galaxies at high redshift are compared to local determinations. Differences in the measurement approach combine with (i) differences in the restframe wavelength of the observations,

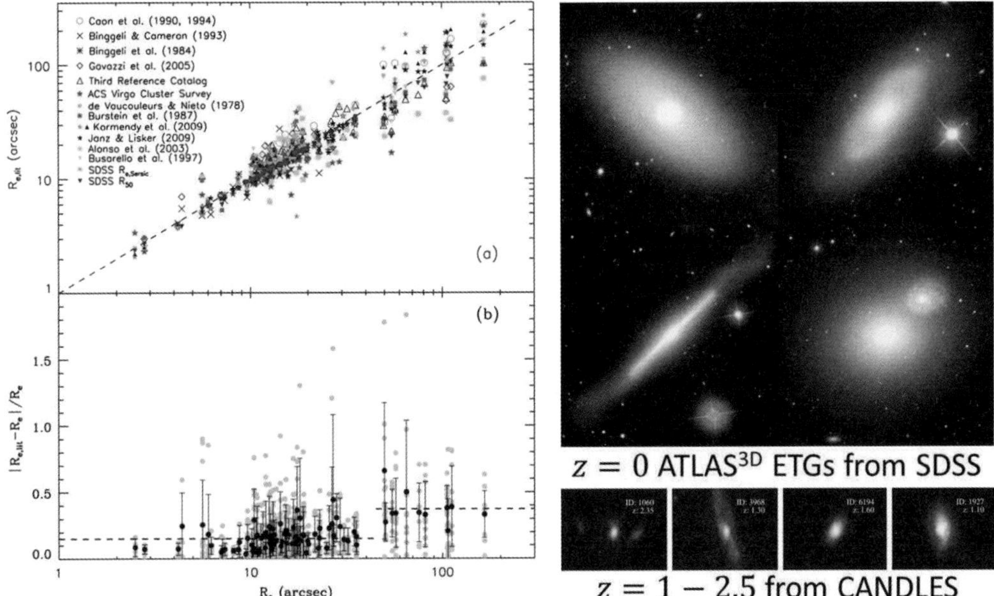

Figure 1. Left: Uncertainty of $R_{\rm e}$ for galaxies in the Virgo cluster. *Top Panel:* Different symbols compare the SDSS half-light radii from Chen *et al.* (2010) with independent determinations of the same quantity from the literature. *Bottom Panel:* Gray dots show the fractional error in the effective radius measurements, plotted as a function of the SDSS effective radii; the mean and standard deviation for individual galaxies are plotted as the black dots with error bars. Differences in $R_{\rm e}$ of a factor 2× are common, especially for the largest galaxies (taken from Chen *et al.* 2010). **Right: Comparing images of nearby and distant ETGs.** *Top Panels:* true colour images of nearby ETGs from the ATLAS³ᴰ sample as imaged by the SDSS survey (Abazajian *et al.* 2009). *Bottom Panels:* true colour images of galaxies at $z = 1 - 2.5$ from the CANDLES survey (taken from Szomoru *et al.* 2012)

(ii) variations in the depth of the photometry (e.g. due to cosmological surface-brightness dimming) and (iii) inferior spatial sampling of the imaging of distant galaxies (see the right panel of Fig. 1), (iv) possible colour gradients and nuclear AGN activity, which are generally stronger in the early Universe.

4. Local benchmark for dynamical scaling relations

An alternative to measuring galaxy sizes to study the evolution of galaxy densities consists of measuring the galaxies velocity dispersion σ. This is more difficult to measure than sizes, as it requires spectroscopic data (resolution $R \approx 1000$) rather than imaging alone. However σ does not suffer from the biases of $R_{\rm e}$ and is closely related to it via the scalar virial equation (e.g. Cappellari *et al.* 2006)

$$\sigma^2 \approx \frac{G\,M_{\rm dyn}}{5.0\,R_{\rm e}}. \tag{4.1}$$

Here $M_{\rm dyn} = 2 \times M_{1/2} \approx M_\star$, where $M_{1/2}$ is the mass within a sphere enclosing half of the total galaxy light, and the last approximation is due to the fact that $M_{1/2}$ is dominated by the stellar mass (Cappellari *et al.* 2012c). The factor 5.0 was calibrated for $R_{\rm e}$ measured in the classic way (Burstein *et al.* 1987; de Vaucouleurs *et al.* 1991; Jorgensen *et al.* 1995) using fixed de Vaucouleurs (1948) $R^{1/4}$ growth curves to extrapolate the outer profiles and using typical photometry of nearby galaxies.

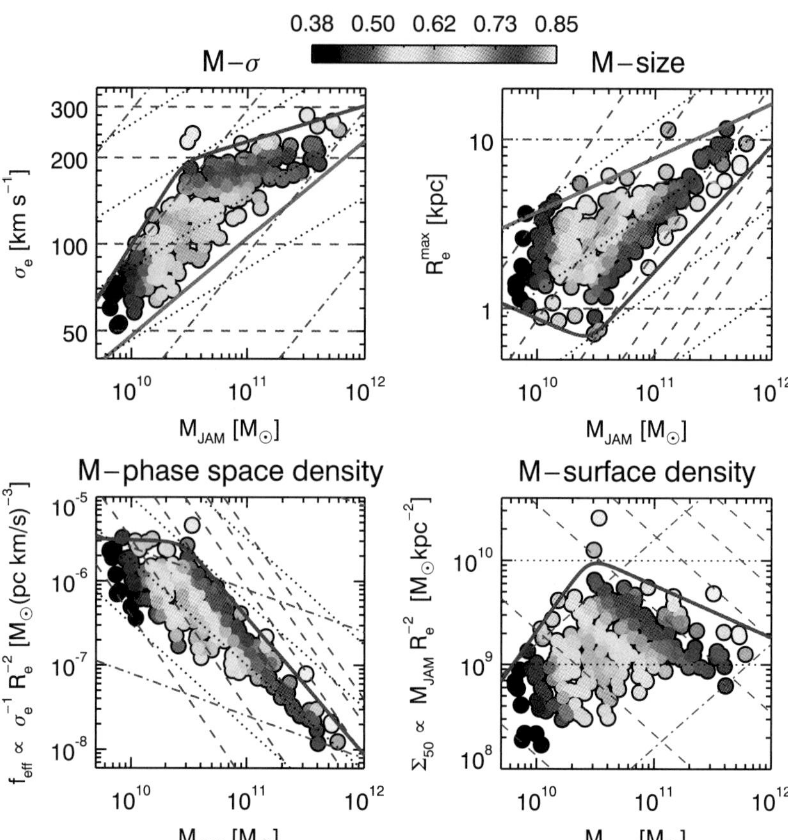

Figure 2. The Mass Plane and it projections. The top two panels show the $(M_{\rm JAM}, \sigma_e)$ and $(M_{\rm JAM}, R_e^{\rm max})$ coordinates. Overlaid are lines of constant $\sigma_e = 50, 100, 200, 300, 400, 500$ km s^{-1} (dashed blue), constant $R_e^{\rm max} = 0.1, 1, 10, 100$ kpc (dot-dashed red) and constant $\Sigma_e = 10^8, 10^9, 10^{10}, 10^{11}$ $M_\odot$ kpc^{-2} (dotted black) *predicted* by the virial equation (4.1). The observed $(M_{\rm JAM}, \sigma_e, R_e^{\rm max})$ points follow the relation so closely that the coordinates provide a unique mapping on these diagram and one can reliably infer characteristics of the galaxies from any individual projection. In each panel the galaxies are coloured according to the (LOESS smoothed) $\log(M/L)_{\rm JAM}$ values, as shown in the colour bar. Moreover in all panels the thick red line shows the same ZOE relation projected via equation (4.1). The green line is the $M - R_e$ relation for late spiral galaxies (equation 3 from Cappellari *et al.* 2011a), which approximately defines the boundary where ETGs disappear (taken from Cappellari *et al.* 2012b).

To study the evolution of galaxy parameters with redshift, a reliable local benchmark is essential. In Fig. 2 we show the (M, σ) and (M, R_e) (and other projections) of the mass plane (M, R_e, σ) of ETGs obtained by the ATLAS$^{\rm 3D}$ project via detailed dynamical modelling of 260 galaxies. The study found that, when accurate and unbiased masses are used, ETGs satisfy equation (4.1) quite accurately. Different two-dimensional projections of the plane contain the same amount of information except for a coordinate transformation. This confirms in particular that one can reliably estimate σ from the knowledge of M and R_e using equation (4.1). The study also found that the zone of avoidance (ZOE) defined by local ETGs in the (M, R_e) plane shows a clear break at a characteristic mass $M_{\rm dyn} \approx 3 \times 10^{10}$ $M_\odot$, with a corresponding break in the ZOE in the (M, σ) plane and other projections. Although indications for a break in the Faber

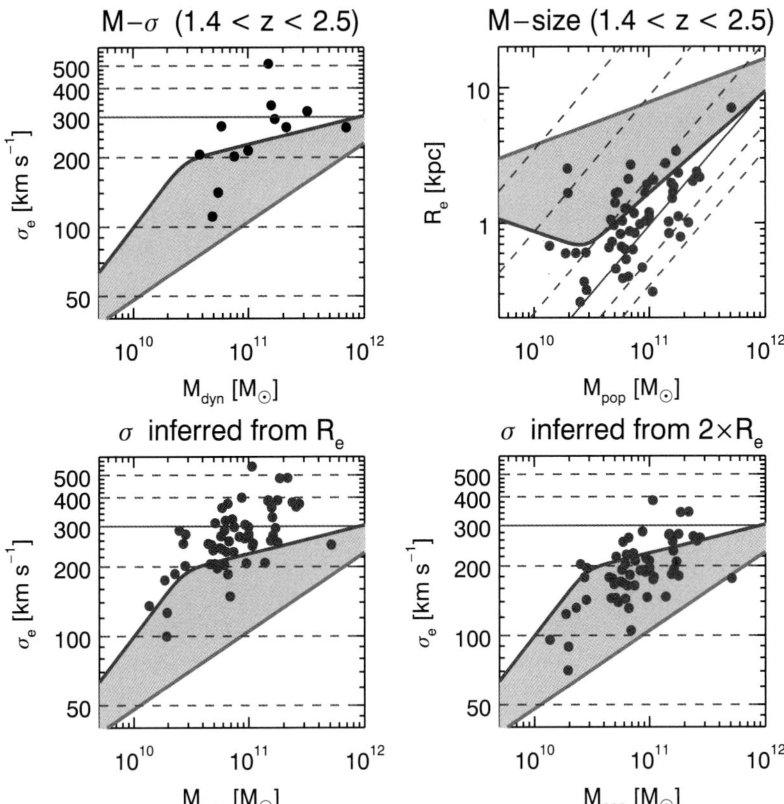

Figure 3. Mass, size and σ distributions at $z \sim 2$. *Top Left:* The observed (M, σ) determinations (black filled circles) for ETGs with $1.4 < z < 2.5$ (from Cenarro & Trujillo 2009; Cappellari *et al.* 2009; van Dokkum *et al.* 2009; van de Sande *et al.* 2011; Newman *et al.* 2010; Onodera *et al.* 2012; Toft *et al.* 2012) is overlaid to the locus of local ETGs from ATLAS^{3D} (grey region). The red and green thick lines are the same as in Fig. 2. *Top Right:* as in the previous panel, for the distribution of (M, R_{e}) determinations (blue filled circles) (from Cassata *et al.* 2011; Szomoru *et al.* 2012). *Bottom Left:* The (M, σ) distribution *predicted* using equation (4.1) from the values in the top-right panel. *Bottom Right:* as in the previous panel, but using $2R_{e}$ to estimate σ instead of the observed R_{e}.

& Jackson (1976) relation between galaxy luminosity and σ (Davies *et al.* 1983) or in the Kormendy (1977) relation between luminosity and surface brightness (Binggeli *et al.* 1984) were found some time ago, this is the first time the break is presented using dynamical masses instead of luminosity, and is shown to represent two projections of the galaxy distribution on the mass plane. This was interpreted as evidence for a change in the galaxy accretion mechanism between bulge growth (internal star formation) and gas-poor merging (external star formation)

5. Importance of σ determinations at high-z

In the top-left panel of Fig. 3 we show the available determinations of σ for galaxies in the redshift interval $1.4 < z < 2.5$ (Cenarro & Trujillo 2009; Cappellari *et al.* 2009; van Dokkum *et al.* 2009; van de Sande *et al.* 2011; Newman *et al.* 2012; Onodera *et al.* 2012; Toft *et al.* 2012) and compare it with the location defined by nearby ATLAS^{3D} ETGs. There is an indication for the high-redshift σ determinations to lie generally above the locus of local galaxies with the same mass, as expected for more compact systems. However, with the exception of one galaxy, the differences are relatively modest and are mostly consistent with local galaxies within the measurement errors.

In the top-right panel of Fig. 3 we show state-of-the-art size determinations of R_e for passive galaxies (Cassata *et al.* 2011; Szomoru *et al.* 2012), in the same redshift range. They were obtained from deep HST/WFC3 observations in the GOODS and CANDLES fields. The photometric samples are much larger than the one with σ determinations, and in this case the difference between the locus of local ETGs and the $z \sim 2$ ones is more dramatic, with the high-z galaxies having much smaller sizes.

In the bottom-left panel of Fig. 3 the observed (M, R_e) distribution of the $z \sim 2$ ETGs is converted into the (M, σ) plane via equation (4.1). This shows that the observed distribution of galaxy sizes implies more extreme σ values than the ones that have been found so far. The expected σ may even be larger than predicted here, due to the fact that high-z galaxies not only are observed to be smaller, but also tend to have steeper profiles than their local counterparts (e.g. Szomoru *et al.* 2012). One way to bring the high-z predicted σ values into better qualitative agreement with the few available determinations, would be to assume the R_e values at $z \sim 2$ are underestimated by about a factor of $2\times$ (bottom-right panel of Fig. 3).

However the comparison between the currently available σ and R_e determinations at $z \sim 2$ is not sufficient to conclude that the high-z R_e determinations are underestimated. The sample of galaxies with available σ is still too small to draw robust conclusions, and it is selected in a very different way than the more complete photometric samples. Moreover the galaxies with available σ generally satisfy equation (4.1) reasonably well. What the comparison does emphasizes is the importance of obtaining more good-quality σ values of high-z galaxies (see e.g. van de Sande in these proceedings), until a fully consistent picture emerges.

The possibility of biases in the R_e determinations of high-z galaxies also emphasizes the danger of using the local virial equation (4.1) for quantitative estimates of M/L, e.g. to study the evolution of the Fundamental Plane or to explore possible variations of the IMF with redshift. There is a clean solution to the problem. It consists of using detailed axisymmetric dynamical models of high-z galaxies (van der Marel & van Dokkum 2007; van der Wel & van der Marel 2008; Cappellari *et al.* 2009), which describe in detail the shape and surface brightness profiles of the galaxies. Contrary to the dynamical M/L inferred from the virial relation, the M/L derived via dynamical models is virtually insensitive to even extreme changes in the surface brightness distribution outside the region where stellar kinematics is available. Although the correctness of this fact is best demonstrated with numerical experiments (e.g. Cappellari *et al.* 2009), the reason for this robustness is similar to the one for which, in the spherical limit, the knowledge of the orbit of a single star is sufficient to infer the mass enclosed inside its orbit, irrespective of the large-scale mass distribution.

A demonstration of this modelling approach is given in the left panel of Fig. 4, which shows a first attempt at measuring the IMF normalization for galaxies at $z \sim 2$ (Cappellari *et al.* 2009). The study used dynamical models to reproduce σ determinations and concluded that some of the high-z ETGs are *required* to have a mass normalization of the IMF as low as Kroupa/Chabrier, like local ETGs. However the study allows for the possibility of some of the studied galaxies to have a heavier Salpeter-like mass normalization. This is not inconsistent with local findings, which indicate the IMF should vary between Kroupa to Salpeter in the interval $\sigma \approx 80 - 260$ km s^{-1} (right panel of Fig. 4; Cappellari *et al.* 2012b). But the current data and samples are still not sufficient to draw strong conclusions on the IMF at high-z.

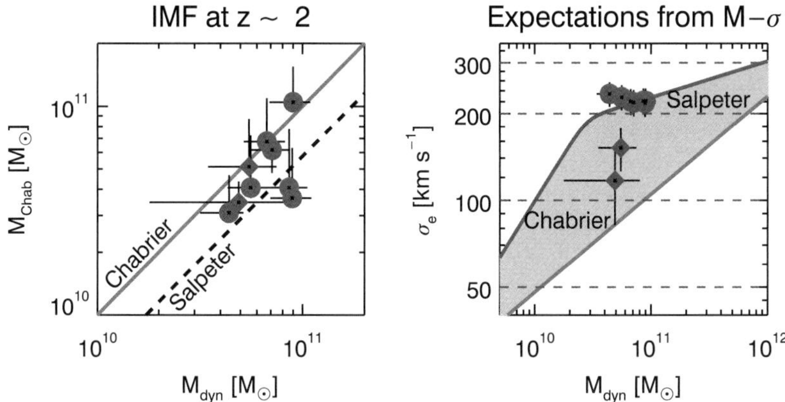

Figure 4. Constraining the IMF at $z \sim 2$. *Left Panel:* dynamical versus population masses determined using a Chabrier IMF. Blue diamonds are galaxies with individually measured σ, while red filled circles are galaxies for which σ from a stacked spectrum was assumed. The green solid line is the one-to-one relation. The dashed line is the one-to-one relation if the IMF was of Salpeter type. The latter is inconsistent with some of the determinations, indicating that some of the object require a Chabrier-like IMF (taken from Cappellari *et al.* 2009). *Right Panel:* Location of the galaxies in the left panel in the (M, σ) plane. The grey area and thick red and green lines are as in Fig. 2. The labels show the IMF normalization expected at different locations on this plane for local ETGs (Cappellari *et al.* 2012b).

The near future looks bright for dynamical studies of galaxies at $z \sim 2$, thanks to the availability of multi-object Spectrometers like MOIRCS on the Subaru telescope, the upcoming MOSFIRE on the Keck telescope and KMOS on the Very Large Telescope. These are optimized for multiplexed near-infrared observations of distant galaxies and should provide in a few years numerous σ determinations at high redshift ETGs.

Acknowledgements

I acknowledge support from a Royal Society University Research Fellowship.

References

Abazajian, K. N., *et al.* 2009, *ApJS*, 182, 543

Auger, M. W., Treu, T., Gavazzi, R., Bolton, A. S., Koopmans, L. V. E., & Marshall, P. J. 2010, *ApJ*, 721, L163

Bastian, N., Covey, K. R., & Meyer, M. R. 2010, *ARAA*, 48, 339

Bezanson, R., van Dokkum, P. G., Tal, T., Marchesini, D., Kriek, M., Franx, M., & Coppi, P. 2009, *ApJ*, 697, 1290

Binggeli, B., Sandage, A., & Tarenghi, M. 1984, *AJ*, 89, 64

Boylan-Kolchin, M., Ma, C.-P., & Quataert, E. 2006, *MNRAS*, 369, 1081

Bruzual, G. & Charlot, S. 2003, *MNRAS*, 344, 1000

Burstein, D., Davies, R. L., Dressler, A., Faber, S. M., Stone, R. P. S., Lynden-Bell, D., Terlevich, R. J., & Wegner, G. 1987, *ApJS*, 64, 601

Caon, N., Capaccioli, M., & D'Onofrio, M. 1993, *MNRAS*, 265, 1013

Cappellari, M., *et al.* 2006, *MNRAS*, 366, 1126

Cappellari, M., *et al.* 2009, *ApJ*, 704, L34

Cappellari, M., *et al.* 2011a, *MNRAS*, 413, 813

Cappellari, M., *et al.* 2011b, *MNRAS*, 416, 1680

Cappellari, M., *et al.* 2012a, *Nature*, 484, 485

Cappellari, M., *et al.* 2012b, *MNRAS*, accepted (arXiv:1208.3523)

Cappellari, M., *et al.* 2012c, *MNRAS*, accepted (arXiv:1208.3522)

Cassata, P., *et al.* 2011, *ApJ*, 743, 96

Cenarro, A. J. & Trujillo, I. 2009, *ApJ*, 696, L43

Chabrier, G. 2003, *PASP*, 115, 763

Chen, C., Côté, P., West, A. A., Peng, E. W., & Ferrarese, L., 2010, *ApJS*, 191, 1

Cimatti, A., *et al.* 2008, *A&A*, 482, 21

Ciotti, L. 2009, Nuovo Cimento Rivista Serie, 32, 1

Conroy, C. & van Dokkum, P. 2012, *ApJ*, 747, 69

Daddi, E., *et al.* 2005, *ApJ*, 626, 680

Davies, R. L., Efstathiou, G., Fall, S. M., Illingworth, G., & Schechter, P. L. 1983, *ApJ*, 266, 41

de Vaucouleurs, G. 1948, Annales d'Astrophysique, 11, 247

de Vaucouleurs, G., de Vaucouleurs, A., Corwin, Jr.H. G., Buta, R. J., Paturel, G., & Fouque, P. 1991, *Third Reference Catalogue of Bright Galaxies* (New York: Springer)

di Serego Alighieri, S., *et al.* 2005, *A&A*, 442, 125

Emsellem, E., *et al.* 2011, *MNRAS*, 414, 888

Faber, S. M. & Jackson, R. E. 1976, *ApJ*, 204, 668

Gallazzi, A. & Bell, E. F. 2009, *ApJS*, 185, 253

Giavalisco, M., *et al.* 2004, *ApJ* (Letters), 600, L93

Hernquist, L., Spergel, D. N., & Heyl, J. S. 1993, *ApJ*, 416, 415

Hopkins, P. F., Bundy, K., Murray, N., Quataert, E., Lauer, T. R., & Ma, C.-P. 2009, *MNRAS*, 398, 898

Jorgensen, I., Franx, M., & Kjaergaard, P. 1995, *MNRAS*, 273, 1097

Kormendy, J. 1977, *ApJ*, 218, 333

Krajnović, D., *et al.* 2011, *MNRAS*, 414, 2923

Kroupa, P. 2001, *MNRAS*, 322, 231

Kroupa, P. 2002, *Science*, 295, 82

Longhetti, M., *et al.* 2007, *MNRAS*, 374, 614

Maraston, C. 2005, *MNRAS*, 362, 799

Maraston, C., Pforr, J., Renzini, A., Daddi, E., Dickinson, M., Cimatti, A., & Tonini, C. 2010, *MNRAS*, 407, 830

Mihos, J. C. & Hernquist, L. 1994, *ApJ* (Letters), 431, L9

Naab, T., Johansson, P. H., & Ostriker, J. P. 2009, *ApJ* (Letters), 699, L178

Newman, A. B., Ellis, R. S., Bundy, K., & Treu, T. 2012, *ApJ*, 746, 162

Newman, A. B., Ellis, R. S., Treu, T., & Bundy, K. 2010, *ApJ* (Letters), 717, L103

Nipoti, C., Londrillo, P., & Ciotti, L. 2003, *MNRAS*, 342, 501

Onodera, M., *et al.* 2012, *ApJ*, 755, 26

Salpeter, E. E. 1955, *ApJ*, 121, 161

Sersic, J. L. 1968, *Atlas de galaxias australes* (Cordoba: Observatorio Astronomico)

Springel, V., *et al.* 2005, *Nature*, 435, 629

Szomoru, D., Franx, M., & van Dokkum, P. G. 2012, *ApJ*, 749, 121

Toft, S., Gallazzi, A., Zirm, A., Wold, M., Zibetti, S., Grillo, C., & Man, A. 2012, *ApJ*, 754, 3

Toft, S., *et al.* 2007, *ApJ*, 671, 285

Trujillo, I., *et al.* 2006, *MNRAS* (Letters), 373, L36

van de Sande, J., *et al.* 2011, *ApJ* (Letters), 736, L9

van der Marel, R. P. & van Dokkum, P. G. 2007, *ApJ*, 668, 756

van der Wel, A. & van der Marel, R. P. 2008, *ApJ*, 684, 260

van Dokkum, P. G. & Conroy, C. 2010, *Nature*, 468, 940

van Dokkum, P. G., *et al.* 2008, *ApJ* (Letters), 677, L5

van Dokkum, P. G., Kriek, M., & Franx, M. 2009, *Nature*, 460, 717

Vazdekis, A., Sánchez-Blázquez, P., Falcón-Barroso, J., Cenarro, A. J., Beasley, M. A., Cardiel, N., Gorgas, J., & Peletier, R. F. 2010, *MNRAS*, 404, 1639

The intriguing life of massive galaxies
Proceedings IAU Symposium No. 295, 2012
D. Thomas, A. Pasquali & I. Ferreras, eds.

© International Astronomical Union 2013
doi:10.1017/S1743921313004225

Gas Accretion and Mergers in Massive Galaxies at $z \sim 2$

C. J. Conselice[1], Jamie Ownsworth[1], Alice Mortlock[1], Asa F. L. Bluck[1,2] and the GNS team

[1]University of Nottingham [2]University of Victoria
email: conselice@nottingham.ac.uk

Abstract. Galaxy assembly is an unsolved problem, with ΛCDM theoretical models unable to easily account for among other things, the abundances of massive galaxies, and the observed merger history. We show here how the problem of galaxy formation can be addressed in an empirical way without recourse to models. We discuss how galaxy assembly occurs at $1.5 < z < 3$ examining the role of major and minor mergers, and gas accretion from the intergalactic medium in forming massive galaxies with $\log M_* > 11$ found within the GOODS NICMOS Survey (GNS). We find that major mergers, minor mergers and gas accretion are roughly equally important in the galaxy formation process during this epoch, with 64% of the mass assembled through merging and 36% through accreted gas which is later converted to stars, while 58% of all new star formation during this epoch arises from gas accretion. We also discuss how the total gas accretion rate is measured as $\dot{M} = 90 \pm 40$ $M_\odot \, \mathrm{yr}^{-1}$ at this epoch, a value close to those found in some hydrodynamical simulations.

Keywords. galaxies: formation, galaxies: evolution

1. Introduction

The formation of galaxies is a difficult problem, and we are still debating the exact history and the physics responsible for establishing the modern galaxy population. One of the issues becoming clear is that this formation and evolution happens in a cosmological context, which ultimately implies that galaxy assembly involves the structure of the universe, as well as the properties of the dark matter and its relationship to baryons.

The major dominant paradigm for understanding galaxy formation theoretically is that the dark matter is cold. This idea when implement in simulations is successful at reproducing the large scale features of the Universe. There are now many simulations and models for how galaxies form within the Cold Dark matter paradigm, however most of these simulations to date have not been able to reproduce the abundances or the formation history of massive galaxies (e.g., Conselice *et al.* 2007; Bertone & Conselice 2009; Marchesini *et al.* 2010; Guo *et al.* 2011).

One example is that when examining the abundances of massive galaxies, typically those with $M_* > 10^{11}$ $M_\odot$, the number densities up to $z \sim 1 - 2$ are similar to what is found in the local Universe (e.g., Conselice *et al.* 2011; Mortlock *et al.* 2011), yet significantly higher than what is predicted in CDM models (e.g., Conselice *et al.* 2007; Marchesini *et al.* 2010). While in CDM models galaxy formation happens hierarchically in the sense that galaxies form largely through mergers (both major and minor), the exact role of this process and its evolution is not yet clear.

To further investigate this problem requires that we examine directly how the formation process of galaxies occurs, and whether models can reproduce this, particularly features such as the merger and measured assembly history. For example, Bertone &

Conselice (2009) compare the merger history of galaxies to the predictions from the CDM Millennium simulation. This comparison shows that this CDM simulation under-predicts the number of major mergers by a similar order of magnitude (factor of 10) that it underpredicts the abundances of galaxies. The reasons for this are unclear, but may relate to either the underlying cosmological assumptions, or the way in which baryons are implemented in these simulations. While this problem, and others, can be alleviated through e.g., radio mode feedback, a full understanding of galaxy assembly cannot be provided by simulations alone, and an empirical method is necessary.

Since the most massive galaxies with $\log M_* > 11$ $M_\odot$ are largely in place by $z \sim 1$ to study their formation we must investigate these systems at higher redshifts, namely at $z > 1$. Recent studies have accomplished this by examining the most massive systems at $1.5 < z < 3$ with a Hubble Space Telescope survey of 80 massive galaxies with $\log M_* > 11$ $M_\odot$, called the GOODS NICMOS Survey (GNS). Using the GNS we have recently made the first measurement of the minor merger history for massive galaxies (Bluck *et al.* 2012), and we have measured the major merger history previously (Bluck *et al.* 2009).

Using the merger history and the star formation history and distribution for the GNS sample (e.g., Bauer *et al.* 2011; Ownsworth *et al.* 2012) we calculate how much gas must be brought in through gas accretion outside of merging. This can be done by examining the imbalance between the amount of gas needed to sustain the star formation rates we observe, and the amount of gas present in-situ within the galaxy, plus that brought in from mergers. As discussed, we find that gas accretion is a necessary and major process in driving the formation of these massive galaxies at $1.5 < z < 3$.

We give a summary of these results here, including an outline for how these features are calculated. Overall we are starting to measure with some certainty how the most massive galaxies assembled during the important epoch $1.5 < z < 3$. A standard cosmology of $H_0 = 70$ km s^{-1} Mpc^{-1}, and $\Omega_{\rm m} = 1 - \Omega_\lambda = 0.3$ is used through out, along with a Salpeter IMF.

2. The Role of Mergers up to $z = 3$

Galaxy assembly is a combination of at least three processes. These are merging with existing galaxies, the accretion of cold gas from the intergalactic medium, and the conversion of in-situ initial gas into stars in a galaxy over time. Understanding the role of these processes is one of the major goals of extragalactic astronomy.

We now have some idea about the role of mergers in galaxy assembly (e.g., Conselice *et al.* 2003). While mergers can dominate much of the assembly of galaxies, including triggering star formation, instigating morphological changes, etc., we are mainly interested here in how merging builds up the stellar masses of galaxies over time.

The amount of stellar mass added to a galaxy due to the merger process is given by the integral over the merger history, based on the fraction of galaxies merging, and the time-scale for mergers (e.g., Bluck *et al.* 2009; 2012). We carry out this integration using the observed merger history measured directly from the GNS sample of massive galaxies at $1.5 < z < 3$, and the modelled time-scale for mergers (e.g., Bluck *et al.* 2009; 2012). The total amount of stellar mass accreted is then a double integral over the redshift range of interest and over the stellar masses which we probe, which for the GNS, sensitive down to $M_* = 10^{9.5}$ $M_\odot$, can be expressed as,

$$M_{*,\rm M} = \int_{z_1}^{z_2} \int_{M_1}^{M_2} M_* \times \frac{f'_m(z, M_*)}{\tau_{\rm m}(M_*)} dM_* dz, \qquad (2.1)$$

where $\tau_{\mathrm{m}}(\mathrm{M}_*)$ is the merger time-scale, which depends on the stellar mass of the merging pair (Bluck *et al.* 2012). The total integration of the amount of mass assembled through merging gives $\mathrm{M}_{*,\mathrm{M}}/\mathrm{M}_*(0) = 0.53 \pm 0.24$, where $\mathrm{M}_*(0)$ is the initial average mass of our sample. This is the fractional amount of stellar mass added due to both major and minor mergers for systems with stellar mass ratios down to 1:100 for the average massive GNS galaxy after following a merger adjusted constant co-moving density (Conselice *et al.* 2012).

However, to fully understand the total baryonic mass assembly of galaxies we need to also account for how much gas mass is brought into these systems through mergers. We calculate this by integrating the amount of gas in these merging systems using an empirical fit to the relationship between the gas mass fraction μ_{gas} and the stellar mass found at $z = 2 - 3$. Overall we find that it is the lower mass galaxies that contribute the bulk of the gaseous mass, whereas most of the stellar mass accreted arises from higher mass mergers. We show this relative role of mergers in Figure 1. We can then use this relation to calculate for the GNS sample how much gas mass is added due to merging, finding $\mathrm{M}_{g,\mathrm{M}}/\mathrm{M}_*(0) = \mathrm{M}_{*,\mathrm{M}}/\mathrm{M}_*(0) \times \mathrm{f_g} = 0.54 \pm 0.24$.

3. Gas Accretion

One important observation of high redshift massive galaxies with $\log \mathrm{M}_* > 11\ \mathrm{M}_\odot$ is that the average star formation rate for this population is around $100\ \mathrm{M}_\odot\,\mathrm{year}^{-1}$ and is constant over the epoch $1.5 < z < 2$ (Bauer *et al.* 2011). This is a large star formation rate, and the amount of gas mass accreted due to merging plus the original amount of gas is not enough to sustain this (Conselice *et al.* 2012).

We show from this that the amount of gas accreted into a massive GNS galaxy is (Conselice *et al.* 2012) $\mathrm{M}_{g,\mathrm{A}}/\mathrm{M}_*(0) = 0.60 \pm 0.46$, such that an amount of mass of the order of the entire initial stellar mass of a massive galaxy is added over time outside of

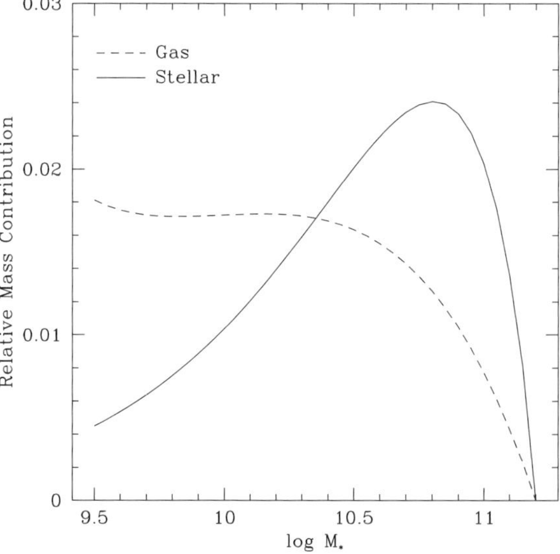

Figure 1. Figure showing the relative amount of gaseous vs. stellar matter accreted from mergers as a function of the merging galaxies stellar mass between $1.5 < z < 3$ for a typical central galaxy of $\mathrm{M}_* = 10^{11.2}\ \mathrm{M}_\odot$ (Conselice *et al.* 2012). The solid line shows the stellar mass contribution and the dashed line the gaseous contribution.

mergers to form stars during $1.5 < z < 3$, a time span of 2.16 Gyr. This reveals a net gas accretion, which is then turned into stars of $55 \pm 40\,\mathrm{M_\odot\,yr^{-1}}$.

We need to also consider that these galaxies have outflows (e.g., Weiner *et al.* 2009) that could easily double the amount of gas mass needed to be accreted from the IGM (e.g., Faucher-Giguere *et al.* 2011). Taking into account these outflows using empirical relations from Weiner *et al.* (2009) we find that the gross inflow rate is:

$$\dot{\mathrm{M}}_{\mathrm{acc}} = \dot{\mathrm{M}}_{\mathrm{outflow}} + \dot{\mathrm{M}}_{\mathrm{g,A}} = 90\,\mathrm{M_\odot\,yr^{-1}},$$

The result of this is that gas accretion accounts for $36\pm31\%$ of the stellar matter added to galaxies from $1.5 < z < 3$. Mergers account for the remainder of the mass assembly, with 1/3 of this minor mergers and 2/3 of this major mergers (Bluck *et al.* 2012). Gas accretion however is responsible for 58% of all new star formation during this epoch. Overall this implies that gas accretion into massive galaxies at early epochs is a major formation method, and dominates over mergers as a formation mechanism for new stars.

Our measured gas accretion rate is roughly consistent with theoretical calculations which predict a similar amount of gas accretion as we find (e.g., Murali *et al.* 2002; Dekel *et al.* 2009). Some of the first measurements of the gas accretion by Murali *et al.* (2002) predict a gas accretion rate of $\dot{M}_{\mathrm{g,A}} \sim 40\ \mathrm{M_\odot\,yr^{-1}}$, while more recent work suggests higher rates of $\dot{M}_{\mathrm{g,A}} \sim 100\ \mathrm{M_\odot\,yr^{-1}}$ (e.g., Dekel *et al.* 2009; Faucher-Giguere *et al.* 2011). Our results are in general agreement with these models.

4. Summary

While we are able to measure how gas accretion and mergers drives the assembly of log $M_* > 11\ \mathrm{M_\odot}$ galaxies by using a deep near infrared HST survey, our results do not apply for all galaxies at all times, but only for these massive galaxies at the epoch $1.5 < z < 3$. The errors are also still large and can be reduced significantly by using larger samples of galaxies. In the future by using surveys such as CANDELS with WFC3 and larger area surveys with e.g., WFIRST we can start to make these kind of measurements for lower mass galaxies and for galaxies at even higher redshifts.

References

Bauer, A. E., *et al.* 2011, *MNRAS*, 417, 289

Bluck, A., *et al.* 2009, *MNRAS* (Letters), 394, 51L

Bluck, A., *et al.* 2012, *MNRAS*, 747, 34

Conselice, C. J., Bershady, M. A., Dickinson, M., & Papovich, C. 2003, *AJ*, 126, 1183

Conselice, C. J., *et al.* 2007, *MNRAS*, 381, 962

Conselice, C. J., *et al.* 2011, *MNRAS*, 413, 80

Conselice, C. J., *et al.* 2013, *MNRAS*, 430, 1051

Dekel, A., *et al.* 2009, *Nature*, 457, 451

Faucher-Giguere, C.-A., Keres, D., & Ma, C.-P. 2011, *MNRAS*, 417, 2982

Guo, Q., *et al.* 2011, *MNRAS*, 413, 101

Marchesini, D., *et al.* 2010, *ApJ*, 725, 1277

Mortlock, A., *et al.* 2011, *MNRAS*, 413, 2845

Ownsworth, J. R., *et al.* 2012, *MNRAS*, 426, 764

Papovich, C., *et al.* 2011, *MNRAS*, 412, 1123

Weiner, B., *et al.* 2009, *ApJ*, 692, 187

The intriguing life of massive galaxies
Proceedings IAU Symposium No. 295, 2012
D. Thomas, A. Pasquali & I. Ferreras, eds.

© International Astronomical Union 2013
doi:10.1017/S1743921313004237

The Morphologies of Massive Galaxies at $1 < z < 3$ in the CANDELS-UDS Field: Compact Bulges, and the Rise and Fall of Massive Disks

V. A. Bruce[1]†, J. S. Dunlop[1], M. Cirasuolo[1,2], R. J. McLure[1],
T. A. Targett[1], E. F. Bell[3], D. J. Croton[4], A. Dekel[5], S. M. Faber[6],
H. C. Ferguson[7], N. A. Grogin[7], D. D. Kocevski[6], A. M. Koekemoer[7],
D. C. Koo[6], K. Lai[6], J. M. Lotz[7], E. J. McGrath[6], J. A. Newman[8]
and A. van der Wel[9]

[1]SUPA‡ Institute for Astronomy, University of Edinburgh, Royal Observatory,
Edinburgh EH9 3HJ
[2]UK Astronomy Technology Centre, Science and Technology Facilities Council,
Royal Observatory, Edinburgh EH9 3HJ
[3]Department of Astronomy, University of Michigan, 500 Church St., Ann Arbor,
MI 48109, USA
[4]Centre for Astrophysics and Supercomputing, Swinburne University of Technology,
PO Box 218, Hawthorn, VIC 3122, Australia
[5]Racah Institute of Physics, The Hebrew University, Jerusalem 91904, Israel
[6]UCO/Lick Observatory, University of California, Santa Cruz, CA 95064, USA
[7]Space Telescope Science Institute, 3700 San Martin Drive, Baltimore, MD 21218, USA
[8]University of Pittsburgh, Pittsburgh, PA 15260, USA
[9]Max-Planck Institut für Astronomie, Königstuhl 17, D-69117 Heidelberg, Germany

Abstract. We have used high-resolution, HST WFC3/IR, near-infrared imaging to conduct a detailed bulge-disk decomposition of the morphologies of $\simeq 200$ of the most massive ($M_* > 10^{11}\,M_\odot$) galaxies at $1 < z < 3$ in the CANDELS-UDS field. We find that, while such massive galaxies at low redshift are generally bulge-dominated, at redshifts $1 < z < 2$ they are predominantly mixed bulge+disk systems, and by $z > 2$ they are mostly disk-dominated. Interestingly, we find that while most of the quiescent galaxies are bulge-dominated, a significant fraction ($25 - 40\%$) of the most quiescent galaxies, have disk-dominated morphologies. Thus, our results suggest that the physical mechanisms which quench star-formation activity are not simply connected to those responsible for the morphological transformation of massive galaxies.

Keywords. galaxies: elliptical and lenticular, cD, galaxies: evolution, galaxies: high-redshift, galaxies: structure

---

1. Introduction

The study of the high-redshift progenitors of today's massive galaxies can provide us with invaluable insights into the key mechanisms that shape the evolution of galaxies in the high-mass regime.

The latest generation of galaxy formation models are now able to explain the number densities and ages of massive galaxies at high redshift. However, this is only part of the challenge, as recent studies have posed new questions about how the morphologies of massive galaxies evolve with redshift.

† E-mail: vab@roe.ac.uk
‡ Scottish Universities Physics Alliance

In addition to the basic question of how high-redshift galaxies evolve in size, there is also still much debate about how these massive galaxies evolve in terms of their fundamental morphological type. Extensive studies of the local Universe have revealed a bimodality in the colour-morphology plane, with spheroidal galaxies typically inhabiting the red sequence and disk galaxies making up the blue cloud (e.g. Baldry *et al.* 2004). However, recent studies at both low (e.g. Bamford *et al.* 2009) and high redshift (e.g. van der Wel *et al.* 2011) have uncovered a significant population of passive disk-dominated galaxies, providing evidence that the physical processes which quench star-formation may be distinct from those responsible for driving morphological transformations. This result is particularly interesting in light of the latest morphological studies of high-redshift massive galaxies by van der Wel *et al.* (2011) and Buitrago *et al.* (2013) who find that, in contrast to the local population of massive galaxies (which is dominated by bulge morphologies), by $z \simeq 2$ massive galaxies are predominantly disk-dominated systems. In this work we attempt to provide significantly improved clarity on these issues.

2. Data and Morphological Fitting

The CANDELS (Grogin *et al.* 2011, Koekemoer *et al.* 2011) near-infrared F160W data provides the necessary combination of depth, angular resolution, and area to enable the most detailed study to date of the rest-frame optical morphologies of massive ($M_* > 10^{11}$ M$_\odot$) galaxies at $1 < z < 3$ in the UKIDSS Ultra Deep Survey (Lawrence *et al.* 2007). For this study we have constructed a sample based on photometric redshifts and stellar mass estimates which were determined using the stellar population synthesis models of Bruzual & Charlot (2003) assuming a Chabrier initial mass function (see Bruce *et al.* 2012 for full details). This provides us with a total mass-complete sample of ~ 200 galaxies.

We have employed the GALFIT (Peng *et al.* 2002) morphology fitting code to determine the morphological properties for all the objects in our sample. To conduct the double component fitting we define three components: a Sérsic index fixed at $n = 4$ bulge, an $n = 1$ fixed disk and a centrally concentrated PSF component to account for any AGN or nuclear starbursts within our galaxies. These three components are combined to generate six alternative multiple component model fits, of varying complexity, for every object in the sample. These models are formally nested, and thus χ^2 statistics can be used to determine the "best" model given the appropriate number of model parameters.

3. Evolution of Morphological Fractions

Armed with this unparalleled morphological information on massive galaxies at high redshift we can consider how the relative number density of galaxies of different morphological type changes during the key epoch in cosmic history probed here.

In Fig. 1 we illustrate this by binning our sample into four redshift bins of width $\Delta z = 0.5$, and consider three alternative cuts in morphological classification as measured by B/T from our bulge-disk decompositions. In the left-hand panel of Fig. 1 we have simply split the sample into two categories: bulge-dominated ($B/T > 0.5$) and disk-dominated ($B/T < 0.5$). In the central panel we have separated the sample into three categories, with any object for which $0.3 < B/T < 0.7$ classed as "Intermediate". Finally, in the right-hand panel we have expanded this Intermediate category to encompass all objects for which $0.1 < B/T < 0.9$.

From these panels it can be seen that $z \simeq 2$ marks a key transition phase, above which massive galaxies are predominantly disk-dominated systems and below which they become increasingly mixed bulge+disk systems. We also note that at the lowest redshifts

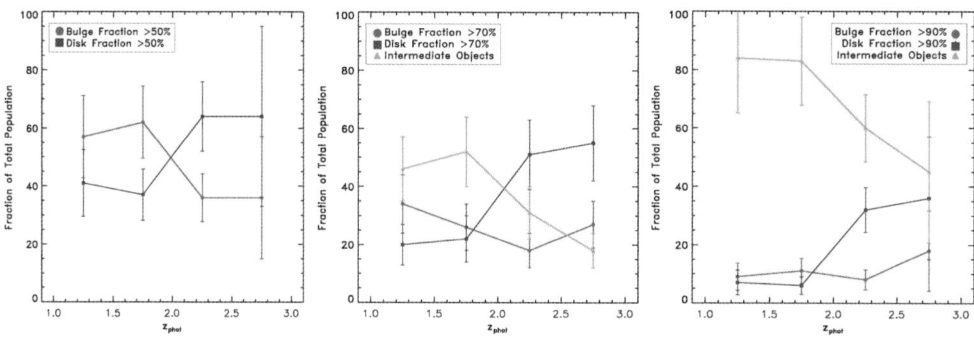

Figure 1. The redshift evolution of the morphological fractions in our galaxy sample, after binning into redshift bins of width $\Delta z = 0.5$ and using three alternative cuts in morphological classification (both to try to provide a complete picture, and to facilitate comparison with different categorisations in the literature).

probed by this study ($z \simeq 1$) it is seen that, while bulge-dominated objects are on the rise, pure-bulge galaxies (i.e. objects comparable to present-day giant ellipticals) have yet to emerge in significant numbers, with $> 90\%$ of these high-mass galaxies still retaining a significant disk component. This is compared with 64% of the local $M_* > 10^{11}$ M$_\odot$ galaxy population, which would be classified as pure-bulges from our definition ($B/T > 0.9$, corresponding to $n > 3.5$) from the sample of Buitrago *et al.* (2013). Thus, our results further challenge theoretical models of galaxy formation to account for the relatively rapid demise of massive star-forming disks, but the relatively gradual emergence of genuinely bulge-dominated morphologies.

4. Star-forming and Passive Disks.

In addition to our morphological decompositions we also make use of the SED fitting already employed in the sample selection to explore the relationship between star-formation activity and morphological type.

Fig. 2 shows specific star-formation rate ($sSFR$) versus morphological type for the massive galaxies in our sample, where morphology is quantified by single Sérsic index in the left-hand panel, and by bulge-to-total H_{160}-band flux ratio (B/T) in the right-hand panel. The values of $sSFR$ plotted are derived from the original optical-infrared SED fits employed in the sample selection, and include correction for dust extinction as assessed from the best fitting value of A_V derived during the SED fitting. As a check of the potential failure of this approach to correctly identify reddened dusty star-forming galaxies, we have also searched for 24 μm counterparts in the *Spitzer* SpUDS MIPS imaging of the UDS, and have highlighted in blue stars those objects which yielded a MIPS counterpart within a search radius of < 2 arcsec.

To first order, our results show that the well-documented bimodality in the colour-morphology plane seen at low redshift, where spheroidal galaxies inhabit the red sequence, while disk galaxies occupy the blue cloud is at least partly already in place by $z \simeq 2$.

Nonetheless, the sample also undoubtedly contains star-forming bulge-dominated galaxies and, perhaps more interestingly, a significant population of apparently-quiescent disk-dominated objects. To highlight and quantify this population we have indicated by a box on both the panels the region occupied by objects with disk-dominated morphologies and $sSFR < 10^{-10}$ yr^{-1}. In the left-hand panel, disk-dominated is defined as $n < 2.5$, and $40\pm7\%$ of the quiescent galaxies lie within this box (if we exclude the 24 μm

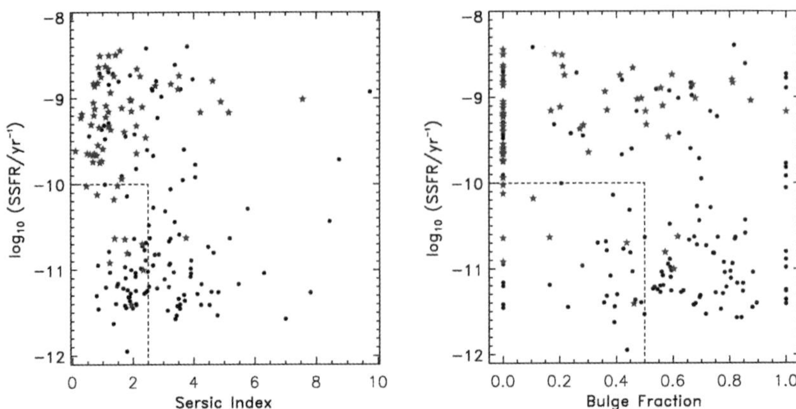

Figure 2. Plots of specific star-formation rate ($sSFR$) versus morphological type as judged by single Sérsic index (left-hand panel) and bulge-to-total H_{160}-band flux ratio (B/T) (right-hand panel).

detections), while in the right-hand panel, disk-dominated is defined by $B/T < 0.5$, in which case $25 \pm 6\%$ of the quiescent objects lie within this region.

The presence of a significant population of passive disks among the massive galaxy population at these redshifts indicates that star-formation activity can cease without a disk galaxy being turned directly into a disk-free spheroid, as generally previously expected if the process that quenches star formation is a major merger.

One possible mechanism for this arises from the latest generation of hydrodynamical simulations (e.g. Kereš *et al.* 2005, Dekel *et al.* 2009a) and analytic theories (e.g. Birnboim & Dekel 2003), which suggest a formation scenario whereby at high redshift star-formation is fed through inflows of cold gas. Another scenario which can account for star-formation quenching, whilst still being consistent with the existence of passive disks, is the model of violent disk instabilities (e.g. Dekel *et al.* 2009b), coupled with "morphology quenching" (Martig *et al.* 2009).

References

Baldry, I. K., Glazebrook, K., Brinkmann, J., Ivezić, Ž., Lupton, R. H., Nichol, R. C., & Szalay, A. S. 2004, *ApJ*, 600, 681

Bamford, S. P., *et al.* 2009, *MNRAS*, 393, 1324

Birnboim, Y. & Dekel, A. 2003, *MNRAS*, 345, 349

Bruce, V. A., *et al.* 2012, *MNRAS*, 427, 1666

Bruzual, G. & Charlot, S. 2003, *MNRAS*, 344, 1000

Buitrago, F., Trujillo, I., Conselice, C. J., & Häussler, B. 2013, *MNRAS*, 428, 1460

Dekel, A., *et al.* 2009a, *Nature*, 457, 451

Dekel, A., Sari, R., & Ceverino, D. 2009b, *ApJ*, 703, 785

Grogin, N. A., *et al.* 2011, *ApJS*, 197, 35

Kereš, D., Katz, N., Weinberg, D. H., & Davé, R. 2005, *MNRAS*, 363, 2

Koekemoer, A. M., *et al.* 2011, *ApJS*, 197, 36

Lawrence, A., *et al.* 2007, *MNRAS*, 379, 1599

Martig, M., Bournaud, F., Teyssier, R., & Dekel, A. 2009, *ApJ*, 707, 250

Peng, C. Y., Ho, L. C., Impey, C. D., & Rix, H. W. 2002, *ApJ*, 124, 266

The intriguing life of massive galaxies
Proceedings IAU Symposium No. 295, 2012
D. Thomas, A. Pasquali & I. Ferreras, eds.

© International Astronomical Union 2013
doi:10.1017/S1743921313004249

High resolution near-infrared imaging of submillimeter galaxies

Paula Aguirre[1,2], Andrew J. Baker[3], Felipe Menanteau[3], Dieter Lutz[4] and Linda J. Tacconi[4]

[1] Pontificia Universidad Católica de Chile, Depto. de Astronomía, Santiago, Chile.
[2] Depto. de Ciencias Físicas, Facultad de Ciencias Exactas, U. Andrés Bello, Santiago, Chile
[3] Department of Physics and Astronomy, Rutgers, the State University of New Jersey.
[4] Max-Planck-Institut für extraterrestrische Physik,d Garching, Germany.

Abstract. We have obtained high-resolution F110W ($\sim J$) and F160W ($\sim H$) band observations of ten submillimeter galaxies (SMGs) with the Hubble Space Telescope's NICMOS camera, in order to resolve their rest-frame optical morphologies, determine the existence of multiple-component, merger-like configurations, and estimate their stellar masses. The selected targets have redshifts in the range $2.2 \leqslant z \leqslant 2.81$ confirmed with millimeter or mid-IR spectroscopy, guaranteeing that the two bands sample the galaxies' rest-frame optical light with the Balmer break falling between them.

Keywords. galaxies: evolution, galaxies: high-redshift, galaxies: peculiar

---

Morphologies: To investigate the possible multi-component, merger-like nature of our targets we used the maximum deblending option in SExtractor to assess whether each system could be resolved into two or more stellar structures coalescing to form a massive SMG. Through our photometric analysis and with the support of previous dynamical evidence, we classify five of our targets as multi-component systems and obtain F110W/F160W magnitudes for individual components. We interpret these system rest-frame optical and/or molecular gas morphologies as signs of ongoing mergers.

The high resolution and long wavelength of our *HST* imaging has also allowed us to characterize our targets with greater confidence than possible until now. As an example, we show in Fig. 1 the F110W and F160W images for SMM J04431+0210; the exceptional spatial resolution of our data allows us to distinguish for the first time two separate components (A and B), which match the positions of two CO$(3-2)$ peaks detected in this z=2.51 SMG by Tacconi *et al.* (2006).

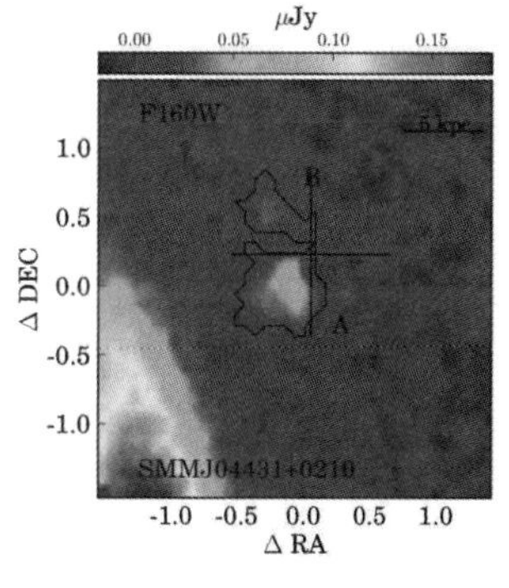

Stellar population synthesis modeling: We have modeled the stellar populations of eight objects with F110W and F160W photometry (including archival *HST* optical data when available) using Hyperzmass (Pozzetti *et al.* 2007) and the stellar synthesis library from Bruzual & Charlot (2003). For multiple systems, we analyzed sub-components separately.

For each component, there are various stellar population models that provide equally good fits to the observed photometry. Thus, we cannot discriminate among star formation histories, but we can constrain stellar masses. After correction for lensing, we obtain a mean $\log(M^*/M_\odot) = 10.88 \pm 0.24$ for our full sample. For multiple systems, mass ratios range from $\sim 3 : 1$ to $\sim 60 : 1$, indicating that both major and minor mergers are represented in the SMG population. The morphologies

Figure 1. *HST*/NICMOS F160W imaging for the submillimeter galaxy SMM J04431+0210.

and merger mass ratios of the least and most massive systems match the predictions of the major-merger and cold accretion (e.g., Davé *et al.* 2010) SMG formation scenarios, respectively, suggesting that both channels have a role in the population's origin.

References

Bruzual, G. & Charlot, S. 2002, *MNRAS* 344, 1000
Davé, R., *et al.* 2010, *MNRAS*, 404, 1355
Ivison, R., *et al.* 2010, *MNRAS*, 404, 198
Pozzetti, L., *et al.* 2007 *A&A*, 474, 443
Swinbank, M., *et al.* 2005, *MNRAS*, 359, 401
Swinbank, M., *et al.* 2010, *MNRAS*, 405,234
Tacconi, L., *et al.* 2006, *ApJ*, 640, 228
Targett, T., *et al.* 2011, *MNRAS*, 412, 295

The intriguing life of massive galaxies
Proceedings IAU Symposium No. 295, 2012
D. Thomas, A. Pasquali & I. Ferreras, eds.

© International Astronomical Union 2013
doi:10.1017/S1743921313004250

The progenitors of the first red sequence galaxies at z ∼ 2

G. Barro[1]†, S. Faber[1], P. Perez-Gonzalez[2], D. Koo[1], C. Williams[3], D. Kocevski[1], J. Trump[1], M. Mozena[1] and CANDELS collaboration

[1]UCO/Lick Observatory, University of California, Santa Cruz, CA 95064

[2]Departamento de Astrofísica y CC. de la Atmósfera, Universidad Complutense de Madrid

[3]University of Massachusetts, Amherst, MA 01003

Abstract. Nearby galaxies come in two flavors: red quiescent galaxies (QGs) with old stellar populations, and blue young star-forming galaxies (SFGs). This color bimodality seems to be already in place at $z = 2 - 3$, presenting also strong correlations with size and morphology. Surprisingly, massive QGs at higher redshifts are ∼5 times smaller than local, equal mass analogs. In contrast, most of the massive SFGs at these redshifts are still relatively large disks. The strong bimodality in both SFR and sizes indicates that some SFGs must experience strong structural transformations accompanied by a rapid truncation of the star-formation to match the observed properties of QGs. Using high-resolution HST/WFC3 F160W imaging from the CANDELS survey in GOODS-S and UDS, along with multi-wavelength ancillary data, we analyze stellar masses, SFRs and sizes of a sample of massive ($M_\star > 10^{10}$ $M_\odot$) galaxies at $z = 1.4 - 3.0$ to identify a population of compact SFGs with similar structural properties as compact QGs at $z \sim 2$. We also find that the number density of QGs increases rapidly since $z = 3$. Among these, the number of compact QGs builds up first, and only at $z < 1.8$ we do start finding a sizable number of extended QGs. This suggests that the bulk of these galaxies are assembled at late times by both continuous migration (quenching) of non-compact SFGs and size growth of cQGs. As a result of this growth, the population of cQGs disappears by $z \sim 1$. Simultaneously, we identify a population of compact SFGs (cSFGs) whose number density decreases steadily with time since $z = 3.0$, being almost completely absent at $z < 1.4$. The number of cSFGs makes up less than 20% of all massive SFGs, but they present similar number densities as cQGs down to $z \sim 2$, suggesting an evolutionary link between the two populations.

Keywords. galaxies: evolution, galaxies: high-redshift, galaxies: statistics

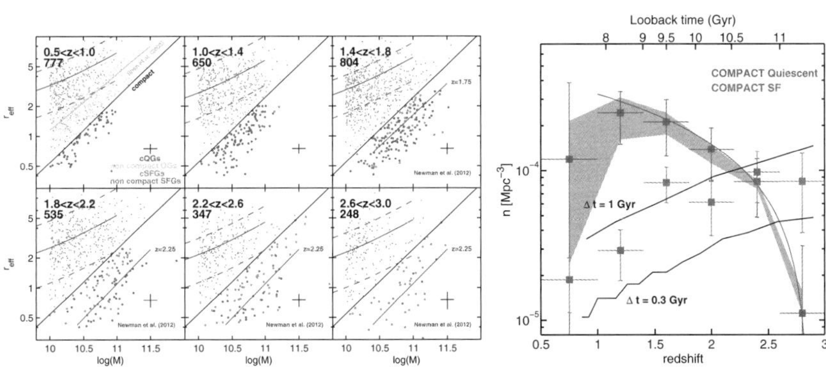

Figure 1. *Left:* Evolution of the mass-size relation for massive galaxies at $z = 3$ to $z = 0.5$. *Right:* Co-evolution of the number density of cQGs and cSFGs, along with a simple model that assigns cSFGs arbitrary life-times for their last burst of star-formation ($\Delta t = 0.3 - 1.0$ Gyr).

† email: gbarro@ucolick.org

The intriguing life of massive galaxies
Proceedings IAU Symposium No. 295, 2012
D. Thomas, A. Pasquali & I. Ferreras, eds.

© International Astronomical Union 2013
doi:10.1017/S1743921313004262

The extended gas halo of QSO host galaxies

R. Falomo[1], E. P. Farina[2], R. Decarli[3], A. Treves[2] and J. Kotilainen[4]

[1]INAF - Osservatorio di Padova, Italy

[2]Max Planck Institute for Astronomy, Heidelberg, Germany

[3]Insubria University, Como, Italy

[4]Finnish Center for Astronomy with ESO (FINCA), University of Turku, Finland

Abstract. We investigate the MgII 2800 and CIV 1540 absorption features of the gas in the halo of a foreground QSO through the absorption imprinting on the spectra of a background QSO that is closely aligned with the nearest quasar. We present the results for 13 QSO pairs ($0.7 < z < 2.2$) that allow us to probe the gas at distances between 60 kpc and 120 kpc from the QSO nucleus. We identify absorption features associated with the foreground QSO in 7 out of 10 systems for MgII , and one out of 3 for CIV (see example in Fig. 1). At variance with the case of inactive and less massive galaxies we find that relatively strong (EW $\sim$ 1 Ang) absorption features are present out to a radius of 100 kpc. This suggests that a large extended halo is associated with massive galaxies.

The comparison of these results with those for inactive (not hosting active black holes) galaxies (see e.g. Chen *et al.* 2010a) shows that the halo of QSOs is similar to that of inactive galaxies. In the observed sample we do not detect a significant enhancement of the absorption strengths, as it could be expected if the QSO nuclear activity were driven by intense gas accretion onto the black hole. Moreover along the line of sight of the QSO we do not detect any Mg II absorbers of the same strength of the transverse one. These results are in agreement with models that consider a non-isotropic emission of the QSO, which are hosted by massive gaseous halos.

Keywords. galaxies: quasars: absorption lines

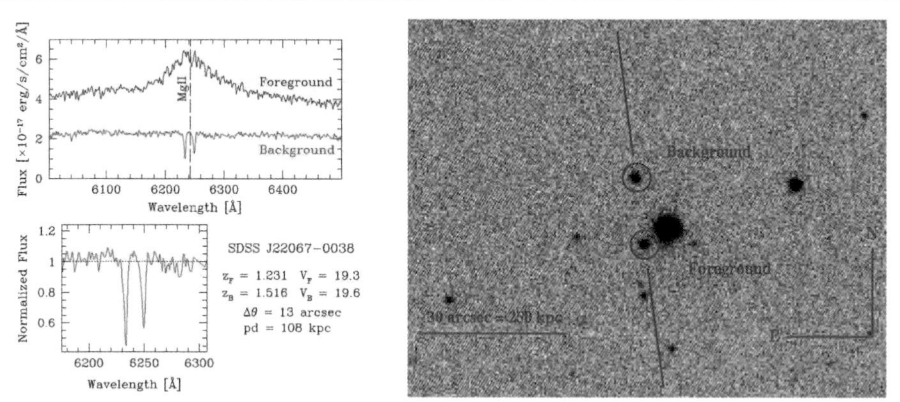

Figure 1. Example of QSO projected pair showing intervening MgII absorption.

References

Chen, H.-W., Helsby, J. E., Gauthier, J. R., Shectman, S. A., Thompson, I. B., & Tinker, J. L. 2010, *ApJ*, 714, 1521

The intriguing life of massive galaxies
Proceedings IAU Symposium No. 295, 2012
D. Thomas, A. Pasquali & I. Ferreras, eds.

© International Astronomical Union 2013
doi:10.1017/S1743921313004274

Sizes of Passively Evolving Galaxies at $z \sim 2$ in CLASH

Lulu Fan[1,2], Yang Chen[3,1], Xinzhong Er[4], Lin Lin[1] and Jinrong Li[1]

[1]Center for Astrophysics, USTC, 230026 Hefei, China
[2]Key Lab. for Research in Galaxies and Cosmology, USTC, CAS, 230026 Hefei,China
[3]Astrophysics Sector, SISSA, Via Bonomea 265, 34136 Trieste, Italy
[4]National Astronomical Observatories, CAS, Beijing 100012, China

Abstract. We select ten passively evolving and massive galaxies at redshift $z \sim 2$ from the Cluster Lensing And Supernova survey with Hubble (CLASH). We derive the stellar properties of these galaxies using the multiwavelength *HST* WFC3 and ACS data, together with Spitzer IRAC observations. We also analyze the optical rest-frame morphology of these high redshift objects by using the GALFIT package (Peng *et al.* 2002). The observed near-IR images, obtained with the *HST*/WFC3 camera with high spatial resolution and amplified by the foreground clusters, provide us with a good chance to study the structures of such systems. Six out of ten galaxies have on average a four times smaller effective radius, in agreement with previous works at redshift $z \sim 2$.

Keywords. galaxies: evolution, galaxies: formation, galaxies: high-redshift

---

1. Introduction and Result

The sizes of passively evolving and massive galaxies at redshift $z \sim 2$ have been found to be ~ 4 smaller than low redshift early-type galaxies of similar stellar mass. The lack of compact massive galaxies at low redshift implies that considerable size evolution must happen between $z = 2$ and $z = 0$. There are two main size evolution mechanisms: (1) dry minor merger (Naab *et al.* 2009); (2) "puff-up" due to gas mass loss (Fan *et al.* 2008, 2010). Recent studies found that the observed size evoultion needed by compact massive galaxies at redshift $z \geqslant 2$ can not be explained well by the present models.

Here we selected ten passively evolving and massive galaxies from the Cluster Lensing And Supernova survey with Hubble (CLASH, Postman *et al.* 2012). The photometric redshift and stellar properties have been derived based on the *HST* ACS and WFC3 observations wih 16 UV-to-NIR filters in total. Spitzer IRAC data, if available, have been also used. With the GALFIT package, we analyze the surface brightness profiles using a Sersic model. We find that galaxy sizes span a wide range: from ~ 2 times larger to ~ 6 times smaller than the local analogs of similar stellar mass. The large size scatter at redshift $z \sim 2$ will give some indications for the size evolution models.

References

Fan, L., Lapi, A., De Zotti, G., & Danese, L. 2008, *ApJ* (Letters), 689, L101
Fan, L., Lapi, A., Bressan, A., *et al.* 2010, *ApJ*, 718, 1460
Naab, T., Johansson, P. H., & Ostriker, J. P. 2009, *ApJ* (Letters), 699, L178
Peng, C. Y., Ho, L. C., Impey, C. D., & Rix, H.-W. 2002, *AJ*, 124, 266
Postman, M., Coe, D., Benitez, N., *et al.* 2012, *ApJS*, 199, 25

The intriguing life of massive galaxies
Proceedings IAU Symposium No. 295, 2012
D. Thomas, A. Pasquali & I. Ferreras, eds.

© International Astronomical Union 2013
doi:10.1017/S1743921313004286

Evolution of the galaxy merger rate since z=2 from the UKIDSS Ultra-Deep Survey

S. Foucaud[1], P.-W. Wang[2], O. Almaini[3], R. Grützbauch[4], W. G. Hartley[3], C. Simpson[5] and C. J. Conselice[3]

[1]National Taiwan Normal University, Taiwan; [2]Laboratoire d'Astrophysique de Marseille, France; [3]University of Nottingham, United Kingdom; [4]University of Lisbon, Portugal; [5]Liverpool John Moores University, United Kingdom.
e-mail: foucaud@ntnu.edu.tw

Abstract. It is now clear that the epoch when the Universe was half of its current age is a crucial period during which galaxies assembled their mass and evolved into the galaxies we observe in the local Universe. However, so far only very few direct studies of mass assembly in action, hence galaxy merging, were conducted over z=1 and usually relied on very small or biased samples. Based on very deep near infrared survey data, the latest UKIDSS-UDS DR8, combined with the optical data conducted by Subaru and CFHT and Spitzer IRAC observations, we explored the evolution of the merging rate up to $z = 2$, over the largest volume of the Universe at $0.4 < z < 2$ ever sampled. The pair fraction is found to decrease by a factor of two during this period, and wet (gas rich) mergers dominate largely. The dry mergers are very rare, ruling it out as the main mechanism for the mass assembly of passive massive galaxies. Also while massive galaxies undergo a decrease of their pair fraction during this period, less massive systems follow an increase during the same period.

Keywords. galaxies: evolution; galaxies: high-redshift

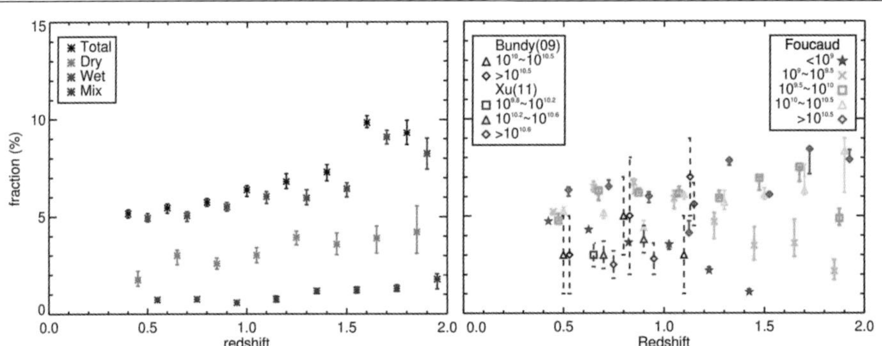

Left: Evolution of close pair fraction between z = 0.4 to z = 2.0, for the total sample, the dry mergers (red-red), wet mergers (blue-blue) and mixed mergers (red-blue). Both corrections for projection effect and close pair incompleteness have been applied. The error bars are generated by Monte-Carlo simulation. **Right:** Evolution of our measured pair fractions with redshift in five different stellar mass bins, compared to recent results (Bundy *et al.* 2009, *ApJ*, 697, 1369; Xu et al. 2012, *ApJ*, 747, 85). Our results imply that merger rates in more massive samples decrease with redshifts while increasing for the less massive samples.

The intriguing life of massive galaxies
Proceedings IAU Symposium No. 295, 2012
D. Thomas, A. Pasquali & I. Ferreras, eds.

© International Astronomical Union 2013
doi:10.1017/S1743921313004298

Clustering of EROs from UKIDSS DXS and Pan-STARRS PS1

J. -W. Kim[1,2], A. C. Edge[1], D. A. Wake[3], V. Gonzalez-Perez[1], C. M Baugh[1] and C. G. Lacey[1]

[1]Institute for Computational Cosmology, Dept. of Physics,
Durham University, Durham DH1 3LE, UK
[2]CEOU, Dept. of Physics & Astronomy, Seoul National University, Seoul,
Korea
email: kjw0704@astro.snu.ac.kr
[3]Dept. of Astronomy, Yale University, New Haven, USA

Abstract. We have measured the angular two-point correlation function of EROs. The halo model is fitted to the observed clustering, and dark matter halo mass, bias and satellite fraction are estimated in three redshift bins. We also compare our results with the semi-analytical galaxy formation model. This work illustrates the power of clustering analysis in providing observational constraints on simulations.

Keywords. galaxies: halos, large-scale structure, infrared: galaxies

1. Introduction & results

Recent wide, deep near-IR surveys provide an opportunity to investigate the clustering of high redshift galaxies. Extremely Red Objects (EROs) are massive galaxies at $z > 1$. We have selected EROs in the Elais-N1 field by merging the UKIDSS DXS with the Pan-STARRS PS1 MDS dataset. Based on halo modeling, the average mass of dark matter haloes hosting EROs is over $10^{13} h^{-1} M_\odot$, and the bias parameter ranges from 2.7 to 3.5 (left and middle panels in Fig. 1). Comparing the halo occupation distributions from the halo model and GALFORM, the GALFORM semi-analytical model predicts too many EROs either as satellites or in less massive haloes (right panel in Fig. 1) than the observations.

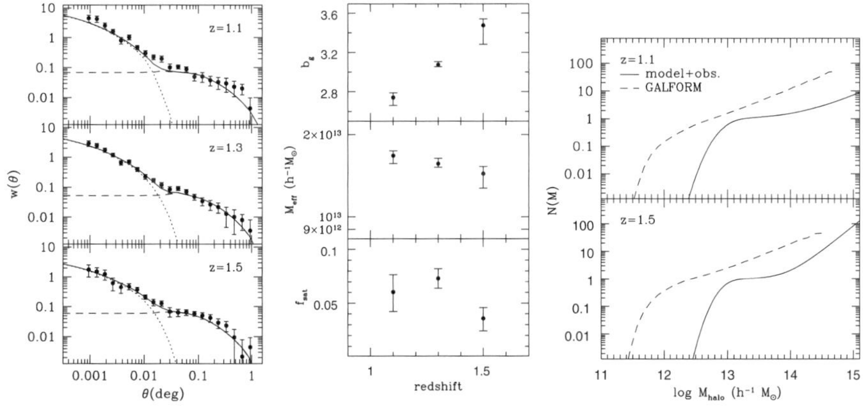

Figure 1. Left : correlation function of EROs and best fit halo models. Middle : derived parameters from the halo model. Right : comparison of HODs from the halo model and GALFORM.

The intriguing life of massive galaxies
Proceedings IAU Symposium No. 295, 2012
D. Thomas, A. Pasquali & I. Ferreras, eds.

© International Astronomical Union 2013
doi:10.1017/S1743921313004304

Cluster galaxies 10 billion years ago

V. Strazzullo[1], E. Daddi[1], R. Gobat[1], and M. Onodera[2]

[1]CEA-Saclay, 91191 Gif sur Yvette, France [2]ETH Zürich, 8093 Zürich, Switzerland
email: veronica.strazzullo@cea.fr

Abstract. Cl J1449+0856 is a spectroscopically confirmed galaxy cluster at $z \sim 2$. The detection of a faint, extended X-ray emission, suggestive of an already evolved, partially relaxed structure, puts this system among the most distant "established" clusters rather than in the realm of $z \gtrsim 2$ proto-clusters. This gives us a chance of studying galaxies in an evolved overdense environment very close to their formation epoch, and in particular to trace the evolution of early-type galaxies in clusters back to ten billion years ago.

Keywords. Galaxies: clusters, Galaxies: evolution, Galaxies: high-redshift

---

Cl J1449+0856 at $z = 2$ is the most distant X-ray detected and spectroscopically confirmed galaxy cluster discovered so far. We studied its galaxy populations with a sample of ~ 100 candidate cluster members ($\gtrsim 10^{10}$ M$_\odot$) within 500kpc of the cluster center. Already at this redshift, the central cluster region ($r < 200$kpc) hosts a population of massive galaxies with quiescent stellar populations and early-type morphologies. At the same time, the cluster core also seems to host massive galaxies still in their active formation phase, with significant star formation and ongoing merging activity. A clear correlation is observed between morphological structure and stellar populations, with quiescent galaxies typically showing a high Sersic index profile, as also observed in the field up to this redshift. In general agreement with many high-redshift studies, quiescent early-type candidate cluster members are smaller than similarly massive early-types in the nearby Universe, although the most compact $z \sim 2$ galaxies in our field are spectroscopically confirmed interlopers, leaving the mass-size relation of cluster early-types a factor $2 - 3$ lower in size than the commonly used Shen *et al.* (2003) local determination.

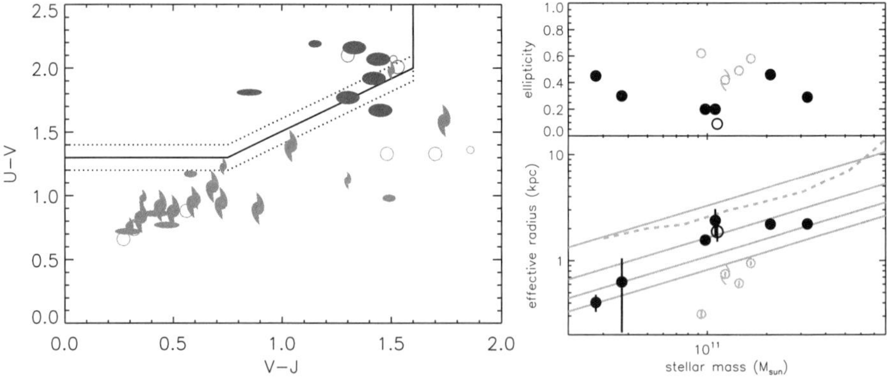

Figure 1. Left: UVJ restframe color plot (Wuyts *et al.* 2007) of candidate cluster members (m$_{F140W}$ <24.5, symbol shape according to Sersic index). Right: effective radius and ellipticity vs stellar mass for candidate quiescent early-type cluster members (dark symbols), and for 4 quiescent spectroscopic interlopers at similar redshift (light symbols). Solid lines show the Shen et al. (2003) local relation scaled down in size by a factor 1,2,3,4.

The intriguing life of massive galaxies
Proceedings IAU Symposium No. 295, 2012
D. Thomas, A. Pasquali & I. Ferreras, eds.

© International Astronomical Union 2013
doi:10.1017/S1743921313004316

Galaxies in most dense environments at redshift 1.4

V. Strazzullo

[1]Irfu/SAp, CEA-Saclay, Orme des Merisiers, 91191 Gif sur Yvette, France
email: `veronica.strazzullo@cea.fr`

Abstract. At a cosmic time when galaxy clusters start showing evidence of a still active galaxy population, the X-ray luminous, massive cluster XMMU J2235-2557 at $z = 1.39$, already hosts massive, quiescent, early-type galaxies on a tight red sequence dominating the cluster core. XMMU J2235-2557 is among the most massive of the very distant clusters, which may explain the evolved status of the system itself, and of its host galaxy populations. It remains a unique laboratory to observe environment-biased galaxy evolution already 9 billion years ago.

Keywords. Galaxies: clusters, Galaxies: evolution, Galaxies: high-redshift

The massive cluster XMMU J2235-2557 at $z = 1.39$ already hosts evolved galaxy populations, with star formation effectively suppressed in massive galaxies, and confined outside of the cluster core. The luminosity/stellar mass functions suggest a stellar mass distribution similar to that of cluster galaxies at lower redshifts. A tight red sequence is already in place, its color and slope in agreement with passive evolution predictions with a formation redshift $z \sim 3$. The red sequence is mainly populated by massive galaxies with no evidence of star formation and early-type morphologies. The smaller sizes of these seemingly passively evolving objects compared to their z=0 counterparts might leave room for later evolution, but its actual relevance for individual galaxies remains unclear due to uncertainties and biases in the masses, sizes, local mass-size relation, and in the comparison of samples at different cosmic epochs (see Strazzullo *et al.* 2010).

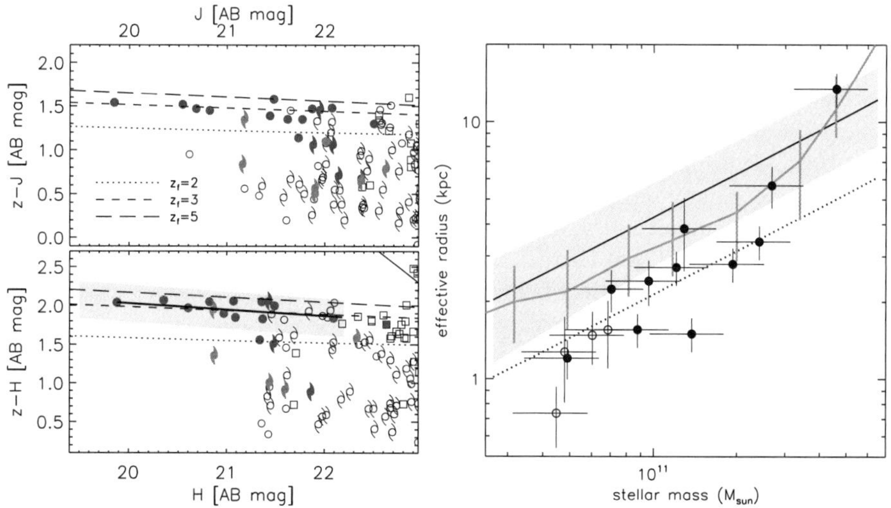

Figure 1. The red sequence (left), and stellar mass vs size relation for early-types (right), in the galaxy cluster XMMU J2235-2557 at z=1.39 (from Strazzullo *et al.* 2010).

The intriguing life of massive galaxies
Proceedings IAU Symposium No. 295, 2012
D. Thomas, A. Pasquali & I. Ferreras, eds.

© International Astronomical Union 2013
doi:10.1017/S1743921313004328

The effects of binary stars on the colors of galaxies at $z \sim 2.0$

Y. Zhang[1], J. Liu[1] and F. Zhang[2]

[1]National Astronomical Observatories/Xinjiang Observatory, Chinese Academy of Sciences,
150 Science 1-Street, Urumqi, Xinjiang 830011, China
email: zhy@xao.ac.cn
[2]National Astronomical Observatories/Yunnan Observatory, CAS, Kunming, 650011, China

Abstract. Using the Monte Carlo simulations and a set of theoretical galaxy templates, which are built on the Yunnan evolutionary population synthesis models with and without binary interactions, we compare the optical color-color (C-C) relations for passive and star-forming galaxies at redshift $z \sim 2.0$, and study the effect of binary interactions on this relation. We find that the influence of binary interactions on the C-C relation is insignificant for passive galaxies at $z \sim 2.0$.

Keywords. binary: general, galaxies: high-redshift, galaxies: stellar content

---

1. Introduction The era at $z \sim 2.0$ is a crucial stage during the cosmic time. During this period, the star forming activity in the Universe and the bulk of stellar mass assembly in galaxies were at their peak levels (Dickinson *et al.* 2003). Binary stars are very common and very important in evolutionary populations synthesis (EPS) models. They can affect the overall shape of the spectral energy distribution (SED) of populations. Especially in the ultraviolet (UV) passbands they make the SED of population bluer by $2 - 3$ magnitudes at $t \sim 1.0$Gyr. The variation in the SED leads to the uncertainties of parameter determinations for stellar population systems (Zhang *et al.* 2012). Therefore, in this work we will investigate the effect of binary interactions on colors for galaxies at $z \sim 2.0$.

2. Method In this study, we use the standard Yunnan EPS (Yunnan models without binary interactions, SSP, Zhang *et al.* 2004; Yunnan models with binary interactions, BSP, Zhang *et al.* 2005) models and exponentially declining star formation rate (SFR), $\psi(t) \propto exp(-t/\tau)$ (τ is the $e-$folding time scale), to build the theoretical galaxy templates. The $e-$folding time scale is short ($\tau \leqslant 1.0$Gyr, $\tau = 0.001 - 1.0$Gyr) for passive galaxies, while long ($\tau > 1.0$Gyr, $\tau = 2.0, 3.0, 5.0, 15.0$, and 30.0Gyr) for star-forming galaxies. Here, we use Monte Carlo simulations to generate model galaxies and their C-C relation with our galaxy templates.

3. Result By comparing the C-C relations between galaxies based on Yunnan-SSP and -BSP models, we find that there is an effect of binary interactions on the C-C relation for passive galaxies, but the effect is negligible for star-forming galaxies at $z \sim 2.0$.

Acknowledgements. This work is supported by Natural Science Foundation (Nos 11103054 and 11273053), the Light in China' Western Region (LCWR) under grant XBBS2011022, and Xinjiang Natural Science Foundation (No.2011211A104).

References
Dickinson, M., Papovich, C., Ferguson, H. C., & Budavári, T. 2003, *ApJ*, 587, 25
Zhang, F., Han, Z., Li, L., & Hurley, J. R. 2004, *MNRAS*, 350, 710
Zhang, F., Han, Z., Li, L., & Hurley, J. R. 2005, *MNRAS*, 357, 1088
Zhang, Y., Han, Z., Liu, J., Zhang, F., & Kang, X. 2012, *MNRAS*, 421, 1678

The intriguing life of massive galaxies
Proceedings IAU Symposium No. 295, 2012
D. Thomas, A. Pasquali & I. Ferreras, eds.

© International Astronomical Union 2013
doi:10.1017/S174392131300433X

The nature of the 8 o'clock arc using Near-IR IFU spectroscopy with SINFONI

M. Shirazi[1], S. Vegetti[2], N. Nesvadba[3], J. Brinchmann[1], S. Allam[4], and D. Tucker[4]

[1] Leiden Observatory, Leiden University, P.O. Box 9513, 2300 RA Leiden, The Netherlands
email: shirazi@strw.leidenuniv.nl
[2] Kavli Institute for Astrophysics and Space Research, Massachusetts Institute of Technology, Cambridge, MA 02139, USA
[3] Institut d'Astrophysique Spatiale, UMR 8617, CNRS, Université Paris-Sud, Bâtiment 121, 91405, Orsay Cedex, France
[4] Fermi National Accelerator Laboratory, P.O. Box 500, Batavia, IL 60510, USA

We present an analysis of the lensed Lyman Break Galaxy (LBG), the 8 o'clock arc, at redshift 2.735. We reduced Near-IR IFU data from SINFONI on VLT covering $\lambda = 2900$ Å to 6500 Å in the rest-frame. From this we recovered the Hβ map and the spatially-resolved Hβ profile which are shown in the right plot in Fig. 1. We can see that Hβ shows different profiles at different spatial pixels and it is composed of multiple components. To study the de-lensed morphology of the galaxy we make use of existing B & H band imaging from the HST. Based on this we constructed a rigorous lens model for the system using the Bayesian grid based lens modeling technique presented by Vegetti & Koopmans (2009, MNRAS, 392, 945). In order to obtain a robust lens model, we first considered the high resolution B band HST image (rest-frame UV, left plot in Fig. 1) and then used this modeling to reconstruct the Hβ line map of the galaxy. We then present the de-lensed Hβ line map, velocity and velocity dispersion maps of this LBG galaxy.

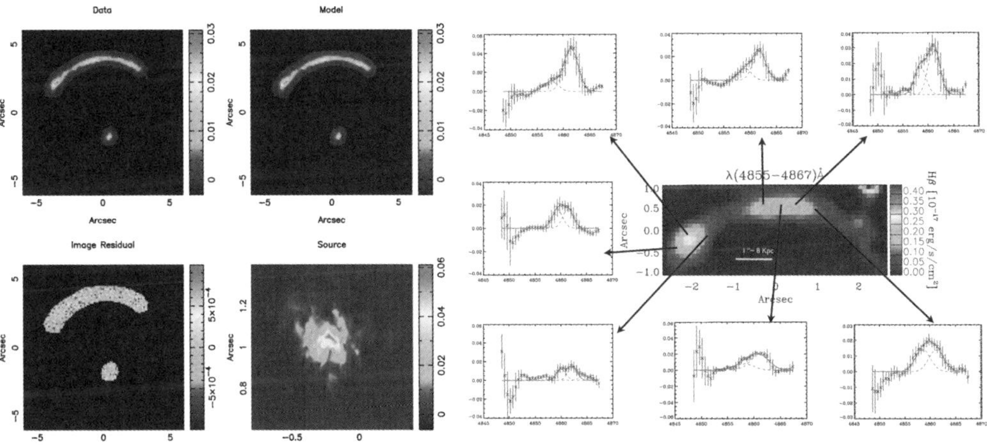

Figure 1. Left plot: The left top panel shows the arc and the counter image in the B band HST image. The elliptical has been removed from this image.The right top panel shows the best-fit model and the next panel shows the residuals after subtracting this model from the data. The reconstructed image is shown on the right bottom panel. From this image we can see that the source is formed of multiple components, two main galaxy components and a clump separated by 0.15 *arcsec*. Right plot: The spatially-resolved asymmetric Hβ profiles of the arc and the Hβ map are shown. The Hβ map is integrated between the wavelengths mentioned in the map.

2.1. *The IR luminosity function at $z \lesssim 2.5$*

The redshift evolution of the average sSFR in main-sequence galaxies out to $z \sim 2$ (e.g. Elbaz *et al.* 2011, Karim *et al.* 2011) can be combined with the evolution of the stellar MF of SFGs to predict the shape of the IR LF if the distribution of sSFR at fixed stellar mass is known. Based on the results of Rodighiero *et al.* (2011) for SFGs at $z \sim 2$ we approximate this distribution by the double log-normal function on the left-hand side of Fig. 1 (identified as being due to main-sequence and – likely interaction-induced – burst-like star formation activity, respectively; see Sargent *et al.* 2012 for details and a discussion of the underlying assumptions). The mapping of the stellar MF to an IR LF is effectively a convolution of the MF and a variable double-Gaussian kernel with (i) normalization fixed by the shape of the MF and (ii) main-sequence peak position that – given the redshift – is uniquely determined by the position of the SF main sequence in the (s)SFR vs. $M_\star$ plane. Thanks to our decomposition, we can then also identify the individual contribution of secular and burst-like SF activity to the IR LF (see Fig. 1, right).

In Fig. 1 we show that the evolution of the stellar MF and of the IR LF (i.e. of the SFR distribution) of star-forming galaxies at $z \leqslant 2$ is self-consistent (see also Bell *et al.* 2007) and that starburst galaxies are the dominant factor shaping the bright end of the IR LFs. We find (cf. Sargent *et al.* 2012) that the fractional contribution of starburst activity to the cosmic SFR density (8–14%) is only weakly redshift-dependent at $z < 2$. We also reproduce the well-known fact that most local ULIRGs are starbursts (e.g., Sanders & Mirabel 1996). At $z > 0.9$ the majority of ULIRGs are main-sequence galaxies. Importantly, however, their high SFR ($>100\,M_\odot/\mathrm{yr}$) is not triggered by merging as in most local ULIRGs but is a secular process linked to large gas reservoirs in high-z disks (e.g., Daddi *et al.* 2010a, Tacconi *et al.* 2010, Geach *et al.* 2011). Local and distant ULIRGs are intrinsically different objects for which direct comparisons should be avoided.

2.2. *IR source counts*

In addition to IR LFs, wavelength-dependent galaxy number counts are another important constraint for evolutionary models of IR-emitters. While purely semi-analytical models (e.g. Lacey *et al.* 2010) struggle to reproduce IR number counts, phenomenological or hybrid models (e.g. Béthermin *et al.* 2011; Gruppioni *et al.* 2011) fare better but are in general descriptive and use an evolution of the IR LF which is not physically motivated. However, these recent models – which reproduce the total counts passably – are excluded at $>3\sigma$ by recent Herschel measurements of counts per redshift slice (Béthermin *et al.* 2012a). Here we present a new model of IR galaxy counts which builds on the 2-SFM framework introduced in Section 2.1.

Having derived SFR distributions (as measured by IR LFs, see Section 2.1) over a continuum of redshifts, the most important ingredient† for predicting IR source counts is the subsequent choice of the IR spectral energy distribution (SED) to assign to individual galaxies. Here we rely on a simple SED library based on Herschel observations: a single SED for the main sequence and another one for starbursts, both growing warmer with redshift (Magdis *et al.* 2012b). We compare the predictions of our model with measurements of differential galaxy counts from $24\,\mu\mathrm{m}$ to $1.4\,\mathrm{GHz}$ in Fig. 2. Overall, the observed

---

† Additional refinements, which are of lesser importance but which lead to an even better match of the IR source counts than is already achieved by the most basic, simplified prediction (cf. Béthermin *et al.* 2012b, and Fig. 2) are: a $M_\star$-dependent extinction law (e.g. Pannella *et al.* 2009), explicit modelling of an AGN component (significant ($>10\%$) only at $24\,\mu\mathrm{m}$ at flux densities $>3\,\mathrm{mJy}$ and negligible ($<2\%$) at longer wavelengths) and a correction for magnification caused by strong lensing.

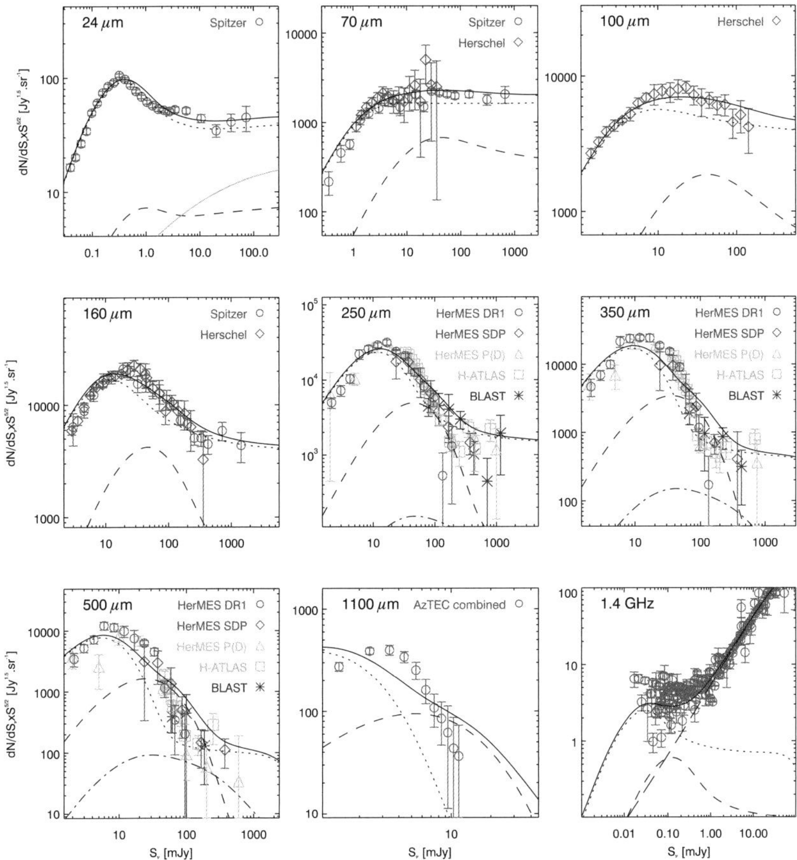

Figure 2. $24\,\mu$m to $1.4\,$GHz number counts (reproduced from Béthermin *et al.* 2012b). Solid line – total counts predicted by the 2-SFM framework, including a contribution for AGN-emission, mass-dependent dust attenuation and lensed sources; gray line – counts predicted when neglecting the three aforementioned refinements; dotted line – main-sequence contribution; short-dashed line – starburst contribution; dot-dashed line – lensed sources; triple-dot-dashed line – difference between counts with and without AGN contribution. At $1.4\,$GHz our model for star-forming galaxies was combined with the model of AGN-driven radio sources by Massardi *et al.* (2010; long-dashed line).

source counts are well reproduced, showing the effectiveness of the core ingredient of the 2-SFM approach, namely the explicit distinction between normal and starbursting galaxies. By distinguishing between main-sequence and burst activity we can explore selection biases toward normal or starburst galaxies in surveys probing various wavelengths and flux density regimes. Main-sequence galaxies dominate the number counts at all flux densities and all wavelengths. However, the relative contribution of starbursts varies a lot with flux density and wavelength and is important ($\sim$30%) around $30\,$mJy at $70\,\mu$m and $50\,$mJy at 350 and $500\,\mu$m.

Despite its simplicity, the 2-SFM predictions provide one of the best fits achieved so far to IR source counts, including counts per redshift slice at SPIRE wavelengths, which were poorly reproduced in earlier models.

redshift-dependent) average values of a galaxy located directly on the main-sequence and compare these to the analogously normalized sSFR (i.e. the offset from the main sequence). This representation of the data reveals that: (1) at all redshifts, the SFE in main-sequence galaxies in our "calibration sample" is almost independent of their position within the main sequence, while their SFR rises in lockstep with f_{gas}; (2) starburst galaxies (here we consider a small sample of local and high-z starbursts with measured α_{CO}) have similarly-sized gas reservoirs as normal galaxies but are characterized by a strongly enhanced SFE. The dashed line in Fig. illustrates the expected variation of SFE ($\propto \mathrm{sSFR}^{-0.17}$) and f_{gas} ($\propto \mathrm{sSFR}^{0.83}$) within the main sequence as it follows from the best-fit integrated K-S relation shown in Fig. 3.b, in agreement with the dust mass-based findings of Magdis *et al.* (2012b).

3.2. *Redshift evolution of gas fractions*

The well-defined relations between $M_\star$ and SFR, and between SFR and H_2-mass allow for a straightforward prediction of the redshift-evolution of f_{gas} in normal galaxies, where $M_{\mathrm{gas}}/M_\star \equiv \mathrm{const.} \times \mathrm{SFR}(M_\star)^{\beta_{\mathrm{K}}-\mathrm{s}}/M_\star$ and $\mathrm{SFR}(M_\star)$ is given by the observed evolution of the sSFR of main-sequence galaxies. The gas fractions of normal galaxies at $z \lesssim 3$ predicted in this manner are in excellent agreement with literature data (Magdis *et al.* 2012a; but note that Narayanan *et al.* (2012) have recently questioned the veracity of the high measured f_{gas} values based on their numerical simulations) suggesting that the gas content of secularly-evolving SFGs is an important driver of the cosmic sSFR-evolution. This is consistent with the observation in Fig. 4 that, also at fixed redshift, variations of sSFR across the main sequence are driven by varying f_{gas}.

3.3. *Molecular gas mass functions and the molecular gas history of the Universe*

To estimate the redshift evolution of the distribution of molecular gas masses in normal galaxies we convert the main-sequence component of the SFR-distribution depicted in Fig. 1 to a gas mass distribution by means of the best-fitting integrated K-S relation of Fig. 3.b. For the starbursting population we assume that the SFE increase (with respect to the value of the main-sequence state prior to the onset of burst activity) in these systems scales with the boost in star-formation rate engendered by the burst (see Sargent *et al.* 2012 for details). This kind of behaviour is suggested by the roughly similar excess in sSFR and SFE observed in Fig. 4.a and also seen in numerical simulations of galaxy mergers (e.g. Di Matteo *et al.* 2007). The total molecular gas MF (i.e. including both normal and starburst galaxies) inferred using this approach for the SFG population at $z \sim 0$, 2 & 5 is shown in Fig. 5. Note that the H_2-MF is only observationally constrained in the local Universe (Keres *et al.* 2003; Obreschkow & Rawlings 2009a) where the measurements are entirely consistent with the 2-SFM predictions. (This is also true of the predicted CO LFs – cf. Fig. 5 – which follow from the H_2-MFs after (a) assigning to main-sequence galaxies α_{CO}-values based on statistically-derived metallicities, computed using the fundamental metallicity relation of Mannucci *et al.* (2010) given the known SFR and $M_\star$, and (b) for starbursts assuming a conversion factor that scales inversely with the SFE-enhancement (see Sargent *et al.* 2012 in preparation for additional details and underlying assumptions/simplifications). The dominant contribution to the predicted H_2-MF stems from main-sequence galaxies which hence also are the main contributors to the evolution of the cosmic abundance of molecular gas which is obtained by integration of the molecular gas mass functions and which is shown in Fig. 6. This indirect measurement of the comoving H_2-density reveals a roughly 5-fold increase of the molecular content of the Universe out to $z \sim 1.5$ where it roughly equals the stellar mass density and the cosmic abundance of atomic hydrogen. The increase traces the rise of

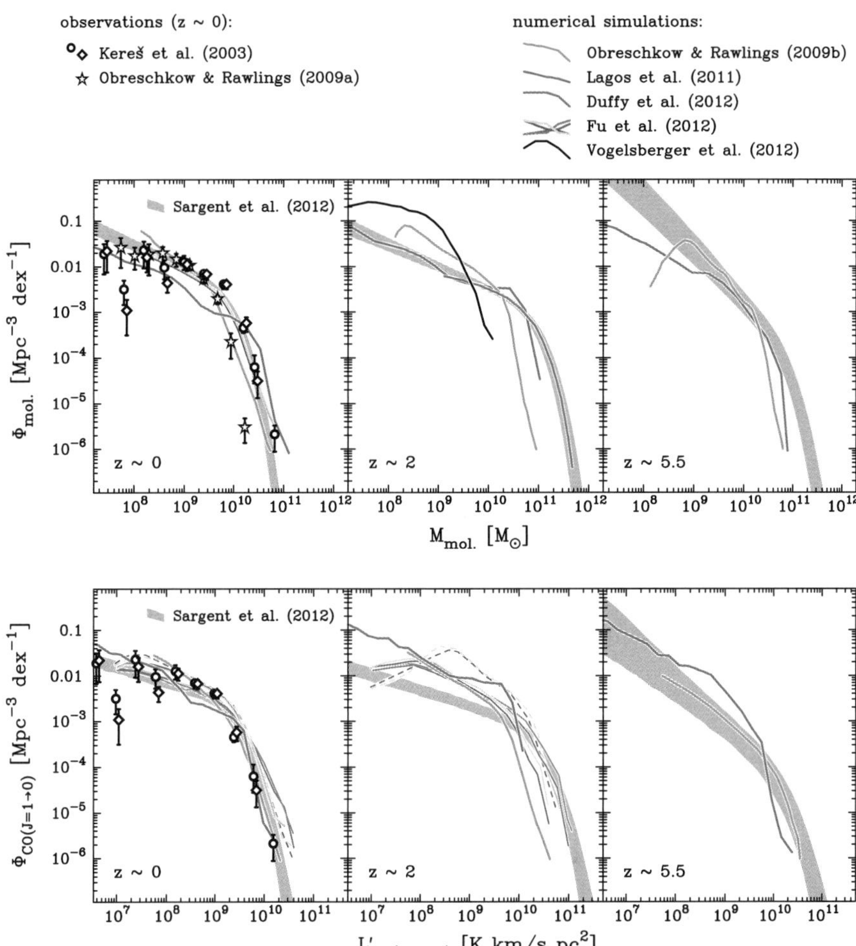

Figure 5. Molecular gas MFs (upper row; grey shading) and CO(1-0) LFs (lower row) based on (1) the evolution of the stellar MF of SFGs, (2) the redshift evolution of the sSFR of main-sequence galaxies, (3) the distribution of main-sequence and starbursting galaxies in the SFR-$M_\star$-plane (see Fig. 1, left), and (4) a metallicity-dependent conversion factor α_{CO}. MFs/LFs include contributions both from 'secular' and burst-like star formation (the latter being characterised by increased SFEs with respect to the secular mode which is assumed to be typical of galaxies residing on the main sequence). Coloured lines –predictions of recent semi-analytical models (see legend above figure) covering a similar redshift range as the empirically-motivated, predictive analysis of Sargent *et al.* (2012, in prep.) . At $z \sim 0$ we show observational constraints on the local MF/LF reported in Keres *et al.* (2003) and Obreschkow & Rawlings (2009b).

the cosmic SFRD but is roughly five times smaller due to the sublinear relation between SFR and H$_2$-mass found in Fig. 3 which implies slowly rising SFEs for the typical SFGs (of stellar mass $\sim 5 \times 10^{10}$ $M_\odot$, cf. Karim *et al.* 2011) that contribute most to the cosmic SFRD density over this time.

4. Summary

Motivated by the homogeneity of the bulk of star-forming galaxies (in particular those occupying the so-called "main sequence" of star-forming galaxies) across a broad range of redshift we have presented a simple framework (2-SFM) for the prediction of

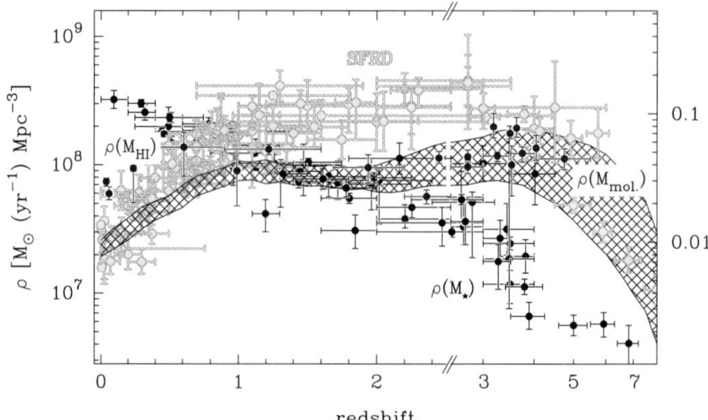

Figure 6. Redshift-evolution of the comoving mass density (cf. scale on left) of molecular gas, $\rho(M_{\rm mol.})$. The hatched region shows the evolution inferred from the integration of the predicted molecular gas MFs in Fig. 5 (extended to $z > 2.5$ with a set of assumptions that successfully match high-z stellar MFs and UV-LFs, e.g. González *et al.* 2010; Bouwens *et al.* 2011). The evolution of $\rho(M_{\rm mol.})$ is compared to compilations of the cosmic evolution of the atomic hydrogen abundance ($\rho(M_{\rm HI})$; e.g. Bauermeister *et al.* 2010, blue symbols) and of $\rho(M_\star)$, the stellar mass density (red symbols; see, e.g., compilation in Marchesini *et al.* 2009). Yellow symbols – SFRD-evolution (cf. scale on right), as constrained by literature data.

basic properties of the star-forming population. By explicitly distinguishing between (1) the population of 'normal' galaxies residing on the star-forming main sequence, and (2) starbursting galaxies for which we assume different characteristics (e.g. IR SEDs or a burst-dependent continuum of SFE enhancements) based on our currently best observational understanding, we show that this simple description is capable of reproducing the evolution of the IR LF out to $z \sim 2.5$ and IR/radio source counts at 24 to 1100 μm and 1.4 GHz. We use the 2-SFM framework to predict the molecular gas mass function and CO(1→0) luminosity function of star-forming galaxies, a fundamental observable which so far has only been measured at $z = 0$ and the extension of which to higher redshift is a major goal of the ALMA era. The strong evolution of the inferred H_2-mass function to higher masses with increasing redshift suggests that the cosmic H_2-abundance was $\sim$5 times larger at the peak of the cosmic SFH than in the local Universe. Furthermore, we provide strong evidence that the higher gas fractions in distant galaxies are directly reflected in the higher sSFR of these systems.

References

Bauermeister, A., Blitz, L., & Ma, C.-P. 2010, *ApJ*, 717, 323
Bell, E. F., *et al.* 2007, *ApJ*, 663, 834
Béthermin, M., Dole, H., Lagache, G., Le Borgne, D., & Penin, A. 2011, *A&A*, 529, 4
Béthermin, M., Le Floc'h, E., Ilbert, O., *et al.* 2012, *A&A*, 542, A58
Béthermin, M., Daddi, E., Magdis, G., *et al.* 2012, *ApJ (Letters)*, 757, L23
Bouwens, R. J., Illingworth, G. D., Oesch, P. A., *et al.* 2011, *ApJ*, 737, 90
Brinchmann, J., Charlot, S., White, S. D. M., *et al.* 2004, *MNRAS*, 351, 1151
Daddi, E., *et al.* 2007, *ApJ*, 670, 156
Daddi, E., *et al.* 2010a, *ApJ*, 713, 686
Daddi, E., *et al.* 2010b, *ApJ (Letters)*, 714, L118
Dannerbauer, H., Daddi, E., Riechers, D. A., *et al.* 2009, *ApJ (Letters)*, 698, L178
Davé, R., Finlator, K., & Oppenheimer, B. D. 2012, *MNRAS*, 421, 98
Di Matteo, P., Combes, F., Melchior, A.-L., & Semelin, B. 2007, *A&A*, 468, 61

Duffy, A. R., Kay, S. T., Battye, R. A., *et al.* 2012, *MNRAS*, 420, 2799

Elbaz, D., *et al.* 2007, *A&A*, 468, 33

Elbaz, D., Dickinson, M., Hwang, H. S., *et al.* 2011, *A&A*, 533, 119

Feldmann, R., Gnedin, N. Y., & Kravtsov, A. V. 2012, *ApJ*, 747, 124

Fu, J., Kauffmann, G., Li, C., & Guo, Q. 2012, *MNRAS*, 424, 2701

Geach, J. E., Smail, I., Moran, S. M., MacArthur, L. A., Lagos, C. d. P., & Edge, A. C. 2011, *ApJ* (Letters), 730, L19

Genzel, R., *et al.* 2010, *MNRAS*, 407, 2091

Genzel, R., Tacconi, L. J., Combes, F., *et al.* 2012, *ApJ*, 746, 69

González, V., Labbé, I., Bouwens, R. J., *et al.* 2010, *A&A*, 713, 115

Goto, T., *et al.* 2011, *MNRAS*, 414, 1903

Graciá-Carpio, J., Sturm, E., Hailey-Dunsheath, S., *et al.* 2011, *ApJ* (Letters), 728, L7

Gruppioni, C., Pozzi, F., Zamorani, G., & Vignali, C. 2011, *MNRAS*, 416, 70

Ilbert, O., Salvato, M., Le Floc'h, E., *et al.* 2010, *ApJ*, 709, 644

Karim, A., *et al.* 2011, *ApJ*, 730, 61

Kennicutt, R. C., Jr. 1998, *ARAA*, 36, 189

Keres, D., Yun, M. S., & Young, J. S. 2003, *ApJ*, 582, 659

Lacey, C. G., Baugh, C. M., Frenk, C. S., *et al.* 2010, *MNRAS*, 405, 2

Lagos, C.d.P., Baugh, C. M., Lacey, C. G., *et al.* 2011, *MNRAS*, 418, 1649

Le Floc'h, E., *et al.* 2005, *ApJ*, 632, 169

Leroy, A. K., *et al.* 2009, *AJ*, 137, 4670

Leroy, A. K., *et al.* 2011, *ApJ*, 737, 12

Magdis, G. E., Daddi, E., Sargent, M., *et al.* 2012, *ApJ* (Letters), 758, L9

Magdis, G. E., Daddi, E., Béthermin, M., *et al.* 2012, *ApJ*, 760, 6

Magnelli, B., Elbaz, D., Chary, R. R., *et al.* 2009, *A&A*, 496, 57

Magnelli, B., Elbaz, D., Chary, R. R., *et al.* 2011, *A&A*, 528, A35

Mannucci, F., Cresci, G., Maiolino, R., Marconi, A., & Gnerucci, A. 2010, *MNRAS*, 408, 2115

Marchesini, D., van Dokkum, P. G., Förster Schreiber, N. M., *et al.* 2009, *ApJ*, 701, 1765

Massardi, M., Bonaldi, A., Negrello, M., *et al.* 2010, *MNRAS*, 404, 532

Narayanan, D., Bothwell, M., & Davé, R. 2012, *MNRAS*, 426, 1178

Obreschkow, D. & Rawlings, S. 2009, *MNRAS*, 394, 1857

Obreschkow, D. & Rawlings, S. 2009, *ApJ* (Letters), 696, L129

Pannella, M., Carilli, C. L., Daddi, E., *et al.* 2009, *ApJ* (Letters), 698, L116

Reddy, N. A., Pettini, M., Steidel, C. C., *et al.* 2012, *ApJ*, 754, 25

Rodighiero, G., *et al.* 2010, *A&A*, 515, A8

Rodighiero, G., *et al.* 2011, *ApJ* (Letters), 739, L40

Rujopakarn, W., Rieke, G. H., Eisenstein, D. J., & Juneau, S. 2011, *ApJ*, 726, 93

Saintonge, A., *et al.* 2011, *MNRAS*, 415, 32

Salmi, F., Daddi, E., Elbaz, D., *et al.* 2012, *ApJ* (Letters), 754, L14

Sanders, D. B., Mazzarella, J. M., Kim, D.-C., Surace, J. A., & Soifer, B. T. 2003, *AJ*, 126, 1607

Sargent, M. T., Béthermin, M., Daddi, E., & Elbaz, D. 2012, *ApJ* (Letters), 747, L31

Schruba, A., Leroy, A. K., Walter, F., *et al.* 2012, *AJ*, 143, 138

Smolčić, V., *et al.* 2009, *ApJ*, 690, 610

Solomon, P. M., Downes, D., Radford, S. J. E., & Barrett, J. W. 1997, *ApJ*, 478, 144

Strazzullo, V., Pannella, M., Owen, F. N., *et al.* 2010, *ApJ*, 714, 1305

Tacconi, L. J., *et al.* 2010, *Nature*, 463, 781

Wuyts, S., Förster Schreiber, N. M., van der Wel, A., *et al.* 2011, *ApJ*, 742, 96

The intriguing life of massive galaxies
Proceedings IAU Symposium No. 295, 2012
D. Thomas, A. Pasquali & I. Ferreras, eds.

© International Astronomical Union 2013
doi:10.1017/S1743921313004353

Mahalo-Subaru: Mapping Star Formation at the Peak Epoch of Massive Galaxy Formation

Tadayuki Kodama[1,2,5], **Masao Hayashi**[2], **Yusei Koyama**[3,2], **Ken-ichi Tadaki**[4,2], **Ichi Tanaka**[1] **and Rhythm Shimakawa**[5]

[1] Subaru Telescope, National Astronomical Observatory of Japan, 650 North A'ohoku Place, Hilo, HI 96720, USA
email: t.kodama@nao.ac.jp
[2] Optical and Infrared Astronomy Division, National Astronomical Observatory, Mitaka, Tokyo 181–8588, Japan
[3] Department of Physics, Durham University, South Road, Durham DH1 3LE, UK
[4] Dept. of Astronomy, Graduate School of Science, Univ. of Tokyo, Tokyo 113–0033, Japan
[5] Dept. of Astronomical Science, Graduate University for Advanced Studies, Mitaka, Tokyo 181–8588, Japan

Abstract. MAHALO-Subaru (MApping HAlpha and Lines of Oxygen with Subaru) is our on-going large programme which aims to investigate how the star forming activities in galaxies are propagated as a function of time, mass, and environment. We are targeting 10 clusters and proto-clusters at $0.4 < z < 2.6$, and two general fields (GOODS-N and SXDF-CANDELS) with Suprime-Cam and MOIRCS by utilizing our unique sets of narrow-band filters. The narrow-band imaging can map out star forming galaxies with the redshifted Halpha and/or [OII] emission lines from our targets, and thus providing relatively unbiased views of star forming activities across time and environment. We have almost completed narrow-band imaging of our targets, and found that star forming activity is very high even in the proto-cluster cores ($z \gtrsim 1.5$), and that the peak of star formation is shifted outwards with time, indicating the inside-out formation of clusters. Moreover, we have identified many "red" emitters especially in high density regions at $z > 2$, which suggests that the mode of star formation and/or the activation of AGN are dependent on environment, and thus holding the key to the environmental effects at the early stage of cluster galaxies formation and evolution.

Keywords. galaxies: clusters, galaxies: evolution, galaxies: formation

---

1. Introduction

The tightness of the colour-magnitude relation of cluster early-type galaxies up to $z \sim 1$–1.5 indicates that star formation in these systems mostly took place long time ago ($z > 1.5$–2) (e.g., Kodama *et al.* 1998). Recent near-infrared imaging of proto-clusters suggests that the massive-end of the red-sequence starts to break down finally at $z \gtrsim 2$ (Kodama *et al.* 2007). It is also suggested that the cosmic star formation rate (SFR) density and the AGN number density both show peaks at $z \sim 2$ (e.g., Bouwens *et al.* 2008; Fan 2006). All these observations clearly indicate that the redshift interval of $1.5 < z < 2.5$ (cosmic look-back time of 9–11 Gyrs) is the critical era for galaxy formation and early evolution. It is therefore essential to systematically map out star forming in the Universe over this period and across various environments. Specifically, we want to know how the star formation in high density regions at high redshifts is intrinsically biased (Cen & Ostriker 1993), how it is affected by galaxy environment, and how the peak activity is shifted outwards to lower density environments with time, and after all what physical processes are involved and responsible for these phenomena. Our previous Subaru observations show that the star formation activity in the cluster core at $z \sim 1.5$

Table 1. Specifications of the narrow-band filters

Camera	Filter	λ_c	FWHM
	NB1190	1.189μm	0.014μm
	NB1550	1.550μm	0.018μm
	NB1657	1.657μm	0.019μm
MOIRCS	NB2071	2.069μm	0.027μm
	NB2095	2.095μm	0.025μm
	NB2288	2.296μm	0.023μm
	NB2315	2.313μm	0.027μm
	NA671	0.6714μm	0.0130μm
Suprime-Cam	NB912	0.9139μm	0.0134μm
	NB921	0.9173μm	0.0132μm
	NB973	0.9755μm	0.020μm

is as high as that in the field, and the peak of star formation is shifted outwards of clusters as time progresses (Koyama *et al.* 2010; Hayashi *et al.* 2010). We need to go further back in time to explore the early stage of galaxy formation and evolution and its environmental dependence.

2. Mahalo-Subaru project

To understand how galaxies form and evolve at the peak epoch, we have been conducting the "Mahalo-Subaru" project (MApping HAlpha and Lines of Oxygen with Subaru), a large part of which was granted as a Subaru open-use intensive program. We are targeting clusters/proto-clusters at $0.4<z<2.53$, and an un-biased general field SXDF/UDS ($z=2.19$ and 2.53 slices) (Table 2). We employ unique sets of narrow-band (NB) filters on the two wide-field cameras, Suprime-Cam (optical; $34'\times27'$) and MOIRCS (NIR; $7'\times4'$) (Table 1). Most of the MOIRCS NB filters are designed and manufactured specifically to our targets, and their wavelengths perfectly match the redshift Hα lines from our $z>1.5$ targets. The Suprime-Cam NB filters in the z'-band are also used for searching for [OII] emitters at $z=1.46$–1.62. The line-of-sight velocity range that fall within the filter FWHM with respect to the cluster center is optimal ($\pm$1000-3000km/s).

Using these NB filters, we have successfully identified lots of star-forming emission line galaxies very efficiently which emit nebular emission lines from ionized star-forming regions, in narrow redshift slices associated with the clusters or in the general field. We are sure about their membership by the presence of emission lines and their SEDs. We have almost completed imaging observations under excellent conditions. Exposure times were typically 1–1.5hrs per broad-band, and 3–4hrs per narrow-band for $z \gtrsim 1.5$ targets.

We have identified well-visible large scale structures traced by the emitters for all of our cluster targets that were analysed so far. For example, the three high redshift clusters, 2215, 0218, and 0332 at $z=1.5$–1.6 are all embedded in extremely large scale structures traced by [OII] emitters which spread over $\sim$20–30 Mpc in co-moving scale. Also at $z \gtrsim 2$, Hα emitters in the vicinity of and physically associated with the radio galaxies are spatially strongly clustered and showing clumpy and/or filamentary structures (Fig. 1). For example, three dense star-bursting groups of galaxies are clearly identified in the proto-cluster USS1558; one including a radio galaxy, another located 3 arcmin away to the Southwest, and the other located in between the two. This system is therefore a clumpy forming cluster to be merged together in the near future.

We find that some of the Hα emitters in the proto-cluster fields show very red colours of $J-K_s>2.3$ (shown by red filled circles/squares), indicating that they are either dusty

Table 2. The list of our targets. S-Cam stands for Suprime-Cam.

environ-ment	target	z	line	λ (μm)	camera	NB-filter	conti-nuum	status as of Oct '12
Low-z cluster	CL0024+1652	0.40	Hα	0.916	S-Cam	NB912	z'	Kodama+'04
	CL0939+4713	0.41	Hα	0.923	S-Cam	NB921	z'	Koyama+'11
	RXJ1716.4+6708	0.81	Hα	1.190	MOIRCS	NB1190	J	Koyama+'10
			[O$_{\text{II}}$]	0.676	S-Cam	NA671	R	observed
High-z cluster	XCSJ2215–1738	1.46	[O$_{\text{II}}$]	0.916	S-Cam	NB912,921	z'	Hayashi+'10,'11
	4C65.22	1.52	Hα	1.651	MOIRCS	NB1657	H	observed
	CL0332–2742	1.61	[O$_{\text{II}}$]	0.973	S-Cam	NB973	y	Hayashi+'13
	CIGJ0218.3–0510	1.62	[O$_{\text{II}}$]	0.977	S-Cam	NB973	y	Tadaki+'12
Proto-cluster	PKS1138–262	2.16	Hα	2.071	MOIRCS	NB2071	K_{s}	Koyama+'13
	4C23.56	2.48	Hα	2.286	MOIRCS	NB2288	K_{s}	Tanaka+'11
	USS1558–003	2.53	Hα	2.315	MOIRCS	NB2315	K_{s}	Hayashi+'12
General field	GOODS-N (70 arcmin^2)	2.19	Hα	2.094	MOIRCS	NB2095	K_{s}	Tadaki+'11
			Hβ	1.551	MOIRCS	NB1550	H	not yet
			[O$_{\text{II}}$]	1.189	MOIRCS	NB1190	J	observed
	SXDF-CANDELS (92 arcmin^2)	2.19	Hα	2.094	MOIRCS	NB2095	K	Tadaki+'13
			Hβ	1.551	MOIRCS	NB1550	H	not yet
			[O$_{\text{II}}$]	1.189	MOIRCS	NB1190	J	not yet
		2.53	Hα	2.315	MOIRCS	NB2315	K_{s}	Tadaki+'13

star forming galaxies or passively evolving galaxies with AGNs. Very interestingly, those red Hα emitters tend to be confined in the densest regions (Hayashi *et al.* 2012; Koyama *et al.* 2013). This suggests that some dramatic events (such as dusty starbursts or AGN feedback) may be occuring in these galaxies in the dense regions due to environmentally driven galaxy-galaxy interactions, for example. In lower redshift clusters (eg., RXJ1716 at z=0.81), such red emitters avoid the cluster centre and are seen only in the surrounding groups or filaments (Koyama *et al.* 2011). Since the location of the red emitters always coincide with the environment where we see a sharp transition in colours or star-formation activity, these must be the key populations under the influence of environmental effects. Star formation histories in the two proto-clusters, 1558 and 1138, are intensively discussed in Hayashi *et al.* (2012) and Koyama *et al.* (2013). These proto-clusters are very likely the sites where early-type galaxies, which will eventually dominate clusters by the present-day, are just in their formation phase. They hence provide excellent samples of star forming galaxies with which we can investigate the physical mechanisms of biased galaxy formation and its early evolution in the dense environment.

We also target a blank field, SXDF-UDS-CANDELS, in order to address environmental dependence of star forming galaxies at high redshifts. This field is unique as lots of coordinated multi-waelength data are avaialble, e.g., Subaru Suprime-Cam deep imaging (Furusawa *et al.* 2008), UKIDSS-UDS (Lawrence *et al.* 2007), Spitzer 3.6–24μm data (SpUDS; PI: J. Dunlop), and in particular deep HST high-resolution optical/NIR imaging with ACS/WFC3 (Grogin *et al.* 2011). Our preliminary analyses show that a large fraction of Hα emitters at z=2.19 and 2.53 show clumpy or merger signatures, while others are compact, centrally concentrated objects. They are probably at different stages of galaxy formation with different modes of star formation.

Our Mahalo-Subaru project is providing a unique dataset of star forming galaxies over the peak epoch of galaxy formation and across various environments. It will tell us how

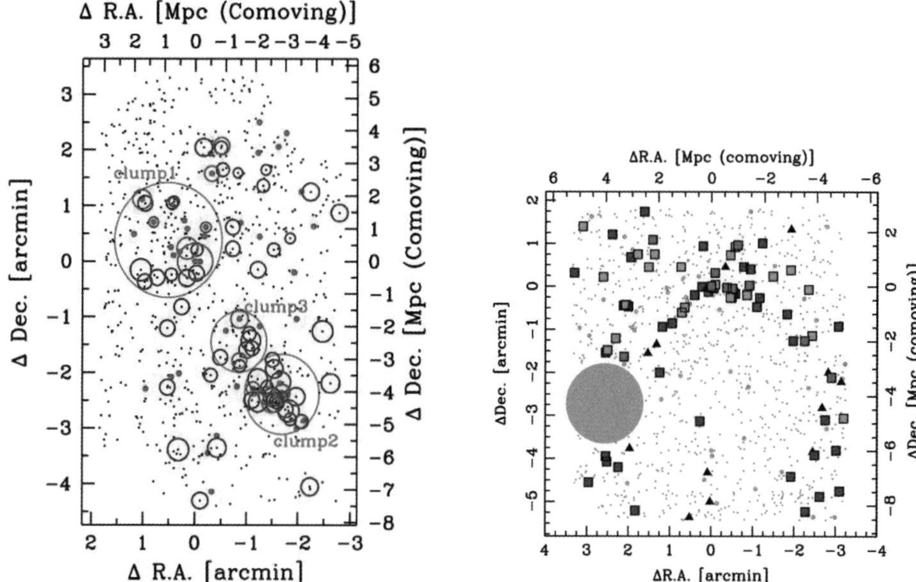

Figure 1. *(Left panel)*: 2-D map of a proto-cluster USS1558–003 at $z=2.53$ (Hayashi *et al.* 2012). The red and blue open circles indicate Hα emitter candidates associated with the proto-cluster. The size of the circles corresponds to a star formation rate. The filled red circles represent galaxies with $(J-K_{\rm s})>1.38$ (DRGs). See the online color version. *(Right panel)*: 2-D map of a proto-cluster PKS1138–262 at $z=2.16$ (Koyama *et al.* 2013). The associated Hα emitter candidates are shown in colored symbols. The red, green and blue squares indicate Hα emitters with $(J-K_{\rm s})>1.38$ (DRGs), $0.8<(J-K_{\rm s})<1.38$ and $(J-K_{\rm s})<0.8$, respectively. The black triangles are the $K_{\rm s}$-undetected emitters. See the online color version.

galaxy formation has progressed with time and how relevant environmental effects are in shaping present-day massive galaxies.

References

Bouwens, R. J., Illingworth, G. D., Franx, M., *et al.* 2008, *ApJ*, 686, 230
Cen, R. & Ostriker, J. P. 1993, *ApJ*, 417, 415
Fan, X. 2006, *NewAR*, 50, 665
Furusawa, H., Kosugi, G., Akiyama, M., *et al.* 2008, *ApJS*, 176, 1
Grogin, N. A., Kocevski, D. D., Faber, S. M., *et al.* 2011, *ApJS*, 197, 35
Hayashi, M., Kodama, T., Koyama, Y., *et al.* 2010, *MNRAS*, 402, 1980
Hayashi, M., Kodama, T., Koyama, Y., *et al.* 2011, *MNRAS*, 415, 2670
Hayashi, M., Kodama, T., Tadaki, K., *et al.* 2012, *ApJ*, 757, 15
Lawrence, A., Warren, S. J., & Almaini, O. 2007, *MNRAS*, 379, 1599
Kodama, T., Arimoto, N., Barger, A. J., *et al.* 1998, *A&A*, 334, 99
Kodama, T., Balogh, M. L., Smail, I., *et al.* 2004, *MNRAS*, 354, 1103
Kodama, T., Tanaka, M., Arimoto, N., *et al.* 2005, *PASJ*, 57, 309
Kodama, T., Tanaka, I., Kajisawa, M., *et al.* 2007, *MNRAS*, 377, 1717
Koyama, Y., Kodama, T., Shimasaku, K., *et al.* 2010, *MNRAS*, 403, 1611
Koyama, Y., Kodama, T., Nakata, F., *et al.* 2011, *ApJ*, 734, 66
Koyama, Y., Kodama, T., Tadaki, K., *et al.* 2013, *MNRAS*, 428, 1551
Tadaki, K., Kodama, T., Koyama, Y., *et al.* 2011, *PASJ*, 63, 437
Tadaki, K., Kodama, T., Ota, K., *et al.* 2012, *MNRAS*, 423, 2617
Tanaka, I., De Breuck, C., Kurk, J., *et al.* 2011, *PASJ*, 63, 415

The intriguing life of massive galaxies
Proceedings IAU Symposium No. 295, 2012
D. Thomas, A. Pasquali & I. Ferreras, eds.

© International Astronomical Union 2013
doi:10.1017/S1743921313004365

Starburst and old population in $z = 3.8$ radio galaxies with Pégase.3

Brigitte Rocca-Volmerange[1] and Guillaume Drouart[2,1]

[1]Institut d' Astrophysique de Paris, UPMC/CNRS, 98bis Bd Arago, F-75014 Paris, France
email: rocca@iap.fr

[2]European Southern Observatory,Karl Schwarzschild Strasse, 85748 Garching bei München, Germany

Abstract. Distant radio galaxies, hosted by massive ellipticals, follow the galaxy evolution process on an extremely large ($0 \geqslant$ z $\geqslant 7$) time-scale $\geqslant 10^{12}$ Gyrs, up to primeval galaxies. The new evolutionary code Pégase.3 predicts on similar time-scales, the coupled stellar and dust emissions of various galaxy types: starbursts and Hubble sequence types. All z=0 templates are fitted on local observations at ages $\simeq 13$ Gyrs (except irregulars at 9 Gyrs). The multi-λ spectral energy distributions (SEDs) of two $z = 3.8$ radiogalaxies, including the most recent *Herschel* data from the *HeRGÉ* consortium, are interpreted in the observer's frame by Rocca-Volmerange *et al.* (2012) with Pégase.3. The apparent SEDs are fitted at best with the sum of a young starburst and an older early-type population, an AGN simple model is taken into account. These results favor massive gas-rich mergers at work in evolved galaxies at $z \simeq 4$. Massive starbursts would be at the origin of galaxy evolution initiated at the earliest epochs ($z_{for} \geqslant 10$). The possible relation with super massive black holes is still debated.

Keywords. galaxies: formation, galaxies: evolution, radio continuum: galaxies, infrared: galaxies, stars: formation

1. Introduction

High-redshift radio galaxies are known as the most distant stellar populations hosting super massive black holes. From their near-IR morphologies, they are identified with massive early-type galaxies (van Breugel *et al.* 1998; Lacy *et al.* 2000; Pentericci *et al.* 2001). Data interpretation with the evolutionary code Pégase.2 (www2.iap.fr/pegase) of the Hubble K-band diagram (De Breuck *et al.* 2002) shows that the brightest luminosity limit corresponds to powerful radio galaxies hosted by massive elliptical galaxies with baryonic masses $M_{\mathrm{bar,max}} \simeq 10^{12} M_\odot$ (Rocca-Volmerange *et al.* 2004). Populations of radio galaxies were likely discovered in the near-infared with IRAS, ISO and Spitzer: the excesses simultaneously observed in faint galaxy counts at 12μm, 15μm with ISO (Seymour *et al.* 2007 and references therein) and at 24μm with Spitzer (Papovich *et al.* 2004) are interpreted as 9% of ultra-bright dusty galaxies evolving as ellipticals, likely AGN hosts (Rocca-Volmerange *et al.* 2007). More recently, distant radio galaxies were observed with *Herschel*/PACS and SPIRE instruments by the HeRGÉ (Herschel Radio Galaxy Evolution) Project (Seymour *et al.* 2012). Together with submmillimeter observations, the continuous SEDs of two $z = 3.8$ radio galaxies 4C 41.17 and TN J2007−1316, selected for their faint AGN contribution and strong star formation signatures from the HeRGÉ sample, are interpreted with the evolutionary code Pégase.3. Section 1 is a presentation of the template models of early types, Section 2 gives the spectral synthesis results of the $z = 3.8$ radio galaxy 4C 41.17 in the observer's frame. Section 3 proposes some preliminary conclusions.

Table 1. The main free parameters of star formation scenarios (hereafter for elliptical and spiral Sa) in the Pégase.3 model. The adopted star formation laws are proportional to the current (neutral and molecular) gas mass M_{gas} with an accretion rate depending on type. The infall time-scale and the age of galactic winds if any, are given. Metal-enrichment and dust amount are computed at all ages from yields and the adopted IMF in the chemical evolution formalism. A variety of IMFs are proposed.

Scenario	Star formation rate	Infall time-scale	Galactic wind age
Elliptical	$3.33\ 10^{-3}\,M_{gas}\,Myr^{-1}$	300 Myrs	1 Gyr
Spiral Sa	$0.71\ 10^{-3}\,M_{gas}\,Myr^{-1}$	2 800 Myrs	no winds

2. Template models of early-type galaxies

Following the readme of the code Pégase.2 (freely accessible at www2.iap.fr/pegase), input parameters for an elliptical and a spiral galaxy are presented in Table 1. Figure 1 illustrates the time evolution of various (total galaxy, stellar and gas) masses for the elliptical galaxy. The parameter set of Table 1 defines the star formation time-scale varying from 1 Gyr to 10 Gyrs from ellipticals to spirals (see also fig.3 of Rocca-Volmerange *et al.* 2004). Synthetic libraries of SEDs evolving from ages of 0 to 20 Gyr are built to simulate the evolution of instantaneous starbursts and of a variety of galaxy types: elliptical (E), lenticular (S0), spiral (Sa , Sb, Sbc, Sc, Sd) and Magellanic irregular (Im) galaxies. Scenarios are robust at $z = 0$, all are fitting local observed colours by types. The star formation process is initiated at the so-called redshift of formation $z_{for}=10$, implying the age of 13 Gyrs (only 9 Gyrs for Irregulars) for local $z=0$ templates. Changing z_{for} to 20 or 30 will only vary ages of local templates by $\leqslant 0.4$ Gyrs according to the z-cosmic time relation, implying small variations in the SED templates. We adopt the following parameters : $H_0 = 70 km.s^{-1} Mpc^{-1}, \Omega_M = 0.3, \Omega_\Lambda = 0.7$. For all scenarios, the

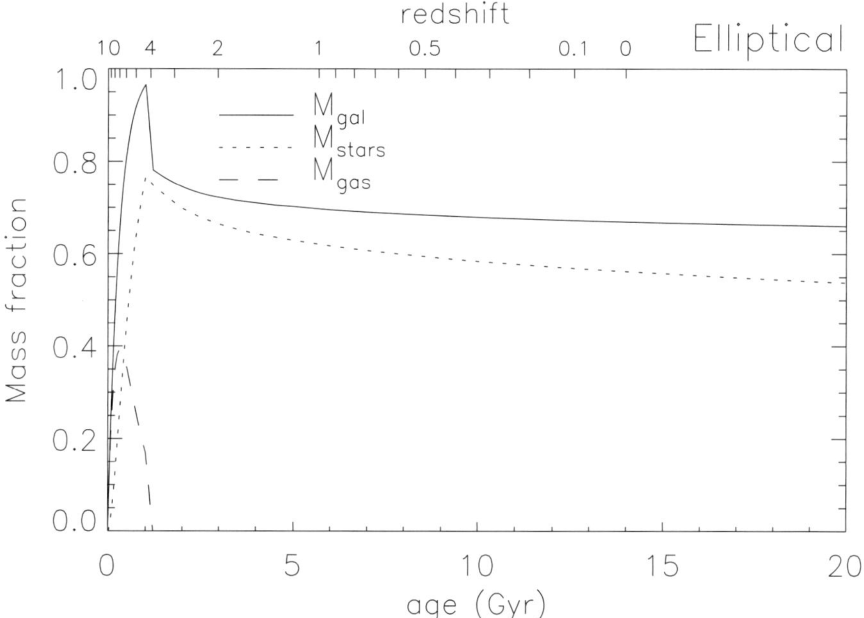

Figure 1. Mass evolution predicted for the elliptical galaxy.

The intriguing life of massive galaxies
Proceedings IAU Symposium No. 295, 2012
D. Thomas, A. Pasquali & I. Ferreras, eds.

© International Astronomical Union 2013
doi:10.1017/S1743921313004389

The intriguing life of star-forming galaxies in the redshift range $1 \leqslant z \leqslant 2$ using MASSIV

P. Amram[1], C. López-Sanjuan[1,2], B. Epinat[1], T. Contini[3],
D. Vergani[4], L. Tasca[1], O. Le Fèvre[1], B. Garilli[5], C. Divoy[3], J.
Queyrel[3], M. Kissler-Patig[6,7], J. Moultaka[3], L. Paioro[5], L. Tresse[1],
V. Perret[1] and F. Bournaud[8]

[1]LAM, AMU, CNRS, Marseille (F), email: `Philippe.Amram@oamp.fr`, [2]FEFCA, Teruel (E),
[3]IRAP, Toulouse (F), [4]INAF, Bologna (I), [5]IASF-INAF, Milano (I), [6]ESO, Garching b.
München (G), [7]GEMINI, Hilo (USA), [8]CEA, SAp, AIM, Saclay (F).

Abstract. MASSIV (Mass Assembly Survey with SINFONI in VVDS) is an ESO large program which consists of 84 star-forming galaxies, spanning a wide range of stellar masses, observed with the IFU SINFONI on the VLT, in the redshift range $1 \leqslant z \leqslant 2$. To be representative of the normal galaxy population, the sample has been selected from a well-defined, complete and representative parent sample. The kinematics of individual galaxies reveals that 58% of the galaxies are slow rotators, which means that a high fraction of these galaxies should probably be formed through major merger processes which might have produced gaseous thick or spheroidal structures supported by velocity dispersion rather than by rotation. Computations on the major merger rate from close pairs indicate that a typical star-forming galaxy underwent ~ 0.4 major mergers in the last ~ 9.5 Gyr, showing that merging is a major process driving mass assembly into the red sequence galaxies. These objects are also intriguing due to the fact that more than one galaxy over four is more metal-rich in its outskirts than in its center.

Keywords. galaxies: evolution, galaxies: fundamental parameters, galaxies: high-redshift, galaxies: kinematics and dynamics

1. Mass assembly of high redshift galaxies

Processes of galaxy mass assembly at early epoch are amongst the largest open issues in galaxy evolution. To tackle this question several mechanisms related to environment should be first understood, e.g. the role of major and/or minor, wet and/or dry mergers versus smooth cold gas accretion along cosmic filaments. On the other hand, the actual impact of feedback from SNe and AGN activity and the secular galaxy evolution should also play a major role in the way the baryonic matter is redistributed within - or ejected from - the galaxies. The knowledge of numerous physical parameters is needed to constrain scenarios of galaxy mass assembly. A number of global parameters could be achieved from global measurements but resolved measurements within individual galaxies extracted from 3D-spectral analysis are necessary to measure the radial distributions of the metallicity, the star formation, the stellar and gaseous masses. With the present-day observational facilities, resolved line measurements could not yet be achieved for galaxies at redshifts higher than $z > 4$. In the redshift range $2 \leqslant z \leqslant 4$, because of strong biases due to color pre-selections, only the brightest star-forming galaxies could be studied which is not the case in the redshift range $1 \leqslant z \leqslant 2$, the 2.6 Gyr-lasting epoch where the stark dichotomy within the galaxy population is rising and is set at $z \sim 1$, and when star-forming galaxies representative of the bulk of galaxy population could be addressed.

MASSIV (Mass Assembly Survey with SINFONI in VVDS) is an ESO large program which consists of 84 star-forming galaxies, in the redshift range $1 \leqslant z \leqslant 2$, observed from 2008 to completion in 2011, with the NIR-IFU SINFONI on the VLT. Spanning a wide range of stellar masses log(M*)=[9,12] and selected from a well-defined, complete and representative parent sample based on accurate spectroscopic redshifts from the VIMOS VLT Deep Survey (Le Fèvre *et al.* 2005), the MASSIV sample is representative of the normal galaxy population. The J- or H-band has been used to target the redshifted $H\alpha$ bright emission line with a high spatial resolution ($< 0.8''$), partly with adaptative optics. A full survey description and global properties of the galaxy sample is given in Contini *et al.* (2012, Paper I). The first epoch sample (50/84 galaxies) leads to several studies about kinematics and close environment classification (Epinat *et al.* 2012, Paper II); evidence for galaxies displaying positive metallicity gradients (Queyrel *et al.* 2012, Paper III); fundamental relations (Vergani *et al.* 2012, Paper IV) and major merger rate from close pairs (López-Sanjuan *et al.* 2012, Paper V).

2. Do high-redshift star-forming galaxies rotate?

The total MASSIV sample of 84 galaxies allows resolved velocity measurements for 76 galaxies. For the kinematic classification, we used several criteria (Epinat *et al.* 2012). One of them is based on the total velocity shear V_{shear} measured from the velocity field not inclination-corrected: 32/76 (42%) galaxies have a high velocity shear ($V_{shear} > 100\ km\ s^{-1}$) and 44/76 (58%) a low velocity shear ($V_{shear} < 100\ km\ s^{-1}$). Another criterion is to distinguish rotators from non-rotating galaxies using a classification based on the disagreement between morphological and kinematic position angles and the mean weighted velocity field residuals normalized by the velocity shear: again 32/76 (42%) galaxies are rotators while 44/76 (58%) are not rotating objects. Finally we consider the nearby environment to distinguish isolated from not isolated galaxies: 58/76 (76%) galaxies are isolated while 18/76 (24%) are not. The galaxies presenting large V_{shear} are mainly rotating disks but also interacting/merging systems. Galaxies showing low V_{shear} may be face-on disks, star-forming spheroids or on-going mergers in transient state. For rotating disks, depending on the total mass of the galaxies, typical maximum rotation velocities V_{rot} range between 100 to 300 $km\ s^{-1}$. Considering these typical maximum rotation velocities $V_{rot} = 200^{+100}_{-100}\ km\ s^{-1}$, if all the disks were pure rotators in planes randomly distributed in the sky, the number of low V_{shear} disks should not exceed 11 galaxies (14%) if all the galaxies were low mass galaxies (for $V_{rot} = 100\ km\ s^{-1}$), 3 galaxies (4%) for $V_{rot} = 200\ km\ s^{-1}$ and 1.5 galaxies (2%) if all the galaxies were high mass galaxies (for $V_{rot} = 300\ km\ s^{-1}$). Furthermore, even if the actual galaxy masses are overestimated, the high fraction of 58% of low velocity shear is incompatible with the hypothesis of face-on rotators.

Finally, it has been shown in Vergani *et al.* (2012) that *(1)* non-rotating galaxies are more compact in their extent of the stellar component than rotators but they are not statistically different in their gas extent; *(2)* marginal evolution in the size-stellar mass and size-velocity relations are observed and *(3)* the large dispersion observed in the stellar and baryonic Tully-Fisher relations is reduced using the $S_{0.5} = \sqrt{(0.5 \times V_{rot}^2 + \sigma_0^2)}$ index instead of V_{rot} (σ_0 being the velocity dispersion). Nevertheless, an intrinsic spread around the median trend remains even when using the $S_{0.5}$ index instead of V_{rot} and slowly rotating galaxies display lower $S_{0.5}$ index and stellar/baryonic masses than fast rotators.

3. One star-forming galaxy over four is more metal-rich in its outskirts than in its centers

Results on the metallicity gradients for the first epoch sample (50 galaxies) have been published in Queyrel *et al.* (2012) while the second epoch sample (35 galaxies) is still under analysis (Divoy *et al.*, in preparation). Among the 50 galaxies of the first epoch, both $H\alpha$ and $[NII]$ lines have been measured for 34 galaxies and metallicity gradients for 26/34. Positive (negative) gradients refers to metallicity being higher (lower) in the center than in the outskirts. We have measured secure negative gradients for 5/26 galaxies and positive ones for 7/26. Preliminary results on the second epoch sample, which contains in average fainter and smaller galaxies than the first epoch one, indicate that integrated metallicities are measurable for $\sim 20/35$ galaxies. Metallicity gradients have been measured for 10/20 galaxies and forthcoming measurements will probably be possible for an extra couple of galaxies. Among those 10 galaxies, 7 of them display negative gradients and 3 positive ones. On the whole MASSIV sample presently available, metallicity gradients have been measured on 36 galaxies; 12 galaxies (33%) show secure classical negative gradients but 10 galaxies (28%) show unexpected positive gradients, the other 14 galaxies (39%) are compatible with a flat radial metallicity distribution. Nine galaxies over 12 (75%) showing a negative gradient are isolated (3/12 or 25% are in interaction) while 6 galaxies over 10 (60%) that show a positive gradient are interacting (4/10 or 40% are isolated). Interactions, mergers or cold gas accretion might be responsible for shallowing and even inverting the abundance gradient.

4. Star-forming galaxy underwent ~ 0.4 major mergers in the last $\sim$9.5 Gyr

The major merger rate at $0.9 \leqslant z \leqslant 1.7$ from IFU-based close pairs in MASSIV is studied in López-Sanjuan *et al.* (2012). A close pair is defined by a couple of galaxies showing *(1)* a projected separation lower than 20 $h^{-1}kpc$ and *(2)* a radial velocity difference lower than 500 $km\ s^{-1}$. When the two galaxies overlap, the two components have been separated both using the morphology and the kinematics (see left panels of Fig. 1). A major (minor) close pair is defined by a couple of galaxies for which the luminosity ratio is $L_2/L_1 > 1/4$ ($L_2/L_1 < 1/4$). N_P is the number of major close pairs over a population of N principal galaxies targeted. The fraction of major merger f_{MM} is basically equal to N_P/N corrected for selection effects and covered area. Within the stellar mass range $10^9 - 10^{10} M_\odot$, the MASSIV sample contains 20 close pair candidates, including $N_P = 13$ major mergers and 7 minors mergers. The major merger rate $R_{MM} \propto f_{MM}/T_{MM}$ at three different epochs, where T_{MM} is the merger time scale extracted from the Millennium simulation, is given in Table 1. Using the MASSIV data combined with data extracted from the literature, Fig. 1 shows that the fraction of major mergers evolves with the redshift z like $f_{MM} \propto (1+z)^{3.91}$ and the major merger rate like $R_{MM} \propto (1+z)^{3.95}$. The average number of major gas-rich mergers per star-forming galaxy between $z = 1.5$ and $z = 0$ is $0.37^{+0.21}_{-0.13}$. Half of the merger activity occurs at high redshift between $z = 1.5$ and $z = 1$ ($\sim 1.6\ Gyr$) and the other half more recently between $z = 1.0$ and $z = 0$ ($\sim 7.7\ Gyr$) which means the average merger activity was higher by a factor 5 at the earliest epoch than at the later one. Accordingly to early type galaxies (ETGs) classification based on fast and slow rotators from Emsellem *et al.* (2011) and using the following assumptions: *(1)* wet major mergers produce fast rotators; *(2)* dry major mergers produce slow rotators, *(3)* dry minor mergers do not change the kinematical state of ETGs and *(4)* all ETGs at z=1.3 are fast rotators, a fraction of slow

Table 1. Fraction of major merger f_{MM} and major merger rate R_{MM}

T_{MM} (Gyr) $[z_1, z_2]$	f_{MM} (%) @ z	R_{MM} (Gyr^{-1}) @ z
1.80 [0.94, 1.06]	21 @ $z = 1.0$	0.12 @ $z = 1.0$
1.37 [0.20, 1.50]	20 @ $z = 1.4$	0.15 @ $z = 1.4$
2.54 [1.50, 1.80]	23 @ $z = 1.6$	0.13 @ $z = 1.6$

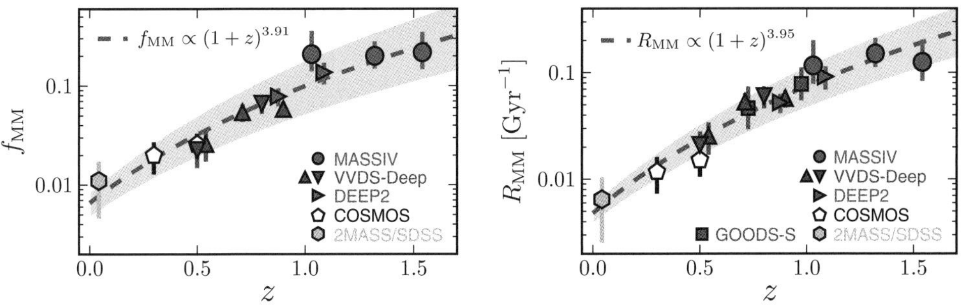

Figure 1. Major merger fraction (left panel) and major merger rate (right panel) of $M_* \sim 10^{10-10.5} M_\odot$ galaxies as a function of redshift. On both panels, diamonds are from this MASSIV data set, triangles from de Ravel *et al.* (2009) and inverted triangles from López-Sanjuan *et al.* (2011), both in VVDS-Deep, pentagons are from Xu *et al.* (2012) in the COSMOS field, and the hexagon is from Xu *et al.* (2012) in 2MASS/SDSS. On the right panel only, squares are from López-Sanjuan *et al.* (2009) in GOODS-S, pentagons are from Xu *et al.* (2012) in the COSMOS field. On both panels, the solid line is the error-weighted least-squares fit of a power-law function, respectively $f_{MM} = 0.0062 \times (1+z)^{3.9}$ and $R_{MM} = 0.0058 \times (1+z)^{3.7}$, to the data. The grey areas mark the 68% confidence interval in the fit.

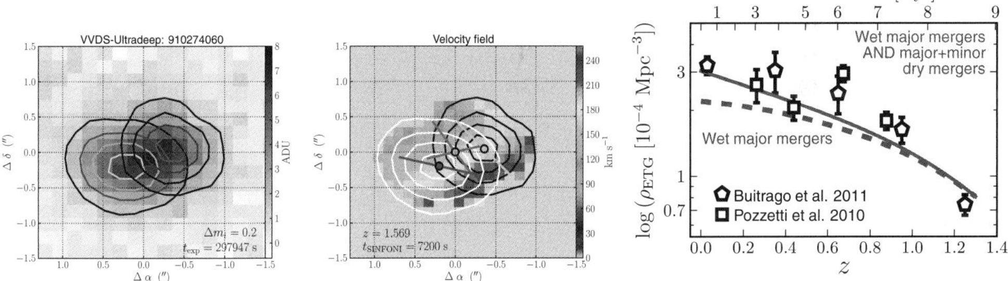

Figure 2. i-band image (left panel) and Velocity field (middle panel) of the MASSIV source 910274060 (major merger). North is up and East is left. The level contours mark the isophotes of the two components obtained with GALFIT in the i-band image. The principal galaxy (red/white) is the one closer to the kinematical centre of the system while the companion (blue/black) is the secondary component. The outer contour was chosen to fit the kinematical maps. (Right panel) Number density evolution of massive ($M_* \geqslant 10^{11.25} M_\odot$) ETGs (E/S0) as a function of redshift from Buitrago *et al.* (2013)(pentagons).

rotator of 55% is computed and leads to the conclusion that merging is a major process driving mass assembly into the red sequence galaxies. The right panel of Fig. 1 shows that the combined effect of gas-rich and dry mergers $f_{tot} = f_{wet} + f_{dry}$ is able to explain the evolution in ETGs since $z \sim 1.3$, with gas rich merging accounting for 2/3. Minor merging is definitively present in the MASSIV sample but due to incompleteness we are not able to assess a minor merger rate.

 P. Amram *et al.*

References

Buitrago, F., Trujillo, I., Conselice, C. J., & Haeussler, B. 2013, *MNRAS*, 428, 1460

Contini, T., Garilli, B., Le Fèvre, O., *et al.* 2012, *A&A*, 539, 91

de Ravel, L., Le Fèvre, O., Tresse, L., *et al.* 2009, *A&A*, 498, 379

Emsellem, E., Cappellari, M., Krajnović, D., *et al.* 2011, *MNRAS*, 414, 888

Epinat, B., Tasca, L., Amram, P., *et al.* 2012, *A&A*, 539, 92

Le Fèvre, O., Vettolani, G., Garilli, B., *et al.* 2005, *A&A*, 439, 845

López-Sanjuan, C., Balcells, M., Pérez-González, P. G., *et al.* 2009, *A&A*, 501, 505

López-Sanjuan, C., Le Fèvre, O., de Ravel, L., *et al.* 2011, *A&A*, 530, 20

López-Sanjuan, C., Le Fèvre, O., Tasca, L. A. M., *et al.* 2012, *A&A*, 553, 78

Queyrel, J., Contini, T., Kissler-Patig, M., *et al.* 2012, *A&A*, 539, 93

Vergani, D., Epinat, B., Contini, T., *et al.* 2012, *A&A*, 546, 118

Xu, C. K., Zhao, Y., Scoville, N., *et al.* 2012, *ApJ*, 747, 85

The intriguing life of massive galaxies
Proceedings IAU Symposium No. 295, 2012
D. Thomas, A. Pasquali & I. Ferreras, eds.

© International Astronomical Union 2013
doi:10.1017/S1743921313004390

Hα Equivalent Widths from the 3D-HST survey: evolution with redshift and dependence on stellar mass†

Mattia Fumagalli[1], Shannon G. Patel[1], Marijn Franx[1], Gabriel Brammer[2], Pieter van Dokkum[3], Elisabete da Cunha[5], Mariska Kriek[4], Britt Lundgren[3], Ivelina Momcheva [3], Hans-Walter Rix[5], Kasper B. Schmidt[5], Rosalind E. Skelton[3], Katherine E. Whitaker[3], Ivo Labbe[1], and Erica Nelson[3]

[1] Leiden Observatory, Leiden University, P.O. Box 9513, 2300 RA Leiden, Netherlands
[2] European Southern Observatory, Alonso de Crdova 3107,Casilla 19001,Santiago, Chile
[3] Department of Astronomy, Yale University, New Haven, CT 06511, USA
[4] Department of Astronomy, University of California, Berkeley, CA 94720, USA
[5] Max Planck Institute for Astronomy (MPIA), Knigstuhl 17, 69117 Heidelberg, Germany

Abstract. We investigate the evolution of the Hα equivalent width, EW(Hα), with redshift and its dependence on stellar mass, using the first data from the 3D-HST survey, a large spectroscopic Treasury program with the HST-WFC3. Combining our Hα measurements of 854 galaxies at $0.8 < z < 1.5$ with those of ground based surveys at lower and higher redshift, we can consistently determine the evolution of the EW(Hα) distribution from z=0 to z=2.2. We find that at all masses the characteristic EW(Hα) is decreasing towards the present epoch, and that at each redshift the EW(Hα) is lower for high-mass galaxies. We find EW(Hα) $\sim (1 + z)^{1.8}$ with little mass dependence. Qualitatively, this measurement is a model-independent confirmation of the evolution of star forming galaxies with redshift. A quantitative conversion of EW(Hα) to sSFR (specific star-formation rate) is model dependent, because of differential reddening corrections between the continuum and the Balmer lines. The observed EW(Hα) can be reproduced with the characteristic evolutionary history for galaxies, whose star formation rises with cosmic time to $z \sim 2.5$ and then decreases to $z = 0$. This implies that EW(Hα) rises to 400 Å at $z = 8$. The sSFR evolves faster than EW(Hα), as the mass-to-light ratio also evolves with redshift. We find that the sSFR evolves as $(1 + z)^{3.2}$, nearly independent of mass, consistent with previous reddening insensitive estimates. We confirm previous results that the observed slope of the sSFR-z relation is steeper than the one predicted by models, but models and observations agree in finding little mass dependence.

Keywords. galaxies: evolution galaxies: formation galaxies: high-redshift

---

† Based on observations made with the NASA/ESA Hubble Space Telescope, obtained at the Space Telescope Science Institute, which is operated by the Association of Universities for Research in Astronomy, Inc., under NASA contract NAS 5-26555. These observations are associated with programs 12177, 12328

The intriguing life of massive galaxies
Proceedings IAU Symposium No. 295, 2012
D. Thomas, A. Pasquali & I. Ferreras, eds.

© International Astronomical Union 2013
doi:10.1017/S1743921313004407

Spatially-Resolved View of High-Redshift Starbursts: the case of Sub-mm Galaxies

Karín Menéndez-Delmestre[1], Andrew W. Blain[2], Mark Swinbank[3], Ian Smail[3], Rob J. Ivison[4,5], and Scott C. Chapman[6]

[1]Observatório do Valongo, Universidade Federal do Rio de Janeiro, Ladeira do Pedro Antônio 43, Rio de Janeiro, RJ 20080-090, Brazil; email: `kmd@astro.ufrj.br`

[2]University of Leicester, University Road, Leicester, LE1 7RH, UK

[3]Institute for Computational Cosmology, Durham University, Durham DH1 3LE, UK

[4]UK Astronomy Technology Centre, Blackford Hill, Edinburgh EH9 3HJ

[5]Institute for Astronomy, Blackford Hill, Edinburgh EH9 3HJ

[6]Institute of Astronomy, Madingley Road, Cambridge, CB3 0HA, UK

Abstract. Ultra-luminous infrared galaxies ($L_{IR} > 10^{12}$ $L_\odot$) are locally rare, but appear to dominate the co-moving energy density at higher redshifts ($z > 2$). Many of these are optically-faint, dust-obscured galaxies that have been identified by the detection of their thermal dust emission at sub-mm wavelengths. Multi-wavelength spectroscopic follow-up observations of these sub-mm galaxies (SMGs) have shown that they are massive ($M_{stellar} \sim 10^{11}$ $M_\odot$) objects undergoing intense star-formation (SFRs $\sim 10^2 - 10^3$ $M_\odot$ yr^{-1}) with a mean redshift of $z \sim 2$, coinciding with the epoch of peak quasar activity. The large fraction of AGNs in SMGs and the derived SMBH masses ($M_\bullet < 10^8$ $M_\odot$) in these galaxies suggest that the submm phase may play an important role in the rapid growth of SMBHs. When both AGN and star-formation activity are present, long-slit spectroscopic techniques face difficulties in disentangling their contributions and may result in SFR and mass overestimates. We present an integral field view of the Hα emission in a sample of 3 SMGs at $z \sim 1.4 - 2.4$ with the IFU instrument OSIRIS on Keck. Designed to be used with Laser Guide Star Adaptive Optics, OSIRIS allows a spatial resolution of up to $10\times$ higher than what has been possible in previous seeing-limited studies of the ionized gas in these galaxies. Our main results are the following: (1) We detect multiple galactic-scale sub-components: the compact, broad Hα emission (FWHM > 1000 km s^{-1}) likely associated with an AGN, the more extended narrow-line Hα emission (FWHM $\lesssim 500$ km s^{-1}) of star-forming regions; the latter are dominated by multiple $1-2$ kpc sized Hα-bright clumps, each contributing 1-25% of the total clump-integrated Hα emission. (2) We derive clump dynamical masses $\sim 1 - 10 \times 10^9 M_\odot$, $1 - 2$ orders of magnitude larger than the kpc-scaled stellar clumps uncovered in optically-selected $z \sim 2$ star-forming galaxies. (3) We determine high star-formation rate surface densities ($\Sigma_{SFR} \sim 1 - 50 M_\odot$yr^{-1} kpc^{-2}, after extinction correction), similar to local starbursts and luminous infrared galaxies. In contrast to these local environments, SMGs undergo such intense activity on significantly larger spatial scales as revealed by extended Hα emission over $4 - 16$ kpc. (4) We find no evidence of ordered global motion as it would be found in a disk, but rather large velocity offsets ($\sim$ few $\times 100$ km s^{-1}) between the distinct stellar clumps. The merger interpretation is likely the most accurate scenario for the SMGs in our sample. However, the final test of whether an underlying disk structure is present will come from studies of the cold gas at the high spatial resolutions possible with ALMA.

We refer the reader to Menéndez-Delmestre *et al.* (2012) for more details.

Keywords. galaxies: high-redshift, galaxies: kinematics and dynamics, galaxies: starburst, galaxies: active, infrared: galaxies, submillimeter, techniques: spectroscopic, instrumentation: adaptive optics

The intriguing life of massive galaxies
Proceedings IAU Symposium No. 295, 2012
D. Thomas, A. Pasquali & I. Ferreras, eds.

© International Astronomical Union 2013
doi:10.1017/S1743921313004419

Connection between the Star Formation Rate and the Gamma-Ray Bursts

Attila Mészáros[1], Zsolt Bagoly[2], Lajos G. Balázs[3] and István Horváth[4]

[1]Faculty of Mathematics and Physics, Astronomical Institute, Charles University, V Holešovičkách 2, CZ 180 00 Prague 8, Czech Republic, email: **meszaros@cesnet.cz**;
[2]Department of Physics of Complex Systems, Eötvös University, H-1117 Budapest, Pázmány P. s. 1/A, Hungary;
[3]Konkoly Observatory, H-1505 Budapest, POB 67, Hungary;
[4]Department of Physics, Bolyai Military University, H-1581 Budapest, POB 15, Hungary

Abstract. It is remarkable that the long gamma-ray bursts, as objects connected with the supernovae - i.e. with the end-stages of massive stars, trace the star formation rate. This connection is discussed in this contribution. The presentation is in essence a recapitulation of the article by Mészáros *et al.* (2006).

Keywords. gamma-ray bursts; supernovae; star-formation rate

1. Overview of the article Mészáros *et al.* (2006)

The BATSE instrument on the Compton Gamma Ray Observatory detected 2704 gamma-ray bursts (GRBs). From this data set it follows that there are two physically different subgroups of GRBs, "short" and "long" ones (Balázs *et al.* 2003); the presence of a further subgroup is not excluded (see Řípa *et al.* 2012, and the references therein).

The question is the following: Can the redshifts of GRBs be distributed in accordance with other objects arising in the star formation regions (Madau 1995 and Dahlen *et al.* 2004)? In addition, this question should be answered separately for any subgroup.

The method is the following: We assume for a given subclass of GRBs that it is distributed in accordance with the redshift distribution of the objects in the star formation regions. Then we compare this theoretical expectation with the observational data from the BATSE Catalog.

The answer is that the redshift distribution of the long bursts may be proportional to star-formation rate (SRF). For the short bursts this can also be the case, but the proportionality is less evident. The connection of the possible third subgroup and of the SFR was not studied yet. All these results are independent on the models of GRBs, and also on the cosmological parameters.

Acknowledgements: This study was supported by the OTKA grant K77795, by the Grant Agency of the Czech Republic grants No. P209/10/0734, and by the Research Program MSM0021620860 of the Ministry of Education of the Czech Republic.

References

Balázs, L. G., Bagoly, Z., Horváth, I., Mészáros, A., & Mészáros, P. 2003, *A&A*, 401, 129
Dahlen, T., Strolger, L. G., Riess, A. G., *et al.* 2004, *ApJ*, 613, 189
Madau, P. 1995, *ApJ*, 441, 18
Mészáros, A., Bagoly, Z., Balázs, L. G., & Horváth, I. 2006, *A&A*, 455, 785
Řípa, J., Mészáros, A., Veres, P., & Park, I. H. 2012, *ApJ*, 756, 44

The intriguing life of massive galaxies
Proceedings IAU Symposium No. 295, 2012
D. Thomas, A. Pasquali & I. Ferreras, eds.

© International Astronomical Union 2013
doi:10.1017/S1743921313004420

Stacking of Interferometric Data

Lukas Lindroos[1] and Kirsten K. Knudsen[1]

[1] Department of Earth and Space Sciences,
Chalmers University of Technology, Onsala Space Observatory, 439 92 Onsala, Sweden
email: `lindroos@chalmers.se`, `kraiberg@chalmers.se`

Abstract. Radio and mm observations play an important role in determining the star formation properties of high redshift galaxies. However, most galaxies at high redshift are too faint to be detected individually at these wavelengths. A way to study this population of galaxies is to use stacking. By averaging the emission of a large number of galaxies detected in optical or near infrared surveys, we can achieve statistical detection.

We investigate methods for stacking data from interferometric surveys. Interferometry poses unique challenges in stacking due to the nature of imaging of this data. We have compared directly stacking the *uv* data with stacking of the imaged data, the latter being the typically used approach. Using simulated data, we find that *uv*-stacking may provide around 50% less noise and that image based stacking systematically loses around 10% of the flux.

Keywords. Stacking, Interferometry

Stacking algorithms and comparison based on simulated data

The aim of our work was to investigate how to best stack data from interferometric surveys. We have evaluated two algorithms. One method based on fully imaged data, similar to the method used by Carilli *et al.* (2008) and one method working directly on the visibility data.

To evaluate the algorithms, we produced a simulated data set with 100 sources at 25 mJy. In order to create a more realistic noise, several brighter sources around a few Jansky were also introduced. Using this data, the image stacking algorithm yielded a 1.5 times higher noise level and systematically missed 10% of the flux. Stacking the *uv*-plane data reproduces the expected result well within the error bars.

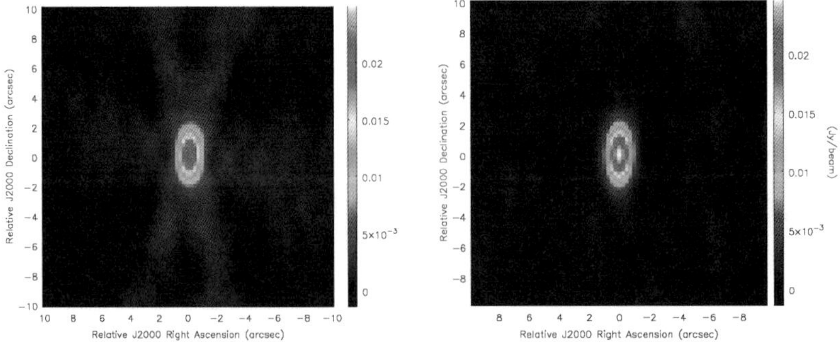

Figure 1. Comparison of stacking done on the image-plane (*left*) and the *uv*-plane (*right*).

References

Carilli, C. L., *et al.* 2008, *ApJ*, 689, 883

The intriguing life of massive galaxies
Proceedings IAU Symposium No. 295, 2012
D. Thomas, A. Pasquali & I. Ferreras, eds.

© International Astronomical Union 2013
doi:10.1017/S1743921313004432

Using the Millennium II simulation to test CDM predictions for the structure of massive galaxies

Andrew P. Cooper[1]†, **Guinevere Kauffmann**[1], **Jing Wang**[1] and **Simon D. M. White**[1]

[1]Max Planck Institut für Astrophysik, Karl-Schwarzschild-Str. 1, 85748 Garching, Germany
email: `acooper@nao.cas.cn`

Abstract. We have combined the semi-analytic galaxy formation model of Guo *et al.* (2011) with a novel particle-tagging technique to predict galaxy surface brightness profiles in a representative sample of ~ 1900 massive dark matter haloes (10^{12}–10^{14} $M_\odot$) from the Millennium II ΛCDM N body simulation. We focus on the outer regions of galaxies and stars accreted in mergers. Our simulations cover scales from the stellar haloes of Milky Way-like galaxies to the 'cD envelopes' of groups and clusters, and resolve low surface brightness substructure such as the tidal streams of dwarf galaxies. We find that the spatial distribution of stars in low surface brightness regions is tightly correlated with DM halo mass and that collisionless merging during the hierarchical assembly of galaxies largely determines the structure of spheroidal stellar components. Our ΛCDM model agrees well with the available data.

Keywords. galaxies: elliptical and lenticular, cD, galaxies: kinematics and dynamics, galaxies: structure, methods: n-body simulations

1. Introduction

Galaxies in a cold dark matter (CDM) universe are assembled both by in situ star formation and by galactic accretion and merging (White & Rees 1978). The balance between these two channels affects how stars and chemical elements are distributed within present-day galaxies. In particular, accretion and merger events are thought to supply most of the stars that contribute to the outer regions of galaxies (e.g. the stellar halo of the Milky Way; Searle & Zinn 1978). The diffuse intracluster light around massive cluster central galaxies may have a similar origin and account for a large fraction of the total stellar mass. Future surveys such as LSST will make possible detailed studies of low surface brightness regions in most massive galaxies at low redshift and thereby provide new ways to test models of galaxy evolution.

The semi-analytic approach treats the fundamental astrophysical processes involved in galaxy formation as a self-consistent system of rate equations, which describe the transfer of mass between different phases (White & Frenk 1991, Kauffmann, White & Guiderdoni 1993, Cole *et al.* 2000). Given cosmological initial conditions and a model for the hierarchical clustering of dark matter, such models provide ab initio predictions for the properties of an entire ΛCDM galaxy population. They allow an efficient exploration of parameter space, leading to a more intuitive understanding of the interplay between different processes, and produce results that can be compared statistically to observations

† Present address: National Astronomical Observatories, Chinese Academy of Scienes, 20A Datun Rd., Beijing 100012, P.R. China

from large galaxy surveys. Our aim is to use the model of Guo *et al.* (2011) to make statistical predictions for the distribution of accreted stars in massive galaxies.

2. Model

Structural predictions from semi-analytic models are currently limited to 1D radial profiles. We have developed a new technique which we call 'particle tagging' (described in detail in Cooper *et al.* 2010; see also Bullock & Johnston 2005) to extend these predictions to all six dimensions of phase space, as in an N body simulation. We 'tag' a set of DM particles in the Millennium II simulation (Boylan-Kolchin *et al.* 2009) for every distinct stellar population (defined by a single age and metallicity) that forms according to the Guo *et al.* (2011) model. Particles are chosen according to their binding energy at the time when their associated population forms, such that they initially occupy a region of phase space appropriate for new stars after dissipative collapse. The phase space evolution of the tagged particles has already been computed, so we use it as a proxy for that of their associated stars, assuming that the gravity of the stars does not alter the evolution of the system. This is likely to be a poor approximation in the centres of massive galaxies, but adequate for accreted stars in their outer regions.

3. Results

Figure 1 shows nine examples of projected stellar mass surface density distributions, from the ~ 1900 massive DM haloes in the Millennium II simulation to which we applied our particle tagging model. We have removed satellite galaxies to show only the light associated with the central galaxy. Galaxies labelled A, E and I are associated with haloes of $M_{200} \sim 10^{12.5} \mathrm{M}_\odot$ while the rest are the central galaxies of massive clusters, $M_{200} \sim 10^{13.5} \mathrm{M}_\odot$†. Black pixels correspond to a surface density of $\sim 10^3 \mathrm{M}_\odot \, \mathrm{kpc}^{-2}$ and white pixels to $\sim 10^9 \mathrm{M}_\odot \, \mathrm{kpc}^{-2}$. The overall light distribution is elongated in most cases, reflecting the triaxial nature of DM haloes. The brightest regions (out to $R \sim 15$ kpc) are dominated by stars formed in situ, while the outer regions are dominated by accreted stars. Streams and shells from recently disrupted companion galaxies are clearly visible.

Figure 2 shows the median surface density profiles of our simulated galaxies obtained by stacking them in bins of M_{200}. The shape of the median profile has a tight correlation with M_{200}. In low mass haloes (comparable to the Milky Way and M31; Li & White 2008) the profile has a clear inflection to a shallower outer slope at ~ 15 kpc; in more massive haloes it is close to a powerlaw over the range $10 < R < 100$ kpc, falling off gradually at larger radius. The trends in our simulations are comparable to those observed, both for individual galaxies and stacks, a number of which we compare in Figure 2.

The behaviour of the average profile can be easily understood by splitting it into in situ and accreted components, as has been emphasised by a number of recent studies (Abadi *et al.* 2006, Cooper *et al.* 2010, Oser *et al.* 2010). Both components can be well-described by Sersic profiles, but they scale differently with M_{200} according both to the mass-dependent efficiency of star formation (e.g. Moster *et al.* 2010) and the mass spectrum, accretion time and orbits of hierarchical progenitors. This is a natural outcome of galaxy formation in CDM (e.g. Cole *et al.* 2000, Purcell *et al.* 2007). Our a priori 6D model of galactic structure is constrained to match the observed galaxy mass function and agrees well with other galaxy population data, and its predictions can therefore be compared directly to observations. The particle tagging technique has

† M_{200} is the mass within a radius enclosing a mean density 200 times the critical density.

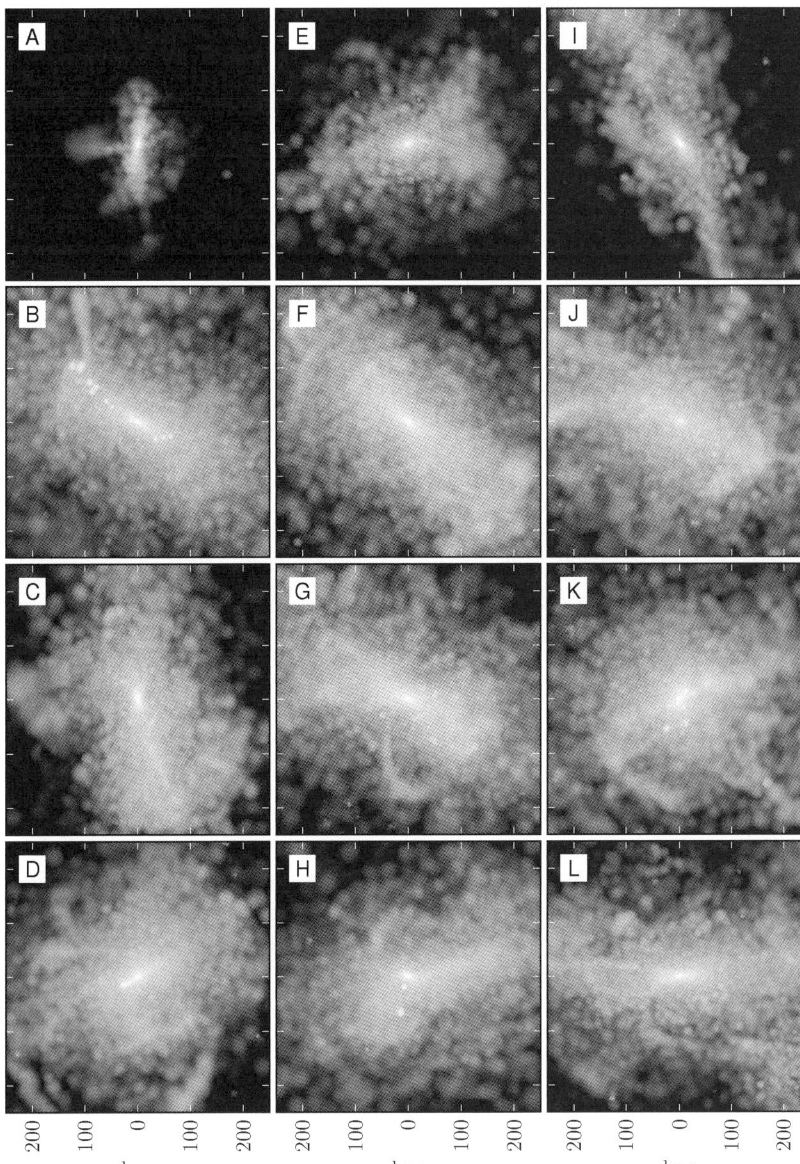

Figure 1. Images of nine of our simulated central galaxies, 250×250 kpc. The logarithmic grey scale corresponds to stellar mass surface density over the range $3 < \log_{10} \Sigma/\mathrm{M}_\odot \, \mathrm{kpc}^{-2} < 9$. The first row (labels A, E, I) shows three galaxies with $M_\star \sim 10^{11} \mathrm{M}_\odot$ and DM halo mass $M_{200} \sim 10^{13} \mathrm{M}_\odot$. The other panels show the central galaxies of cluster-mass haloes $M_{200} \sim 10^{14} \mathrm{M}_\odot$. Satellite galaxies are *not* shown. Simulation particles have been smoothed with an $N = 64$ cubic spline kernel; the remaining lumpiness in these images is due to shot noise in the particle distribution.

a number of interesting applications that we will present in forthcoming papers, include stellar population gradients in early-type galaxies; the kinematics of diffuse light; the effects of interactions on the structure of satellite galaxies and environmental trends; the frequency of tidal features and their correlations with other galaxy properties; and the intracluster light of massive clusters.

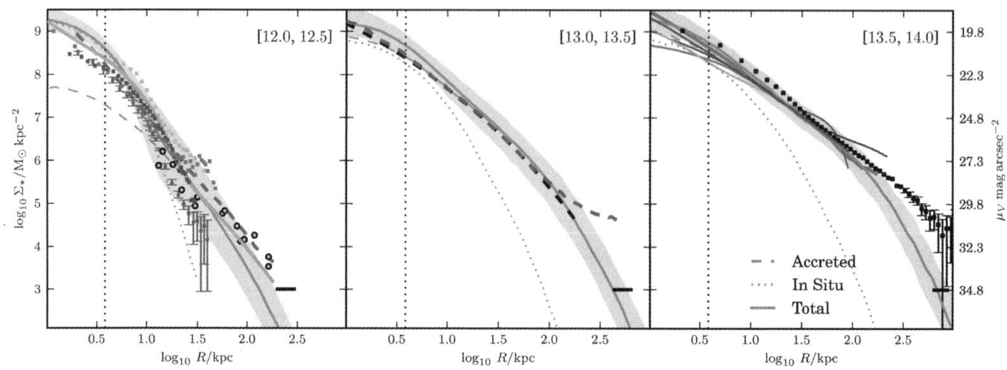

Figure 2. Median surface density profiles (shaded region shows 10-90 percentile range) for stacks of simulated galaxies in three bins of $\log_{10} M_{200}$, compared with observations. Simulated profiles are split accreted (dashed) and in situ (dotted) contributions. Black bar shows the range of virial radii R_{200} in each mass bin. Data is taken from the following authors. Left: Barker *et al.* (2009), Gilbert *et al.* (2009), Courteau *et al.* (2011), Font *et al.* (2011; GIMIC simulations), Barker *et al.* (2012), Bakos & Trujillo (2012). Centre: van Dokkum *et al.*(2010), Tal & van Dokkum (2011). Right: Zibetti *et al.*(2005), Seigar *et al.* (2007), Kormendy *et al.* (2009).

References

Abadi, M. G., Navarro, J. F., & Steinmetz, M. 2006, *MNRAS*, 365, 747

Barker, M. K., Ferguson, A. M., Irwin, M., *et al.* 2009, *AJ*, 138, 1469

Barker, M. K., Ferguson, A. M., Irwin, M. J., *et al.* 2012, *MNRAS*, 419, 1489

Bakos, J. & Trujillo, I. 2012, *ApJ*, submitted (arXiv:1204.3082)

Boylan-Kolchin, M., Springel, V., White, S. D. M., & Jenkins, A., Lemson G. 2009, *MNRAS*, 398, 1150

Bullock, J. S. & Johnston, K. V. 2005, *ApJ*, 635, 931

Cole, S. 1991, *ApJ*, 367, 45

Cole, S., Lacey, C. G., Baugh, C. M., & Frenk, C. S. 2000, *MNRAS*, 319, 168

Cooper, A. P., *et al.* 2010, *MNRAS*, 406, 744

Courteau, S., *et al.* 2011, *ApJ*, 739, 20

Font, A. S., *et al.* 2011, *MNRAS*, 416, 2802

Gilbert, K. M., Font, A. S., Johnston, K. V., & Guhathakurta, P. 2009, *ApJ*, 701, 776

Guo, Q., *et al.* 2011, *MNRAS*, 413, 101

Kauffmann, G., White, S. D. M., & Guiderdoni, B. 1993, *MNRAS*, 264, 201

Kormendy, J., Fisher, D. B., Cornell, M. E., & Bender, R. 2009, *ApJS*, 182, 216

Li, Y.-S. & White, S. D. M. 2008, *MNRAS*, 384, 1459

Moster, B. P., *et al.* 2010, *ApJ*, 710, 903

Searle, L. & Zinn, R. 1978, *ApJ*, 225, 357

Seigar, M. S., Graham, A. W., & Jerjen, H. 2007, *MNRAS*, 378, 1575

Somerville, R. S. & Primack J. R. 1999, *MNRAS*, 310, 1087

Tal, T. & van Dokkum, P. G. 2011, *ApJ*, 731, 89

van Dokkum, P. G., *et al.* 2010, *ApJ*, 709, 1018

White, S. D. M. & Frenk, C. S. 1991, *ApJ*, 379, 52

White, S. D. M. & Rees, M. J. 1978, *MNRAS*, 183, 341

Zibetti, S., White, S. D. M., Schneider, D. P., & Brinkmann, J. 2005, *MNRAS*, 358, 949

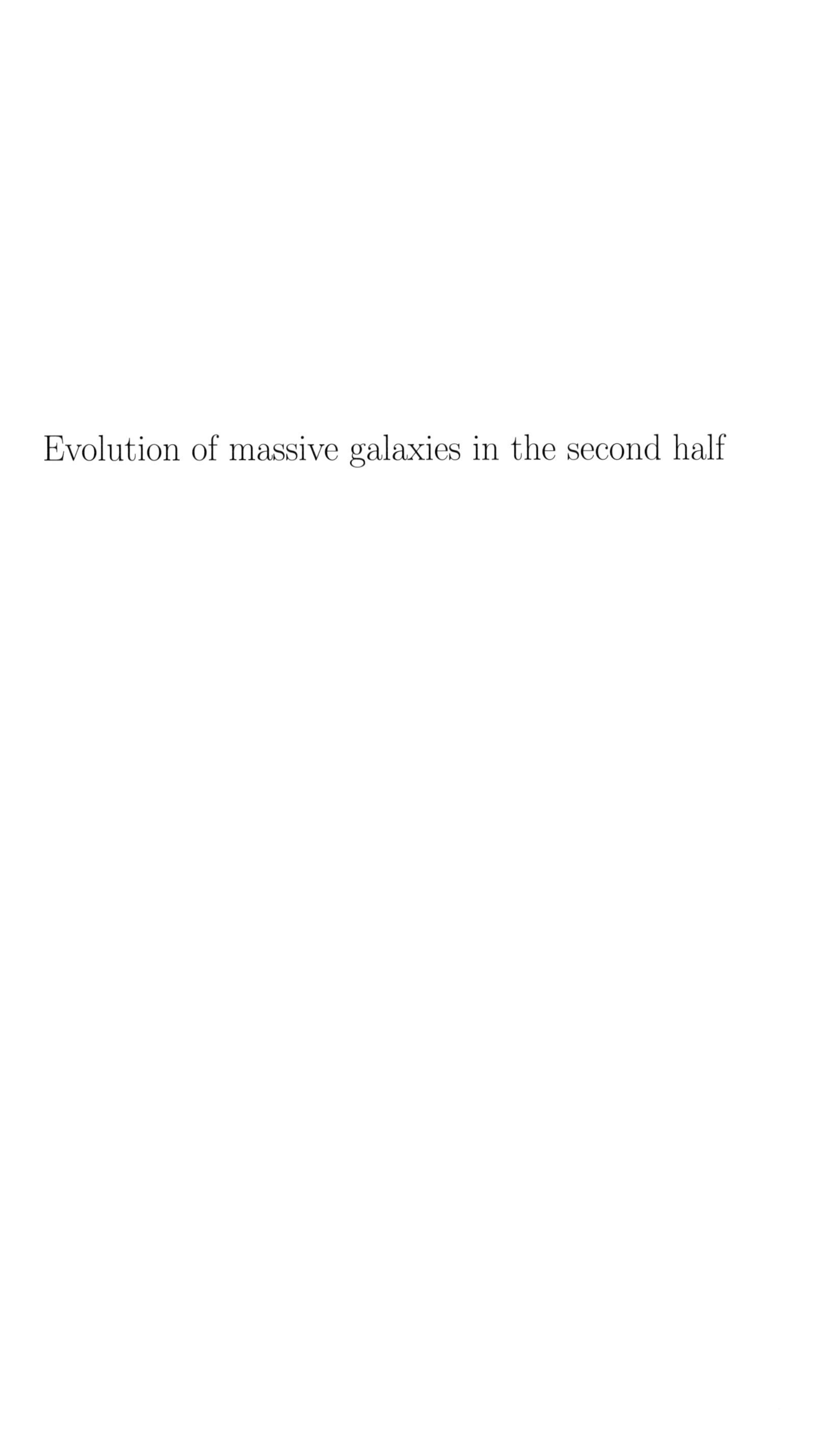

Evolution of massive galaxies in the second half

The intriguing life of massive galaxies
Proceedings IAU Symposium No. 295, 2012
D. Thomas, A. Pasquali & I. Ferreras, eds.

© International Astronomical Union 2013
doi:10.1017/S1743921313004444

The evolving structure of massive quiescent galaxies

Ivana Damjanov[1] and the GDDS team

[1]Harvard-Smithsonian Center for Astrophysics,
60 Garden St., MS-20, Cambridge, MA 02138, USA
email: idamjanov@cfa.harvard.edu

Abstract. The evolution of size and shape of massive quiescent galaxies over cosmic history has been challenging to explain within standard models of galaxy assembly. Several mechanisms have been proposed to explain the size growth of these systems, including major mergers, expansion, and late accretion via a series of minor mergers. The central mass density is shown to be an excellent tool for discriminating between different evolutionary scenarios. We present here the analysis performed on a spectroscopic sample of ~ 500 quiescent systems with stellar masses $M_* > 10^{10}$ M$_\odot$ spanning the redshift range $0.2 < z < 2.7$ for which we calculate stellar mass densities within central 1 kpc and show that this quantity evolves linearly with redshift. Our results do not change when only systems at constant number density are considered in order to account for the mass growth during mergers and to relate progenitors to their descendants. Discrepancy between our findings and other recent studies performed on an order of magnitude smaller samples emphasizes the need for larger homogeneous spectroscopic samples to be used in such analysis.

Keywords. galaxies: evolution, galaxies: fundamental parameters, galaxies: high-redshift, galaxies: stellar content, galaxies: structure

1. Introduction

The recent discovery of a population of compact massive quiescent galaxies ('Red Nuggets') at redshift $1 < z < 3$ (e.g., Szomoru *et al.* 2012 and references therein) has posed profound challenges to standard models of galaxy formation and evolution. Compared to present-day galaxies of similar (stellar) mass ($M_* \sim 10^{11}$ M$_\odot$), these high-z compacts are (depending on z) a factor of $\sim 2 - 5$ smaller (e.g., Damjanov *et al.* 2009) with half-light or effective radii $R_e \lesssim 1$ kpc. Furthermore, while their ellipticities resemble those of massive spheroids at $z \sim 0$, their Sérsic indices are better matched to the local disk-dominated population of massive galaxies, suggesting that 'red nuggets' may constitute a new class of objects (Chevance *et al.* 2012). Several mechanisms have been proposed to explain observed structural evolution of these systems, including major mergers, expansion, and satellite accretion. Low fraction of close pairs among quiescent galaxies observed at $0 < z < 2$ (e. g., Man *et al.* 2012) and the small number of near equal mass mergers produced in N-body simulations (e. g., Shankar *et al.* 2010) suggests that major mergers can only be partly responsible for the observed size growth and that additional secular processes, such as adiabatic expansion (Fan *et al.* 2010, Damjanov *et al.* 2009) and/or a series of minor mergers (Naab *et al.* 2009, Hopkins *et al.* 2010), may be needed to expand these compact systems.

Recent studies that have been tracing massive galaxies at constant number density over the redshift range $0 < z < 3$ in order to link progenitors and their descendants (van Dokkum *et al.* 2010, Patel *et al.* 2013) seem to confirm the predictions of a two-phase model of massive galaxy assembly (Oser *et al.* 2012). The two phases of this model involve (1) in situ star formation phase at $z \gtrsim 2$ in which star-formation is the dominant mechanism driving the structural evolution of massive galaxies and (2) the accretion phase at $z \lesssim 2$ when a series of minor mergers becomes the controlling factor in shaping these systems. However, the low inferred rate of mergers at $1 < z < 2$ (Newman *et al.* 2012) and the fine tuning of these stochastic processes needed to explain the tightness of local scaling relations such as the size-luminosity relation (Nair *et al.* 2011) suggest that a series of minor mergers may not be the only mechanisms driving the size growth of massive galaxies after their star formation ceases. We focus here on the change with time (redshift) of the stellar mass confined within the central 1 kpc region of a massive quiescent galaxy as a powerful tool for testing the minor mergers scenario.

2. Central mass density evolution

One of the parameters that can be effectively used to discriminate between different evolutionary scenarios is the central stellar mass density of massive quiescent galaxies, which is conveniently described by the mass density within the central kiloparsec, $\rho(R_e < 1\ \text{kpc})$. If a massive system grows by accumulating low density material in its outskirts (through accretion of low surface brightness satellites), its central stellar density will not change. On the other hand, if these systems increase their size by blowing out the baryonic material from central regions (via AGN feedback or stellar mass loss), their central mass density will decrease sharply (Hopkins *et al.* 2010).

To test these predictions, we used a compilation of ~ 500 quiescent massive galaxies with confirmed spectroscopic redshifts in the range $0.2 < z < 2.7$ (described in detail in Damjanov *et al.* 2011) and a combination of their stellar masses (based on the spectral energy distribution fitting) and the parameters of their Sérsic profiles. We calculated central mass densities by assuming that the total stellar mass is following the light profile and that systems in our sample have spherical symmetry. The resulting values are in excellent agreement with the results of a previous analysis where a slightly different method was used for deprojecting galaxy light profile (MYSIC subsample, Bezanson *et al.* 2009). Upper panels of Figure 1 show the steady evolution in the central mass density as a function of redshift for our sample of galaxies with high resolution HST imaging providing spatial sampling of $\sim 1\,\text{kpc}$ at $z \sim 2$. Within our uncertainties, this growth of central stellar mass density with redshift can be parameterized rather simply as almost linear growth with redshift: $\rho \propto (1 + z)^{0.96 \pm 0.29}$ †. The decrease in the central stellar mass density by a factor of ~ 3.5 since $z \sim 2.5$ may prove challenging to explain in a scenario in which the structural evolution is driven by a succession of minor mergers. Taking dynamical friction into account, the expected change in ρ is a factor of 1.5 over our redshift range (Naab *et al.* 2009), which seems insufficient to explain the rather dramatic changes in size and, consequently, central density observed in our sample.

The resulting steady evolution of the central mass density is based on the assumption that the mass of passive galaxies does not change in the redshift range we are probing. However, it has been shown that the number density of massive quiescent galaxies evolves with redshift (e.g., Brammer *et al.* 2011). In order to take into account the change of mass during mergers, we selected a subsample of quiescent galaxies that follow the change in

---

† The best fit is obtained by fitting the median values in the six redshift bins, i.e. giving each redshift range equal weight. The range of $1\ \sigma$ errors is obtained by using the bootstrap resampling method. This fit is shown in red in the upper panels of Figure 1, with the corresponding uncertainty shown as a gray band.

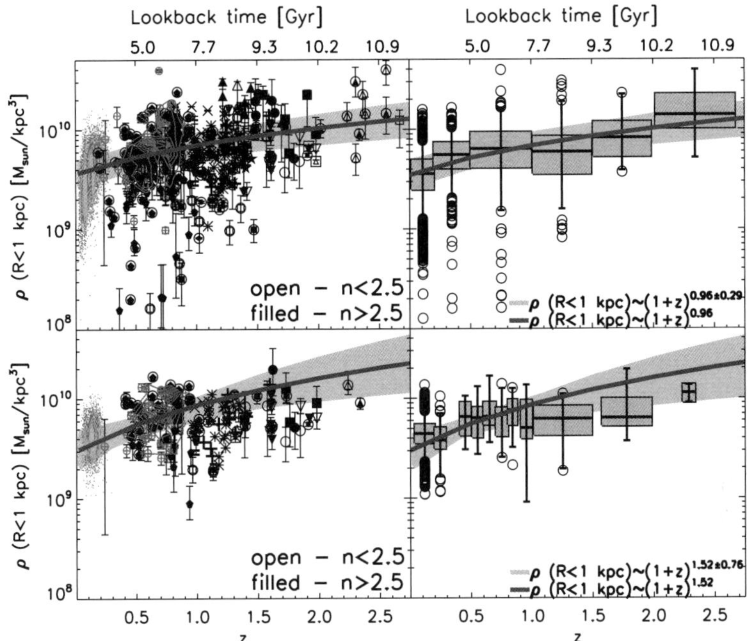

Figure 1. Central stellar mass density (within 1 kpc) of massive quiescent galaxies as a function of redshift. Upper panels show the complete sample from Damjanov *et al.* (2011), while the lower panels present the objects at constant number density (based on the analysis given in Patel *et al.* 2013). *Left panels:* Each symbol type corresponds to a different survey, while blue (red) contours denote the regions of constant density of $z \sim 0$ ($0.2 < z \lesssim 0.9$) galaxies in size-redshift parameter space. A legend mapping data points to individual surveys is provided in Damjanov *et al.* (2011). *Right panels:* The box-and-whisker diagram for $\rho(R_e < 1\,\mathrm{kpc})$ divided into redshifts bins. The top and bottom of the box show the 25th and 75th percentile of the distribution. The horizontal line bisecting the box is the median. The top and bottom of the error bars correspond to the 9th and 91st percentile. Circles are outliers. The red line and the grey shaded area in all four panels show the best fit to the median redshift points and the $\pm 1\sigma$ errors of the best relation, respectively. Note that in the lower panels fitted medians lie at $z < 1$. See text for details.

mass with redshift at constant number density by employing Eq. 2 from Patel *et al.* (2013) and selecting the bin width of $(\log M_*(n_c = 1.4 \times 10^{-4}\,\mathrm{Mpc}^{-3})/\mathrm{M}_\odot)^{+0.2}_{-0.2}$ to account for the uncertainties in mass estimates within our compilation. Furthermore, we select to fit the median values of central mass densities for galaxies in eight $\Delta z = 0.1$ redshift bins at $z < 1$ since at higher redshifts, unlike in our sample, the fraction of star-forming objects among massive galaxies at constant number density is not negligible (Patel *et al.* 2013, van Dokkum *et al.* 2010). The change in the central mass density of massive systems at $z < 1$ selected in this fashion can be presented by a power law that is even steeper than the one obtained using the complete sample ($\rho \propto (1 + z)^{1.52 \pm 0.76}$, lower panels of Figure 1), confirming that at the constant number density central mass density of quiescent galaxies decreases ~ 2.8 times from $z = 1$ to $z = 0$, a factor that may be difficult to explain by invoking (only) the minor mergers scenario.

3. Conclusions

Central mass density evolution is an excellent indicator for discriminating between different scenarios proposed to explain the structural evolution of massive quiescent galaxies. Using a compilation of ~ 500 passive systems from Damjanov *et al.* (2011) we find that the observed evolution of $\rho(R_e < 1\,\mathrm{kpc})$ is a linear function of redshift, showing an increase of at least a factor of 3.5 over the redshift range $0 < z < 2.5$. These results are not altered if, in order to trace the same galaxy population, we select only massive galaxies at the constant number density at redshifts $z < 1$, where a large fraction of systems selected in this way are passive. Our findings are in a good agreement with the central densities found at high redshift (e.g, Bezanson *et al.* 2009). However, recent results based on a small sample of 34 quiescent galaxies spanning the redshift range $0.9 < z_\mathrm{spec} < 2$ (Saracco *et al.* 2012) suggest that the central density of massive quiescent systems is independent of redshift. Similar conclusions are reached in a study of 23 massive ($M_* \sim 10^{11}\,\mathrm{M_\odot}$) galaxies drawn from the Sloan Digital Sky Survey, where the central mass densities based on De Vaucouler profiles are compared to the central densities at $z > 1$ (Tiret *et al.* 2011). This discrepancy highlights the need to perform a self-consistent analysis of the central mass density on a large homogeneous sample of massive galaxies covering a wide range of spectroscopically confirmed redshifts. Furthermore, current theoretical models do not explore in detail the effects of dynamical friction during minor mergers. For example, one of the factors that needs to be considered is the trajectory of accreted systems (Naab, priv. comm.). More detailed theoretical predictions on the possible decrease in ρ during minor mergers would allow us to fully explore the indicative potential of the observed evolution in the central stellar mass density of massive galaxies.

References

Bezanson, R., Van Dokkum, P. G., Tal, T., Marchesini, D., Kriek, M., Franx, M., & Coppi, P. 2009, *ApJ*, 697, 1290

Brammer, G., Whitaker, K., van Dokkum, P., Marchesini, D., Franx, M., Kriek, M., Labbe, I., Lee, K., Muzzin, A., Quadri, R., Rudnick, G., & Williams, R. 2011, *ApJ*, 739, 24

Chevance, M., Weijmans, A.-M., Damjanov, I., Abraham, R. G., Simard, L., van den Bergh, S., Caris, E., & Glazebrook, K. 2012, *ApJ* (Letters), 754, L24

Damjanov, I., McCarthy, P. J., Abraham, R. G., Glazebrook, K., Yan, H., Mentuch, E., Le Borgne, D., Savaglio, S., Crampton, D., Murowinski, R., Juneau, S., Carlberg, R. G., Jørgensen, I., Roth, K., Chen, H.-W., & Marquee, R. O. 2009, *ApJ*, 695, 101

Damjanov, I., Abraham, R. G., Glazebrook, K., McCarthy, P. J., Caris, E., Carlberg, R. G., Chen, H. W., Crampton, D., Green, A. W., Jørgensen, I., *et al.* 2011, *ApJ* (Letters), 739, L44

Fan, L., Lapi, A., Bressan, A., Bernardi, M., De Zotti, G., & Danese, L. 2010, *ApJ*, 718, 1460

Man, A. W. S., Toft, S., Zirm, A. W., Wuyts, S., & van der Wel, A. 2000, *ApJ*, 744, 85

Naab, T., Johansson, P. H., & Ostriker, J. P. 2009, *ApJ* (Letters), 699, L178

Nair, P., van den Bergh, S., & Abraham, R. G. 2011, *ApJ* (Letters), 734, L31

Newman, A. B., Ellis, R. S., Bundy, K., & Treu, T. 2012, *ApJ*, 746, 162

Oser, L., Naab, T., Ostriker, J. P., & Johansson, P. H. 2012, *ApJ*, 744, 63

Patel, S. G., van Dokkum, P. G., Franx, M., Quadri, R. F., Muzzin, A., Marchesini, D., Williams, R. J., Holden, B. P., & Stefanon, M. 2013, *ApJ*, 766, 15

Saracco, P., Gargiulo, A., & Longhetti, M. 2012, *MNRAS*, 422, 3107

Shankar, F., Marulli, F., Bernardi, M., Dai, X., Hyde, J. B., & Sheth, R. K. 2010 *MNRAS*, 403, 117

Szomoru, D., Franx, M., & van Dokkum, P. G. 2012 *ApJ*, 749, 121

Tiret, O., Salucci, P., Bernardi, M., Maraston, C., & Pforr, J. 2011, *MNRAS*, 411, 1435

van Dokkum, P. G., Whitaker, K. E., Brammer, G., Franx, M., Kriek, M., Labbé, I., Marchesini, D., Quadri, R., Bezanson, R., Illingworth, G. D., Muzzin, A., Rudnick, G., Tal, T., & Wake, D. 2010, *ApJ*, 709, 1018

The intriguing life of massive galaxies
Proceedings IAU Symposium No. 295, 2012
D. Thomas, A. Pasquali & I. Ferreras, eds.

© International Astronomical Union 2013
doi:10.1017/S1743921313004456

The emergence of the red sequence at $z \sim 2$ seen through galaxy clustering in the UKIDSS UDS

William G. Hartley[1], Omar Almaini[1], Alice Mortlock[1], Chris Conselice[1] and the UDS team

[1]School of Physics and Astronomy, University of Nottingham, University Park, Nottingham NG7 2RD
email: will.hartley@nottingham.ac.uk

Abstract. We use the UKIDSS Ultra-Deep Survey, the deepest degree-scale near-infrared survey to date, to investigate the clustering of star-forming and passive galaxies to $z \sim 3.5$. Our new measurements include the first determination of the clustering for passive galaxies at z$>$ 2, which we achieve using a cross-correlation technique. We find that passive galaxies are the most strongly clustered, typically hosted by massive dark matter halos with $M_{halo} > 10^{13}$ $M_\odot$ irrespective of redshift or stellar mass. Our findings are consistent with models in which a critical halo mass determines the transition from star-forming to passive galaxies.

Keywords. Galaxies, Evolution, Large-scale structure, etc.

1. Introduction

One of the most fundamental observational results in extragalactic astrophysics is the bimodal form of the galaxy population. Galaxies with spheroidal morphologies tend to have very low star-formation rates (hereafter, 'passive') and are the dominant population at high stellar masses. The more abundant disky, star-forming objects, meanwhile, typically have lower stellar mass. However, the origin of this bimodality remains unclear. A gas-rich merger or the collapse of an unstable disk is thought to result in very vigorous star-formation, possibly triggering a quasar phase. Either of these could have sufficient energy to expel the remaining gas from the galaxy, or heat it sufficiently to prevent further star formation (Silk & Rees 1998, Fabian 1999). Without cold gas present, the galaxy can no longer form stars and it is observed as a passive object.

Recent theoretical models, however, suggest an alternative mechanism by which galaxies end their star formation. In the so called 'hot halo' model, a high mass dark matter halo can sustain a reservoir of hot gas. New gas infall is shock heated as it encounters this gas halo and is therefore not immediately available for star formation. Henceforth, only a modest amount of energy input, possibly from radio AGN feedback, is required to maintain a high gas temperature. In this way, galaxies within high mass halos become starved of star-forming gas with the result that ongoing star-formation rates become negligible. A clear consequence of the hot halo model is that there should be a characteristic halo mass scale, above which the relative fraction of passive objects (at fixed stellar mass) increases sharply. Current models suggest that this halo mass scale is a few $\times 10^{12}$ $M_\odot$ and becomes maximally efficient at $M_{halo} > 10^{13}$ $M_\odot$ (Croton et al. 2006, Cen 2011).

Estimates of the typical host dark matter halo of a galaxy sample can be achieved through their clustering statistics. In order to use the large-scale structure to estimate halo masses at $z > 1$ we require selection at near-infrared wavelengths. A large contiguous field is clearly necessary to measure the bias at large scales (r $\sim$ few Mpc) and find rare objects at high redshift.

Our sample is drawn from the K-band image of the UKIRT Infrared Deep Sky Survey (UKIDSS, Lawrence *et al.* 2007), Ultra-Deep Survey† (UDS) data release 8 and matched multi-wavelength photometry. The UDS covers 0.77 square degrees in J, H and K-bands, reaching depths of J = 24.9, H = 24.2 and K = 24.6 (AB magnitudes, estimated from the RMS between $2''$ apertures).

In addition to the three UKIDSS bands, the field is covered by comparable data from CFHT Megacam u-band, optical Subaru Suprime-cam data (B, V, R, i' and z'-bands) and Spitzer IRAC channels 1 and 2. Finally, we make use of existing 24μm, X-ray and radio data to remove obvious AGN.

Photometric redshift probability distributions were computed for each source with dispersion of $\Delta z/(1+z) \sim 0.031$ after excluding outliers ($\Delta z/(1+z) > 0.15$, $< 4\%$ of objects). The stellar masses and rest-frame luminosities used in this work are measured using a multicolour stellar population fitting technique. We fit the 11 photometric bands to a large grid of synthetic spectral energy distributions (SEDs) constructed from the stellar population models of Bruzual & Charlot (2003), assuming a Chabrier initial mass function.

Our separation of passive and star-forming galaxies is based initially on the U, V and J-Bessel band rest-frame luminosities. Objects with passive UVJ colours but SSFR $> 10^{-8}$ yr^{-1} derived from the template fits were re-assigned to the star-forming sample. Furthermore, the 24μm data were used to identify objects that harbour dust-enshrouded star-formation by a nearest match to the K-band catalogue within $6''$.

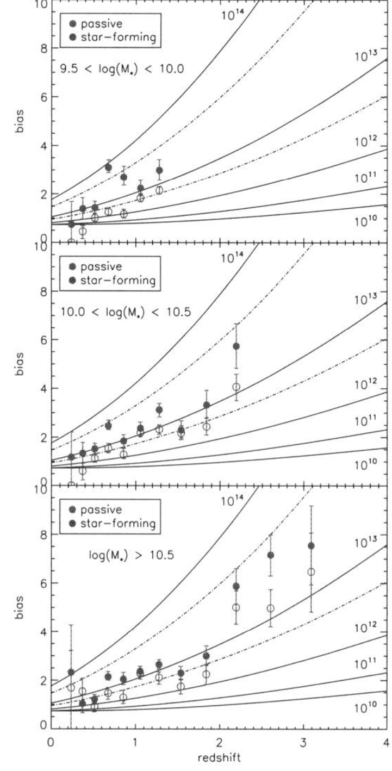

Figure 1. Galaxy bias inferred from our cross-correlation analysis against redshift for passive and star-forming samples. Filled red points represent passive samples while the open blue points are for our star-forming samples. The stellar mass range of galaxies in each panel is indicated in the upper left. Solid and dot-dashed lines show the bias for dark matter halos of various masses (in solar masses, as labelled). The dot-dashed lines are 5×10^{13} M$_\odot$ and 5×10^{12} M$_\odot$, upper and lower respectively.

2. Clustering measurements

We estimate the host dark matter halo masses of passive and star-forming sub-samples as a function of their stellar mass and redshift in intervals of 500 Mpc across $0.3 < z < 3.36$. Using cross-correlation statistics, we can cross a small sample for which we want to know the bias (e.g. a passive sample at high redshift) with a full mass-limited tracer population and achieve much smaller statistical uncertainties on their bias, b_2, than we would through an auto-correlation analysis. The bias can be recovered as follows:

† http://www.nottingham.ac.uk/astronomy/UDS/

$$b_2 = \frac{b_{12}^2}{b_1}, \qquad (2.1)$$

where b_1 is the bias computed from the auto-correlation function of the tracer population and b_{12}^2 is the bias measured from the cross-correlation function. In order to find b_{12}^2 we first compute the cross-correlation function using the estimator of Landy & Szalay (1993)

We then fit a single bias parameter to the galaxy clustering measurements at large scales using the non-linear dark matter power spectrum routine of Smith *et al.* (2003),

$$w_{obs}(\theta) = b_{12}^2 \times w_{dm}(\theta), \qquad (2.2)$$

minimising the χ^2. Once we have computed the bias of a sample, b_2, we obtain a typical host halo mass using the models of Mo & White (2002) (and references therein).

In Figure 1 we display the bias determined from our cross-correlation method as a function of redshift for passive and star-forming objects, with three panels representing the three bins in stellar mass. Red points correspond to our passive galaxy sample, while blue points represent the star-forming objects. Any subsample in which the lower mass limit falls below the 95% mass completeness limit is not considered.

From Figure 1 a number of trends are apparent. Firstly, passive galaxies are, on average, clearly more strongly clustered than their star-forming counterparts at similar stellar mass. This result is most apparent at lower stellar masses (M $<$ 10^{10} M$_\odot$) and for subsamples at $z < 1$.

A second feature of these results is the mass dependence within samples. The least massive of the passive galaxies appear to be the most strongly clustered (and therefore in the most massive dark matter halos). The bias typically decreases as the masses of the subsamples increase. Despite this mass dependence, our passive galaxy samples appear to have consistently high parent halo masses with typical masses $M_{\rm halo} \gtrsim 10^{13}$ M$_\odot$ or higher.

For star-forming sub-samples, galaxies of different stellar masses but similar redshifts do not show any clear trends. The measurements are consistent with star-forming galaxies being typically hosted by similar mass halos irrespective of stellar mass. This absence of a clear correlation for star-forming galaxies was identified in the luminosity-selected subsamples in Hartley *et al.* (2010). Similarly, the downsizing-like behaviour we found in Hartley *et al.* is also present in this work, with subsamples of equal stellar mass typically hosted by less massive halos at low redshift.

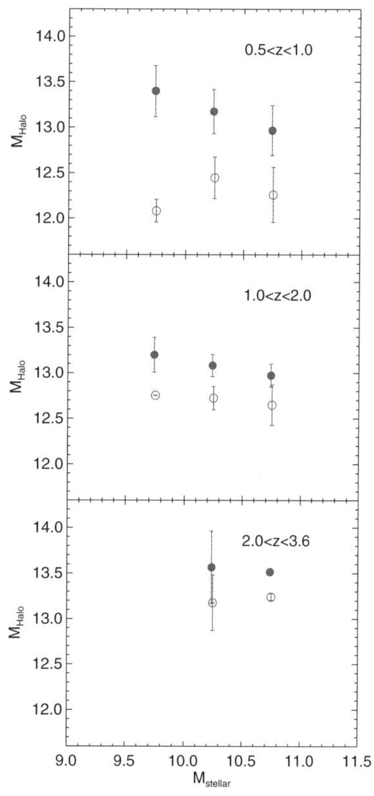

Figure 2. Average halo masses implied by our bias measurements in three redshift intervals (as labelled). The data represent mean halo masses for samples of fixed stellar mass, while the uncertainties are the standard error on the mean. As in previous Figures, filled red points represent passive galaxy samples and open blue points are for star-forming objects.

The raw measurements presented in Figure 1 are difficult to fully digest. We therefore condense our results into a more intuitive set of plots in Figure 2. We plot the mean halo mass implied by our clustering results as a function of stellar mass for passive and star-forming subsamples in three redshift ranges: $0.5 < z < 1.0$, $1.0 < z < 2.0$ and $2.0 < z < 3.36$. Each point in Figure 1 is converted to a typical halo mass, based on the measured bias, and these are then averaged over the indicated redshift intervals. The uncertainties on the data are the standard error on the mean.

The behaviour described above can be seen more clearly in Figure 2: the strong clustering of low-mass passive samples, the lack of a stellar mass dependence on halo mass for star-forming galaxies and the 'downsizing' of halo masses with redshift are all apparent. The averaged halo masses of all passive galaxy samples are greater than or equal to 10^{13} $M_\odot$ and within each redshift range the lowest-mass galaxies are found in the most massive halos. This result is clearest at low redshifts, indicating a build up of relatively low mass passive objects in high mass halos. Although there is no clear stellar mass dependence on halo mass for star-forming galaxies, the downsizing-like behaviour is clearly apparent. Typical halo masses in our lowest redshift interval are $10^{12-12.5}$ $M_\odot$, while they exceed 10^{13} $M_\odot$ at the highest redshifts.

Taken at face value, the average parent halo masses for passive galaxy subsamples shown in Figure 2 are in agreement with a hot halo model and consistent with a limiting mass scale for passivity. Passive galaxies of all masses are typically hosted by halos of $M_{\mathrm{halo}} \geqslant 10^{13}$ $M_\odot$ or greater. The typical host halo masses for star-forming galaxies show no consistent stellar mass dependence. However, at $z > 2$ our star-forming samples are typically also found in halos with mass, $M_{\mathrm{halo}} \geqslant 10^{13}$ $M_\odot$. It is therefore possible that the efficiency with which a hot halo can shut-off star-formation evolves with redshift. We will conduct an interpretation of our results in the context of the halo occupation framework in future work.

References

Bruzual, G. & Charlot, S. 2003, *MNRAS*, 344, 1000
Cen, R. 2011, *ApJ*, 741, 99
Croton, D. J., *et al.* 2006, *MNRAS*, 365, 11
Fabian, A. C. 1999, *MNRAS*, 308, L39
Hartley, W. G., *et al.* 2010, *MNRAS*, 407, 1212
Landy, S. D. & Szalay, A. S. 1993, *ApJ*, 412, 64
Lawrence, A., *et al.* 2007, *MNRAS*, 379, 1599
Mo, H. J. & White, S. D. M. 2002, *MNRAS*, 336, 112
Silk, J. & Rees, M. J. 1998, *A&A*, 331, L1
Smith, R. E., *et al.* 2003, *MNRAS*, 341, 1311

The intriguing life of massive galaxies
Proceedings IAU Symposium No. 295, 2012
D. Thomas, A. Pasquali & I. Ferreras, eds.

© International Astronomical Union 2013
doi:10.1017/S1743921313004468

The Co-Evolution of Supermassive Black Holes and Galaxies: Observational Constraints

Xian Zhong Zheng

Purple Mountain Observatory, Chinese Academy of Sciences,
2 West-Beijing Road, Nanjing 210008, P. R. China
email: xzzheng@pmo.ac.cn

Abstract. The connection between the growth of supermassive black holes (SMBHs) and the assembly of their host galaxies is termed 'co-evolution'. Understanding co-evolution is one of the most fundamental issues in modern astrophysics. In this contribution, we review recent progress in addressing how the growth of SMBHs is linked to the properties of their host galaxies in the context of galaxy evolution, from the observational point of view. Although a coherent picture has not yet emerged, multiple pathways of co-evolution appear to be favored with a probable dependence on AGN luminosity and redshift.

Keywords. galaxies: general, galaxies: evolution, galaxies: nuclei, quasars: general

1. Introduction

In the past two decades great effort had been made to understand the origin of the tight correlation between the mass of supermassive black holes (SMBHs) and velocity dispersion/bulge stellar mass in nearby massive galaxies. The SMBHs are often believed to couple only with bulges and have no relationship with the mass of the underlying dark matter halo (Kormendy & Bender 2011; Beifiori *et al.* 2012) or with pseudobulges/disks (Kormendy *et al.* 2011). The energetic output of growing SMBHs, seen as AGN, can affect their surrounding galaxies/environments (Fabian 2012 and references therein). AGN feedback from SMBHs is taken as a more effective mechanism of quenching gas cooling and star formation in the framework of galaxy formation and evolution (Somervile *et al.* 2008; Silk & Mamon 2012), although its physical processes are poorly understood. Recently, observational evidence accumulates towards a non-universal $M - \sigma_*$ relation (e.g., Gultekin *et al.* 2009; Greene *et al.* 2010; McConnell *et al.* 2011), implying complications for the interaction between SMBHs and galaxies. These findings and constraints from observations provide new insights in exploring multiple aspects of the co-evolution and its processes (Schawinski 2012).

2. Statistical Links between AGN and Star Formation Activity

Deep multi-wavelength surveys over a large area provide a complete census of galaxy and AGN populations out to $z \sim 4$. This is particularly true for $z < 1$. It has been pointed out by a number of studies that the global BH accretion history traces the global galaxy star formation history. The latter is indeed the history of galaxy stellar mass growth. An up-to-date comparison between the two is presented in Zheng *et al.* (2009) (Fig. 1). In particular, the two activities also trace each other in intensity, following a mode of so-called 'downsizing'. The match between BH accretion rate and star formation rate

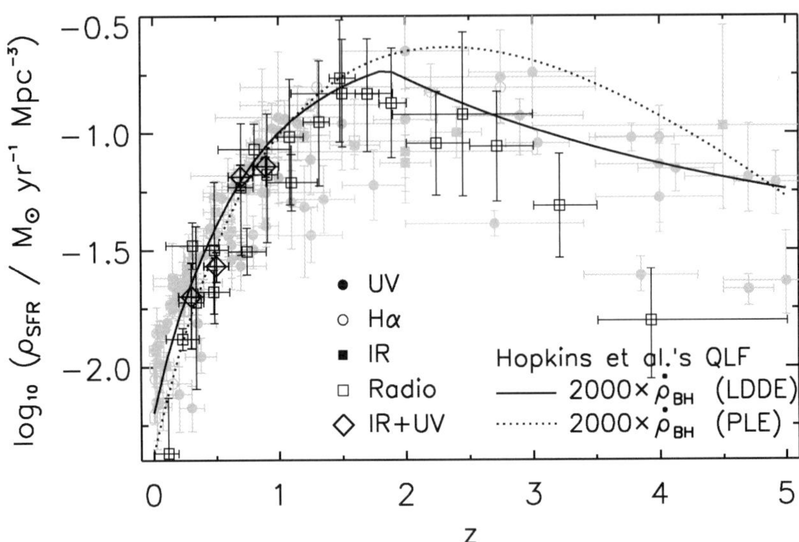

Figure 1. The global similarity of cosmic star formation history and cosmic BH accretion history. The proportion of 2000 between the two is consistent with the local SMBH-bulge mass relation accounting for the gas recycling of a factor 2 due to stellar evolution (Zheng *et al.* 2009).

suggests that the duty cycle of AGN is about two orders of magnitude shorter than that of starbursts. The integral of the growth of SMBHs and the growth of stellar mass through star formation results in a BH-bulge mass ratio consistent with the local $M - \sigma_*$ relation.

The global and differential similarities of AGN and starburst activity are only valid in a statistical sense. They are not necessarily linked on a one-to-one basis. The star-forming galaxies hosting the vast majority of cosmic star formation are indeed dominated by disk galaxies out to $z \sim 2$ (Wuyts *et al.* 2011). The enhanced star formation induced by mergers represents only a small fraction of the overall SFR at $z \sim 2$ (Rodighiero *et al.* 2011) and $z < 1$ (Guo *et al.* 2011). This is also confirmed by direct measurement of SFRs in close pairs and mergers (Robaina *et al.* 2009; Jogee *et al.* 2009). It becomes clear that starburst events mostly take place in isolated disks at least out to $z \sim 2$. These facts rule out the simplest co-evolution picture that BH accretion and bulge growth through star formation take place simultaneously in the same event. If a delay exists between BH accretion and starburst in a galaxy (Hopkins 2012), one would not see a direct link between the two processes from observations but SMBHs and bulges could grow in a proportional manner. Such a picture is also ruled out because starbursts are mostly associated with isolated disks other than growing bulges. The similarity between the global accretion history of SMBHs and the star formation history of galaxies is likely controlled by gas supply, which declines rapidly with cosmic time since $z \sim 2$. However, processes triggering both AGN activity and star formation remain to be understood (Kauffmann *et al.* 2007).

The local $M - \sigma_*$ relation involves mass and morphology (i.e., bulges/spheroids). A bulge can be built up through star formation and/or merging stars from, e.g., its surrounding disk. The bulk of stars formed in isolated disks needs to be transformed into bulges via certain physical processes (e.g., mergers, disk instability or secular processes) in order to meet the local $M - \sigma_*$ relation. The buildup of bulges has been associated with the quenching of star formation (Cheung *et al.* 2012). What are the processes governing the growth of bulges in different galaxy populations at different cosmic epochs?

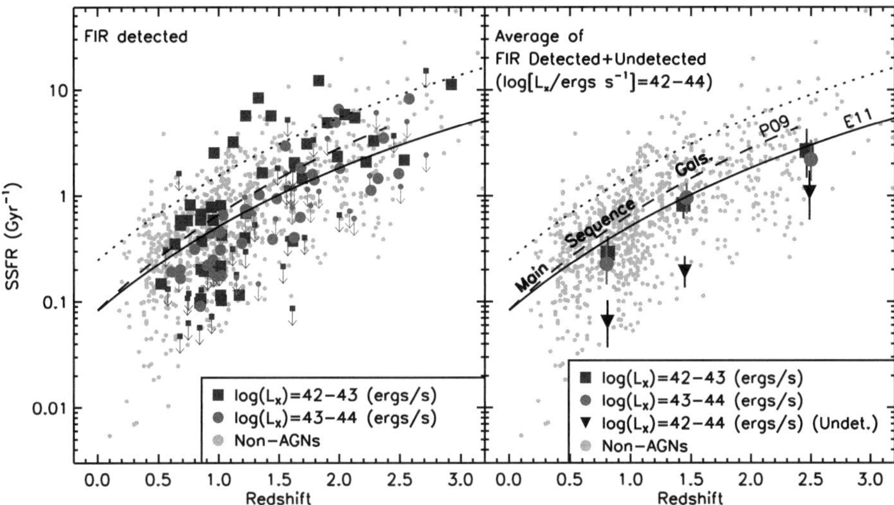

Figure 2. Comparison of the specific SFR between the host galaxies of moderate-luminosity X-ray selected AGN and normal star-forming galaxies over $0 < z < 3$ (Mullaney *et al.* 2012). The increase in sSFR of the AGN host galaxies with redshift follows the increase in the global sSFR of all galaxies. No correlation is seen between AGN luminosity and sSFR.

Are these processes universally responsible for the ignition of AGN activity and the establishment of the $M - \sigma_*$ relation? What are the key mechanisms over galactic scales for transporting materials to the center of a galaxy and feeding of the central black hole? The relationship between AGN activity and the properties of their host galaxies hold the key to answering these questions.

3. The Properties of AGN Host Galaxies

The presence of AGN affects the determination of the properties of the host galaxy, especially for the most luminous ones. Relevant studies often target moderate-luminosity AGN in order to minimize contamination from the AGN. On the other hand, the diversity of the AGN population induces uncertainties in the selection of AGN samples hindering the derivation of a consistent picture

Star formation and stellar mass. The AGN host galaxies are found to exhibit a star formation activity similar to the entire galaxy population in a statistical manner (Fig. 2; Mullaney *et al.* 2012; see also Shao *et al.* 2010). Silverman *et al.* (2009) noted that the star formation rate of the host galaxies does not correlate with AGN luminosity (Fig. 3). This is mostly true for the low- and moderate-luminosity AGN. In contrast, the most luminous AGN have star formation rates likely enhanced by merger events (Rosario *et al.* 2012). The incidence of AGN strongly depends on galaxy stellar mass in the sense that AGN are preferentially found in massive galaxies, but the BH accretion mode seems independent of both the stellar mass and color of the host galaxy (Aird *et al.* 2012; Bongiorno *et al.* 2012). All together, one may conclude that the occurrence of an AGN involves a massive galaxy and sufficient gas supply to fuel both the AGN and star formation; the presence of an AGN seems neither to be related to quenching of star formation nor to have significant influence on the properties of the host galaxy.

Morphologies. Large studies of the morphologies of AGN host galaxies are attributed to the low-/moderate-luminosity AGN. These AGN are mostly found in disk-dominated

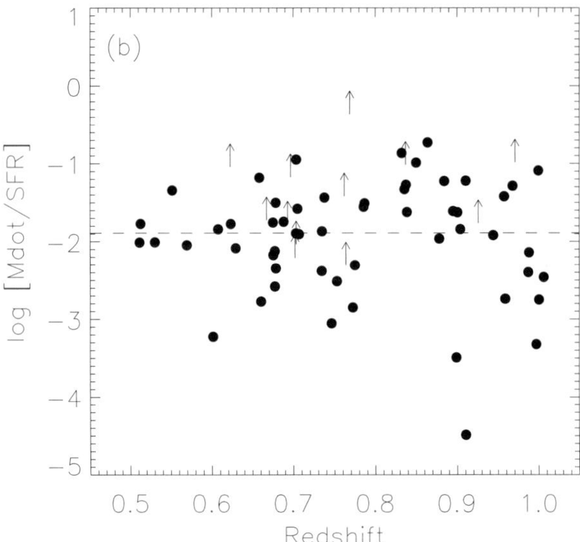

Figure 3. Comparison between BH accretion rate and star formation rate of the host galaxies of X-ray-selected AGN from zCOSMOS (Silverman *et al.* 2009). It is clearly seen that there is no correlation or anti-correlation of intensity between AGN and star formation activity over the redshift range examined.

galaxies other than mergers/interacting galaxies out to $z \sim 2$. The AGN host galaxies tend to be systematically more concentrated than normal galaxies at $z < 1$ (Grogin *et al.* 2005; Geogakakis *et al.* 2008; Cisternas *et al.* 2011), and indistinguishable from the galaxy population at $z \sim 2$ (Fig. 4; Schawinski *et al.* 2011). The latter is confirmed by Kocevski *et al.* (2012) using HST/WFC3 imaging from the CANDELS survey. It is now convincingly established that stochastic accretion of gas, instead of violent dynamical processes (major mergers/interactions), plays the major role regarding the fuelling of low-/moderate-luminosity AGN over cosmic time at least out to $z \sim 2$. The stochastic accretion is argued to be driven by secular processes in the host galaxies. The morphologies of galaxies hosting the most luminous AGN, known as quasars, are less explored because of glare from the central AGN over the host galaxy (e.g., Bahcall *et al.* 1997). Major mergers are believed to play a crucial role in triggering the most luminous AGN (Treister *et al.* 2012), although the role of violent events is possibly less important at $z > 1$ (Rosario *et al.* 2012).

The BH-galaxy mass ratios. One critical question to understand the co-evolution of SMBHs and galaxies is which one grew first. This is vital to ascertaining which of the two components regulates the other (Volonteri 2012). Only indirect measurements of the BH-galaxy mass ratio are feasible for AGN over cosmic distances. SMBHs in distant AGN are often found to be more massive at fixed galaxy mass than the predictions based on the local BH-bulge mass relation, suggesting SMBHs to be in place at earlier times during the galaxy formation process (Fig. 5, Bennert *et al.* 2011; Merloni *et al.* 2010). The scatter in the BH-galaxy mass ratio appears to be rather large for galaxies with growing SMBHs, illustrating the complexities of the pathways that govern the growth of SMBHs and bulges (Bennert *et al.* 2011, Greene *et al.* 2010).

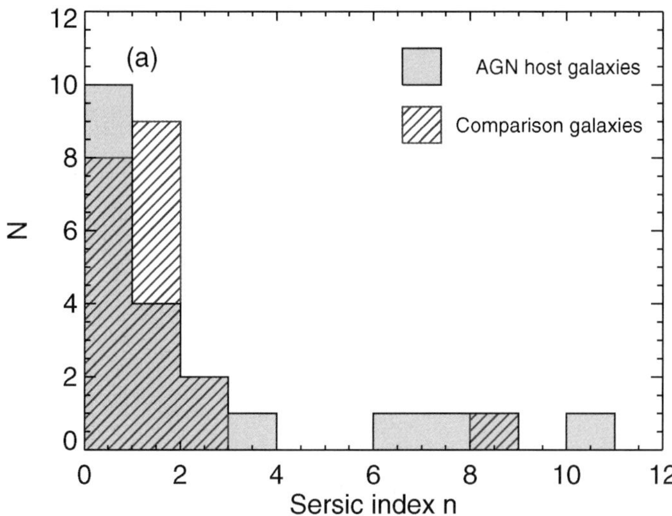

Figure 4. The distribution of Sersic index derived from rest-frame optical imaging for AGN hosts and comparison galaxies at $z \sim 2$, suggesting that the AGN host galaxies are dominated by late-type systems with low Sersic indices and indistinguishable from the parent galaxy population (Schawinski *et al.* 2011).

4. Multiple Evolutionary Pathways

The links between AGN and star formation activity, along with the connections between AGN activity and the properties of AGN host galaxies provide important observational constraints on the SMBH-galaxy co-evolution. In particular,

- the gas supply is key to the fuelling of AGN and star formation within galaxies;
- stochastic accretion of gas is commonly seen in disk-dominated galaxies over cosmic time. BH accretion does not require a bulge;
- secular processes are predominantly responsible for triggering and feeding moderate-luminosity AGN;
- the growth of SMBHs preceding the assembly of galaxies hints that processes feeding AGN are easier to happen at $z \gtrsim 2$;
- disk disruption plays an important role in driving bulge growth and the evolution of the BH-bulge mass relation;
- major mergers are preferentially associated with the most luminous AGN.

The observational evidence points to multiply evolutionary pathways for massive galaxies satisfying the local $M - \sigma_*$ relation. The SMBH-galaxy co-evolution is neither universally a "hand-in-hand dance" nor a growth in lock step with a delay in occurrence time.

Merger-driven scenario. Most luminous AGN are exclusively hosted by major mergers in contrast to moderate- and low-luminosity AGN, which are mostly associated with normal galaxies (Treister *et al.* 2012), suggesting that the triggering and fuelling mechanism is likely a function of AGN luminosity. The mechanisms involving violent mergers may be more efficient in producing luminous AGN compared to stochastic fuelling of SMBHs in disk galaxies. This is supported by the finding of distinct BH accretion modes with respect to the dependence on luminosity among local AGN (Kauffmann *et al.* 2009) and AGN at intermediate redshifts (Rosario *et al.* 2012). Regarding merger-driven evolution in self-regulated models, the growth of the SMBH and the formation of the bulge are closely linked through feedback or feeding (Di Matteo *et al.* 2005; Hopkins *et al.* 2008;

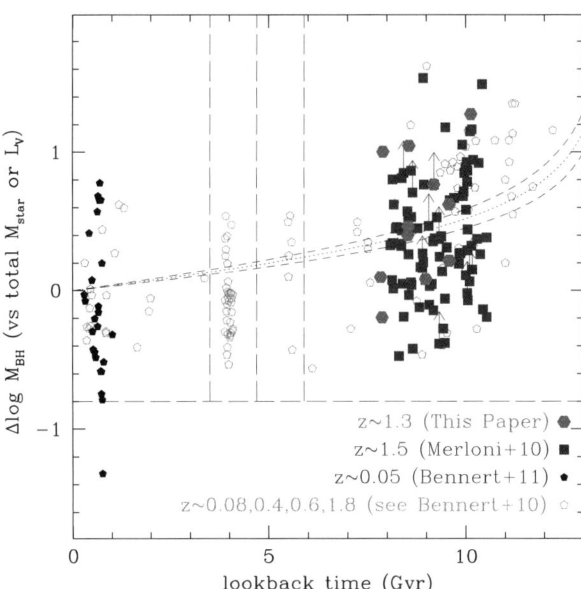

Figure 5. The evolution of BH-to-galaxy mass ratio of AGN with respect to the local relation of AGN (Bennert *et al.* 2011). The higher BH-to-galaxy mass ratio at increasing redshift suggests that BH growth precedes galaxy assembly.

Somerville *et al.* 2008). How AGN feedback affects the quenching of star formation in bulges remains an open question. This picture represents one possible pathway in the formation of massive galaxies.

Secular evolution. The majority of moderate-luminosity AGN is fuelled by internal mechanisms rather than violent mergers, suggesting that secular evolution plays a crucial role in regulating the growth of SMBHs in disk-dominated galaxies. The disk instabilities could be an important AGN feeding mechanism at $z > 1$. The AGN activity at $z \sim$ 2 are mostly regulated by secular processes and supposed to be ideal for probing the conditions in these processes. In this scenario, the growing SMBHs reside in bulgeless galaxies (Simmons *et al.* 2012; Greence *et al.* 2010).

Non-causal evolution. The SMBH-galaxy correlation can, arguably, be caused as the natural product of hierarchical galaxy evolution, regardless of the merger history of the galaxy (Peng 2007; Jahnke & Maccio 2011). This correlation can be obtained even without an apparent causation between the growth of SMBHs and galaxy assembly. Hence, in this scenario, SMBH growth and galaxy growth need not be coupled through direct feedback processes (Cen 2012).

Implications from the assembly of massive galaxies

It is likely that outliers on the $M - \sigma_*$ relation provide the key clues to understanding the connection between SMBHs and galaxies. It is found that Brightest Cluster Galaxies (BCGs) show systematic offsets from the correlation in Fig.6 (McConnell *et al.* 2011). This could point to unique evolutionary processes near the highest density peaks in the Universe. The growth of the most massive galaxies can be characterized by two stages: an early one that forms the central part at $z > 2$, and a later phase that builds up the outer stellar halo via continuously accreting satellite galaxies (van Dokkum *et al.* 2010). The formation of the outer halo is unlikely to fuel BH accretion and to disturb significantly the kinematics of stars. Therefore, these BCGs with a higher BH-to-bulge mass ratio at

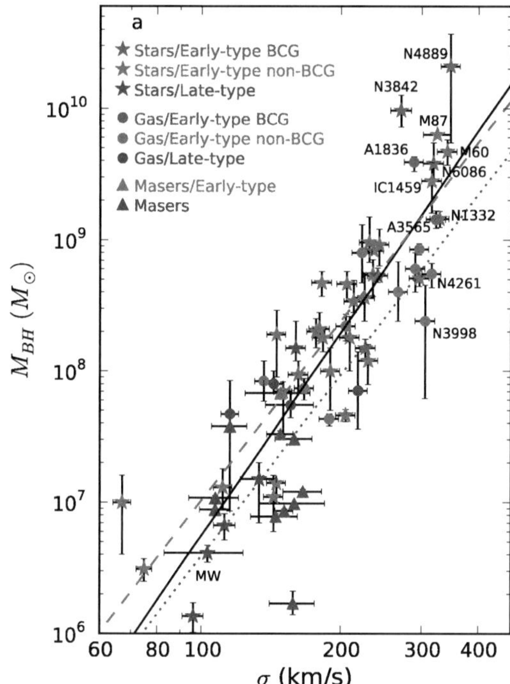

Figure 6. SMBH mass versus velocity dispersion of host galaxies from McConnell *et al.* (2011). The Brightest Cluster Galaxies (BCGs) appear to contain SMBHs much more massive than the predicted by the $M - \sigma_*$ relation.

$z \sim 2$ could evolve little in the $M - \sigma_*$ relation since then, appearing as the outliers with respect to the less massive galaxies.

Acknowledgments

I am very thankful to the IAU Symposium 295 organizers, in particular Daniel Thomas, Anna Pasquali and Ignacio Ferreras, for inviting me to give the targeted talk, and for organizing such a wonderful meeting. It has been a great experience for me.

References

Aird, J., Coil, A. L., Moustakas, J., *et al.* 2012, *ApJ*, 746, 90
Alexander, D. M. & Hickox, R. C. 2012, *New Astronomy*, 56, 93
Bahcall, J. N., Kirhakos, S., Saxe, D. H., & Schneider, D. P. 1997, *ApJ*, 479, 642
Beifiori, A., Courteau, S., Corsini, E. M., & Zhu, Y. 2012, *MNRAS*, 419, 2497
Bennert, V. N., Auger, M. W., Treu, T., Woo, J.-H., & Malkan, M. A. 2011, *ApJ*, 742, 107
Bogdán, Á., Forman, W. R., Zhuravleva, I., *et al.* 2012, *ApJ*, 753, 140
Bongiorno, A., Merloni, A., Brusa, M., *et al.* 2012, *MNRAS*, 427, 3103
Cen, R. 2012, *ApJ*, 755, 28
Cheung, E., Faber, S. M., Koo, D. C., *et al.* 2012, *ApJ*, 760, 131
Cisternas, M., Jahnke, K., Inskip, K. J., *et al.* 2011, *ApJ*, 726, 57
Di Matteo, T., Springel, V., & Hernquist, L. 2005, *Nature*, 433, 604
Fabian, A. C. 2012, *ARAA*, 50, 455
Georgakakis, A., Coil, A. L., Laird, E. S., *et al.* 2009, *MNRAS*, 397, 623
Greene, J. E., Peng, C. Y., Kim, M., *et al.* 2010, *ApJ*, 721, 26

Digital Sky Survey (SDSS, York *et al.* 2000), and the availability of population synthesis models with comparable spectral resolution (Bruzual & Charlot 2003). The spectra of massive galaxies ($M_s > 10^{11} M_\odot$) at $z \sim 0$, which mainly correspond to an early-type morphology, are dominated by old, metal-rich populations (see, e.g., Thomas et al. 2005). Detailed analyses based on spectral fitting reveal an early, intense, short-lived process of star formation, with 80% of the stellar mass already formed within the first 3 Gyr (de la Rosa *et al.* 2011). In addition, we find an abrupt transition in the star formation history of galaxies with stellar mass below $\gtrsim 3 \times 10^{10} M_\odot$, suggesting a different channel of formation (Trevisan *et al.* 2012).

These studies focus on the central regions of the galaxies (the fiber-fed SDSS spectra only probe an aperture of 3 arcsec). It is more challenging to explore radial gradients of stellar populations with spectroscopy out to several effective radii. Some studies on a small set of targets find very small age radial gradients and a significant change in the metallicity gradient with respect to galaxy mass (Sánchez-Blázquez *et al.* 2007, Spolaor *et al.* 2009). A photometric optical-NIR analysis of the radial profiles of the populations in a large sample of massive ETGs from SDSS reveal a metallicity-dominated trend (decreasing outwards), with homogeneously old populations out to an effective radius (La Barbera *et al.* 2012), and a steepening of the metallicity gradient, along with a trend towards older populations at distances 4–$8 \mathrm{R}_e$ (La Barbera *et al.* 2012). These results are consistent with numerical simulations, where the infall of small structures, formed early, would populate the outskirts of massive galaxies (Lackner *et al.* 2012).

3. CASE B: Massive, early-type galaxies at moderate redshift

At higher redshifts, massive galaxies span a wider range of morphologies in constrast to the ETG-dominated population at low redshift. We focus here on the $z \sim 0.7$ Universe, where ETGs constitute a fraction $\gtrsim 60\%$ of all massive galaxies (Buitrago *et al.* 2013). So far, the best spectra for the analysis of the stellar populations in this range has been taken by the slitless grisms of ACS and WFC3, instruments onboard *HST*. The superb flux calibration of these data allows for a robust constraint of the ages and metallicities of galaxies at moderate redshift (Ferreras *et al.* 2009a), confirming the quiescent star formation histories obtained in the low-redshift sample. Photometric analyses covering a much wider range in redshift give a similar picture, with the bulk of the stellar populations of massive galaxies forming at redshifts $z_F \gtrsim 3$ (see, e.g., Pérez-González *et al.* 2005). The analysis of radial gradients in the properties of the underlying populations at $z \sim 0.7$ is currently limited to photometric data, and consistently show a nearly-flat colour gradient out to an effective radius, with the main trend driven by metallicity (Ferreras *et al.* 2005). The presence of younger populations in spheroidal galaxies at those redshifts is only found in the form of blue cores, at stellar masses $\lesssim 10^{10} M_\odot$ (Pasquali *et al.* 2006, Ferreras *et al.* 2009b).

4. CASE C: Size evolution and stellar populations

In order to understand the growth in size of massive galaxies between $z \sim 2$ and 0, we compared the SDSS-based samples presented in cases A and B above, using the same methodology to extract their star formation histories (Trujillo *et al.* 2011). The analysis found no significant difference in the population ages on the mass-size plane, ruling out growth scenarios where a significant amount of star formation is expected, as in the puffing-up mechanism of Fan et al. (2008), or the emergence hypothesis of van der Wel *et al.* (2008), where a contribution in the "emergent" systems created from

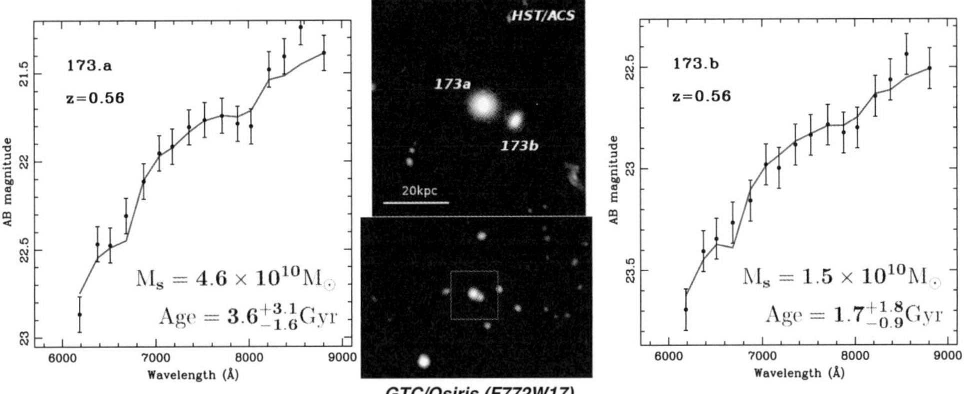

Figure 1. An example of the analysis of stellar populations in close pairs involving a massive galaxy, with data from the SHARDS collaboration (Pérez-González *et al.* 2013). ACS/*HST*, and OSIRIS/*GTC* images are shown in the central top and bottom panels, respectively. The left/right diagrams correspond to the data (points) and best SSP model (line) for the centreal (left) and satellite galaxy, respectively (Ferreras *et al.*, in preparation).

late-type galaxies to the $z \sim 0$ census of massive galaxies would have shown up as a measurable fraction of young stars in the analysis of de la Rosa et al. (2011). Hence, the most plausible scenario involves dry minor mergers over this redshift range (Naab *et al.* 2009). A simple analysis of the evolution of galaxies on the mass-size plane shows that a relatively small number of minor mergers is needed to account for the growth between $z \sim 1$ and 0 (Trujillo *et al.* 2011), consistent with the fraction of observed satellites around massive spheroidal galaxies (Mármol-Queraltó *et al.* 2012). This study was followed up by an analysis of the stellar populations using broadband photometry, finding similar stellar ages between the populations of the central (massive) and satellite galaxies at high redshift, with a trend towards an age difference $t_C - t_S \gtrsim 1.5\,\mathrm{Gyr}$ at lower redshifts (Mármol-Queraltó *et al.* 2013).

We are continuing this project of characterization of the populations of (massive) centrals and satellites at $z \lesssim 1$ with the SHARDS survey, involving deep photometry of the GOODS North 130 arcmin^2 field with 24 contiguous medium-width filters in the spectral range 5000–9500Å (Pérez-González *et al.* 2013). The dataset effectively provide $R \sim 50$ spectra down to AB=26.5 mag (3σ), enabling us to achieve an accurate determination of the redshifts (essential to select close pairs), and to probe in detail the stellar populations of the companions of massive galaxies at $z \lesssim 1$. Figure 1 gives an example for a close pair at $z = 0.56$, with a simple comparison with SSP models from Bruzual & Charlot (2003).

5. CASE D: A different formation channel at early times?

Observational evidence, therefore, points towards a two-stage formation process in massive ETGs, involving an initial phase of rapid, early star formation between $z_F \sim 5 - 10$ and 2, followed by a general quiescent behaviour along with stellar mass growth of mostly gas-poor minor mergers. The physical conditions of star formation during the rapid early phase must have been significantly different from the bulk of star forming galaxies at lower redshifts. Detailed spectroscopic analyses of the populations of massive galaxies at $z \gtrsim 2$ (see, e.g. van Dokkum & Brammer 2010, Ferreras *et al.* 2012, Toft *et al.* 2012) will enable us to quantify the relative importance of mergers and cold accretion to the process.

Preliminary studies of the IMF under those conditions suggest a trend towards a bottom-heavy IMF in strong star-bursting systems (Hopkins 2012), similar to the observed trend in gravity-sensitive line strengths that reveal an excess of low-mass stars in the most massive galaxies (van Dokkum & Conroy 2010, Spiniello *et al.* 2012, Ferreras *et al.* 2013). Although those studies still need to improve, to be able to robustly constrain the shape of the low-mass end of the IMF, independent work based on dynamical modelling, or strong gravitational lensing seem to suggest a similar scenario (Auger *et al.* 2010, Thomas *et al.* 2011, Cappellari *et al.* 2012).

References

Auger, M. W., *et al.* 2010, *ApJ Lett.*, 721, L163

Bruzual, G. & Charlot, S. 2003, *MNRAS*, 344, 1000

Buitrago, F., *et al.* 2013, *MNRAS*, 428, 1460

Cappellari, M., *et al.* 2012, *Nature*, 484, 485

de la Rosa, I. G., *et al.* 2011, *MNRAS Lett.*, 418, L74

Dekel, A. & Birnboim, Y. 2006, *MNRAS*, 368, 2

Fan, L., Lapi, A., De Zotti, G., & Danese, L. 2008, *ApJ Lett*, 689, L101

Ferreras, I., *et al.* 2005, *ApJ*, 635, 243

—. 2009a, *ApJ*, 706, 158

—. 2009b, *MNRAS*, 396, 1573

—. 2012, *AJ*, 144, 47

—. 2013, *MNRAS Lett.*, 429, L15

Hopkins, P. 2012, *MNRAS*, submitted (arXiv:1204.2835)

La Barbera, F., *et al.* 2011, *ApJL*, 740, L41

—. 2012, *MNRAS*, 426, 2300

Lackner, C., *et al.* 2012, MNRAS, 425, 641

Mármol-Queraltó, E., *et al.* 2012, *MNRAS*, 422, 2187

—. 2013, *MNRAS*, 429, 792

Naab, T., Johansson, P. H., & Ostriker, J. P. 2009, *ApJ Lett.*, 699, L178

Oser, L., *et al.* 2010, *ApJ*, 725, 2312

Pasquali, A., *et al.* 2006, ApJ, 636, 115

Pérez-González, P., *et al.* 2005, *ApJ*, 630, 82

—. 2013, *ApJ*, 762, 46

Sánchez-Blázquez, P., *et al.* 2007, *MNRAS*, 377, 759

Spiniello, C., *et al.* 2012, *ApJ Lett*, 753, L32

Spolaor, M., *et al.* 2009, *ApJ Lett.*, 691, L138

Thomas, D., *et al.* 2005, *ApJ*, 621, 673

Thomas, J., *et al.* 2011, *MNRAS*, 415, 545

Toft, S., *et al.* 2012, ApJ, 754, 3

Trevisan, M., *et al.* 2012, *ApJ Lett.*, 752, L27

Trujillo, I., *et al.* 2007, MNRAS, 382, 109

Trujillo, I., Ferreras, I., & de la Rosa, I. G. 2011, MNRAS, 415, 3903

Valentinuzzi, T., *et al.* 2010, *ApJ*, 712, 226

van Dokkum, P., *et al.* 2008, *ApJ Lett.*, 677, L5

van Dokkum, P. & Brammer, G. 2010, *Nature*, 468, 940

van Dokkum, P. & Conroy, C. 2010, *Nature*, 468, 940

Vazdekis, A., *et al.* 2012, *MNRAS*, 424, 157

White S. D. M., Rees, M. J. 1978, *MNRAS*, 183, 341

York, D., *et al.* 2000, AJ, 120, 1579

The intriguing life of massive galaxies
Proceedings IAU Symposium No. 295, 2012
D. Thomas, A. Pasquali & I. Ferreras, eds.

© International Astronomical Union 2013
doi:10.1017/S174392131300450X

Stellar velocity dispersions and emission line properties of SDSS-III/BOSS galaxies

D. Thomas[1,2], O. Steele[1], C. Maraston[1,2], J. Johansson[1,3], A. Beifiori[1,4], J. Pforr[1,5], G. Strömbäck[1], C. A. Tremonti[6], D. Wake[7] and the BOSS collaboration

[1]ICG, University of Portsmouth, Dennis Sciama Bldg, Burnaby Road, PO1 3FX, UK
[2]SEPnet, South East Physics Network, (www.sepnet.ac.uk)
[3]Max-Planck-Institut für Astrophysik, D-85748 Garching, Germany
[4]MPI für extraterrestrische Physik, Giessenbachstraße, D-85748 Garching, Germany
[5]NOAO, 950 North Cherry Ave., Tucson, AZ 85719, USA
[6]Department of Astronomy, University of Wisconsin-Madison, Madison, WI 53706, USA
[7]Yale Center for Astronomy and Astrophysics, Yale University, New Haven, CT, USA

Abstract. We perform a spectroscopic analysis of 492,450 galaxy spectra from the first two years of observations of the Sloan Digital Sky Survey-III/Baryonic Oscillation Spectroscopic Survey (BOSS) collaboration. This data set has been released in the ninth SDSS data release, the first public data release of BOSS spectra. We show that the typical signal-to-noise ratio of BOSS spectra is sufficient to measure stellar velocity dispersion and emission line fluxes for individual objects. The typical velocity dispersion of a BOSS galaxy is 240 km/s, with an accuracy of better than 30 per cent for 93 per cent of BOSS galaxies. The distribution in velocity dispersion is redshift independent between redshifts 0.15 and 0.7, which reflects the survey design targeting massive galaxies with an approximately uniform mass distribution in this redshift interval. The majority of BOSS galaxies lack detectable emission lines. We analyse the emission line properties and present diagnostic diagrams using the emission lines [OII], H_β, [OIII], Halpha, and [NII] (detected in about 4 per cent of the galaxies). We show that the emission line properties are strongly redshift dependent and that there is a clear correlation between observed frame colours and emission line properties. Within in the low-z sample around $0.15 < z < 0.3$, half of the emission-line galaxies have LINER-like emission line ratios, followed by Seyfert-AGN dominated spectra, and only a small fraction of a few per cent are purely star forming galaxies. AGN and LINER-like objects, instead, are less prevalent in the high-z sample around $0.4 < z < 0.7$, where more than half of the emission line objects are star forming. This is a pure selection effect caused by the non-detection of weak H_β emission lines in the BOSS spectra. Finally, we show that star forming, AGN and emission line free galaxies are well separated in the $g - r$ vs $r - i$ target selection diagram.

Keywords. surveys, galaxies: general, galaxies: evolution, galaxies: kinematics and dynamics, galaxies: ISM, galaxies: active evolution

1. Introduction

Recent large galaxy surveys have opened a new avenue in observational galaxy evolution, by providing large data sets that allow statistical studies of large galaxy samples in the nearby universe. In the last decade an overwhelming number of studies based on data from the Sloan Digital Sky Survey (York *et al.* 2000) have led to significant progress in our understanding of the local galaxy population. The Baryon Oscillation Spectroscopic Survey (hereafter BOSS, cite[Dawson *et al.* 2012]Dawson12), part of the Sloan Digital Sky Survey III collaboration Eisenstein *et al.* (2011), is extending this database to higher redshifts obtaining spectra of 1.5 million luminous galaxies up to redshifts $z \sim 0.7$ by

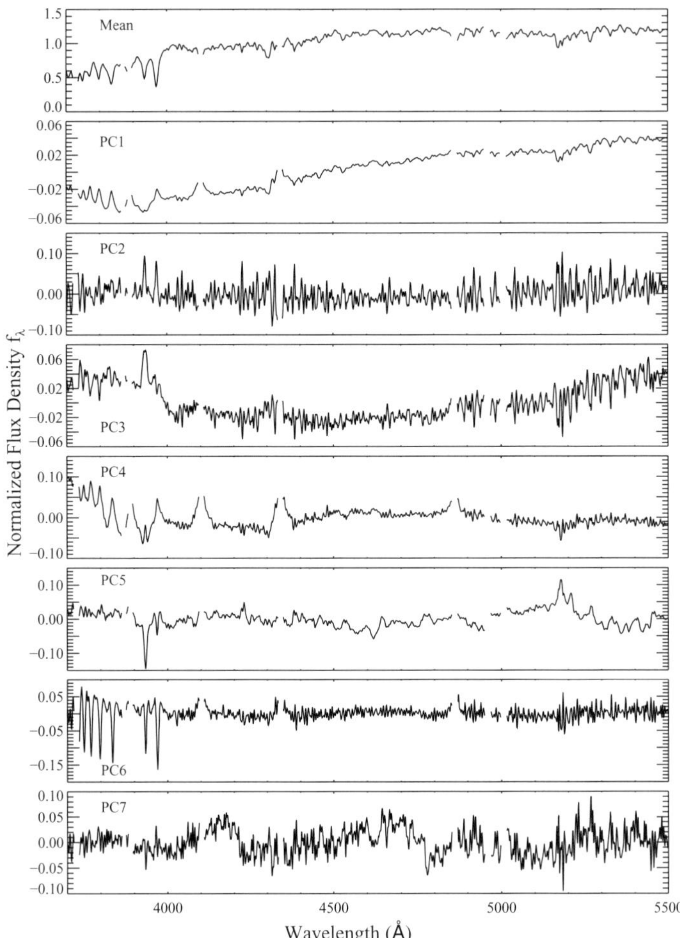

Figure 1. From top to bottom: the mean spectrum of the model library followed by the seven most significant eigenspectra.

of the model spectra is convolved to a velocity that is uniformly distributed over the range of values from 50 to 400 km s^{-1} to mimic stellar velocity dispersion.

Second, we run the PCA code on the model library to identify the significant principal components (PCs) of the spectral library. Figure 1 presents the mean spectrum and the first seven PCs for our input model library. As expected, the mean spectrum is typical of that of a galaxy with an intermediate age stellar population. The first PC provides a first-order measure of the age of the stellar population and it is strongly correlated with both 4000Å break and Balmer absorption line strengths. The second and third PC contain information about velocity dispersion and metallicity. The fourth PC clearly recovers information contained in the Balmer absorption lines. CaII (H+K) and Mgb absorption lines are clearly visible in the fifth PC, this component carries information about galaxy metallicity.

Third, we decompose each model and observed spectrum into its PC representation. Figure 2 is an example, the black spectrum is BOSS data, the red is the best fit. It is a linear combination of PCs:

$$S_\lambda = M_\lambda + \sum_\alpha C_\alpha \ E_{\alpha,\lambda}, \tag{2.1}$$

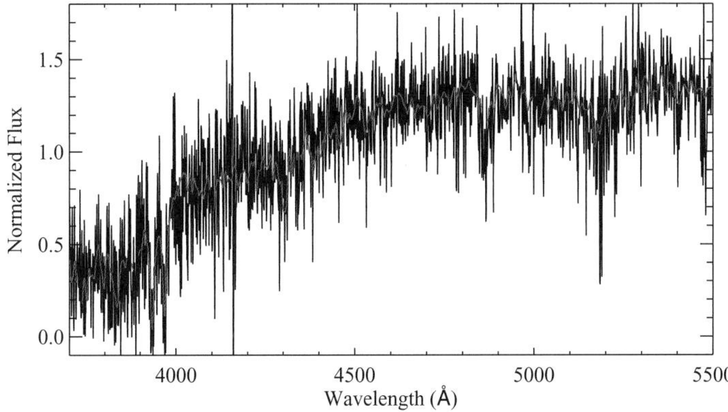

Figure 2. Example of the projection of a spectrum from BOSS data (black line) onto PCs (represented by the red central line).

where M_λ is the mean spectrum of the model library. C_α is the amplitude of the α-th PC $E_{\alpha,\lambda}$ (Note that α ranges from 1 to 7). Through this decomposition, we get the coefficient C_α.

Finally, in the seven-dimensional C_α space, the observed galaxy is a point. In the ideal case, the 25,000 models should be uniformly distributed in PC-space. Let's take the observed galaxy as the center, the errors of C_α as the radius in the seven-dimensional space. We get physical parameters of the observed galaxy by finding all the models located within the chosen radius, and assuming the physical parameters of the observed galaxy are similar to these models.

3. Implications

Evolution of the most massive galaxies to $z \sim 0.6$. We select a low-redshift galaxy sample from DR7: $14.5 < r < 17.6$, $0.055 < z < 0.3$ and a high-redshift sample from BOSS ("CMASS") with $0.55 < z < 0.7$. With the SFHs, we define a parameter, $F(> 10\%)$, fraction of galaxies with more than 10% of their stellar masses formed in the last Gyr. For simplicity, we will refer to it as the fraction of actively star-forming galaxies. The left panel of figure 3 shows how this fraction changes with stellar mass and redshift. Two conclusions from the left panel of figure 3: (1) the fraction of actively star-forming galaxies with $\log M_* > 11.4$ has evolved strongly since redshift 0.6; (2) at $z \sim 0.6$ the fraction of active star forming galaxies flattens above $\log M_* > 11.7$.

The link between radio AGN activity and star formation. We cross-match the SDSS DR7 (BOSS) spectroscopic galaxy sample with the FIRST and NVSS surveys to generate the low (high) redshift radio-loud AGN sample. Any galaxy that meets our radio flux cut is termed "radio-loud". We create samples of radio-quiet galaxies, which are matched as closely as possible to the radio-loud AGN hosts. For each galaxy in the radio-loud sample, we find another galaxy located in the FIRST and NVSS survey area without a radio detection that is closely matched in redshift ($|\Delta z| < 0.005$), stellar mass ($|\Delta \log M_*| < 0.05$), and velocity dispersion ($|\Delta \sigma_*| < 10 \mathrm{km\ s^{-1}}$). In order to study evolutionary effects, we finally match the DR7 and CMASS galaxies by stellar mass. The right panel of figure 3 shows the fraction of actively star forming galaxies as a function radio luminosity. Black lines show results for DR7 (low-z), while red lines are for CMASS (high-z). The dashed and solid lines represent the radio-loud galaxies and the radio-quiet controls, respectively. Note that for the radio-quiet galaxies, we use the radio luminosity from their radio-loud

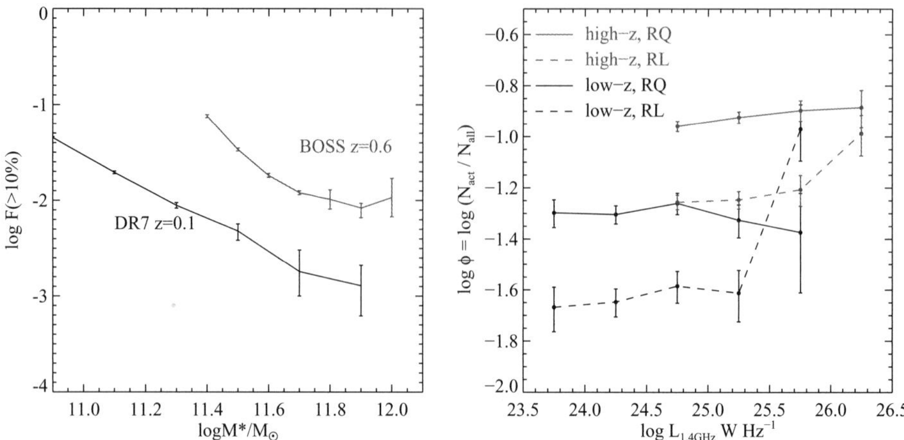

Figure 3. The left panel shows how this fraction of actively star forming galaxies changes with stellar mass and redshift; the right panel shows the fraction of actively star forming galaxies as a function radio luminosity. Black lines show results for DR7 (low-z), while red lines are for CMASS (high-z). The dashed and solid lines represent the radio-loud galaxies and the radio-quiet controls, respectively.

twins as the x-axis quantity. RL and RQ represent radio-loud and radio-quiet respectively. The main point of this plot is that the fraction of actively star forming galaxies is about 2 times higher in the radio quiet samples. This fraction remains constant over a wide range in radio luminosity. At the very highest radio luminosities, this fraction suddenly increases in radio loud samples.

References

Brinchmann, J., Charlot, S., White, S. D. M., Tremonti, C., Kauffmann, G., Heckman, T., & Brinkmann, J. 2004, *MNRAS*, 351, 1151
Bruzual, G. & Charlot, S. 2003, *MNRAS*, 344, 1000
Charlot, S. & Fall, S. M. 2000, *ApJ*, 539, 718
Chen, Y. M., Kauffmann, G., Tremonti, C. A., *et al.* 2012, *MNRAS*, 421, 314
Kauffmann, G., Heckman, T. M., White, S. D. M., *et al.* 2003, *MNRAS*, 341, 33
Maraston, C., & Stromback, G. 2011 *MNRAS*, 418, 2785
Salim, S., Charlot, S., Rich, R. M., *et al.* 2005, *ApJ*, 619, L39

The intriguing life of massive galaxies
Proceedings IAU Symposium No. 295, 2012
D. Thomas, A. Pasquali & I. Ferreras, eds.

© International Astronomical Union 2013
doi:10.1017/S1743921313004523

Galaxy formation and evolution with the Dark Energy Survey

Diego Capozzi[1], Daniel Thomas[1],
Claudia Maraston[1] and Luke J. M. Davies[1,2]

[1] Institute of Cosmology and Gravitation, University of Portsmouth
Dennis Sciama Building, Burnaby Road
Portsmouth, PO1 3FX, UK
email: diego.capozzi@port.ac.uk
email: daniel.thomas@port.ac.uk
email: claudia.maraston@port.ac.uk
[2] Department of Physics, University of Bristol
H. H. Wills Physics Laboratory, Tyndall Avenue
BS8 1LT, UK
email: luke.davies@bristol.ac.uk

Abstract. The Dark Energy Survey (DES) will be the new state-of the-art in large-scale galaxy imaging surveys. With $5,000$ deg^2, it will cover an area of the sky similar to SDSS-II, but will go over two magnitudes deeper, reaching 24th magnitude in all four optical bands (*griz*). DES will further provide observations in the redder *Y*-band and will be complemented with VISTA observations in the near-infrared bands *JHK*. Hence DES will furnish an unprecedented combination of sky and wavelength coverage and depth, unreached by any of the existing galaxy surveys. The very nature of the DES data set – large volume at intermediate photometric depth – allows us to probe galaxy formation and evolution within a cosmic-time range of ~ 10 Gyr and in different environments. In fact there will be many galaxy clusters available for galaxy evolution studies, given that one of the main aims of DES is to use their abundance to constrain the equation of state of dark energy. The X-ray follow up of these clusters, coupled with the use of gravitational lensing, will provide very precise measures of their masses, enabling us to study in detail the influence of the environment on galaxy formation and evolution processes. DES will leverage the study of these processes by allowing us to perform a detailed investigation of the galaxy luminosity and stellar mass functions and of the relationship between dark and baryonic matter as described by the Halo Occupation Distribution.

Keywords. Astronomical data bases: surveys, galaxies: abundances, galaxies: formation, galaxies: evolution, galaxies: clusters: general, cosmology: observations

1. Introduction

In the past decades, the astrophysical community has put much effort into understanding how galaxies and the structures in which they assemble over cosmic time, form and cluster together. Within the currently accepted standard cosmological model, Dark Matter and Dark Energy (both of them only having been indirectly detected) play a fundamental role, which can be inferred by the outstanding results obtained by comparing the predictions of N-body simulations (e.g., Springel *et al.* 2005; Klypin *et al.* 2011) – which aim at reproducing the Universe's dark matter distribution – with observations performed with large galaxy surveys like the Sloan Digital Sky Survey (SDSS), using galaxies as dark matter tracers (e.g., Springel *et al.* 2006). The current picture predicts that structures (or haloes) form hierarchically due to the dark matter clustering. However, the picture portrayed for dark matter might not be applicable to baryonic matter (constituting the galaxy stellar and gas content), because it is not subject only to gravity. In fact, despite the general belief that galaxies form hierarchically – most of semi-analytic models are built on this premise (e.g., De Lucia *et al.* 2006) – several are the observations this scenario struggles to reproduce (e.g., downsizing, see Cowie *et al.* 1996; Thomas *et al.* 2005; Capozzi *et al.* 2010, Pozzetti *et al.* 2010 and references therein). Furthermore, the influence of the physics driving galaxy formation on galaxy-structure properties is still unresolved (e.g., Peacock & Smith 2000; Berlind & Weinberg 2002; Lin *et al.* 2004; Capozzi et al. 2012). The Dark Energy Survey (DES) will cover a niche of medium depth with a sky coverage that will allow us to probe the build-up of galaxy mass, and the dark-baryonic matter relation in galaxy structures out to $z \sim 1.3$.

2. Abundance of massive galaxies

The way the massive ($\gtrsim 10^{11}$ M$_\odot$) end of the stellar mass function (SMF) builds up over cosmic time is currently a matter of debate. On the one hand, theoretical (semi-analytic) models (e.g., De Lucia *et al.* 2006; De Lucia & Blaizot 2007), constructed on the hierarchical structure build-up, predict that this part of the SMF evolved significantly in the last ~ 10 Gyr, during which galaxies accreted between 20 and 50% of their final mass; on the other hand, observations hint at a mass-dependent galaxy number-density evolution with z, i.e. the higher the mass the less the number-density evolution (Cimatti *et al.* 2006; Pozzetti *et al.* 2010). These discrepancies between observations and theoretical models put uncertainties on galaxy formation, which could be reduced by having a survey able to access a volume large enough to detect a representative sample of the most massive galaxies in the Universe (the rarest part of the galaxy population), and deep enough to identify a mass-complete galaxy sample out to $z \gtrsim 1$. The recent results of the Baryonic Oscillations Spectroscopic Survey (Maraston et al. 2012) at $0.4 < z < 0.7$ (a crucial range for the late mass build up predicted theoretically) showed a mild evolution of the massive end of the SMF. However, the z range explored was too narrow to include the highest-z window necessary for fully probing the mass build-up process. Studying the SMF with the DES dataset will allow us to do so, given its outstanding photometric quality, depth and spanned volume (the surveyed area will be 5,000 sq. deg).

3. Detection of massive galaxies at $z > 3$

An intriguing byproduct of the DES will be the identification of very massive high redshift ($z > 3$) galaxies (Davies et al. 2012). Cold Dark Matter simulations (scaled to DES accessible volume) predict that no massive ($M > 10^{12}$ M$_\odot$) sources should exist at $z > 3$, while passively evolving the low-z mass function predicts such sources should be rare but nonetheless present at early times. However – given the large integration time and accessible volume needed to identify high-z sources – rare, massive ($M > 10^{12}$ M$_\odot$) galaxies, which may be already passively evolving at $z \sim 3$, have probably missed detection in past and current surveys (Fig. 1). DES depth and sky coverage will allow the identification of such galaxies, should they exist, helping us to constrain galaxy formation models. By modeling (Davies *et al.* 2012) these sources we find that they must be sufficiently young, massive and relatively dust free, to be detected.

4. Halo occupation distribution over cosmic time

DES will detect a large number of galaxy clusters out to $z \sim 1.3$. Hence, it will provide a unique opportunity to study the evolution of the Halo Occupation Distribution

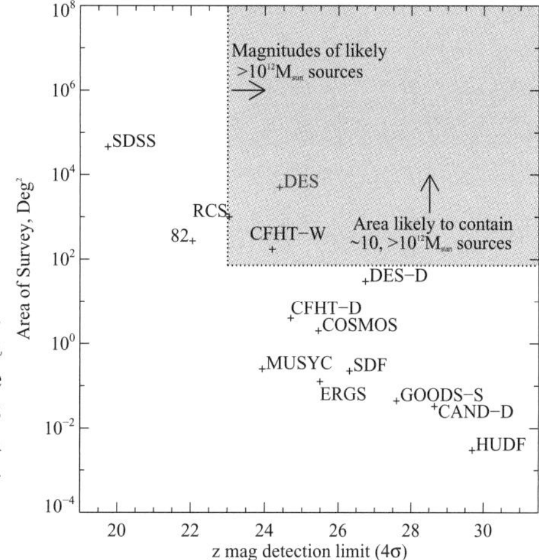

Figure 1. Figure from Davies *et al.* (2012). Comparison of z-band magnitude limit vs. area for a sample of surveys. The grey region is the parameter-space area which is likely to detect $M > 10^{12}$ M$_\odot$ galaxies at $z > 3$. DES falls in this region as it covers a large area and is relatively deep in z.

(HOD, Peacock & Smith 2000; Berlind & Weinberg 2002; Kravtsov *et al.* 2004) over cosmic time. Within the halo model, we focus on two aspects of the HOD framework: i) the richness-mass $(N - M)$ relation; ii) the relation between galaxy and dark matter spatial distributions within haloes, as parametrized by the density-profile concentration parameter (c).

i) For haloes with masses $> 10^{13}$ M$_\odot$ (galaxy groups and clusters), the $N - M$ relation can be approximated by a power law, $N \propto M^\beta$. The β parameter at low z has been deeply investigated in the literature (e.g., Lin et al. 2004; Muzzin *et al.* 2007) and the majority of the results point at a value $\beta < 1$. This means that more massive structures have a lower number of galaxies per unit mass compared to their less massive counterparts, a fact that could be due to galaxy structures being less rich at high z. By studying a sample of some of the highest-z $(0.8 < z < 1.5)$ X-ray-selected clusters (Capozzi *et al.* 2012), we found evidence that clusters were indeed poorer in the past compared to local ones of the same mass (Fig. 2, left panel). However, this fact cannot be considered as the cause of the local $\beta < 1$, unless corroborated by a significant evolution of the latter towards a value of 1 at high z. This possibility could be studied only by using mock clusters, with our results showing β remaining ~ 0.8 between $0 < z < 1.3$. However, a detailed study of this issue should be carried out observationally over a wide z range.

ii) The importance of c resides in its relation with structure formation time (Navarro *et al.* 1997; Neto et al. 2007; Gao *et al.* 2008). However, this somewhat direct correspondence is driven by gravity only. Differences between the dark-matter measured c and the one measured using baryonic matter (e.g., galaxies or intra-cluster medium) could break this correspondence and shed light on the influence that galaxy formation has on the formation and evolution of structures over comic time. Hints of such differences (Fig. 2, right panel) have been found using simulations (Capozzi *et al.* 2012). However this result could not be associated with a physical process, nor could it be probed observationally, due to the lack of adequate data-sets.

Due to the heterogeneity of the analysis methods and cluster-sample characteristics, the observation-based studies in the literature also provide a blurred picture of the HOD evolution with z, leaving us with the need of a systematic and homogeneous study of $N - M$ and c to be carried out on a statistically-rich cluster sample over a large z-range (out to $z \gtrsim 1$). DES will provide such a sample.

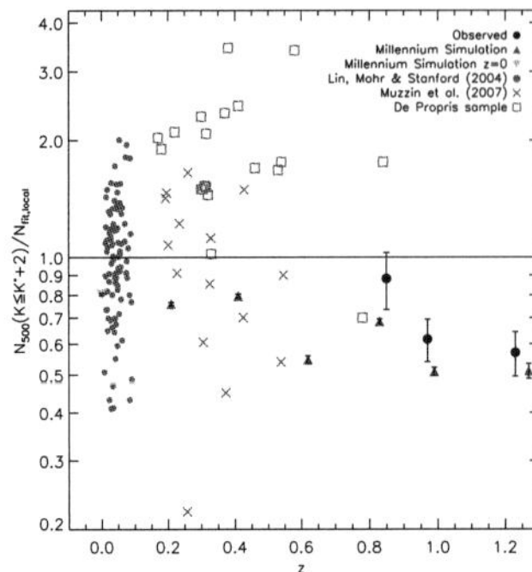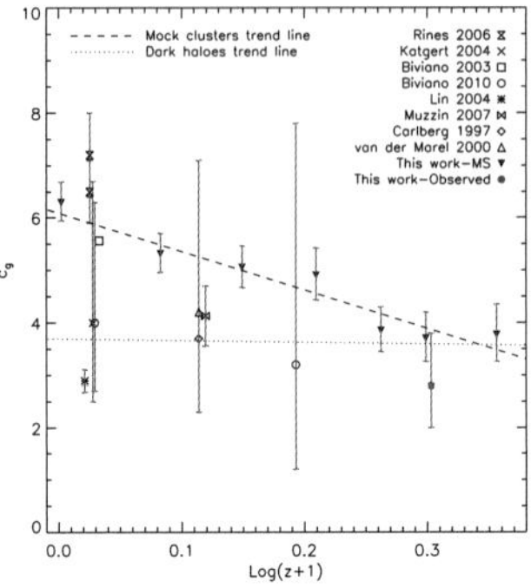

Figure 2. Figures from Capozzi *et al.* (2012) (see references therein). *Left panel*: ratio of measured N_{500} over the prediciton of the low-z best-fitting relation of Lin *et al.* (2004), vs. z. *Right panel*: average galaxy c vs. $\log(z + 1)$. Data points are observations, except full downward triangles (mock clusters in Capozzi et al. 2012). Dashed and dotted lines are trend lines for mock clusters and haloes with same mass by Capozzi et al. (2012) and Gao *et al.* (2008).

5. Conclusions

With its characteristics – area of $5,000$ deg^2; depth of $\sim 24^{th}$ in four optical bands (*griz*) – DES will allow us to study in detail the galaxy mass build-up process since $z \sim 1.3$ and to possibly detect, should they exist, $M > 10^{12}$ M$_\odot$ galaxies at $z > 3$. This will help us to constrain galaxy formation models. Finally, the large number of galaxy clusters that DES will detect between $0 < z < 1.3$ will allow us to study the prospective evolution of the HOD in detail, hence giving us the chance to probe the relation between dark and baryonic matter over such a large redshift range.

References

Berlind, A. A. & Weinberg, D. H. 2002, *ApJ*, 575, 587

Capozzi, D., Collins, C. A., & Stott, J. P. 2010, *MNRAS*, 403, 1274

Capozzi, D., Collins, C. A., Stott, J. P., & Hilton, M. 2012, *MNRAS*, 419, 2821

Cimatti, A., Daddi, E., & Renzini, A. 2006, *A&A*, 453, L29

Cowie, L. L., Songaila, A., Hu, E. M., & Cohen, J. G. 1996, *AJ*, 112, 839

Davies, L. J. M., Maraston, C., Thomas, D., & Capozzi, D., the DES collaboration 2012, *submitted*

De Lucia, G., Springel, V., White, S. D. M., Croton, D., & Kauffmann, G. 2006, *MNRAS*, 366, 499

De Lucia, G. & Blaizot, J. 2007, *MNRAS*, 375, 2

Gao, L., *et al.* 2008, *MNRAS*, 387, 536

Klypin, A. A., Trujillo-Gomez. S., & Primack, J. 2011, *ApJ*, 740, 102

Kravtsov, A. V., *et al.* 2004, *ApJ*, 609, 35

Lin, Y., Mohr, J. J., & Stanford, S. A. 2004, *ApJ*, 610, 745

Maraston, C., *et al.* 2012, *MNRAS*, submitted (arXiv:1207.6114)

Muzzin, A., Yee, H. K. C., Hall, P. B., & Lin, H. 2007, *ApJ*, 663, 150

Navarro, J. F., Frenk, C. S., & White, S. D. M. 1997, *ApJ*, 490, 493

Neto, A. F., *et al.* 2007, *MNRAS*, 381, 1450

Peacock, J. A. & Smith, R. E. 2000, *MNRAS*, 318, 1144

Pozzetti, L., *et al.* 2010, *A&A*, 523, A13

Springel V., *et al.* 2005, *Nature*, 435, 629

Springel, V., Frenk, C. S., & White, S. D. M. 2006, *Nature*, 440, 1137

Thomas, D., Maraston, C., Bender, R., & Mendes de Oliveira, C. 2005, *ApJ*, 621, 673

The intriguing life of massive galaxies
Proceedings IAU Symposium No. 295, 2012
D. Thomas, A. Pasquali & I. Ferreras, eds.

© International Astronomical Union 2013
doi:10.1017/S1743921313004535

A phenomenological approach to the evolution of galaxies

Simon J. Lilly[1], **Yingjie Peng**[1], **Marcella Carollo**[1] **and Alvio Renzini**[2]

[1]Institute of Astronomy, ETH Zurich, 8093 Zurich, Switzerland
[2]INAF Osservatorio Astronomico di Padova, vicolo dell'Osservatorio 5, I- 35122 Padova, Italy,
and Department of Physics and Astronomy Galileo Galilei, Universita degli Studi di Padova,
via Marzolo 8, I-35131 Padova, Italy

Abstract. Increasingly good statistical data on the galaxy population at high and low redshift enable the development of new phenomenological approaches to galaxy evolution based on application of the simplest continuity equations. This has given new insights into the different ways in which star-formation in galaxies is quenched, the role of merging in the population, and in to the control of star-formation in star-forming galaxies and the links with chemical evolution. The continuity approach provides a self-consistent view of the evolving population and exposes linkages between different aspects of galaxy evolution.

Keywords. Galaxy evolution, star-formation rate, chemical evolution

1. Introduction: the continuity approach to galaxy evolution

Large surveys of the extragalactic Universe, both nearby and at high redshifts, are producing high quality statistical data on the galaxy population over a wide range of epochs. These allow the possibility of a new phenomenological approach to galaxy evolution that is built around the simple continuity equation(s) that link the galaxy population at different epochs. In this approach, we first identify, both locally and at high redshift, a few underlying simplicities, or symmetries, exhibited by the galaxy population and then explore analytically the implications of these via the continuity equations.

This approach is in a sense reversed from the usual theoretical approach of the semi-analytic models, which start with the hierarchical build-up of dark matter haloes, within which baryons evolve subject to a large number of physical processes that are described by analytic formulae. The models end up involving a large number of overt free parameters (at least 30) plus many additional assumptions that are buried in the implementation. State of the art semi-analytic models are quite successful in reproducing the galaxy population, but the inevitable complexity of the models means that there is considerable uncertainty as to the uniqueness of any particular implementation, and agreement with observations is never perfect.

Our continuity approach steps back from physical inputs (apart from the most basic continuity equations) and focuses directly on what the *data* appear to require. This approach produces phenomenological descriptions of the evolution of galaxies that may help constrain more physically motivated models. It can also identify hidden connections between different aspects of galaxy evolution that appear at first sight to be disjoint. At the very least, this approach should produce a simple self-consistent framework for interpreting different data, provide a common language for discussion, and allow well-posed questions to guide future observational programs.

We here review our recent work in this area, and discuss some of the insights that have been gained in a series of recent papers. In Section 2 (based on Peng *et al.* 2010, 2012, hereafter P10 and P12), we focus on the inter-relationships between star-formation, stellar mass and environment and the "quenching" of star-formation in galaxies. In Section 3 (based on Peng *et al.* in preparation) we briefly discuss the merging of galaxies which is also discussed in another contribution to these Proceedings. Finally, in Section 4, which is based on Lilly *et al.* (2013, hereafter L13), we examine the control of star-formation in star-forming galaxies, and the connections between metallicity, the cosmic star-formation history of galaxies and the efficiency with which dark-matter haloes produce stars.

2. The quenching of galaxies

Most star-forming galaxies have an SFR that is well-correlated with their existing stellar mass, producing a tight "Main Sequence" (MS) on which the specific SFR (sSFR) depends only weakly on stellar mass, $sSFR \propto m^{\beta}$ with $\beta \sim -0.1$, with a scatter of order 0.3 dex. A few percent of galaxies have significantly elevated sSFR, but these contribute only about 10% of the overall star-formation over a range of redshifts $0 < z < 2$ (Rodighiero *et al.* 2011, Sargent *et al.* 2012). Locally, these star-bursts are associated with major mergers of galaxies, and the same may be true at high redshifts also. Finally, there is a substantial population of galaxies with much reduced sSFR. These "quenched" galaxies are not forming a significant number of stars, i.e. $sSFR^{-1} >> \tau_H$. The physical process that turns off star-formation in galaxies is not well understood. Plausible ideas include the action of AGN or extreme merger-induced star-bursts, the inability of some haloes to deliver cool gas to the central galaxy, plus a whole host of ways in which satellite galaxies may interact with the environment.

2.1. *Key observational axioms*

Our analysis of quenching is based two key observational facts, which we use as axioms:

• **Separability of the red fraction**: As shown in P10, the SDSS red fraction $f_{red}(m,\rho)$ is separable in mass and environment ("environment" is here a nearest neighbour measure of local density, but could equally well be a central-satellite division etc.). This means that the fraction of "surviving" blue galaxies can be written as the product of two functions, one of mass but not environment, and the other of environment but not mass.

$$f_{blue}(m,\rho) = (1 - \epsilon_m(m)) \times (1 - \epsilon_\rho(\rho)) \tag{2.1}$$

Separability was already implicit in one of the fitting formulae for the SDSS $f_{red}(m,\rho)$ given by Baldry *et al.* (2007). We argued in P10 that this separability suggests that there are two "probabilistic" channels for quenching galaxies: one we call "mass-quenching", which is linked to the mass of galaxies but not environment, and the other, "environment-quenching", determined by the environment of a galaxy but not by its stellar mass. Of course, galaxies are unlikely to evolve probabilistically, and so there are presumably hidden parameters at work, but these must be independent of environment for mass-quenching, and vice versa.

• **The shape of the mass function of star-forming galaxies stays the same back to at least $z \sim 2$ and probably back to $z \sim 4$.** : Both the characteristic mass M* and the faint-end slope α_S of the Schechter mass function of star-forming galaxies show very little evolution between SDSS locally and the high redshift population. In contrast, the density normalization $\phi*$ increases by a factor of about 3 since $z \sim 2$, and by 20 since $z \sim 4$. This "vertical" evolution in $\phi_{SF}(m)$ is quite different from the "horizontal" evolution of the mass-function of dark matter haloes, which changes primarily in $M*$ with more or less constant $\phi*$. The surprising constancy of $M*$ was first highlighted by

Bell *et al.* (2005) to $z \sim 1$, was evident in Gonzales *et al.* (2010) to $z \sim 4$ using photo-z to $z \sim 4$, and is firmly established in COSMOS with both photo-z and spec-z (Ilbert *et al.* 2010, Pozzetti *et al.* 2010, Ilbert *et al.* 2013).

Other important observations that simplify the analytic treatment, but are not really required per se, are (i) the small dispersion in sSFR on the Main Sequence, (ii) the weak dependence of the Main Sequence sSFR on mass, $\beta \sim 0$, and (iii) the evidence that the environment term $\epsilon_\rho(\rho)$ is more or less constant back to $z \sim 0.8$ (P10). It is also convenient to assume that quenching is instantaneous and one-way. The characteristic sSFR of the MS gives the rate at which galaxies increase their logarithmic mass, and the evolving sSFR(t) therefore "sets the clock" of the evolving population but does not affect the outcome.

These most basic observational results are of course inter-linked. The dramatic increase in sSFR to high redshift implies that a given galaxy that stays on the Main Sequence will have increased its mass by a factor of about 50 since $z \sim 2$. This makes the constancy of $M*$ and α_S all the more striking, and emphasizes the importance of fluid-like continuity equations to track the evolving galaxy population.

2.2. *Environment-quenching as satellite-quenching*

In P12, it was shown that the environmental effects in SDSS that we highlighted in P10 are confined to satellite galaxies. Unlike the satellites, the properties of central galaxies do not depend on ρ (at fixed mass). The satellite-quenching efficiency, ϵ_{sat}, is the probability that a blue star-forming galaxy is quenched because it becomes the satellite of another galaxy. This is empirically independent of stellar mass in SDSS (as in van den Bosch *et al.* 2008) with a mean value of $\epsilon_{sat} \sim 0.4$. As predicted from P10, ϵ_{sat} increases with the local density ρ which may be taken as an (imperfect) proxy for the location within a group/cluster. It is largely independent of group mass (at fixed ρ). In other words, the environment-quenching of P10 is simply "satellite-quenching". Similar results are also evident in the detailed study of 170 low redshift ZENS groups (Carollo *et al.* in preparation).

Satellite-quenching is independent of stellar mass, and so it cannot change the shape of the mass-function of star-forming galaxies. Instead, it will produce a mass-function of satellite-quenched passive galaxies that has the same $M*$ and α_S as the star-forming population, but a relative $\phi*_{EQ}/\phi*_{SF}$ that increases with the local density.

2.3. *Mass-quenching*

Mass-quenching acts on all galaxies, independent of environment, i.e. on both centrals and satellites. Since only mass-quenching is linked to galaxy mass, it must be this process alone that controls the shape of the mass function of the (surviving) star-forming galaxy population. Indeed, "mass-quenching" is usefully thought of as *that* process which controls the mass function of star-forming galaxies, and the star-forming mass-function therefore gives us as clear a view of the action of mass-quenching as the more traditional study of the red fraction.

The observed constancy of $M*$ for the star-forming population places a strong requirement on the form of mass-quenching. This can be expressed either as a quenching rate λ_m (the chance of a given galaxy to be mass-quenched per unit time) or, equivalently, as a survival probability for a given galaxy to reach a particular mass m without being quenched, $P(m)$. Both versions include a constant μ that is the inverse of the (time-independent) $M*$, i.e. $\mu = M*^{-1}$. It is easy to show analytically from the continuity equation that $\lambda_m(m,t)$ and $P(m)$ must have the following form

$$\lambda_m = \mu \times SFR \iff P(m) = exp(-\mu m) \tag{2.2}$$

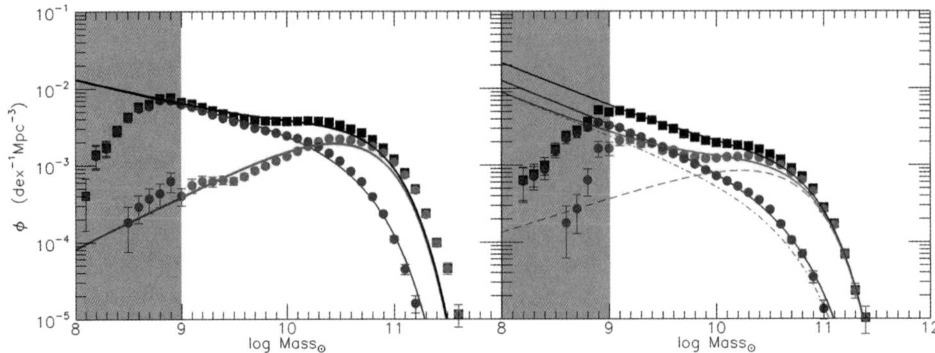

Figure 1. The mass functions of SDSS centrals (left panel) and satellites (right panel) that are observed in the SDSS, split into blue and red populations (as indicated), plus the overall population (in black). In each case, the mass functions of the red (plus overall populations) that are *predicted* on the basis of the *observed* blue population using the P10/P12 formalism are shown as continuous lines (split in the case of satellites into the mass-quenched and satellite-quenched components as the dashed lines). These mass-functions provide an excellent representation of the data. The excess of high mass red central galaxies is interpreted as a modest amount of mass-increase through merging after quenching. Figure is taken from P12.

2.4. *The origin of the Schechter function(s)*

The action of the mass-quenching described by Equation (2.2) not only maintains the M* of the (single) Schechter function of star-forming galaxies (as it must, by design) but it will also *establish* a precise Schechter shape starting from a more general power-law mass-function of the right slope. The mass function of the mass-quenched passive galaxies also has a particularly simple form: it too is a single Schechter function, with exactly the same M* as the star-forming population, but with a faint end slope that differs by $\Delta\alpha_S = (1 + \beta) \sim 1$. Given that $M*$ for the star-forming population is observed to be constant, the M* of the (mass-quenched) passive galaxies will also evolve with constant M* as the population builds up over time. The continuity equation also requires that the $\phi*$ of the two populations should increase in step, with $\phi*_{MQ}/\phi*_{SF} = -(1+\alpha_{S,SF})^{-1} \sim 2.5$.

The superposition of these different Schechter functions for the mass-quenched and environment-quenched passive galaxies produces a double Schechter function for the overall passive population. The second component depends on environment and will be absent in centrals. Adding the star-forming population, the overall population will also be a double Schechter function. Interestingly, the strength of the two components in this overall double Schechter function does not depend on the amount of environment quenching and is given by the same relation as before, $\phi*_{1+\alpha_S}/\phi*_{\alpha_S} = -(1+\alpha_S)^{-1} \sim 2.5$

These quantitative relations between the Schechter parameters are seen, to impressive precision, within the SDSS sample as a function of environment (P10), and for centrals/satellites (P12, see Figure 1). Not least, the characteristic inflection of the double Schechter function due to the superposition of the red and blue populations, is clearly visible in the SDSS, GAMA and zCOSMOS mass-functions (Pozzetti *et al.* 2010, P10, P12, Baldry *et al.* 2012).

2.5. *Interpretation of mass-quenching*

We stress that the λ form of Equation (2.2) does not necessarily require that mass-quenching is physically caused by star-formation (or by stellar mass, as sometimes assumed), only that the quenching rate must *mimic* the mass- *and* time- dependencies of the SFR in MS galaxies.

Given the known observational similarities between SFR and black hole accretion in galaxies (Silverman 2009), quenching due to AGN activity could satisfy this requirement. Furthermore, given the close coupling between m_{star} and m_{halo} for central galaxies, the second, $P(m)$, form could conceivably be connected to a limit in the m_{halo} that can sustain star-formation in its central galaxy. Note, though, that the $M*$ of central and satellite star-forming galaxies are essentially identical in SDSS (P12) and that the satellite $M*$ is constant over $12.5 < \log(M_{halo}/M_\odot) < 15$ (see P12), so the mass-quenching process must operate identically in centrals and in satellites over a very wide range of halo mass. Also, the observation that the f_{red} of centrals is correlated with halo mass, at fixed stellar mass (e.g. Woo et a 2012), does not necessarily imply a causal link between quenching and halo mass as it can arise by the continued increase of halo mass, but not of stellar mass, after a galaxy has been quenched (see P12). Finally, we stress that mass-quenching need not be physically caused by stellar mass either - indeed it is clear that morphology (e.g. central mass density, concentration or Sersic parameter) is a better "predictor" of the sSFR than m_{star}, even though causal basis of this link is unclear. The only thing we know for sure is that mass-quenching must satisfy Equation (2.2).

2.6. *Quenching at higher redshifts*

Densely sampled spectroscopic surveys like zCOSMOS and DEEP allow the construction of group catalogues similar to those in SDSS at redshifts out to $z \sim 1$. The ϵ_{sat} appears to be still independent of stellar mass at $z \sim 0.8$ (Knobel *et al.* 2013) with a mean value of $\epsilon_{sat} \sim 0.5$ close to that seen in SDSS. The mass functions of star-forming and passive centrals and satellites at high z also follow the interrelations expected from the simple P10/P12 model. The form of the mass- and environment- quenching efficiencies look similar, the role of satellites in driving the latter appears to be the same, and the form of $\epsilon_{sat}(\rho)$, all seem to be consistent at $z \sim 0.8$ with what is seen locally (Kovac *et al.* in preparation). We conclude that the action of the environment on satellite galaxies appears to be rather similar when the Universe was a half its current age, as today. This provides another interesting constraint on the physical processes involved.

2.7. *Overall evolution of the population*

The action of the simple quenching laws is well seen in a movie that is available from http://iopscience.iop.org/0004-637X/721/1/193/fulltext/apj351504f13.mov.

At early times, almost all galaxies are star-forming and the population evolves primarily by increasing in mass at a rate given by the rsSFR of the population. We call this "horizontal" evolution of $\phi(m)$ as Phase 1 of the evolution. As galaxies approach what will become the characteristic $M*$, mass-quenching becomes increasingly important and starts to produce massive quenched galaxies. The star-forming $\phi(m)$ stabilizes as a single Schechter function (if it is not one already) and thereafter increases in $\phi*$ with constant $M*_{SF}$. The mass-quenched passive population also increases in $\phi*$ at the same rate (with the same constant $M*$ but different α_S). This "vertical" evolution of $\phi(m)$ is Phase 2. Finally, the quenching of satellite galaxies becomes important, producing the second component of passive galaxies which increasingly comes to dominate at lower masses. We call this Phase 3 although it proceeds in parallel with Phase 2. The transition between Phase 1 and 2 is marked by the establishment of constant $M*_{SF}$ and the appearance of the first massive passive galaxies, and represents a divergence in evolution between the overall galaxy and halo mass functions. This probably occurs sometime around $z \sim 3-4$ and quite possibly at different times in different large scale environments.

An important point is that the environment- (or satellite-) quenched population of passive galaxies, which dominates at lower galaxy masses, emerges later compared to the population of mass-quenched galaxies, which dominate the higher masses around $M*$. The mass-quenching rate was twenty times higher at $z \sim 2$ than now, whereas the environment-quenching rate has evolved more gradually as groups and clusters were assembled. This difference accounts for a correlation between mean stellar age and stellar mass for today's passive population, as well as the over-abundance of α-elements in the more massive galaxies (see P10 for details). It also naturally accounts for the apparent mass-*dependent* differential build up of the passive population (see Ilbert *et al.* 2013), even though both channels, individually, build up the passive population *independently* of mass. These observational effects are both sometimes taken as a signature of a "down-sizing" process, but this is rather misleading since we are dealing with two distinct processes proceeding at different rates.

The continuity approach also clarifies two other widespread misconceptions: It is clear from the mass-functions that there is today a threshold mass above which most galaxies are passive and below which most are still actively forming stars. It is often assumed that, in order to produce today's massive passive galaxies in the early Universe, this threshold must have been shifted to higher masses at earlier times (at variance with the observed constant M*). Alternatively, it is assumed that massive passive galaxies were generally produced at lower masses and then moved to higher masses via widespread dry-merging. The continuity analysis emphasizes that both star-forming and passive populations have very similar values of M*, and the different shapes of $\phi(m)$ come about purely from the different α_S of the two populations, that are themselves a result of the particular, required, form of the mass-quenching process. The observed (constant-shape) star-forming population can make all except the most massive of today's passive population directly.

Finally, the continuity formalism makes a clear prediction for the mass function of any set of objects that are seen at some stage in the process of being quenched. Even if they are still actively forming stars, these transition systems should have the same $M*$ and α_S of the *passive* population, but a $\phi*$ given by the $\phi*$ of the *star-forming* population multiplied by $\tau_T \times sSFR$, where τ_T is the duration of the relevant phase of quenching, see Equation (28) of P10.

3. Merging of galaxies

The merging of galaxies enters into the picture in at least two ways. This is discussed in Peng *et al.* (in preparation), and also in the paper by Peng *et al.* in these Proceedings. Here we very briefly summarize the results, for completeness:

• **Mass-addition after-quenching**: Substantial addition of stellar mass to passive galaxies after they have been quenched would perturb the quantitative relations between the Schechter functions of star-forming and passive galaxies outlined above. This limits the average mass increase for passive galaxies to be quite modest, i.e. $< 25\%$ for passive galaxies generally, and $< 35\%$ for passive centrals.

• **Maintaining constant** α_S: If the slope of the sSFR-mass relation is negative, then low mass galaxies will increase in logarithmic mass faster than more massive ones, potentially producing a steepening of the mass function, which is not seen. The destruction of low mass galaxies through merging provides an obvious way of countering this. If we define a *specific merger mass rate* ($sMMR$) that is the mass increase of a galaxy through merging with smaller galaxies (as opposed to the in situ star-formation described by the rsSFR), then it turns out that an sMMR ~ 0.1 sSFR will compensate for the steepening effect of associated with $\beta \sim -0.1$. An obvious physical explanation for the apparent link between merging and cosmic sSFR is that both are being driven by the specific growth rate of dark matter haloes, as discussed further in Section 4 below. The fact that both the merger rate of galaxies and the specific star-formation rate follow the growth of haloes produces the "symmetry" or "simplicity" of constant α_S in the overall population.

- **Merger quenching**: The introduction of "merger-quenching" as a potential quenching channel in P10 was one of the few ad hoc aspects of that paper, which we have now clarified: If $\epsilon_\rho(\rho)$ is to remain more or less constant (which was not imposed as a condition in P10) then the κ_- term in Equation (18) of P10 must have a specific form: $\kappa_-(\rho) = \epsilon_\rho(\rho) \times sSFR$. This link with $\epsilon_\rho(\rho)$ makes it attractive to view this term as part of the general "environment-quenching" process, which may or may not be linked to merging *per se*, rather than as a distinct third quenching channel.

4. The evolution of cosmic star-formation rate, haloes and metallicity

A similar phenomenological approach to star-forming galaxies also reveals linkages between different aspects of galaxy evolution - namely the epoch dependence of the characteristic sSFR of MS galaxies, the chemical evolution of galaxies, and the varying efficiencies with which dark matter haloes have converted gas into stars (see Lilly *et al.* 2013, hereafter L13, for details).

We again start with a striking fact about the galaxy population: The increase in the sSFR of Main Sequence galaxies back to $z \sim 2$ (Noeske 2007, Daddi 2007, Elbaz 2007 and Panella *et al.* 2009) is very similar to the change in the specific mass increase rate of dark matter haloes (sMIR$_{DM}$) over the same interval (see Neistein & Dekel 2009). Above $z \sim 2$ the situation becomes less clear: Initial evidence for a more or less constant $sSFR(z)$ (Perez-Gonzalez *et al.* 2009) has been disputed (Stark *et al.* 2012, Schaerer *et al.* 2012). Note that we must use a "reduced" sSFR, i.e. ($rsSFR = (1 - R) \times sSFR$) to give the mass-doubling timescale of the long-lived stellar population (see L13). These similarities are shown in Figure (2). There are, however, important differences in detail: note that the rsSFR for typical mass galaxies is is consistently significantly higher, especially at high redshifts, implying a shorter doubling time, and that the weak mass-dependence is reversed, with $\beta_{DM} \sim +0.1$ compared to $\beta \sim -0.1$ for star-formation.

4.1. *Gas-regulated evolution of galaxies*

A very simple model for galaxies can be constructed (see also Bouche *et al.* 2010) in which the star-formation rate in a galaxy is regulated by the gas content of the galaxy, via some star-formation efficiency $\epsilon = SFR/m_{gas}$, so that ϵ^{-1} is the gas depletion time-scale τ_{gas}^{-1}. This implies that the $sSFR$ and the gas ratio $\mu = m_{gas}/m_{star}$ are linked by $sSFR = \epsilon\mu$. The star-formation may drive a wind $\lambda \times SFR$, where λ is the mass-loading. The inflow can be parameterized to be a fraction f_{gal} of the baryonic inflow into the halo. If f_{gal} and the internal parameters ϵ and λ are all independent of time, then this model has the attractive feature of quickly setting and maintaining the $rsSFR$ to be exactly equal to the $sMIR_{DM}$ of the halo. It does so by forming a constant fraction f_{star} of the inflowing baryons into stars - the actual fraction depending on λ and the ($\epsilon \times sSFR$) product. If these parameters of the regulator change with time, either directly or because they are functions of the stellar mass, then this equality will be perturbed, as discussed below.

4.2. *Chemical evolution*

For as long as the gas consumption timescale is shorter than the timescales on which λ and μ ($= \epsilon \times sSFR$) are changing, gas can be considered to be continuously flowing

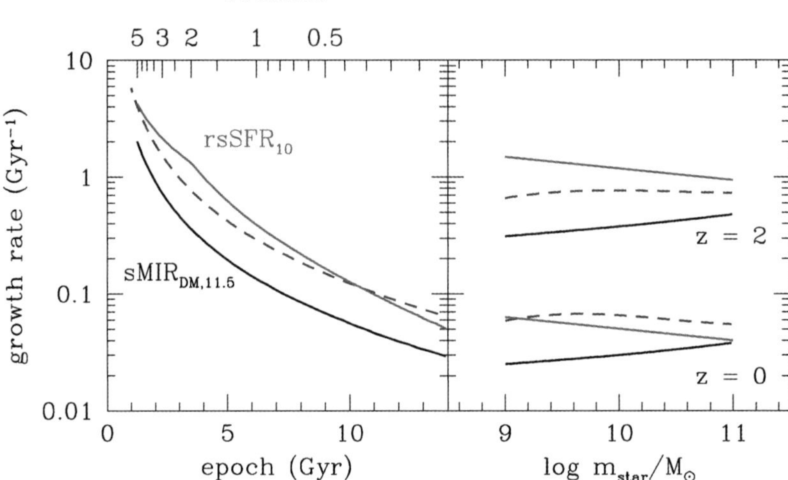

Figure 2. Comparison of the specific growth rate of dark matter haloes, sMIR$_{DM}$ (in black) from simulations, and of the stellar populations of galaxies, given by the observed rsSFR (in red), as a function of epoch for two particular masses (10^{10} and $10^{11.5}$ M$_\odot$ respectively in the left panel, and as a function of mass at $z = 0$ and $z = 2$ in the right panel. These curves show the systematic offset of the sMIR$_{DM}$ to higher values (shorter doubling times) and the reversal of the mass-dependence of the $sMIR$ relative to the rsSFR. The blue dashed curves show the boosting of the rsSFR relative to the $sMIR_{DM}$ expected from the $f_{star}(m)$ dependence - see text.

through the system. The metallicity of the gas in this quasi-steady-state flow is given by

$$Z = Z_0 + \frac{y_R}{(1 + \lambda(1 - R)^{-1} + \epsilon^{-1}(sSFR + (1 - R)^{-1}d\ln\mu/dt)} \tag{4.1}$$

where Z_0 is the metallicity of the inflowing gas (assumed to be small) and y_R is the yield of metals per unit mass locked up in long-lived stars. The metallicity of the system is to a large degree set "instantaneously" by the regulator. The only knowledge of the previous history is in the (small) $\epsilon^{-1}d\ln\mu/dt$ term. As would be expected in a continuous flow model, Equation (4.1) can be written simply in terms of the fraction f_{star} of incoming baryons that are transformed into stars,

$$Z = Z_0 + y_R \times f_{star}, \tag{4.2}$$

without knowledge of the quantities λ and $\mu = \epsilon^{-1}sSFR$ which actually set the value of f_{star}. These two Equations (4.1 and 4.2) highlight several inter-connections between different aspects of galaxy evolution.

4.3. *The existence of SFR as a second parameter in the mass-metallicity relation*

Equation (4.1) requires that the SFR should enter as a second parameter in the mass-metallicity relation, as has been claimed observationally in several papers (Ellison & Kewley 2008, Manucci *et al.* 2010, Lara-Lopez *et al.* 2010, Yates *et al.* 2012). Furthermore, because the metallicity in the regulator is set "instantaneously", the $Z(m, SFR)$ relation should only evolve with time in so far as the parameters ϵ and λ themselves change with time. Equation (4.1) provides a natural explanation for a more or less epoch-independent "fundamental" $Z(m, SFR)$ relation (e.g. Manucci *et al.* 2010).

The derivation of gas-phase metallicities from emission line data is fraught with uncertainties (e.g. Kewley & Dopita 2002) and there has been a lot of debate as to the

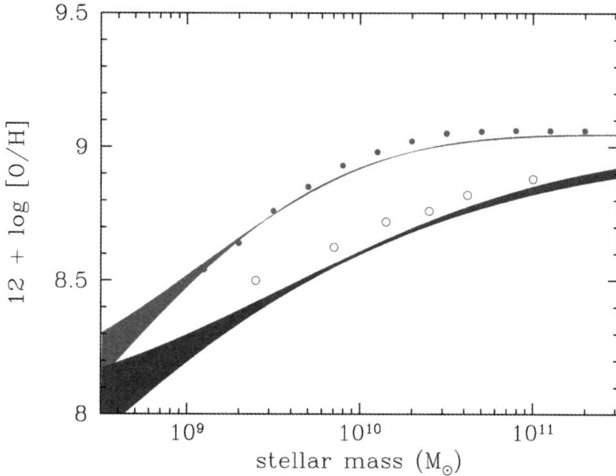

Figure 3. The mean mass-metallicity relation at $z \sim 2$ (blue) and $z \sim 0$ (red). The data is from Mannucci *et al.* (2010) and Erb *et al.* (2004) - the latter adjusted by 0.3 dex, see L13 for details. The simple model described in the text and given by Equation (4.1) qualitatively reproduces the observed evolution, which is primarily driven by the much higher sSFR at high redshift. The Figure is taken from L13.

form of the $Z(m, SFR)$ relation, even at low redshift (see e.g. Yates *et al.* 2011). With this significant caveat in mind, we can fit Equation (4.1) to the SDSS $Z(m, SFR)$ data tabulated by Mannucci *et al.* (2010) and examine the recovered $\epsilon(m)$ and $\lambda(m)$. For simplicity, y_R and Z_0 are assumed to be independent of mass. Intriguingly, the $\epsilon(m)$ and $\lambda(m)$ are not unreasonable from independent astrophysical arguments: $\epsilon(m)$ implies a (total = atomic+molecular) gas depletion timescale at $m \sim 10^{10} \mathrm{M}_\odot$ of 2-3 Gyr (c.f. Genzel *et al.* 2010, Tacconni *et al.* 2012), with a weak inverse mass dependence, while the nominal wind-loading factor at this mass is $0.3 < \lambda < 0.5$ with a quite strong inverse dependence on stellar mass, $m^{-0.8}$.

Finally, we can use Equation (4.1) to predict a mean mass-metallicity relation for MS galaxies at $z \sim 2$ (assuming $\epsilon \propto (1 + z)$). This is in good qualitative agreement with the available data from Erb *et al.* (2006), as shown in Figure 3.

4.4. *The stellar and dark matter content of haloes and the boost to the rsSFR*

The product $f_{star} f_{gal}$ in this simple treatment gives the fraction of baryons that enter the halo and are transformed into stars, which, integrated up, will give m_{star}/m_{halo}. If we define ζ as the mass-dependence of this quantity, $f_{star} f_{gal} \propto m_{star}^{\zeta}$, then $\zeta \sim 0.5$ is required to reconcile the faint end slope of the galaxy stellar mass function, $(\alpha_{S,star} \sim -1.45)$, and the low mass end slope of the dark matter halo mass function, $\alpha_{S,halo} \sim -1.85$. From Equation (4.2), the mean $Z(m)$ of MS galaxies gives (for constant y_R) the variation of $f_{star}(m)$. Estimates of the logarithmic slope of $Z(m)$ at low masses range from $Z \propto m^{0.35}$ (Tremonti *et al.* 2004) to $Z \propto m^{0.5}$ (Andrews & Martini 2012, Manucci *et al.* 2010). This simple analysis suggests that the mass-dependence of f_{star} that is implied from the mass-metallicity relation of MS galaxies (and therefore reflecting baryonic processes operating within galaxies) may be sufficient to produce the dependence of m_{star} on m_{halo} required to reconcile the different faint end slopes - see also Peeples & Shankar (2012).

Secondly, as noted above, a constant $f_{star} f_{gal}$ in the regulator sets the rsSFR to be exactly the sMIR$_{DM}$. Conversely, if $f_{star} f_{gal}$ increases with stellar mass, then the $rsSFR$ will be boosted relative to the $sMIR_{DM}$. Using the ζ parameter defined above, this boost to the rsSFR will be given by $rsSFR \sim (1 - \zeta)^{-1} sMIR_{DM}$. It can be seen that the value of $\zeta \sim 0.5$, that is indicated from the mass-metallicity relation and from the abundance matching of mass-functions, would give a boost of the $rsSFR$ of a factor of about two above the $sMIR_{DM}$. This is about what is seen on Figure (2), especially at lower redshifts. This boost effect also explains the reversed mass-dependences of the rsSFR relative to the underlying sMIR$_{DM}$, because of the reduction in ζ associated with the flattening of the $Z(m_{star})$ relation.

Acknowledgements. Our work has made extensive use of the COSMOS and zCOSMOS data sets, in addition to SDSS, and many of the ideas developed therein arose in discussions with collaborators in those projects, whose contributions are gratefully acknowledged.

References

Andrews, B. H. & Martini, P. 2012, *arXiv:1211.3418*.
Baldry, I. K., *et al.* 2012, *MNRAS*, 421, 621.
Bouche, N., *et al.* 2010, *ApJ*, 718.
Daddi, E. *et al.* 2007, *ApJ*, 670, 156.
Dave, R., Finlator, K., & Oppenheimer, B. D., 2012, *MNRAS*, 421, 98.
&Elbaz, D. *et al.* , 2007, *A&A*, 468, 33.
Ellison, S. L. & *et al.* , 2008, *ApJL*, 672, L107.
Erb, D. & *et al.* , 2006, *ApJ*, 647, 128.
Genzel *et al.* 2010, *MNRAS*, 407, 2091.
Gonzalez, V., *et al.* 2010, *ApJ*, 713, 115.
Ilbert, O. & *et al.* , 2010, *ApJ*, 709, 644.
Ilbert, O., *et al.* 2013, *arXiv1301.3157*.
Kewley, L. J. & Dopita, M. A. 2002, *ApJS*, 142, 35.
Knobel, C., *et al.* 2012, *arXiv1211.5607*
Lara-Lopez, M. A., *et al.* 2010, *A&A*, 521, L53.
Lilly, S. J., *et al.* 2013, *ApJ*, submitted.
Lotz, J., *et al.* 2011, *ApJ*, 742, 103.
Mannucci, F., *et al.* 2010, *MNRAS*, 408, 2115.
Neistein, E. & Dekel, A. 2008, *MNRAS*, 388, 1792.
Noeske, K. G. *et al.* 2007, *ApJ*, 660L, 43.
Pannella, M., *et al.* 2009, *ApJ*, 698, L116.
Peeples, M. S. & Shankar, F., 2011, *MNRAS*, 417, 2962.
Peng, Y.-J., *et al.* 2010, *ApJ*, 721, 193 (P10)
Peng, Y.-J. & *et al.* , 2012, *ApJ*, 757, 4 (P12)
Perez-Gonzalez, P. G. & *et al.* , 2008, *ApJ* 675, 243.
Pozzetti, L. & *et al.* , 2010, *A&A*, 523, 13.
Rodighiero, G. & *et al.* , 2011, *ApJ*, 739, 40.
Sargent, M., Bethermin, M., Daddi, E., & Elbaz, D., 2012, *ApJ*, 747, 31.
Schaerer, D., de Barros, S., & Sklias, P., 2012, *arXiv:1207.3074*.
Silverman, J. D. & *et al.* , 2009, *ApJ*, 743, 2.
Stark, D. P., *et al.* 2012, *arXiv:1208.3529*.
Tacconi, L., *et al.* 2012, *arXiv:1211.5743*.
Tremonti, C. A. *et al.* 2004, *ApJ*, 613, 898.
Woo, J., *et al.* 2012, *MNRAS*, 428, 3306.
Yates, R. A. M., Kauffmann, G., & Guo, Q. 2011, *MNRAS*, 422, 215.

The intriguing life of massive galaxies
Proceedings IAU Symposium No. 295, 2012
D. Thomas, A. Pasquali & I. Ferreras, eds.

© International Astronomical Union 2013
doi:10.1017/S1743921313004547

The evolution of galaxy sizes

Bianca M. Poggianti[1], Rosa Calvi[2], Daniele Bindoni[2], Mauro D'Onofrio[2], Alessia Moretti[2], Tiziano Valentinuzzi, Giovanni Fasano[1], Jacopo Fritz[3], Gabriella De Lucia[4], Benedetta Vulcani[2], Daniela Bettoni[1], Marco Gullieuszik[1] and Alessandro Omizzolo[1]

[1] INAF-Astronomical Observatory of Padova, vicolo dell'Osservatorio 5, 35122 Padova, Italy
email: bianca.poggianti@oapd.inaf.it
[2] Department of Astronomy, University of Padova
[3] Sterrenkundig Observatorium, University of Gent, Belgium
[4] INAF-Astronomical Observatory of Trieste

Abstract. We present a study of galaxy sizes in the local Universe as a function of galaxy environment, comparing clusters and the general field. Galaxies with radii and masses comparable to high-z massive and compact galaxies represent 4.4% of all galaxies more massive than $3 \times 10^{10} M_\odot$ in the field. Such galaxies are 3 times more frequent in clusters than in the field. Most of them are early-type galaxies with intermediate to old stellar populations. There is a trend of smaller radii for older luminosity-weighted ages at fixed galaxy mass. We show the relation between size and luminosity-weighted age for galaxies of different stellar masses and in different environments. We compare with high-z data to quantify the evolution of galaxy sizes. We find that, once the progenitor bias due to the relation between galaxy size and stellar age is removed, the average amount of size evolution of individual galaxies between high- and low-z is mild, of the order of a factor 1.6.

Keywords. galaxies: evolution — galaxies: structure — galaxies: fundamental parameters

1. Introduction

Many recent observational studies have found a population of high-z ($z = 1 - 2.5$), massive ($M_\star > 10^{10} M_\odot$), compact ($R_e < 2\,\mathrm{kpc}$) galaxies (e.g. Daddi *et al.* 2005, Trujillo *et al.* 2006, Cimatti *et al.* 2008, van Dokkum *et al.* 2008, Saracco *et al.* 2009, Cassata *et al.* 2011 and others). In the great majority of these studies, galaxies are selected to be already passively evolving (devoid of ongoing star formation) at the redshift they are observed. Passive galaxies at high-z have been found to display a wide range in sizes, from extremely compact to those whose sizes are comparable to normal galaxies in the local Universe (Saracco *et al.* 2009, Mancini *et al.* 2010, Cassata *et al.* 2011).

Most high-z works use as local comparison the r-band Sloan median mass-size relation for galaxies with Sersic index $n \geqslant 2.5$ from Shen *et al.* (2003). Trujillo *et al.* (2009) and Taylor *et al.* (2010) searched for massive compact galaxies in Sloan and both found very few such galaxies in the general field at low redshift.

Here we present a search for massive and compact galaxies in different environments at low redshift using two galaxy samples: for the general field we use the Padova Millennium Galaxy and Group Catalogue at $z = 0.04 - 0.1$ (PM2GC, Calvi *et al.* 2011), and for galaxy clusters the WIde-field Nearby Galaxy cluster Survey at $z = 0.04 - 0.07$ (WINGS, Fasano *et al.* 2006).

"Superdense" galaxies, with galaxy stellar masses and effective sizes similar to those of high-z galaxies ($\Sigma_{50} = (0.5 * M_\star)/(\pi * R_e^{2}) > 3 \times 10^9 M_\odot \mathrm{kpc}^{-2}$), were found to represent $\sim 20\%$ of all galaxies with $3 \times 10^{10} < M_\star < 4 \times 10^{11} M_\odot$ in WINGS clusters based on

V-band images (Valentinuzzi *et al.* 2010, hereafter V10). Given the paucity of similar galaxies in the field studies mentioned above, this suggested a strong environmental dependence in the frequency of massive and compact galaxies at low-z (V10).

V10, and a number of other works (Saracco *et al.* 2011, Cassata *et al.* 2011, Cappellari *et al.* 2013 and these proceedings, to name a few) also found that galaxy sizes depend on the age of stars: galaxies with older luminosity-weighted ages (hereafter, LW ages) are smaller. As a consequence, selecting the oldest galaxies at high-z (those that have already stopped forming stars at $z = 1 - 2.5$) means selecting the most compact ones. At lower redshifts, larger galaxies stop forming stars and enter the passive galaxy samples, causing the median mass-size relation to move to larger radii at fixed mass. Hence, in order to measure the correct amount of size evolution, the sizes of high-z galaxies should be compared only with those of galaxies whose LW age is so old that they were already passive at high-z (V10).

Moreover, in clusters at $z = 0.4 - 0.8$, superdense galaxies are 40% of the galaxy population, and the mass-size relation in clusters evolves very little between $z \sim 0.6$ and $z = 0$ (Valentinuzzi *et al.* 2010b).

2. Results in the field

Recently (Poggianti *et al.* 2012, submitted), we have searched for massive compact galaxies in the field using the PM2GC, a sample of low-z galaxies representative of the general field population drawn from the Millennium Galaxy Catalogue (MGC, Liske *et al.* 2003, Driver *et al.* 2005). We measured galaxy radii and galaxy morphologies with GASPHOT (Pignatelli & Fasano 2006) and MORPHOT (Fasano *et al.* 2012), respectively, from the MGC B-band images. Size and stellar mass estimates are in very good agreement with other existing measurements.

After inspecting each superdense candidate, we found 44 superdense galaxies out of 995 galaxies in the PM2GC, corresponding to a fraction of 4.4%±0.7 among galaxies with $M_\star > 3 \times 10^{10} M_\odot$. The corresponding fraction in clusters from B-band WINGS images is 11.8%±1.7 (about half of the V-band percentage found by V10). This is due to the small but systematic change of galaxy radii with wavelength, with radii increasing using bluer passbands.

Most of the PM2GC superdense galaxies are red, early-type galaxies (70% S0s and 23% ellipticals), quite flattened, with a mean Sersic index $n = 2.8$, and have intermediate-to-old stellar populations, with a mean LW age =5.5 Gyr and a mean mass-weighted age =9.3 Gyr.

Figure 1 illustrates the relation between size and LW age for galaxies in different mass bins and different environments (for each galaxy mass, the lower line is the PM2GC, and the upper line is WINGS). At given size and environment, more massive galaxies have older ages, both in clusters and in the field, following downsizing. The plot shows that, at given mass and environment, galaxies with smaller radii (more compact) are older. This effect is present both in clusters and in the field, although our analysis (not shown) finds that the dependence of the median mass-size relation on stellar age is stronger in clusters than in the field.

As a consequence, the population of passive galaxies at low-z includes both the old, on average small, galaxies and those galaxies that have stopped forming stars at later times that have on average larger sizes. This needs to be taken into account when trying to quantify the evolution of the high-z samples.

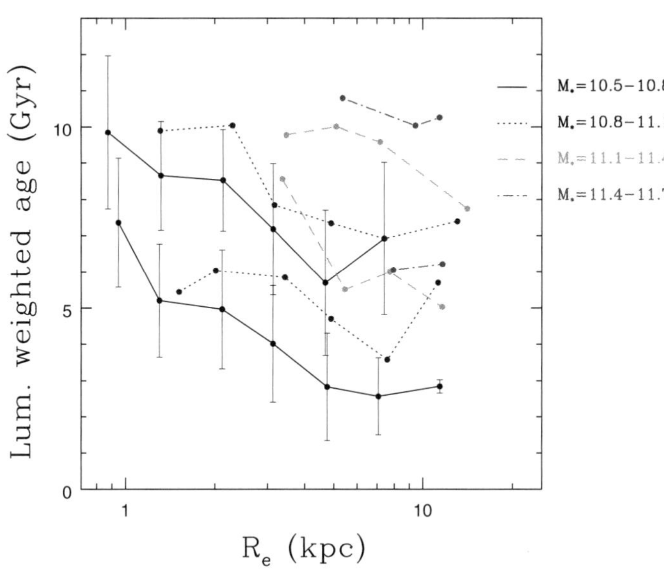

Figure 1. Median LW age of galaxies as a function of effective radius for four bins of galaxy masses (see legend). For each mass bin, the lowest line refers to the field and the upper line to clusters.

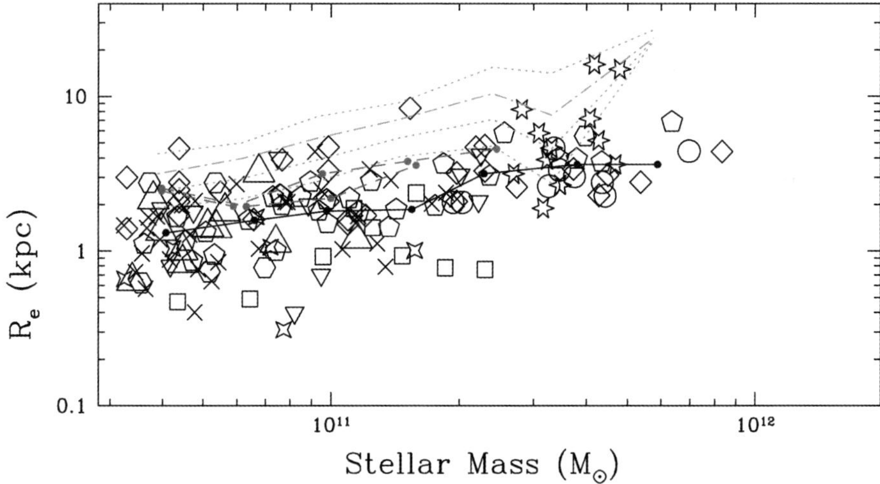

Figure 2. Median mass-size relation (solid line) of high-z galaxies (large symbols) compared to the relation of PM2GC and WINGS galaxies with LW ages $\geq$ 10 Gyr (dotted) and of all PM2GC galaxies (the dashed-dotted line is the median, dotted lines are 1 and 2σ).

2.1. *Size evolution: comparison with high-z studies*

Figure 2 shows a compilation of various datasets of galaxies at $z \sim 1 - 2.5$ (different symbols, Daddi *et al.* 2005, Trujillo *et al.* 2006, van Dokkum et al. 2008, Cimatti *et al.* 2008, van der Wel *et al.* 2008, Saracco *et al.* 2009, Damjanov *et al.* 2009, Mancini *et al.* 2010, Cassata *et al.* 2011) and their median mass-size relation (solid line).

To quantify the evolution in size, we compare this relation with the median mass-size relation of *old* (LW age $\geq$ 10 Gyr) galaxies in WINGS and the PM2GC (two dotted lines in Fig. 2). We find an average evolution of 0.2-0.25 dex, a factor 1.6-1.8, using WINGS and PM2GC, respectively. This is half of what is found using all, passive or $n \geqslant 2.5$ galaxies in the PM2GC, as usually done in the literature. The evolution of galaxy sizes is therefore much smaller than would be (erroneously) inferred at face value considering as local descendants the whole passive galaxy population regardless of age.

Using old WINGS or old PM2GC galaxies does not significantly change the results. It is however important to stress that our simulations (Millennium Simulation + semi-analytic model presented in De Lucia & Blaizot 2007) predict that 60% of all galaxies that at $z \sim 2$ are massive ($\geqslant 10^{11} M_\odot$) and passive end up in haloes with masses above $10^{14} M_\odot$ by $z = 0$, therefore in WINGS-like clusters in the local Universe. Hence, a large fraction of the high-z passive and massive galaxies have evolved into today's cluster galaxies. This may explain the strong environmental dependence of the superdense incidence we find at low-z, and indicates that galaxies in nearby clusters are the most appropriate local counterparts to the high-z studies.

References

Calvi, R., Poggianti, B. M., & Vulcani, B. 2011, *MNRAS*, 416, 727
Cappellari, M., *et al.* 2013, *MNRAS*, 432, 1862
Cassata, P., *et al.* 2011, *ApJ*, 743, 96
Cimatti, A., *et al.* 2008, *A&A*, 482, 21
Daddi, E., *et al.* 2005, *ApJ*, 626, 680
Damjanov, I., *et al.* 2009, *ApJ*, 695, 101
De Lucia, G. & Blaizot, J. 2007, *MNRAS*, 375, 2
Driver, S. P., Liske, J., Cross, N. J. G., De Propris, R., & Allen, P. D. 2005, *MNRAS*, 360, 81
Fasano, G., *et al.* 2006, *A&A*, 445, 805
Fasano, G., *et al.* 2012, *MNRAS*, 420, 926
Liske, J., Lemon, D. J., Driver, S. P., Cross, N. J. G., & Couch, W. J. 2003, *MNRAS*, 344, 307
Mancini, C., *et al.* 2010, *MNRAS*, 401, 933
Saracco, P., Longhetti, M., & Andreon, S. 2009, *MNRAS*, 392, 718
Saracco, P., Longhetti, M., & Gargiulo, A. 2011, *MNRAS*, 412, 2707
Shen, S., *et al.* 2003, *MNRAS*, 343, 978
Taylor, E. N., Franx, M., Glazebrook, K., Brinchmann, J., van der Wel, A., & van Dokkum, P. G. 2010, *ApJ*, 720, 723
Valentinuzzi, T., *et al.* 2010, *ApJ*, 712, 226
Trujillo, I., *et al.* 2006, *MNRAS*, 373, L36
Valentinuzzi, T., *et al.* 2010, *ApJ*, 712, 226
Valentinuzzi, T., *et al.* 2010b, *ApJL*, 721, L19
van der Wel, A., *et al.* 2008, *ApJ*, 688, 48
van Dokkum, P., *et al.* 2008, *ApJL*, 677, L5

The intriguing life of massive galaxies
Proceedings IAU Symposium No. 295, 2012
D. Thomas, A. Pasquali & I. Ferreras, eds.

© International Astronomical Union 2013
doi:10.1017/S1743921313004559

The GAMA Panchromatic Survey

Simon P. Driver[1,2]

[1]International Centre for Radio Astronomy Research, University of Western Australia,
Crawley, WA 6009, Australia,
[2]Scottish Universities Physics Alliance (SUPA), School of Physics and Astronomy, University
of St Andrews, St Andrews, Scotland

Abstract. The Galaxy And Mass Assembly Survey (GAMA) has now been operating for almost
5 years gathering spectroscopic redshifts for five regions of sky spanning 300 sq degrees in total
to a depth of $r < 19.8$ mag. The survey has amassed over 225,000 redshifts making it the
third largest redshift campaign after the SDSS and BOSS surveys. The survey has two novel
features that set it apart: (1) complete and uniform sampling to a fixed flux limit ($r < 19.8$ mag)
regardless of galaxy clustering due to multiple-visits to each sky region, enabling the construction
of high-fidelity catalogues of groups and pairs, (2) co-ordination with diverse imaging campaigns
which together sample an extremely broad range along the electro-magnetic spectrum from
the UV (GALEX) through optical (VST KIDs), near-IR (VISTA VIKING), mid-IR (WISE),
far-IR (Herschel-Atlas), 1m (GMRT), and eventually 20cm continuum and rest-frame 21cm
line measurements (ASKAP DINGO). Apart from the ASKAP campaign all multi-wavelength
programmes are either complete or in the final stages of observations and the UV-far-IR data
are expected to be fully merged by the end of 2013. This article provides a brief flavour of
the coming panchromatic database which will eventually include measurements or upper-limits
across 27 wavebands for 380,000 galaxies. GAMA DR2 is scheduled for the end of January 2013.

1. Introduction

The Galaxy And Mass Assembly Survey (GAMA; Driver *et al.* 2009, Driver *et al.* 2011)
will contain spectroscopic redshifts at $> 98\%$ completeness for 380,000 galaxies drawn
from five distinct sky regions (2^h, 9^h, 12^h, 15^h, and 23^h). Details of the survey concept,
input catalogue, tiling strategy and first data release are provided in Driver *et al.* (2009),
Baldry *et al.* (2010), Robotham *et al.* (2010), and Driver *et al.* (2011) respectively. The
construction of the first group catalogue is given by Robotham *et al.* (2011) and the
measurement of robust stellar masses and Sérsic profiles by Taylor *et al.* (2011) and
Kelvin *et al.* (2012) respectively. Current early science includes the measurement of the
nearby galaxy luminosity functions from $UV - K$ Driver *et al.* (2012a) and their evolution
in *ugriz* to $z < 0.5$ by Loveday *et al.* (2012). The nearby galaxy stellar mass function,
which exhibits a distinctive upturn at intermediate masses, is given by Baldry *et al.* (2012)
and upcoming papers by Taylor *et al.* (2013), Lopez-Cruz *et al.* (2013), Gunawardhana
et al. (2013) and Robotham *et al.* (2013) explore galaxy colour bimodality, the evolution
of the mass-metallicity relation, the evolution of the Hα luminosity function, and the life
and times of L* galaxies. Numerous (> 100) other papers are currently in progress. Fig. 1
shows the end-of-year 2012 update to the survey of surveys table given in Driver (2012b)
and provides some indication of GAMAs emerging status amongst established leading
surveys from the past three decades. Modern surveys like GAMA, BOSS and VIPERS
are all expected to rise rapidly up this league table over the coming years and updates
will be reported annually.

Table 1. Major extragalactic optical/near-IR surveys and their impact in terms of refereed papers and citations

Survey	Papers	Citations
Sloan Digital Sky Survey (SDSS)	3803	163288
Hubble Deep Field (HDF)	594	49360
Two Micron All Sky Survey (2MASS)	1076	32506
Great Observatories Origins Deep Survey (GOODS)	567	30331
Two degree Field Galaxy Redshift Survey (2dFGRS)	206	25301
Centre for Astrophysics (CfA)	342	19883
Galaxy Evolution Explorer (GALEX)	563	15532
Cosmic Evolution Survey (COSMOS)	394	13060
Ultra Deep Field (UDF)	202	9169
Canada France Redshift Survey (CFRS)	90	8797
Deep Evolutionary Exploratory Probe (DEEP/DEEP2)	154	7767
Classifying Objects by Medium-Band Observations (COMBO-17/COMBO17)	89	6882
UKIRT Infrared Deep Sky Survey (UKIDSS)	135	4661
VIMOS VLT Deep Survey (VVDS)	83	4531
Las Campanas Redshift Survey (LCRS)	96	4197
Southern Sky Redshift Survey (SSRS/SSRS2)	89	4151
APM Galaxy Survey (APMGS)	57	4109
Point Source Catalogue Survey (PSCz)	95	3553
2dF QSO redshift Survey (2QZ)	57	3305
Canadian Network for Observational Cosmology Field Galaxy Redshift Survey (CNOC2)	65	3169
CFHT Legacy Survey (CFHTLS)	87	3151
Galaxy Evolution from Morphologies and SEDs (GEMs)	68	3109
NOAO Deep Wide Field Survey (NDFWS)	76	2772
zCOSMOS	60	2218
Millennium Galaxy Catalogue (MGC)	38	1854
The 2dF-SDSS LRG And QSO Survey (2SLAQ)	36	1636
AGN and Galaxy Evolution Survey (AGES)	34	1575
Gemini Deep Deep Survey (GDDS)	20	1479
Stromlo-APM Redshift Survey (SARS)	19	1402
Baryonic Acoustic Oscillations Survey (BOSS)	45	980
ESO Slice Project (ESP)	24	971
The 6dF Galaxy Survey (6dFGS)	26	749
WiggleZ	26	646
Galaxy And Mass Assembly (GAMA)	33	468
2MASS Redshift Survey (2MRS)	19	413
VIMOS Public Extragalactic Redshift Survey (VIPERS)	3	3

Note these numbers were determined on 28th Dec 2012 using the SOA/NASA ADS Astronomy Query Form by searching for refereed papers which contained abstract keywords based on the following boolean logic: (galaxy or galaxies) and ("<long survey name>" or <short survey name>). The table is purely indicative and not weighted by survey age or effective cost and is updated on an annual basis. My apologies in advance for the many surveys not included.

2. The GAMA panchromatic campaign

As stated earlier a number of major imaging surveys are underway on a variety of internationally leading facilities which include within their survey footprints the GAMA sky regions. These complimentary imaging surveys are a mixture of GAMA motivated/led programs (e.g., GALEX, GMRT), those in which GAMA team members are playing a leading role (e.g., VISTA VIKING, ASKAP DINGO), to those in which GAMA is peripheral or playing a supporting role (e.g., VST KIDs, WISE, Herschel-Atlas). Regardless of the motivation these data collectively form the GAMA panchromatic campaign and need to be assembled, astrometrically matched, and for photometric measurements to be conducted which manages highly variable flux limits and spatial resolutions. Typically only 10-20% of the GAMA galaxies have clear unambiguous counterparts in all bands and spatial resolutions which vary with wavelength from 0.5″ to 30″. The significance of managing these two issues (flux limits and spatial resolutions), cannot be overstated. The correct strategy to manage this process is not yet entirely clear, the wrong strategy

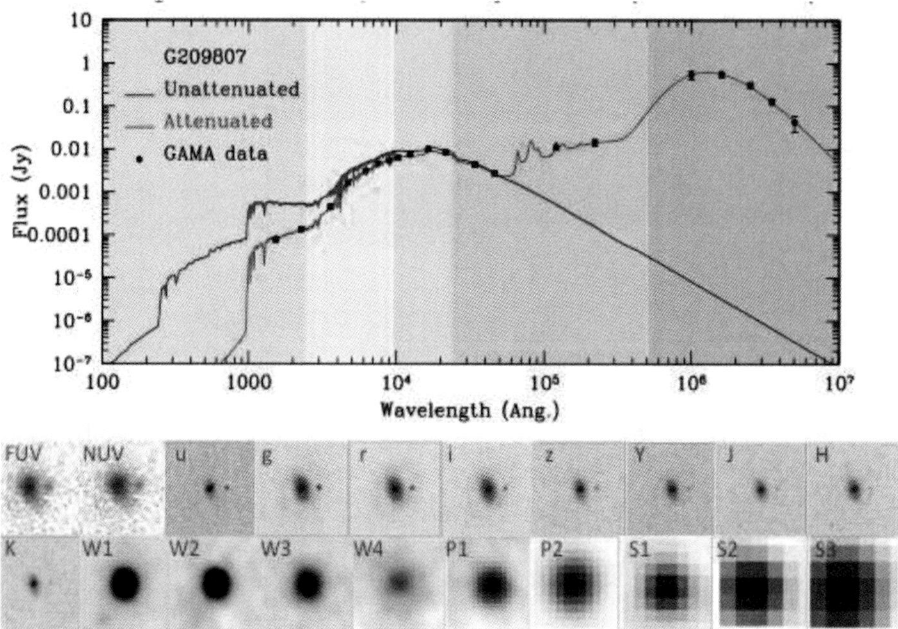

Figure 1. (*upper plot*) MAGPHYS modelling of a single GAMA galaxy detected in GALEX, SDSS, UKIDSS, WISE, HERSCHEL PACS and HERSCHEL SPIRE. (*lower panels*) The postage stamps images are shown in the lower panels highlighting the varied spatial resolutions of the contributing surveys.

however is. To simply conduct independent source finding in each band and then attempt to table match the resulting catalogues leads to a highly biased (ill defined) sample with high levels of blended contamination in the poorer resolution bands. The problem is particularly acute when matching data from the optical to the far-IR where the incidence of high-z interlopers and lensed sources becomes a significant problem. Fig. 1 highlights some of these issues by showing the assembled data for a single object which is clearly detected and unambiguous but also highlights the varying spatial resolution across the wavebands sampled. The red curve on Fig. 1 shows the simple best-SED fit to these data from the MAGPHYS energy balance software of da Cunha, Charlot & Elbaz (2008). The blue curve shows the actual energy production within this galaxy. In particular it is worth noting the significant change in the slope of the SED in the blue optical region where the observed colour is often erroneously used to divide the galaxy population into *intrinsically* red and blue populations. Indeed this object is observationally red with a high-Sérsic index yet also dusty and star-forming (i.e., *intrinsically* blue). This system alone highlights that galaxies do not simply come in two flavours but are a medley of intermingled parameters with all possibilities existing in colour/star-formation/profile-shape/opacity/halo/stellar mass parameter space. In order to complete this analysis not only are broad spectrum measurements vital but in the case of non-detections robust upper limits with meaningful errors. This demands the definition of the sample in a single band from which measurements/upper-limits are garnered across the full electromagnetic range. A key question is which band to start from, possibly the most logical is that most closely aligned to stellar mass, i.e., H or K band.

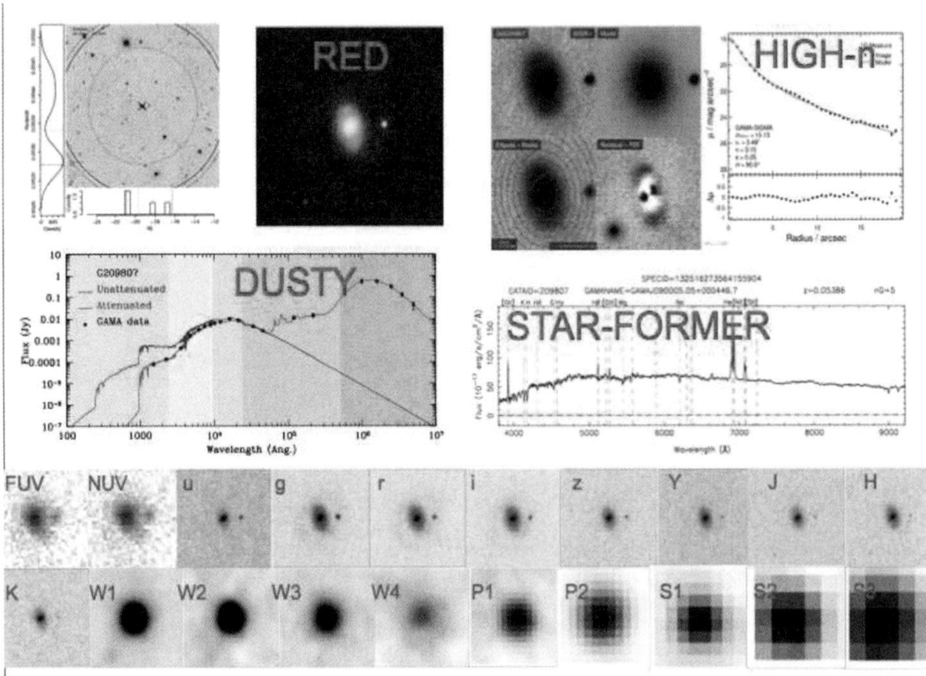

Figure 2. The information available for our 380,000 galaxies. Which includes (clockwise from top-right): halo and group information, colours, 2D profiles (single Sérsic and bulge-disc decompositions), spectroscopic line analysis, panchromatic imaging, and UV-farIR SED modelling.

3. Summary

Unlocking the interdependency of galaxy properties as a function of environment and redshift requires a comprehensive panchromatic survey such as GAMA. Fig. 2 shows the type of data we are now assembling for 380,000 galaxies which will enable us to move beyond simple two-dimensional descriptions of the galaxy population and allow us to explore the rich interdependencies of the many properties which we will measure. Extensions to fainter flux-limits, higher spatial resolutions, and higher redshifts are underway.

References

Baldry I. K., *et al.*, 2010, *MNRAS*, 404, 86
Baldry I. K., *et al.*, 2012, *MNRAS*, 421, 621
da Cunha E., Charlot S., & Elbaz, D., 2008, *MNRAS*, 388, 1595
Driver S. P., *et al.*, 2009, *A&G*, 50, 12
Driver S. P., *et al.*, 2011, *MNRAS*, 413, 971
Driver S. P., *et al.*, 2012a, *MNRAS*, 427, 3244
Driver S. P., *et al.*, 2012b, *Panchromatic properties of galaxies in wide-field optical spectroscopic and photometric surveys*, IAU Symposium 284, p268-278
Gunawardhana M., *et al.*, 2013, in press
Kelvin L. S., *et al.*, 2012, *MNRAS*, 421, 1007
Lopez-Cruz M., *et al.*, 2013, in press
Loveday J., *et al.*, 2012, *MNRAS*, 420, 1239
Robotham A. S. G., *et al.*, 2010, *PASA*, 27. 76
Robotham A. S. G., *et al.*, 2011, *MNRAS*, 416, 2640
Robotham A. S. G., *et al.*, 2011, *MNRAS*, in press
Taylor E., *et al.*, 2011, *MNRAS*, 418, 1587
Taylor E., *et al.*, 2013, *MNRAS*, in press

The intriguing life of massive galaxies
Proceedings IAU Symposium No. 295, 2012
D. Thomas, A. Pasquali & I. Ferreras, eds.

© International Astronomical Union 2013
doi:10.1017/S1743921313004560

GAMA: The effect of environment on galaxy emission line properties

Oliver Steele[1], Daniel Thomas[1], Claudia Maraston[1], James Etherington[1] and the GAMA collaboration

[1] Institute of Cosmology and Gravitation, University of Portsmouth, Dennis Sciama Building, Burnaby Road, Portsmouth, PO1 3FX, UK
email: `oliver.steele@port.ac.uk`

Abstract. We study the influence of environment on emission line properties using the Galaxy And Mass Assembly (GAMA) survey, taking care to disentangle the role of mass and environment. We look at the role of local density separating galaxies into classifications star forming, AGN, and SF/AGN composite using the BPT diagnostic diagram. We find that environment is generally less important as a driving factor than galaxy mass. The presence of emission lines, whether driven by star formation or central supermassive black hole activity mostly depends on galaxy mass consistently for all galaxy types.

Keywords. galaxies: active, galaxies: evolution, galaxies: general

1. Introduction

The impact of environment on galaxy evolution is a much debated and researched topic best visible through the well-known morphology-density relationship (e.g. Oemler 1974, Dressler 1980), through which there is a strong preference for comparable galaxy morphologies to appear in similar density environments. Studies into the importance of environment at a given morphology have proven indecisive, with some favouring environment as the major driving factor (e.g. Trager *et al.* 2000, Mendes de Oliveira *et al.* 2005) whilst others claim that mass is a more significant parameter (e.g. Thomas *et al.* 2010, Wake *et al.* 2005, Bundy *et al.* 2006).

Whilst some papers have investigated the impact of environment on star formation rates (SFR) (e.g. Peng *et al.* 2010) and generally found that it is negligible compared to that of mass, the impact on general emission line properties has not yet been studied in great detail.

In this paper we use the Galaxy And Mass Assembly (GAMA) survey to investigate the impact of environment on emission line properties.

2. Data

The GAMA survey is a multi-wavelength galaxy survey that upon completion will cover over 360 square degrees and consist of roughly 400,000 galaxies for $z \lesssim 0.4$ (Driver *et al.* 2011). It combines optical spectroscopy taken at the AAT with photometry taken at a variety of telescopes (CST, VISTA, Herschel, ASKAP and GALEX) ranging from radio to ultraviolet wavelengths. In Phase-I of GAMA, objects were spectroscopically observed up to an r-band magnitude limit of 19.4 for two of the GAMA fields, whilst the third field had a deeper 19.8 mag limit, all to a 98% completeness level. This depth makes it an excellent complement to SDSS as it probes down to lower stellar masses. A summary of the spectroscopic processing applied to the data can be found in Hopkins *et al.* (2013).

Our sample was created by combining $\sim 153,600$ objects spectroscopically observed by GAMA, with $\sim 15,000$ SDSS Data Release 7 objects that fall into the GAMA fields.

3. Analysis Tools

3.1. *Spectroscopic measurements*

We use the publicly available codes pPXF (Cappellari & Emsellem 2004) and GANDALF (Sarzi *et al.* 2006) to calculate stellar kinematics and derive emission line properties. GANDALF fits stellar population templates and Gaussian emission line templates to the observed spectra simultaneously, allowing for the separation of stellar continuum and absorption from ionised gas emission, whilst pPXF extracts stellar kinematics from the data in pixel space. Dust reddening is accounted for by adopting a Calzetti (2001) obscuration curve and applying it to the data. Using this method we are able to extract stellar kinematics, gas kinematics, emission line fluxes and emission line equivalent widths (EWs) from the data.

We adopt the stellar population templates from Maraston & Strömbäck (2011) based on the MILES stellar library (Sánchez Blázquez *et al.* 2006), extended in the blue for $\lambda < 3500$ Å using theoretical spectra from the model of Maraston *et al.* (2009) based on the theoretical library UVBLUE (Rodríguez-Merino *et al.* 2005).

Details of this analysis tool and its calibration with SDSS data can be found in Thomas *et al.* (2013). A thorough presentation of the emission line measurements in GAMA is provided by Hopkins *et al.* (2013).

3.2. *Environment*

We calculated the Adaptive Gaussian Environment (AGE) Density of all of the galaxies in the sample with a redshift below 0.18 to define a pseudo-volume-limited population. See Schawinski *et al.* (2007a) and Thomas *et al.* (2010) for details on how this method was performed. We measured the environment for $\sim 37,000$ spectra.

4. Emission line properties

The equivalent widths and amplitude-over-noise (AoN) values of the lines [OIII]5007, [NII]6583, Hβ and Hα are measured for all objects with redshift below ~ 0.45, representing 96.8% of all objects. These lines allow for the usage of the well known Baldwin, Phillips and Terlevich (BPT) diagnostic diagram introduced in Baldwin, Phillips & Terlevich (1981) and Veilleux & Osterbrock (1987), and widely used in galaxy studies. This method compares the line ratios [OIII]/Hβ and [NII]/Hα to determine the ionisation source in galaxies.

We use the empirical separation between star forming galaxies and AGN from Kauffmann *et al.* (2003) and the theoretical extreme starburst line from Kewley *et al.* (2001) to identify pure star forming and pure AGN emission. We adopt the common practise of assuming the area between these lines is populated by galaxies with composite star forming and AGN spectra. We further use the dividing line defined by Schawinski *et al.* (2007b) to distinguish between LINER and Seyfert emission based on SDSS galaxy classifications obtained through the [SII]/Hα ratio. We find that 18.4% of the total sample have all four BPT lines with AoN > 2, of which 54.2% are classified as star forming, 28.7% as composite, 9.5% as Seyfert and 6.5% as LINER.

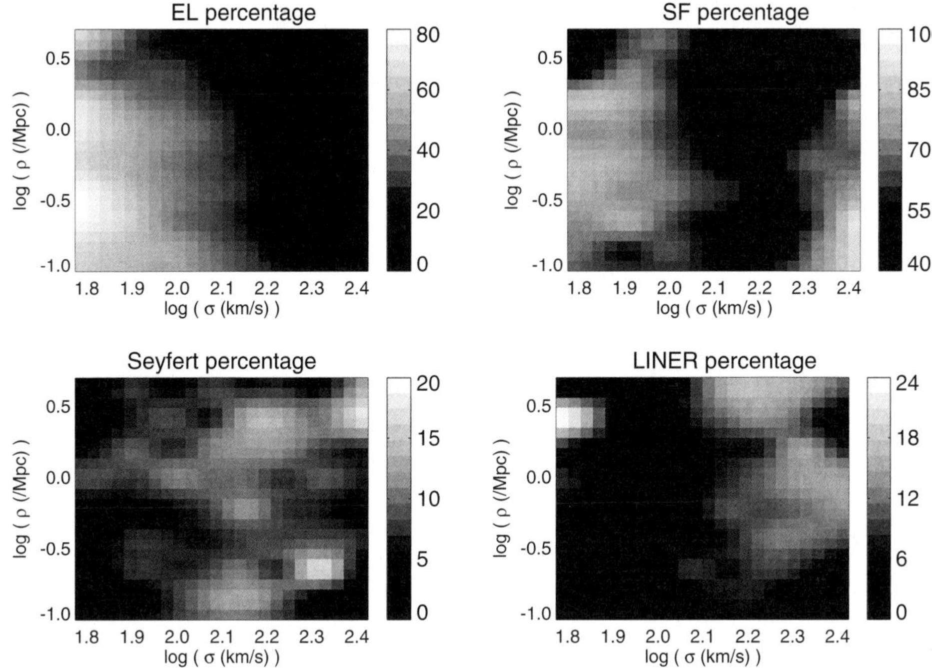

Figure 1. Percentage of emission line galaxies (top left panel) as a function of galaxy velocity dispersion and environmental density. The percent of star forming galaxies, Seyferts, and LINERs within the emission line population are shown by the top-left, bottom-left, and bottom-right panels, respectively. The high LINER fraction at very low sigma is an artefact caused by low number statistics, as is the high fraction of star forming galaxies at high sigma and low density.

For the purposes of this paper we consider galaxies with all four lines having a value of AoN > 2 as having emission lines, and all other galaxies as 'passive'.

5. Results & Discussion

After applying cuts requiring that all objects investigated have 65 km s^{-1} < σ < 400 km s^{-1}, $d\sigma/\sigma$ < 0.3 and 0 Mpc^{-1} < ρ < 50 Mpc^{-1}, we were left with 19,000 objects, of which 26% had emission lines.

Fig. 1 shows the percentage of emission line galaxies out of the whole sample, and star forming galaxies, composite galaxies, Seyferts and LINERs out of the emission line sample as functions of stellar velocity dispersion and local density. There are a few artefacts to note that were caused by low number statistics, specifically the high LINER percentage bubble at low sigma and the high star forming percentage at the high-sigma low-density corner.

We find that emission lines have a far stronger dependence on velocity dispersion (a proxy for galaxy mass) than environmental density. The top-left panel clearly shows that the emission line fraction decreases with increasing mass and only depends mildly on environment. There is a second order effect of slightly increasing emission line fraction with decreasing density at a given velocity dispersion.

Regarding specific BPT types, in general we find that none of the various emission line classes have a strong dependence on density, with differences being driven primarily by mass. Most of our sample consists of star forming galaxies at low velocity dispersion and across a wide range of densities. The fraction of star forming galaxies shows a similar pattern. The top-right panel shows that the fraction of star forming galaxies is highest at low galaxy masses and, again, only depends mildly on galaxy environment. The secondary fingerprint from the environment can be seen from a minimum of star forming galaxy fraction at high galaxy masses and environmental densities.

Finally, the bottom-right panel shows that the fraction of galaxies with LINER-like emission line properties again mostly depends on mass with the largest fraction of LINERs being found in massive galaxies, a trend largely independent of galaxy environment.

These results are in agreement with stellar population studies (e.g., Thomas *et al.* 2010), and show that galaxy mass is the primary driving factor of galaxy properties, with environmental density playing a secondary role.

These findings are inkeeping with current theory and observations, as late-type galaxies are more likely to have emission lines, are less massive and are found in less dense environments than their early-type counterparts, leading to a morphologically driven explanation for these results. A detailed analysis based on GAMA DR2 will be presented in a forthcoming paper (Steele *et al.*, in preparation).

Acknowledgments

The Science, Technology and Facilities Council is acknowledged for their financial support. Numerical computations were done on the Sciama High Performance Computer cluster which is supported by the ICG, SEPNet and the University of Portsmouth.

References

Baldwin J. A., Phillips M. M., & Terlevich R. 1981, *PASP*, 93, 5
Bundy K. *et al.* 2006, *ApJ*, 651, 120
Calzetti, D. 2001, *PASP*, 113, 1449
Cappellari M. & Emsellem E. 2004, *PASP*, 116, 138
Dressler, A. 2000, *ApJ*, 236, 351-365
Driver S. P. *et al.* 2011, *MNRAS*, 413, 971
Hopkins A. M. *et al.* 2013, *MNRAS*, 430, 2047
Kauffman G., *et al.* 2003, *MNRAS*, 346, 1055
Kewley L. J., Dopita M. A., Sutherland R. S., Heisler C. A., & Trevena J. 2001, *ApJ*, 556, 121
Maraston C., Nieves Colmenárez L., Bender R., & Thomas D. 2009, *A&A*, 493, 425
Maraston C. & Strömbäck G. 2011, *MNRAS*, 418, 2785
Mendes de Oliviera C., Coelho P., González J. J., & Barbuy B. 2005, *AJ*, 130, 55
Oemler, A., Jr. 1974, *ApJ*, 194, 1-20
Peng Y.-j. *et al.* 2010, *ApJ*, 721, 193
Rodríguez-Merino L. H., Chavez M., Bertone E., & Buzzoni A. 2005, *ApJ*, 626, 411
Sánchez Blázquez P., Gorgas J., Cardiel N., & González J. J. 2006, *A&A*, 457, 809
Sarzi M. *et al.* 2006, *MNRAS*, 366, 1151
Schawinski K. *et al.* 2007, *ApJS*, 173, 512
Schawinski K., Thomas D., Sarzi M., Maraston C., Kaviraj S., Joo S.-J, Yi S. K., & Silk J. 2007, *MNRAS*, 382, 1415
Thomas D., Maraston C., Schawinski K., Sarzi M., & Silk J. 2010, *MNRAS*, 404, 1775
Thomas D., Steele O., & Maraston C., *et al.* 2013, *MNRAS*, 431, 1383
Trager S. C., Faber S. M., Worthey G., & González J. J. 2000, *AJ*, 120, 165
Veilleux S. & Osterbrock D. E. 1987, *ApJS*, 63, 295
Wake D., Collins C. A., Nichol R. C., Jones L. R., & Burke D. J. 2005, *ApJ*, 627, 186

The intriguing life of massive galaxies
Proceedings IAU Symposium No. 295, 2012
D. Thomas, A. Pasquali & I. Ferreras, eds.

© International Astronomical Union 2013
doi:10.1017/S1743921313004572

Quenching star formation at intermediate redshifts: downsizing of the mass flux density in the green valley

Thiago S. Gonçalves[1], D. Christopher Martin[2], Karín Menéndez-Delmestre[1], Ted K. Wyder[2], and Anton Koekemoer[3]

[1] Observatório do Valongo, Universidade Federal do Rio de Janeiro,
Ladeira Pedro Antonio, 43, Saúde, Rio de Janeiro–RJ, CEP 22240-060
email: `tsg@astro.ufrj.br`

[2] California Institute of Technology, [3] Space Telescope Science Institute

Abstract. The bimodality in galaxy properties has been observed at low and high redshift, with a clear distinction between star-forming galaxies in the blue cloud and passively evolving objects in the red sequence; the absence of galaxies with intermediate properties indicates that the quenching of star formation and subsequent transition between populations must happen rapidly. In this work, we present a study of over 100 transiting galaxies in the so-called "green valley" at intermediate redshifts ($z \sim 0.8$). By using very deep spectroscopy with the DEIMOS instrument at the Keck telescope, we are able to infer the star formation histories of these objects and measure the stellar mass flux density transiting from the blue cloud to the red sequence when the Universe was half its current age. Our results indicate that the process happened more rapidly, affecting more massive galaxies in the past, suggesting a top-down scenario whereby the massive end of the red sequence assembles first. This represents another aspect of downsizing, with the mass flux density moving towards smaller galaxies in recent times.

Keywords. galaxies: evolution, galaxies: stellar content, galaxies: high-redshift

1. Introduction

The distinction between blue star-forming galaxies and red, passively evolving ones has been known for a long time. However, the existing bimodality in the galaxy distribution, with a clear distinction between blue spirals and red spheroids, and its establishment still remain a puzzle. This bimodality in galaxy colors has been observed at low redshifts ($z \sim 0.1$; Wyder *et al.* 2007), but also extends to earlier times, at least out to $z \sim 1.0$ (Willmer *et al.* 2006). If the color distribution of galaxies can be described as two distinctive peaks, then one can denominate the minimum at intermediate color values the "green valley". The question then arises: why does such a minimum exist, instead of a homogeneous distribution across the color-magnitude diagram?

The low number density (or number deficit) of galaxies in the green valley indicates that the transition between both groups occurs rapidly. To infer how rapidly galaxies are moving across the green valley, Martin *et al.* (2007) have used spectroscopic features in green valley galaxies to obtain information on their star formation histories. Along with measurements of typical galaxy masses and number densities in the color-magnitude diagram, the authors have been able to determine the mass flux across the green valley at low redshifts ($z \sim 0.1$). We apply here the methods introduced in that work to a sample of galaxies at higher redshift. This will allow us to measure the mass flux in the green valley at intermediate redshifts, and to compare it with results found for galaxies in the low-redshift Universe.

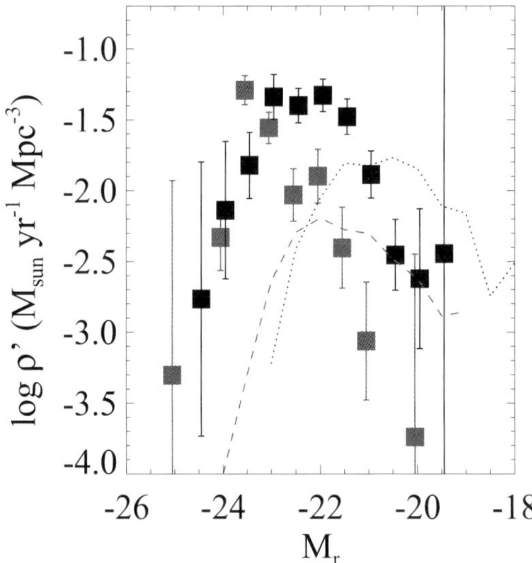

Figure 3. Mass flux density as a function of r magnitude. The flux at intermediate redshifts is shown as solid symbols; red squares for the extinction-corrected sample, black squares, no extinction correction. The lines represents the values found for the local Universe; the red dashed line indicates the extinction corrected sample at $z = 0.1$. In both cases, the peak at high redshift is shifted towards brighter magnitudes by $\gtrsim 1$ mag.

the red sequence forming "from the top down": in the past, more massive star-forming galaxies were being formed and subsequently quenched, forming the more massive red sequence galaxies. At later times, star formation shifts to less massive galaxies; these are then quenched as well, and the fainter end of the red sequence is created.

Since we have calculated the quenching timescales of individual objects, the next logical step is to compare our results with other observable properties in the green valley galaxies, in particular merger signatures and AGN activity. We expect this analysis will help clarify the role of each of these processes in the total quenching of star formation, both at low and at high redshifts.

References

Bruzual, G. & Charlot, S. 2003, *MNRAS*, 344, 1000

Faber, S. M., Willmer, C. N. A., Wolf, C., *et al.* 2007, *ApJ*, 665, 265

Gonçalves, T. S., *et al.* 2012, *ApJ*, 759, 67

Martin, D. C., Wyder, T. K., Schiminovich, D. *et al.* 2007, *ApJS*, 173, 342

Newman, J. A., Cooper, M. C., Davis, M., *et al.* 2012, *ApJS*, submitted (arXiv:1203.3192)

Willmer, C. N. A., Faber, S. M., Koo, D. C., *et al.* 2006, *ApJ*, 647, 853

Wyder, T. K., Martin, D. C., Schiminovich, D., *et al.* 2007, *ApJS*, 173, 293

The intriguing life of massive galaxies
Proceedings IAU Symposium No. 295, 2012
D. Thomas, A. Pasquali & I. Ferreras, eds.

© International Astronomical Union 2013
doi:10.1017/S1743921313004584

The Intriguing Life of Massive Galaxies: The Connections between α_s, β and Merging

Ying-jie Peng[1], **Simon J. Lilly**[1], **Alvio Renzini**[2] **and Marcella Carollo**[1]

[1]Institute of Astronomy, ETH Zurich, 8093 Zurich, Switzerland

[2]INAF - Osservatorio Astronomico di Padova, Vicolo dell'Osservatorio 5, 35122, Padova, Italy

Abstract. In this proceeding paper, we discuss the important underlying connections between the faint end slope α_s of the stellar mass function of star-forming galaxies, the logarithmic slope β of the sSFR-mass relation and merging through the continuity approach, as we introduced in Peng *et al.* (2010, hereafter P10) and (2012, hereafter P12).

Keywords. Galaxy Evolution, Mass Function, Star Formation

1. Introduction

The galaxy population appears to be composed of infinitely complex different types and properties at first sight. However, when large samples of galaxies are studied, it appears that the majority of galaxies just follow simple scaling relations while the outliers represent some minority. We demonstrate the astonishing underlying simplicities of the galaxy population emerged from large surveys and take a new approach to the topic of galaxy evolution and derive the analytical forms for the dominant evolutionary processes that control the galaxy evolution. This model successfully explained the observed evolution of the galaxy stellar mass functions (GSMF) and the origin of the Schechter form of the GSMF. All of these detailed quantitative relationships are indeed seen, to very high precision, in SDSS, leading strong support to our simple empirically-based model. This model also offers natural explanations for the "anti-hierarchical" age-mass relation and the α-enrichment patterns for passive galaxies and makes many other testable predictions. Although still purely phenomenological, the model makes clear what the evolutionary characteristics of the relevant physical processes must in fact be. This model thus offers a new powerful analytical framework to study galaxy evolution from low to high redshifts.

2. α_s of the stellar mass function of star-forming galaxies

In Figure 1 we show the evolution of the Schechter parameter α_s determined from observed stellar mass function of star-forming galaxies as a function of cosmic time from various literatures. Although the values of α_s change systematically from sample to sample, the α_s derived from a single homogeneous survey in a self-consistent way, such as the results from Pérez-González *et al.* (2008) and Ilbert *et al.* (2010), supports that the α_s of the stellar mass function of the star-forming galaxies is constant over a broad range of redshifts from z $\sim$ 0 up to z $\sim$ 2.

There might be a hint that the α_s is getting steeper at higher redshifts of z $>$ 4 from some observations as shown in Figure 1, but these measurements are highly uncertain since the completeness mass of a given survey progressively increases with increasing

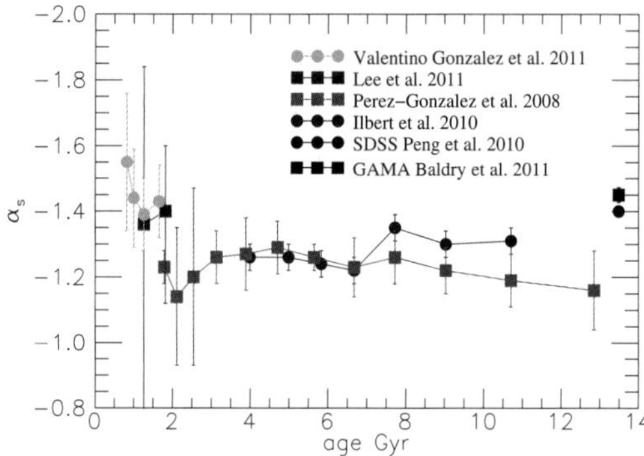

Figure 1. Evolution of the Schechter parameter α_s of the observed stellar mass function of star-forming galaxies with cosmic time in various literatures. Although the value of α_s changes systematically from sample to sample, for the α_s derived from a given homogeneous survey in a self-consistent way such as the results from Pérez-González *et al.* (2008) and Ilbert et al. (2010), it implies that the α_s of the stellar mass function of the star-forming galaxies is constant over a broad range of cosmic time.

redshift. The value of α_s can only be determined reliably if the completeness mass is well below the M* at a given epoch. To make it more difficult, the M* of the mass function at higher redshifts of z > 4 is expected to be smaller than the M* in the local universe (see the discussion in P10), thus it will require a lower completeness mass to reliably determine the α_s comparing to the mass function at low redshifts. Also it is still unclear that whether the galaxy main sequence has already been in place at z > 4 and thus the slope β is essentially unconstrained at these redshifts. There are some indirect evidences that the α_s of the SFR functions show no clear evolution with cosmic time from z $\sim$ 0 to z $\sim$ 7 (Smit *et al.* 2012). If we naively convert the SFR function to the mass function of the star-forming galaxies by assuming that the galaxy main sequence is already in place at z $\sim$ 7 and β is constant over cosmic time, it would imply that the α_s of the mass function of the star-forming galaxies is also constant from z $\sim$ 0 to z $\sim$ 7. In this work we will mainly focus on the redshift range where the α_s is observed to be more or less constant with time and explore its consequence through the continuity equation.

3. The connection between α_s and β

As discussed in P10, a negative value of β implies that the low mass galaxies grow faster than the more massive ones. If we do not consider any quenching and merging process, a negative β will apparently lead to the steepening of the faint end slope of the star-forming mass function, as shown in Figure 2. This will also cause the steepening of the global mass function as well, since the low mass end of the mass function is dominated by star-forming galaxies.

It can be shown analytically from the continuity equations that for negative β, α_s quickly diverges with time and for positive β, α_s converges towards the value of $1+\beta$. Therefore, for negative β, the mass function of the star-forming galaxies is dynamically unstable and α_s will quickly steepen with time, unless $\alpha_s = -(1+\beta)$. The differential steepening will also quickly destroy the Schechter form of the mass function. Since essentially all the observations favor a negative value of β up to z $\sim$ 2, there is an apparent conflict between the observed negative value of β and the fact that the mass function of the star-forming galaxies is a Schechter function with α_s being more or less constant with epoch up to at least of z $\sim$ 2.

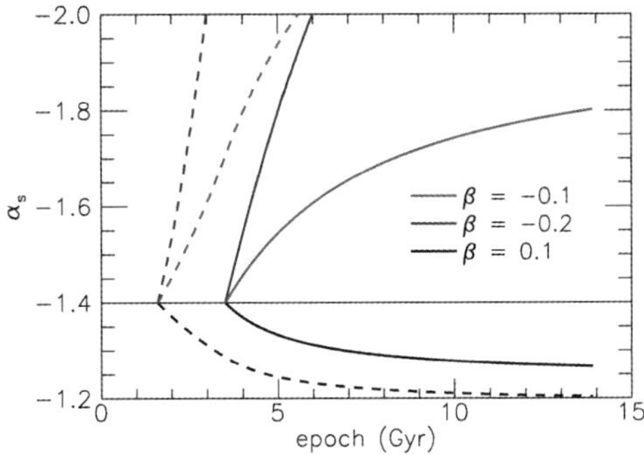

Figure 2. For non-zero β, the predicted evolution of the Schechter parameter α_s of the stellar mass function of the star-forming galaxies with cosmic time. The solid lines show if we start the evolution with $\alpha_s = -1.4$ at $z = 2$, the steepening (for negative β) or the flattening (for positive β) of the α_s. The dashed lines show that the divergence of α_s is even faster if we start with the same α_s and β, but at an earlier epoch of $z = 4$.

Destroying low mass galaxies by merging them into larger galaxies and increasing the mass of the larger galaxies, will however act to counter this steepening effect and keep the α_s constant over time. It becomes more efficient for high mass ratio mergers, i.e. minor mergers or mini mergers, to control the evolution of α_s. Since one massive galaxies can merge and destroy many low mass galaxies with only a small increase of its own stellar mass, this will help to reduce the number density of the low mass galaxies and thus to flatten α_s.

4. Constraints on mergers through the destruction of galaxies

In this section we explore the constraints on mergers via destruction of galaxies to keep the α_s constant and demonstrate the underlying connections between α_s, β and merging.

We introduce κ as the destruction rate of the galaxies by mergers. Analogous to the sSFR, we introduce the sMMR as the specific rate of mass added to a given galaxy through mergers. One key question in studying the galaxy stellar mass assembly is to understand the relationship between the sSFR and sMMR. We find the sMMR/sSFR ratio to be around 0.1 for the often observed values of $\alpha_s \sim -1.4$ (e.g. P10; Baldry *et al.* 2008 and 2012) and $\beta \sim -0.1$ (e.g. Elbaz *et al.* 2007; Daddi *et al.* 2007; Dunne *et al.* 2009; P10), i.e. sMMR ~ 0.1 sSFR. This implies that on average across the whole population, mergers bring about 10% of the stellar mass that is assembled through star-formation. In other words, stellar mass assembly through merging is significant but not dominant relative to the in situ star-forming as indicated by the observed sSFR.

We also find that there is a strong dependence of the sMMR/sSFR ratio on β. More negative values of β make it increasingly hard to keep α_s constant. It becomes impossible to control α_s when $\alpha_s + \beta + 2 \leqslant 0$, i.e. for $\beta \leqslant -(\alpha_s + 2)$. This thus strongly argues against any claim of the possibility of having the logarithmic slopes of the sSFR-m relation that are significantly steeper than -0.1.

The galaxy merger rate at a given stellar mass m, defined as the number of galaxies that have been merged into the primary galaxy of mass m per unit time, is given by

$$R(m) = x*\text{sMMR}(m),$$

where x is the mass ratio between the merging galaxies pair.

With sMMR ~ 0.1 sSFR for $\alpha_s \sim -1.4$ and $\beta \sim -0.1$, it implies a merger rate of order $0.1x$ sSFR. For the major merger defined as having a mass ratio of $1 \leqslant x \leqslant 3$, this implies a major merger rate of order $0.1\,\mathrm{sSFR} \sim 0.3\,\mathrm{sSFR}$. If we simply take the average mass ratio of $\langle x \rangle \sim 2$, we find an average major merger rate of order $0.2\,\mathrm{sSFR}$. Encouragingly, these predicted major merger rates show excellent agreement with the observed values from various literatures (e.g. Bridge *et al.* 2010; Lotz *et al.* 2011).

5. The connection between the sMMR and the assembly of dark matter halos

In the previous section we show that $\mathrm{sMMR} \sim 0.1\,\mathrm{sSFR}$ will be able to keep α_s constant over cosmic time for $\beta \sim -0.1$, which implies a major merger rate of order $\sim 0.2\,\mathrm{sSFR}$. As mentioned in P10 and P12, it is important to notice that both the average sSFR of star-forming main sequence galaxies and the specific dark matter Mass Increase Rate (sMIR) of typical haloes are only weakly dependent on mass, with $\beta_{DM} \sim +0.1$ (Faucher-Giguere *et al.* 2011) and $\beta \sim -0.1$. Both quantities have similar time dependence of $\sim (1 + z)^{2.2}$, at least in the redshift range from $z = 0$ to $z \sim 2$. The similarity of the sSFR and the sMIR suggests that the sSFR of a star-forming galaxy is closely linked to the assembly of its dark matter halo and the star-forming galaxy may have been built up through the conversion of some fraction of the baryonic mass in the halo into stars. Thus it is not surprising to find that our phenomenological analysis requires a merger rate that is closely linked to the cosmic sSFR. The connection between the sSFR and sMIR will be further discussed in our future papers.

References

Baldry, I. K., *et al.* 2012, *MNRAS*, 421, 621
Baldry, I. K., Glazebrook, K., & Driver, S. P. 2008, *MNRAS*, 388, 945
Bridge, C. R., Carlberg, R. G., & Sullivan, M. 2010, *ApJ*, 709, 1067
Daddi, E., *et al.* 2007, *ApJ*, 670, 156
Dunne, L., Ivison, R. J., & Maddox, S. 2009, *MNRAS*, 394, 3
Elbaz, D., *et al.* 2007, *A&A*, 468, 33
Faucher-Giguere, C., Keres, D., & Ma, C. 2011, *MNRAS*, 417, 2982
Gonzĺez, V., *et al.* 2011, *ApJ*, 735, 34
Ilbert, O., *et al.* 2010, *ApJ*, 709, 644
Lee, K., *et al.* 2012, *ApJ*, 752, 66
Lotz, J. M., *et al.* 2011, *ApJ*, 742, 103
Peng, Y., Lilly, S. J., Kovac, K., *et al.* 2010, *ApJ*, 721, 193
Peng, Y., Lilly, S. J., Renzini, A., & Carollo, M. 2012, *ApJ*, 757, 4
Pérez-González, P. G., *et al.* 2008, *ApJ*, 675, 234
Smit, R., *et al.* 2012, *ApJ*, 756, 14

The intriguing life of massive galaxies
Proceedings IAU Symposium No. 295, 2012
D. Thomas, A. Pasquali & I. Ferreras, eds.

© International Astronomical Union 2013
doi:10.1017/S1743921313004596

Massive galaxies: born as disks, dead as spheroids

Fernando Buitrago[1,2]

[1]Institute for Astronomy, University of Edinburgh, Royal Observatory, Edinburgh,
EH9 3HJ, U.K.
email: fb@roe.ac.uk

[2]University of Nottingham, School of Physics & Astronomy, Nottingham, NG7 2RD, U.K.

Abstract. Present day massive ($M_{stellar} > 10^{11} M_\odot$) galaxies are composed mostly of early-type objects. To ascertain whether this was also the case at higher redshift, we have compiled over a thousand massive galaxies at $0 < z < 3$ with HST imaging and spectroscopic redshifts for the majority of them. We have also analyzed using 3D spectroscopy another sample of 10 massive galaxies at $z = 1.4$. Both works highlight the progressive change between a late-type/peculiar nature at $z > 2$ and a predominance of early-type morphologies only since $z = 1$.

Keywords. galaxies: evolution, galaxies: high-redshift, galaxies: structure

1. Description of the present communication

Massive galaxies change dramatically their observational properties from low to high redshift, namely their sizes (e.g. Buitrago *et al.* 2008), other structural parameters and star formation rates (e.g. Bell *et al.* 2012). This must be reflected in their morphology. To shed light into this transformation we have analysed in Buitrago *et al.* (2013) a representative (> 1000) sample of massive galaxies at $0 < z < 3$ both quantitatively (using the Sérsic index) and qualitatively (by visual classification). The rise of galaxies displaying large Sérsic indices is connected with the development of the galaxy outskirts, and possibly their bulges. Massive galaxies also increase their number density by an order of magnitude, and the mechanism in charge is most efficient creating early-type objects.

The evidence for this morphological transformation is based so far on the detailed morphological analysis of these objects which ultimately rests on the shape of their surface brightness profiles. To explore the consistency of this scenario, it is necessary to have access to the dynamical status of these galaxies. We have studied in Buitrago *et al.* (2012, in preparation) 10 massive galaxies at $z \sim 1.4$ with the SINFONI integral field spectrograph in the H-band by mapping the Hα emission line. Most of these objects show large rotational field maps, with half of them being consistent with rotationally supported disks. We conclude that massive galaxies acquire more rapidly a given morphology and gravitational equilibrium than less-massive objects. Mass is crucial for understanding the galaxy morphological development, whereby massive galaxies progressively join the observational properties of their local Universe counterparts.

References

Bell, E., *et al.* 2012, *ApJ*, 753, 167
Buitrago, F., *et al.* 2008, *ApJ* (Letters), 687, L61
Buitrago, F., *et al.* 2013, *MNRAS*, 428, 1460

The intriguing life of massive galaxies
Proceedings IAU Symposium No. 295, 2012
D. Thomas, A. Pasquali & I. Ferreras, eds.

© International Astronomical Union 2013
doi:10.1017/S1743921313004602

Evolution in cluster cores since z ∼ 1

Claire Burke[1], Chris Collins[1], John Stott[2] and Matt Hilton[3]

[1] Astrophysics Research Institute, Liverpool John Moores University, UK.
email: `cb@astro.livjm.ac.uk`
[2] Extragalactic & Cosmology Group, Durham University, UK.
[3] Astrophysics & Cosmology Research Unit, University of KwaZulu-Natal, South Africa.

Abstract. A large fraction of the stellar mass in galaxy clusters is thought to be contained in the diffuse low surface brightness intracluster light (ICL). Being bound to the gravitational potential of the cluster rather than any individual galaxy, the ICL contains much information about the evolution of its host cluster and the interactions between the galaxies within. However due its low surface brightness it is notoriously difficult to study. We present the first detection and measurement of the flux contained in the ICL at z ∼ 1. We find that the fraction of the total cluster light contained in the ICL may have increased by factors of 2–4 since z ∼ 1, in contrast to recent findings for the lack of mass and scale size evolution found for brightest cluster galaxies. Our results suggest that late time build-up in cluster cores may occur more through stripping than merging and we discuss the implications of our results for hierarchical simulations.

Keywords. galaxies: clusters: general, galaxies: elliptical and lenticular, cD, galaxies: evolution, galaxies: halos, galaxies: interactions

---

The central regions of galaxy clusters are usually dominated by a massive brightest cluster galaxy (BCG), which generally sits at the centre of mass of the cluster; and a diffuse, extended halo of intracluster light (ICL). The study of the ICL can reveal details of the evolution histories and processes occurring within galaxy clusters. A significant fraction of the total stellar mass and luminosity of galaxy clusters is thought to be contained within the ICL, however since it has very low surface brightness it is often difficult to detect. Using a sample of 6 X-ray selected clusters at $0.8 \leqslant z \leqslant 1.22$ with deep J-band imaging from HAWK-I on the VLT we have performed the first detection and measurement of the ICL at high redshift. We detect the ICL down to a surface brightness of $\mu_J \sim 23$ mag/arcsec^2 and, using a cautious but robust method, find it to contain 1–4% of the total cluster light. When compared to nearby clusters, we find the ICL to have grown by 2–4 times as a fraction of the total cluster light since z = 1.

We also examine the surface brightness profiles of the BCGs in our sample. We find that up to 50% of the light from the BCG is outside of a 50 kpc radius, suggesting that when measuring photospheric light from BCGs a significant fraction will be contaminated by the ICL, therefore BCG masses measured observationally will include some contribution from the ICL. Recent hierarchical simulations (e.g. De Lucia & Blaizot 2007) under-predict the masses of BCGs at high redshift by as much as 50%, therefore we suggest that the inclusion of the ICL in the mass budgets of simulated BCGs may ease this discrepancy.

When considering recent results for the lack of scale size evolution of BCGs (in contrast to the size growth seen for field ellipticals (see Stott *et al.* 2011), we suggest that evolution and interactions in the central regions of clusters may involve stripping rather than merging causing the ICL to be built up while the BCG remains relatively unchanged (Burke *et al.* 2012).

References

Burke, C., *et al.* 2012, *MNRAS*, 425, 2058
De Lucia, G. & Blaizot, J. 2007, *MNRAS*, 375, 2
Stott, J. P., *et al.* 2011, *MNRAS*, 414, 445

The intriguing life of massive galaxies
Proceedings IAU Symposium No. 295, 2012
D. Thomas, A. Pasquali & I. Ferreras, eds.

© International Astronomical Union 2013
doi:10.1017/S1743921313004614

Colour Properties of Group Galaxies in Pan-STARRS MD Survey Fields

C. W. Chen[1], L. Lin[1], H. Y. Jian[2] and S. Foucaud[3]

[1]Institute of Astronomy and Astrophysics, Academia Sinica, Taipei, Taiwan
email: `chinwei.chen@asiaa.sinica.edu.tw`

[2]Dept. of Physics, National Taiwan University, Taipei, Taiwan
[3]Dept. of Earth Science, National Taiwan Normal University, Taipei, Taiwan

Abstract. How environment shapes galaxies on groups scale in the early universe is poorly constrained. Here we carry out a study of the colour properties of galaxies against environment using Pan-STARRS medium deep (PS1 MD) survey data. We first focus on the MD04 field which overlays with the COSMOS field with published photo-z and group catalogs. Photo-z with the accuracy of $\Delta z/(1 + z) \sim 0.058$ for 0.36 million galaxies are derived based on PS1 photometry and CFHT-u. Together with a probabilistic-based group finder, PFOF, we are able to identify galaxy groups and find a 83%-86% matching rate with the X-ray groups (George *et al.* 2011) or spectroscopically selected groups (Knobel *et al.* 2012) with intermediate redshift ($z \sim 0.7$). Among the matched samples (see Fig. 1), we found the colours of BCGs to be indistinguishable from other members, but group galaxies tend to be redder than those in the field for a given range of z-band magnitude, suggesting environmental effects on the evolutionary history of galaxies. This is qualitatively consistent with the X-ray study (George *et al.* 2011). The rest-frame quantities, e.g. colours, stellar mass and star formation rates, will be included and expanded to larger MD fields ($\sim 70\ deg^2$) to probe the cosmic evolution of galaxy properties in a forthcoming study.

Keywords. galaxies: evolution, galaxies: elliptical and lenticular, cD

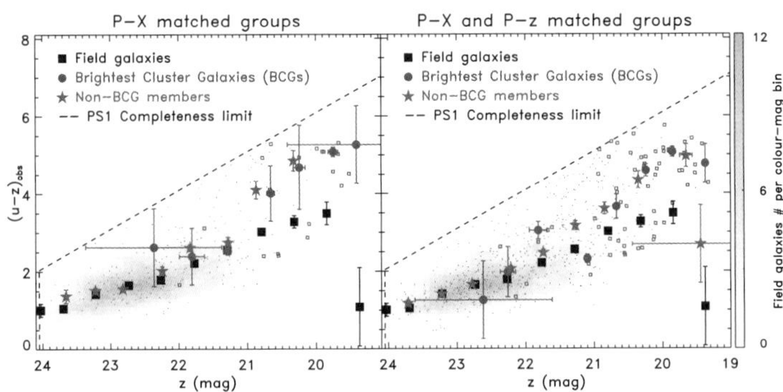

Figure 1. The left hand panel shows the colour properties of 25 PFOF-X-ray matched groups, serving as the most secure group sample in MD04. In addition, there are spec-z selected zCOSMOS groups in the field, allowing us to increase the sample size with groups fairly reliably identified. Therefore, a sample of 75 groups is collected by combining "PFOF-X-ray matched" with "PFOF-X-ray unmatched but PFOF-zCOSMOS matched" groups (right panel). Both panels show a significant difference between galaxy colour in groups and field at the bright end.

References

George, M. R., *et al.* 2011, *ApJ*, 742, 125
Knobel, C., *et al.* 2012, *ApJ*, 753, 121

The intriguing life of massive galaxies
Proceedings IAU Symposium No. 295, 2012
D. Thomas, A. Pasquali & I. Ferreras, eds.

© International Astronomical Union 2013
doi:10.1017/S1743921313004626

Evolution of the Fundamental Plane for early–type galaxies up to z = 1.2

M. Fernández Lorenzo[1], J. Cepa[2,3], A. Bongiovanni[2,3], A. M. Pérez García[2,3], A. Ederoclite[4], M. A. Lara-López[5], M. Pović[1] and M. Sánchez-Portal[6]

[1]Instituto de Astrofísica de Andalucía (IAA-CSIC), email: mirian@iaa.es, [2]Instituto de Astrofísica de Canarias (IAC), [3]Universidad de la Laguna (ULL), [4]Centro de Estudios de Física del Cosmos de Aragón (CEFCA), [5]Australian Astronomical Observatory (AAO), [6]European Space Astronomy Centre (ESAC)/ESA

Abstract. We studied the evolution of the Fundamental Plane in B– and g–bands, for two samples of early–type galaxies at $0.2 < z < 0.35$ and $0.35 < z < 1.2$. We found a difference in the intercept that can be interpreted as galaxies at $<z> \sim 0.7$ being 0.68 and 0.52 mag brighter in the B– and g–band respectively, than their local counterparts. From the study of the Kormendy relation, we found the existence of a population of very bright $(-21.5 > M_g > -22.5)$ and compact (Re < 2 kpc) galaxies of which only a small fraction of 0.4% exists at $z = 0$, and that would be responsible for the apparent evolution in the Kormendy relation (Fernández Lorenzo *et al.* 2011). These compact objects would have evolved mainly in size by the action of "dry" minor mergers.

We checked the effect of these bright and compact objects in the FP by comparing the galaxies with $-21.5 > M_g > -22.5$ of our high–redshift sample and the SDSS early–type galaxies of Bernardi *et al.* (2003) in Fig. 1. The previous evolution found in the FP seems to be caused mainly by these galaxies, which have virtually disappeared at $z = 0$.

Keywords. galaxies: evolution, galaxies: fundamental parameters

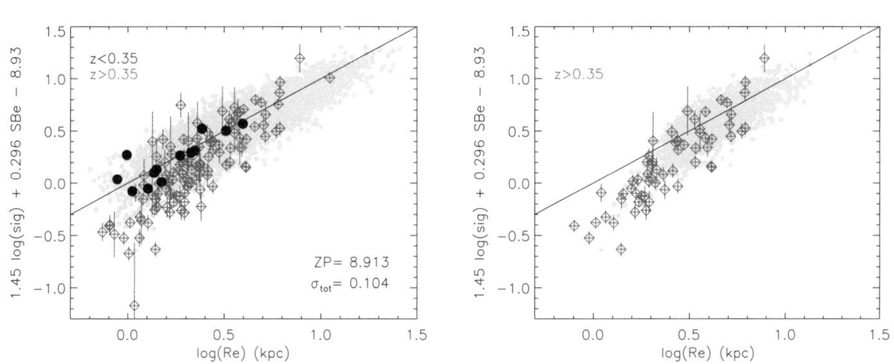

Figure 1. Edge–on projection of the FP in the g–band for all the galaxies in our local (black) and high–z (red) sample, and for the SDSS sample (yellow) (left). In the right panel we represented only galaxies with $-21.5 > M_g > -22.5$.

References

Bernardi, M., *et al.* 2003, *AJ*, 125, 1866

Fernández Lorenzo, M., Cepa, J., Bongiovanni, A., *et al.* 2011, *A&A*, 526, 72

The intriguing life of massive galaxies
Proceedings IAU Symposium No. 295, 2012
D. Thomas, A. Pasquali & I. Ferreras, eds.

© International Astronomical Union 2013
doi:10.1017/S1743921313004638

Evolutionary paths among different red galaxy types at $0.3 < z < 1.5$ and the build-up of massive E-S0's

Jesús Gallego[1]**, Mercedes Prieto, M. Carmen Eliche-Moral, Marc Balcells, David Cristóbal-Hornillos, Peter Erwin, David Abreu, Lilian Domínguez-Palmero, Angela Hempel, Carlos López-Sanjuan, Rafael Guzmán, Pablo G. Pérez-González, Guillermo Barro and Jaime Zamorano**

[1] Departamento de Astrofísica, Universidad Complutense de Madrid,
Facultad de CC. Físicas, Ciudad Universitaria, E-28040 Madrid, Spain
email: `j.gallego@fis.ucm.es`

Abstract. Some recent observations seem to disagree with hierarchical theories of galaxy formation on the role of major mergers in a late build-up of massive early-type galaxies. We re-address this question by analysing the morphology, structural distortion level, and star formation enhancement of a sample of massive galaxies ($M_* > 5 \times 10^{10} M_\odot$) lying on the Red Sequence and its surroundings at $0.3 < z < 1.5$. We have used an initial sample of $\sim$1800 sources with $K_s < 20.5$ mag over an area $\sim 155\,\mathrm{arcmin}^2$ on the Groth Strip, combining data from the Rainbow Extragalactic Database and the GOYA Survey. Red galaxy classes that can be directly associated to intermediate stages of major mergers and to their final products have been defined. For the first time we report observationally the existence of a dominant evolutionary path among massive red galaxies at $0.6 < z < 1.5$, consisting in the conversion of irregular disks into irregular spheroids, and of these ones into regular spheroids. This result points to: 1) the massive red regular galaxies at low redshifts derive from the irregular ones populating the Red Sequence and its neighbourhood at earlier epochs up to $z \sim 1.5$; 2) the progenitors of the bulk of present-day massive red regular galaxies have been blue disks that have migrated to the Red Sequence majoritarily through major mergers at $0.6 < z < 1.2$ (these mergers thus starting at $z \sim 1.5$); 3) the formation of E-S0's that end up with $M_* > 10^{11} M_\odot$ at $z = 0$ through gas-rich major mergers has frozen since $z \sim 0.6$. Our results support that major mergers have played the dominant role in the definitive build-up of present-day E-S0's with $M_* > 10^{11} M_\odot$ at $0.6 < z < 1.2$, in good agreement with the hierarchical scenario proposed in the Eliche-Moral *et al.* (2010a) model (see also Eliche-Moral *et al.* 2010b). This study is published in Prieto *et al.* (2012).

Supported by the Spanish Ministry of Science and Innovation (MICINN) under projects AYA2009-10368, AYA2006-12955, AYA2010-21887-C04-04, and AYA2009-11137, by the Madrid Regional Government through the AstroMadrid Project (CAM S2009/ESP-1496), and by the Spanish MICINN under the Consolider-Ingenio 2010 Program grant CSD2006-00070: "First Science with the GTC" (http://www.iac.es/consolider-ingenio-gtc/). S. D. H. & G.

Keywords. galaxies: elliptical and lenticular, cD, galaxies: evolution, galaxies: formation, galaxies: interactions, galaxies: luminosity function, galaxies: mass function, galaxies: morphologies

---

References

Eliche-Moral, M. C., *et al.* 2010, *A&A*, 519, 55

Eliche-Moral, M. C., Prieto, M., Gallego, J., & Zamorano, J. 2010, *ApJ*, submitted (arXiv:1003.0686)

Prieto, M., *et al.* 2012, *MNRAS*, 428, 999

The intriguing life of massive galaxies
Proceedings IAU Symposium No. 295, 2012
D. Thomas, A. Pasquali & I. Ferreras, eds.

© International Astronomical Union 2013
doi:10.1017/S174392131300464X

Which Galaxy Property Best Predicts Quiescence?

Joel Leja[1], Pieter van Dokkum[1] and the 3D-HST Collaboration

[1]Department of Astronomy, Yale University, New Haven, CT 06520-8101
email: joel.leja@yale.edu
email: pieter.vandokkum@yale.edu

Abstract. It is generally accepted that local elliptical galaxies assembled most of their mass in a burst of star formation between $1 < z < 3$, yet today, their star formation has been almost entirely quenched. In order to constrain this quenching mechanism, we measure Hα line emission in galaxies sorted by multiple galaxy properties as a function of redshift to what galaxy parameter best predicts quiescence. This is done for samples of the most massive, most luminous, and galaxies with the highest velocity dispersion both locally ($0.05 < z < 0.07$ in the SDSS) and at high redshift ($0.7 < z < 1.5$ in 3D-HST). It is demonstrated through spectral stacking that velocity dispersion results in the lowest Hα line equivalent width both locally and at high redshift. The spatial distribution of the emission line flux is available from grism spectroscopy: the line flux from the high dispersion stack is centrally peaked and thus likely associated with AGN activity rather than star formation, strengthening this conclusion. Since velocity dispersion may also be the best predictor of halo mass (Wake *et al.* 2012), this may imply that the quenching mechanism is directly related to halo mass.

Keywords. galaxies: evolution, galaxies: high-redshift

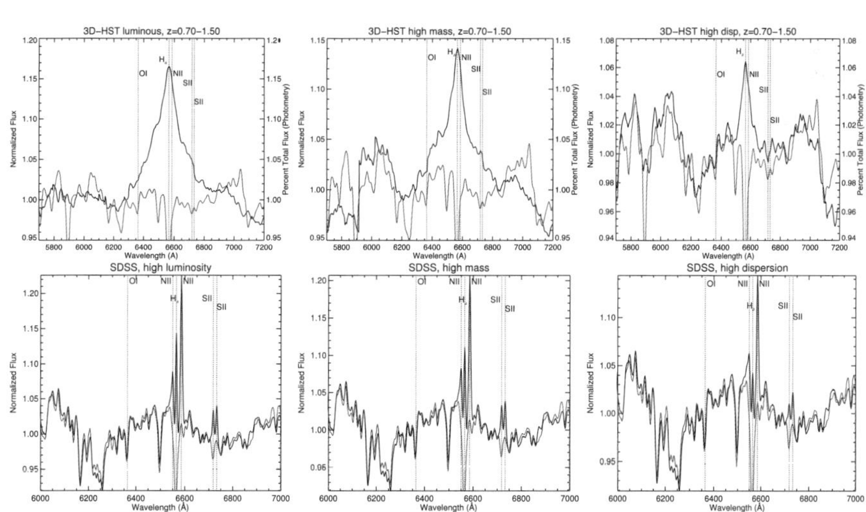

Figure 1. Stacks of SDSS spectra and 3D-HST grism spectra with stellar population synthesis models convolved to the appropriate resolution (Vazdekis *et al.* 2010).

References

Wake, D. A., Franx, M., & van Dokkum, P. G. 2012, *ApJ*, submitted (arXiv:1201.1913)
Vazdekis, A., Sánchez-Blázquez, P., Falcón-Barroso, J., Cenarro, A. J., Beasley, M. A., Cardiel, N., Gorgas, J., & Peletier, R. F. 2010, *MNRAS*, 404, 1639

The intriguing life of massive galaxies
Proceedings IAU Symposium No. 295, 2012
D. Thomas, A. Pasquali & I. Ferreras, eds.

© International Astronomical Union 2013
doi:10.1017/S1743921313004651

Stellar Populations in the Most Luminous Obscured Quasars at $z > 0.5$

Xin Liu[1,2]

[1]Harvard College Observatory, 60 Garden Street, Cambridge, MA 02138, USA
Email: xinliu@cfa.harvard.edu

[2]Einstein Fellow

There is evidence that the growth of stellar populations and supermassive black holes (SMBHs) are coupled across cosmic time: the redshift evolution of star formation rate and quasar number density are similar; SMBH masses in local inactive galaxies are correlated with the velocity dispersion of their stellar bulges. Models predict that SMBHs predominantly grow in brief quasar phases accompanied by starbursts, but on-going starbursts in luminous quasars have been difficult to quantify. There have been extensive photometric studies of quasar host galaxies. However, spectroscopic studies that provide crucial information on stellar populations such as age and velocity dispersion are scarce, especially at the highest luminosities, where the stars are vastly outshone by the quasar.

To study stellar populations in luminous quasars, we target obscured (type 2) quasars. The circum-nuclear obscuring material works as a natural coronagraph, allowing us to study the host galaxies in a luminosity regime virtually inaccessible for unobscured quasars. Our targets were selected to have high intrinsic luminosities ($M_V < -26\,$mag), and they are likely to be accreting at close to Eddington rates. Our goal is to search for direct evidence for ongoing or recent starburst that may accompany peak quasar activity.

In our pilot study of nine SDSS luminous obscured quasars (Liu *et al.* 2009), we found a substantial contribution from very young stellar populations ($< 0.1\,$Gyr) in all cases. Furthermore, we detected Wolf-Rayet emission in three objects, indicating the presence of an extremely young post-starburst phase ($\sim 5\,$Myr; e.g., Brinchmann *et al.* 2008). Population synthesis modeling of the stellar continuum lends further support to the young stellar ages. The scattered light component was carefully constrained using broad Hβ, if any. The inferred fraction of infant stellar populations in luminous quasars directly tested the link between on-going starburst and luminous quasar activity.

To get more statistically significant results, we conducted deep Gemini/GMOS spectroscopy of an additional 21 luminous obscured quasars. With improved sensitivity and an enlarged sample, we have (i) confirmed the high frequency of infant starbursts from Wolf-Rayet emission features and population synthesis modeling of the stellar continuum, (ii) quantified old stellar populations from stellar absorption features; (iii) measured stellar velocity dispersions and estimated BH masses, and (iv) found a significant contribution from scattered quasar light to the observed continuum. With a sample size of 30, we correlated the quasar and galaxy properties to probe the physical link and mutual influences between starburst and quasar activity, as an important test of our understanding of the nature and evolution of the coupled growth of bulge stellar populations and SMBHs.

References

Brinchmann, J., Kunth, D., & Durret, F. 2008, *A&A*, 485, 657
Liu, X., Zakamska, N. L., Greene, J. E., Strauss, M. A., Krolik, J. H., & Heckman, T. M. 2009, *ApJ*, 702, 1098

The intriguing life of massive galaxies
Proceedings IAU Symposium No. 295, 2012
D. Thomas, A. Pasquali & I. Ferreras, eds.

© International Astronomical Union 2013
doi:10.1017/S1743921313004663

The Age-Redshift Relation For LRGs

**Gaochao Liu[1,2], Youjun Lu[1], Xuelei Chen[1],
Yongheng Zhao[1], Wei Du[1], and Xianmin Meng[1]**

[1] Key Laboratory of Optical Astronomy, National Astronomical Observatories, Chinese
Academy of Science, Beijing 100012, China
email: `luyj@nao.cas.cn`

[2] College of Science, China Three Gorges University, YiChang 443002, China
email: `gcliu@nao.cas.cn`

Abstract. The relative age of galaxies at different redshifts can be used to infer the Hubble parameter and put constraints on cosmological models. We select 23,883 quiescent luminous red galaxies (LRGs) from the SDSS DR7, dividing them into four sub-samples according to their velocity dispersions. Each sub-sample was further divided into 12 redshift bins. The spectra of the LRGs in each redshift and velocity bin were co-added. Adopting the GalexEV/SteLib model, we estimate the mean ages of the LRGs from these combined spectra by the full-spectrum fitting method and find it is consistent with the well known "downsizing" formation of galaxies. Based on the age–redshift relation, we further estimate the Hubble parameter.

Keywords. cosmological parameters – cosmology:evolution – galaxies:stellar content

---

1. Sample and Results

We select 23,883 quiescent LRGs from the SDSS DR7. These LRGs are divided into four sub-samples according to their velocity dispersions (σ_v) and each LRG sub-sample is further divided into 12 redshift bins. We co-add the spectra of galaxies in each sub-sample (and each red-shift bin) to obtain combined spectra with significantly higher S/N (> 40). These combined LRG spectra are then fitted by a single stellar population by using the GalexEV/SteLib model in the software ULySS, which implements the full spectrum fitting method. The combined spectra, with higher S/N, enable us to obtain the LRG mean age (t_{age}) with considerable accuracy. Our results show that t_{age} decreases with increasing redshift, and the age of LRGs with higher σ_v tends to be older than those with lower σ_v, consistent with the well known "downsizing" formation scenario of galaxies. Assuming that the $t_{\mathrm{age}} - \sigma_v$ relation follows a power-law, i.e., $t_{\mathrm{age}} \propto \sigma_v^\gamma$, we find that the mean value of γ is $\simeq 0.77$ with a standard deviation of 0.25.

The age–redshift relation obtained from observations can be directly fitted by $t_{\mathrm{age}} = t_{\mathrm{U}} - t_{\mathrm{form}}$, where t_{form} is the mean formation time of the quiescent galaxies and assumed to be a constant for each sub-sample, and t_{U} is the age of the Universe. Assuming a spatially flat ΛCDM model, then $t_{\mathrm{U}} = \frac{1}{H_0} \int_z^\infty \frac{dz}{(1+z)\sqrt{\Omega_{\mathrm{m}}(1+z)^3 + \Omega_\Lambda}}$. Using the standard χ^2 statistic, we find the best fit (and its uncertainty) for the Hubble constant at the present time is $H_0 = 65^{+7}_{-3} \mathrm{km\ s^{-1}\ Mpc^{-1}}$ if using the age-redshift relation obtained from the LRG sub-sample with the highest σ_v, or $H_0 = 74^{+5}_{-4} \mathrm{km\ s^{-1}\ Mpc^{-1}}$ if using the age-redshift relations obtained from the two LRG sub-samples with high σ_v simultaneously.

See Liu *et al.* (arXiv:1208.6502) for more details,

The intriguing life of massive galaxies
Proceedings IAU Symposium No. 295, 2012
D. Thomas, A. Pasquali & I. Ferreras, eds.

© International Astronomical Union 2013
doi:10.1017/S1743921313004675

Discriminating Quasars from Stars Based on SDSS and UKIDSS Databases

He Ma[1,2], **Yanxia Zhang**[1], **Yongheng Zhao**[1] **and Bo Zhang**[2]

[1] Key Laboratory of Optical Astronomy, National Astronomical Observatories,
Chinese Academy of Sciences, 100012 Beijing, China
[2] Hebei Normal University, 050016 ShiJiaZhuang, Hebei, China

Abstract. In this work, two different algorithms: Linear Discriminant Analysis (LDA) and Support Vector Machines (SVMs) are combined for the classification of unresolved sources from SDSS DR8 and UKIDSS DR8. The experimental result shows that this joint approach is effective for our case.

Keywords. Classification, Astronomical databases: miscellaneous, Catalogs, Methods: Data Analysis, Methods: Statistical

1. Sample

The data are retrieved from SDSS DR8 and UKIDSS DR8. The training sample and test sample for stars comprises 334474 and 61338, respectively. The number of quasars in the two samples is 12061 and 12061, separately.

2. Algorithm and Result

The main idea of LDA is to find a projection in a low dimensional space for different classes to be well separated. The main principle of SVMs is to transfer input vectors into a high-dimensional feature space and find the optimal separating hyperplane in this space. Seven kinds of colors ($u - g, g - r, u - r, g - z, u - z, g - z, Y - K$) are taken as the input of SVMs because of their better color-color diagram separation. Anothr six kinds of LDA output for $u - r$ vs. $g - J$, $u - r$ vs. $r - Y$, $u - z$ vs. $r - z$, $u - Y$ vs. $r - Y$, $u - J$ vs. $u - r$, $g - Y$ vs. $i - K$ are also used as the input of SVMs simultaneously. In addition, r magnitude and i magnitude are used as the input of SVMs. As a result, the precision and completeness of stars is 99.60% and 99.80%, respectively, while that of quasars is 98.93% and 97.99%, respectively.

3. Conclusion

Multi-band data provide a large amount of information about objects. We use multi-band data to classify objects effectively. On the other hand, we combine traditional color-color, magnitudes and outputs of LDA as the input pattern of SVMs to create a new classifier. The classification result is rather satisfactory.

This paper is funded by National Natural Science Foundation of China under grant No.10778724, 11178021 and No.11033001, the Natural Science Foundation of Education Department of Hebei Province under grant No. ZD2010127 and by the Young Researcher Grant of National Astronomical Observatories, Chinese Academy of Sciences.

The intriguing life of massive galaxies
Proceedings IAU Symposium No. 295, 2012
D. Thomas, A. Pasquali & I. Ferreras, eds.

© International Astronomical Union 2013
doi:10.1017/S1743921313004687

Satellites of massive galaxies: the infalling pieces of the puzzle

E. Mármol-Queraltó[1,2], I. Trujillo[1,2], P. G. Pérez-González[3], G. Barro[4], J. Varela[5], and V. Villar[3]

[1]Instituto de Astrofísica de Canarias, c/Vía Láctea s/n, E-38205, La Laguna, Spain

[2]Dpto. de Astrofísica, Universidad de La Laguna, E-38205, La Laguna, Spain

[3]Dpto. de Astrofísica, CC. Físicas, Universidad Complutense de Madrid, E-28040, Spain

[4]UCO/Lick Observatory, University of California, Santa Cruz, CA 95064

[5]Centro de Estudios de Física del Cosmos, Plaza de San Juan, E-44001, Teruel, Spain

Abstract. Accretion of minor satellites has been postulated as the most likely mechanism to explain the significant size evolution of massive galaxies over cosmic time. A direct way of probing this scenario is to measure the frequency of satellites around massive galaxies at different redshifts. Here we present our study of satellites around massive galaxies ($M_{\rm star} \sim 10^{11}$ M$_\odot$) up to $z \sim 2$. We find (Fig. 1) that the fraction of massive galaxies with satellites down to 1:10 mass ratio is ∼15% (∼30% down to 1:100), not varying with redshift (Mármol-Queraltó et al. (2012)). We also find that our satellites are younger than their central galaxies at low z (Mármol-Queraltó et al., 2013). Then, if minor merging is acting to form massive galaxies, their ourtskirts should be younger than their cores. The challenge to find this age gradient in nearby massive galaxies is opened.

Keywords. galaxies: massive galaxies, galaxies: evolution galaxies: high-redshift galaxies:formation

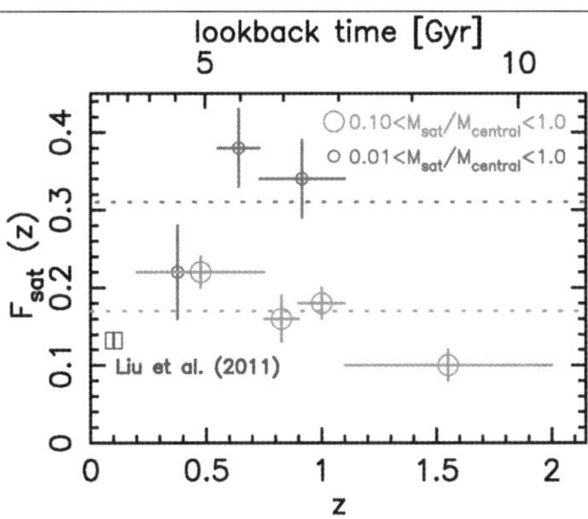

Figure 1. Fraction of massive galaxies with satellites within a projected distance of 100 kpc for different redshifts. The magenta small circles indicate the fraction of massive galaxies with satellites down to 1:100 mass ratio, and the orange big circles down to 1:10 mass ratio. The blue cross indicates the fraction of satellites around massive galaxies in the nearby universe (∼ 12%, Liu et al. (2011)), in agreement with our results at higher redshifts.

References

Liu, L., Gerke, B. F., et al. 2011, ApJ, 733, 62

Mármol-Queraltó, E., Trujillo, I., Pérez-González, P., Varela, J., & Barro, G. 2012, *MNRAS*, 422, 2187

Mármol-Queraltó, E., Trujillo, I., Villar, V., Barro, G., & Pérez-González, P. 2013, *MNRAS*, 429, 792

The intriguing life of massive galaxies
Proceedings IAU Symposium No. 295, 2012
D. Thomas, A. Pasquali & I. Ferreras, eds.

© International Astronomical Union 2013
doi:10.1017/S1743921313004699

Byurakan-IRAS galaxies as massive galaxies with nuclear and starburst activity

Areg M. Mickaelian and Gohar S. Harutyunyan

Byurakan Astrophysical Observatory (BAO), Byurakan 0213, Aragatzotn province, Armenia
email: aregmick@aras.am, goharutyunyan@gmail.com

Byurakan-IRAS Galaxies (BIG) (Mickaelian 1995) are the result of optical identifications of IRAS PSC sources at high-galactic latitudes using the First Byurakan Survey (FBS) low-dispersion spectra (Markarian *et al.* 1989). Among the 1577 targets, 1178 galaxies have been identified. Most are dusty spiral galaxies and there is a number of ULIRGs among these objects. Our spectroscopic observations, carried out with three telescopes (Byurakan Astrophysical Observatory 2.6m, Russian Special Astrophysical Observatory 6m and Observatoire de Haute Provence 1.93m; Mickaelian & Sargsyan 2010), for 172 galaxies, as well as the SDSS DR8 spectra for 83 galaxies make up the list of 255 spectroscopically studied BIG objects. The classification regarding activity type for narrow-line emission galaxies has been carried out using the diagnostic diagrams by Veilleux & Osterbrock (1987). All possible physical characteristics have been measured and/or calculated, including radial velocities and distances, angular and physical sizes, absolute magnitudes and luminosities (both optical and IR). IR luminosities and star-formation rates have been calculated from the IR fluxes (Duc *et al.* 1997).

In our spectroscopic sample of 255 BIG objects, we have 43 (17%) AGN, 25 (10%) composite spectrum objects, 157 (62%) starbursts, 29 emission-line galaxies without a definite type, and 1 absorption-line galaxy. There are 3 Ultra-Luminous InfraRed Galaxies (ULIRGs). The average redshift is $z = 0.06743$ and the average distance, 269 Mpc.

The masses of BIG objects have been estimated based on mass-luminosity relations for spiral galaxies. As it appears, most of these objects are giant massive galaxies. The average optical luminosities (absolute magnitudes) in B and R are -18.89^m and -20.51^m, respectively, average physical sizes, $D = 17$ kpc, average FIR and IR luminosities, $L_{fir}=1.36E+11L_o$ and $L_{ir}=2.31E+11L_o$ respectively.

Various multiwavelength (MW) data have been retrieved for the full sample of 1178 objects from recent catalogs from X-ray to radio (ROSAT, GALEX, APM, MAPS, USNO, GSC, SDSS, 2MASS, WISE, IRAS, AKARI, NVSS, FIRST, etc.) to make a complete study of these galaxies possible. MW SEDs have been built, which have been matched to their optical classifications. Star-formation rates have been calculated to compare to their other physical characteristics, such as morphology, activity type, UV, optical, IR and radio luminosities.

Keywords. surveys, infrared: galaxies, galaxies: active, galaxies: starburst

References

Duc, P. A., Mirabel, I. F., & Maza, J. 1997, *A&AS*, 124, 533
Markarian, B. E., Lipovetski, V. A., Stepanian, J. A., Erastova, L. K., & Shapovalova, A. I. 1989 *ComSAO*, 62, 5
Mickaelian, A. M. 1995, *Ap*, 38, 625
Mickaelian, A. M. & Sargsyan, L. A. 2010, *Ap*, 53, 483
Veilleux, S. & Osterbrock, D. E. 1987, *ApJS*, 63, 295

The intriguing life of massive galaxies
Proceedings IAU Symposium No. 295, 2012
D. Thomas, A. Pasquali & I. Ferreras, eds.

© International Astronomical Union 2013
doi:10.1017/S1743921313004705

Dry minor mergers and the size evolution of high-z compact massive early-type galaxies

Taira Oogi[1] and Asao Habe[2]

Department of Comosciences, Graduate School of Science, Hokkaido University,
Kita-ku Kita 10 Nishi 8, Sapporo, Hokkaido 060-0810, Japan
[1]email: oogi@astro1.sci.hokudai.ac.jp
[2]email: habe@astro1.sci.hokudai.ac.jp

1. Summary

Recent observations show evidence that high-z ($z \sim 2 - 3$) early-type galaxies (ETGs) are more compact than those with comparable mass at $z \sim 0$ (e.g. Trujillo *et al.* 2007; Buitrago *et al.* 2008). Such a size evolution is most likely explained by the 'dry merger scenario'. However, previous studies based on this scenario are not able to consistently explain both the properties of the high-z compact massive ETGs and the local ETGs (Nipoti *et al.* 2009). We investigate the effect of multiple sequential minor mergers on the size evolution of the compact massive ETGs. From an analysis of the Millennium Simulation Database (Springel *et al.* 2005; De Lucia & Blaizot 2007), we show that such minor (stellar mass ratio $M_2/M_1 < 1/4$) mergers are extremely common during hierarchical structure formation. We perform N-body simulations of sequential dry minor mergers with parabolic and head-on orbits, including a dark matter component and a stellar component. Typical mass ratios of the minor mergers are $1/20 < M_2/M_1 \leqq 1/10$.

We show that sequential dry minor mergers of compact satellite galaxies are the most efficient at promoting size growth and decreasing the velocity dispersion of the compact massive ETGs in our simulations. In particular, we show that the sequential minor mergers of the compact satellites have the most efficient size growth efficiency, $\alpha \simeq 2.7$ ($R_e \propto M_*^\alpha$). The change of stellar size, density, and stellar velocity dispersion of the merger remnants is consistent with recent observations (e.g. Bezanson *et al.* 2009).

Furthermore, we construct the merger histories of candidates for the high-z compact massive ETGs using the Millennium Simulation Database, and estimate the size growth of the galaxies in the dry minor merger scenario. We can reproduce the mean size growth of the galaxies between $z = 2$ and $z = 0$, assuming the efficient size growth, $\alpha \gtrsim 2.3$, obtained during sequential dry minor mergers in our simulations. However, we note that our numerical result is only valid for merger histories with typical mass ratios between $1/20$ and $1/10$ with parabolic and head-on orbits, and that our most efficient size growth efficiency is likely to serve as an upper limit. In future studies, we will investigate various mass ratios and merger orbits for a more robust prediction.

References

Bezanson, R., *et al.* 2009, *ApJ*, 697, 1290
Buitrago, F., *et al.* 2008, *ApJ*, 687, L61
De Lucia, G. & Blaizot, J. 2005, *MNRAS*, 375, 2
Nipoti, C., *et al.* 2009, *ApJ*, 706, L86
Springel, V., *et al.* 2005, *Nature*, 435, 629
Trujillo, I., *et al.* 2007, *MNRAS*, 382, 109

The intriguing life of massive galaxies
Proceedings IAU Symposium No. 295, 2012
D. Thomas, A. Pasquali & I. Ferreras, eds.

© International Astronomical Union 2013
doi:10.1017/S1743921313004717

Merger rates for early-type galaxies: combining clustering and luminosity function measurements

Nelson D. Padilla[1] Eric Gawiser[2] Daniel Christlein[3] and Danilo Marchesini[3]

[1]Universidad Católica de Chile, Chile, email: npadilla@astro.puc.cl [2]Rutgers University, USA; [3]Tufts University, USA; [4]Max Planck Institute for Astrophysics, Garching, Germany.

Abstract. We present a study of the evolution of early-type galaxies (ETGs) that combines luminosity function and clustering measurements. This technique shows that ETGs at a given redshift evolve into brighter galaxies in the rest-frame passively evolved optical at lower redshifts. Notice that this indicates that a stellar-mass selection at different redshifts does not necessarily provide samples of galaxies in a progenitor-descendant relationship. The comparison between high redshift ETGs and their likely descendants at $z = 0$ points to a higher number density for the progenitors by a factor 3 to 11, implying the need for mergers to decrease their number density by today. Because the progenitor-to-descendant ratios of luminosity density are consistent with the unit value, our results show no need for strong star-formation episodes in ETGs since $z = 1$, which indicates that the needed mergers are dry, i.e. gas free.

Keywords. galaxies: evolution, interactions, luminosity function, mass function

1. Methods and results

We present the analysis of the evolution of MUSYC (MUlti-wavelength Survey by Yale- Chile, Gawiser *et al.* 2006) early type galaxies (ETG) by Padilla *et al.* (2011, P11). This analysis uses luminosity and correlation functions for MUSYC ETG galaxies with rest-frame $M_R(0) < 19.7$ corrected by passive evolution, which corresponds to $M_{\mathrm{stellar}} > 10^{10} M_\odot/h$. The former are extracted from Christlein *et al.* (2009), and the latter from Padilla *et al.* (2010).

The method consists of selecting ETGs at a given redshift to then use their measured correlation function to obtain the dark-matter (DM) halo mass of their hosts. DM haloes of this mass are followed to $z = 0$ in a numerical simulation. At $z = 0$, their average DM mass or, equivalently, clustering amplitude are measured. We then find observational samples of $z = 0$ ETGs with the same clustering amplitude, and define these as the descendants of the high redshift MUSYC ETG. Using the luminosity function of the high and low redshift samples, we measure their number and luminosity densities. The former are lower in the descendant samples which indicates the need for mergers, whereas the latter show no change with redshift, indicating no need for star formation, which suggests the mergers would be dry.

References

Christlein D., Gawiser E., Marchesini D., & Padilla N., 2009, *MNRAS*, 400, 429

Gawiser E., *et al.* 2006, *ApJS*, 162, 1

Padilla N., Christlein D., Gawiser E., & Marchesini D., 2011, *A&A*, 531, 142

Padilla N., Christlein D., Gawiser E., Gonzalez R., Guaita L., & Infante L., 2010, *MNRAS*, 409, 184

The intriguing life of massive galaxies
Proceedings IAU Symposium No. 295, 2012
D. Thomas, A. Pasquali & I. Ferreras, eds.

© International Astronomical Union 2013
doi:10.1017/S1743921313004729

Studying Luminous Red Galaxies to probe H(z) at high redshift

A. Ratsimbazafy[1], C. Cress[1,2], S. Crawford[3] and SCALPEL team

[1] Physics Dept, University of the Western Cape, Private Bag X17. Cape Town 7535,
South Africa
email: `raljha.a@gmail.com`

[2] Centre of High Performance Computing, 15 Lower Hope St. Cape Town 7700, South Africa
[3] South African Astronomical Observatory, PO box 9. Cape Town 7935, South Africa

Abstract. Luminous Red Galaxies (LRGs) have old, red stellar populations often interpreted as evidence of a formation scenario in which these galaxies form in a single intense burst of star formation at high redshift. By measuring the average age of LRGs at two different redshifts, one can potentially measure the redshift interval corresponding to a time interval and thus measure the Hubble parameter $H(z) \approx -(1+z)^{-1}\Delta z/\Delta t$ (as in Jimenez & Loeb). The goal of this project is to measure directly the expansion rate of the universe at the redshift range $0.1 < z < 1.0$ within 3% precision. We explore the age-dating of Sloan Digital Sky Survey LRGs using the stellar population models of Lick absorption line indices after stacking spectra in redshift bins to increase the signal-to-noise. We also use the method of full spectral fitting to measure the ages of LRGs observed with the Southern Africa Large Telescope (SALT).

Keywords. galaxies: evolution , cosmology: cosmological parameters , cosmology: observations

---

Below, we show one of the SALT spectra of an LRG at $z = 0.40$ which we will use, along with many other LRG spectra at $z = 0.4$ and $z = 0.55$ to calculate Δt associated with $\Delta z = 0.15$ at $z_{av} \approx 0.47$. See Crawford *et al.* (2010) for more details.

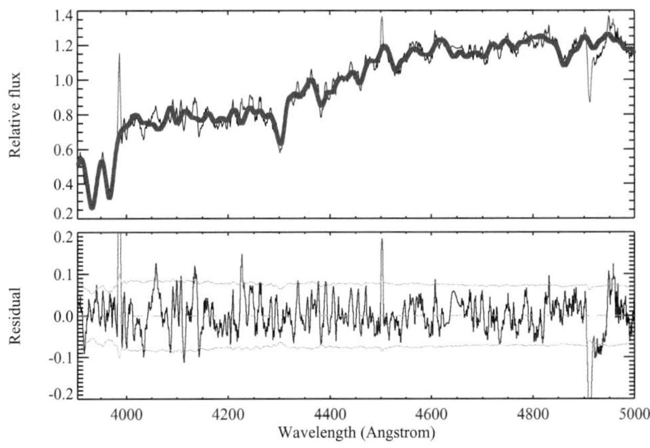

Figure 1. Preliminary result: Full spectrum fitting of SALT LRG at $z = 0.40$ using Ulyss package (Vazdekis models, Miles library). The thin black line is the observed SALT spectrum, the thick blue line is the best fit. Age = 8.165 ± 1.897 Gyr.

References

Crawford, S., Ratsimbazafy, A., Cress, C., *et al.* 2010, *MNRAS*, 406, 2569
Jimenez & Loeb 2002, *ApJ*, 573, 37

The intriguing life of massive galaxies
Proceedings IAU Symposium No. 295, 2012
D. Thomas, A. Pasquali & I. Ferreras, eds.

© International Astronomical Union 2013
doi:10.1017/S1743921313004730

Is there evolution in the black hole - bulge relation?

Andreas Schulze[1] and Lutz Wisotzki[2]

[1]Kavli Institute for Astronomy and Astrophysics, Peking University, 100871 Beijing, China
email: aschulze@pku.edu.cn

[2]Leibniz-Institut für Astrophysik Potsdam (AIP), An der Sternwarte 16,
14482 Potsdam, Germany

Abstract. We present a framework to investigate sample selection effects on the observed BH - bulge relation and its evolution with redshift. We particularly discuss an active fraction bias and a luminosity bias. Applying our framework to literature studies, we cannot find statistically significant evidence for an evolving BH-bulge relation.

Keywords. Galaxies: active - Galaxies: nuclei - quasars: general

---

The existence of the observed black hole - bulge relation implies a close link between the growth of black holes and the evolution of galaxies. The redshift evolution of this correlation provides constraints on the origin of this link.

In Schulze & Wisotzki (2011) we presented an investigation of sample selection effects that influence the observed black hole - bulge relation and its evolution with redshift. Our framework can be used to investigate all kinds of selection effects on the black hole-bulge relation, but we focus on the consequences of using broad-line AGN and their host galaxies to search for evolution in the black hole - bulge relation.

We particularly consider two important effects. (i) The active fraction among SMBH locally decreases with increasing black hole mass (Schulze & Wisotzki 2010). In connection with intrinsic scatter of the black hole - bulge relation this effect implies a bias towards a low BH mass at given bulge property. (ii) The flux limited nature of observational AGN samples leads to a bias towards a high black hole mass at given bulge property, as already discussed by others (Salviander *et al.* 2007, Lauer *et al.* 2007).

A quantitative prediction of these biases requires (i) a realistic model of the sample selection function, and (ii) knowledge of the relevant underlying distribution functions. At higher redshift, we have currently limited observational constraints on the latter.

Using the available constraints, we estimate the expected magnitude of sample selection biases for a number of recent observational attempts to study the BH-bulge evolution (e.g. Salviander *et al.* 2007, Merloni *et al.* 2010). In no case do we find statistically significant evidence for an evolving BH-bulge relation. While the observed apparent offsets in the BH-bulge ratio from the local relation can be quite large, the sample selection bias estimated from our formalism is typically of the same magnitude.

References

Lauer, T. R., Tremaine, S., Richstone, D., & Faber, S. M. 2007, *ApJ*, 670, 249
Merloni, A., Bongiorno, A., Bolzonella, M., *et al.* 2010, *ApJ*, 708, 137
Salviander, S., Shields, G. A., Gebhardt, K., & Bonning, E. W. 2007, *ApJ*, 662, 131
Schulze, A. & Wisotzki, L. 2010, *A&A*, 516, A87+
Schulze, A. & Wisotzki, L. 2011, *A&A*, 535, A87

The intriguing life of massive galaxies
Proceedings IAU Symposium No. 295, 2012
D. Thomas, A. Pasquali & I. Ferreras, eds.

© International Astronomical Union 2013
doi:10.1017/S1743921313004742

Dynamically Close Pairs of Galaxies Selected in the NIR

Ryan C. Keenan,[1] **Sebastien Foucaud,**[2] **Roberto De Propris,**[3,4] **and Jing-Hua Lin**[1,5]

[1] Academia Sinica Institute for Astronomy & Astrophysics, Taipei, Taiwan
[2] National Taiwan Normal University, Department of Earth Sciences, Taipei, Taiwan
[3] Cerro Tololo Inter-American Observatory, La Serena, Chile
[4] European Southern Observatory, Santiago, Chile
[5] National Taiwan University, Graduate Institute of Astrophysics, Taipei, Taiwan

Abstract. Studies of dynamically close pairs of galaxies can serve as a powerful probe of the galaxy merger rate and its evolution. Here we present a large sample of dynamically close pairs of galaxies selected in the $K-$band from the UKIDSS LAS. These data span ~ 175 deg^2 on the sky in the 2dFGRS equatorial region ($10^h < \mathrm{RA} < 14^h$). Combining the 2dFGRS redshifts with those from the SDSS, our $K-$band selected catalog is $> 90\%$ spectroscopically complete at $K_{\mathrm{AB}} < 16.4$. In this study, we focus on quantifying the relative contributions of wet, dry, and mixed mergers to the stellar mass buildup of galaxies over the past $1 - 2\,\mathrm{Gyr}$.

Keywords. Galaxies, Mergers, Near-Infrared

Synopsis: The galaxy merger rate and its evolution are important quantities for theories of galaxy formation. In hierarchical ΛCDM models, galaxies are expected to accrete most of their stellar mass via mergers, with at least 50% of the total stellar mass growth occurring at $z < 1$. Mergers should have a profound influence on galaxy properties such as morphology, star formation rate, and nuclear activity, among others.

Mergers are often categorized into three types: "wet", "dry", and "mixed". These labels describe whether the galaxies involved are star-forming, and contain significant cold gas, or are passively evolving. Wet mergers involve two galaxies that are actively forming stars and contain significant reservoirs of cold gas. Wet mergers are expected to trigger additional star formation, and potentially nuclear activity as well. On the contrary, in "dry" mergers both galaxies are passively evolving and contain little cold gas. Dry mergers are not expected to enhance star formation in the remnant. A "mixed" merger is one in which a star-forming galaxy with cold gas merges with a passive galaxy.

Of critical importance to understanding the stellar mass assembly of galaxies is what fraction of mergers trigger enhanced star formation. As noted above, roughly 50% of the stellar mass growth of galaxies takes place at $z < 1$. During the same epoch, the star formation rate density is declining dramatically, and thus, dry mergers must play an increasingly important role in the stellar mass buildup of galaxies over the last half of the age of the universe.

In our sample, we find a similar total merger rate ($\sim 1.5\%$) to $K-$band selected samples at lower redshifts. However, we find that the majority of the pairs in our sample are dry merger candidates, nearly a factor of three higher than the fraction (25%) found by other studies at $z \sim 0.1$. Some of this discrepancy could come from the fact that other groups have performed optical selections, while ours is in the NIR, or from a difference in classifying "blue" vs. "red" galaxies. In a forthcoming series of papers we will present a detailed study of stellar mass assembly of galaxies inferred from our dynamically close pair sample.

The intriguing life of massive galaxies
Proceedings IAU Symposium No. 295, 2012
D. Thomas, A. Pasquali & I. Ferreras, eds.

© International Astronomical Union 2013
doi:10.1017/S1743921313004754

Updated catalog of 132,684 galaxy clusters and evolution of brightest cluster galaxies

Z. L. Wen and J. L. Han

National Astronomical Observatories, Chinese Academy of Sciences, 20A Datun Road,
Chaoyang District, Beijing 100012, China; zhonglue@bao.ac.cn.

Abstract. We identified 132,684 clusters in the redshift range of $0.05 < z < 0.8$ from SDSS DR8. The spectroscopic redshifts of 52,683 clusters have been included in the catalog using SDSS DR9 data. We found that BCGs are more luminous in richer clusters and at higher redshifts.

From the Sloan Digital Sky Survey Data Release 8 (SDSS DR8), we identified 132,684 clusters in the redshift range of $0.05 < z < 0.8$ (Wen *et al.* 2012). Using photometric redshifts of galaxies, we recognized a cluster when the richness $R_{L*} = L_{\text{total}}/L^* \geqslant$ 12 and the number of member galaxy candidates within a photometric redshift gap of $z \pm 0.04(1 + z)$ and a radius of r_{200}, $N_{200} \geqslant 8$. Here, r_{200} is the radius within which the mean density of a cluster is 200 times of the critical density of the universe, L_{total} is the total luminosity of member galaxies, L^* is the characteristic luminosity. Monte Carlo simulations show that for rich clusters with a mass of $M_{200} > 1.0 \times 10^{14}\ M_\odot$ in the redshift range of $0.05 \leqslant z < 0.42$, the sample has a purity of $> 95\%$ and a completeness of $> 95\%$. The spectroscopic redshifts of 52,683 clusters have been included in the catalog based on the spectra data of the SDSS DR9. The updated catalog of galaxy clusters is now available at http://zmtt.bao.ac.cn/galaxy_clusters/.

Brightest cluster galaxies (BCGs) are luminous elliptical galaxies located at the potential centers of galaxy clusters. Because of the dominant role inside clusters and their unusual properties, the formation and evolution of BCGs are very intriguing. The BCGs of the clusters in our catalog are recognized as the brightest member galaxies within a radius of 0.5 Mpc from the number density peaks. We found that BCGs are more luminous in richer clusters and at higher redshifts, in the form of

$$M_r = (-21.25 \pm 0.01) - (1.75 \pm 0.03)z - (1.10 \pm 0.03)\log R_{L*}.$$

The color evolution of BCGs was investigated using a higher redshift cluster sample identified from the Canada-France-Hawaii Telescope Deep Survey and the Cosmic Evolution Survey using photometric redshifts (Wen & Han 2011). There are 294 clusters in the redshift range of $0.5 < z < 1.6$. The colors $r' - z'$ and $r^+ - m_{3.6\mu m}$ of most BCGs are consistent with a stellar population synthesis model (Bruzual & Charlot 2003) in which the BCGs are formed at redshift $z_f > 2$ and are evolved passively.

We also checked if the BCGs are luminous red galaxies (LRGs). We found that 25% of LRGs are the BCGs of our clusters, and 36% of LRGs are cluster member galaxies. In our cluster sample, 63% of BCGs satisfy the SDSS LRGs selection criteria for magnitude of $r_{\text{petro}} < 19.5$.

References

Bruzual, G. & Charlot, S. 2003, *MNRAS*, 344, 1000
Wen, Z. L. & Han, J. L. 2011, *ApJ*, 734, 68
Wen, Z. L., Han, J. L., & Liu, F. S. 2012, *ApJS*, 199, 34

The intriguing life of massive galaxies
Proceedings IAU Symposium No. 295, 2012
D. Thomas, A. Pasquali & I. Ferreras, eds.

© International Astronomical Union 2013
doi:10.1017/S1743921313004766

LAMOST 2D pipeline

BAI Zhongrui

National Astronomic Observatories, Chinese Academy of Sciences,
A20 Road Datun, Chaoyang Distinct, Beijing, China
email: `zrbai@bao.ac.cn`

Abstract. This paper describes the 2-dimensional data reduction of the LAMOST pilot survey.

Keywords. LAMOST, 2D pipeline, data reduction

The Large Sky Area Multi-Object Fiber Spectroscopic Telescope (LAMOST, also called the Guoshoujing Telescope) has the capability of taking 4000 spectra with resolution (R=1800) simultaneously in a single exposure (Cui *et al.* 2012). The LAMOST 2D pipeline aims to extract spectra from CCD images, and to calibrate them, including wavelength calibration, sky subtraction, flux calibration and co-addition. A number of procedures from the spectro2d pipeline of SDSS are used (Stoughton *et al.* 2012).

The pipeline has been run successfully on all spectroscopic data obtained during LAMOST Pilot Surveys, from October 24, 2010 to June 17, 2012(Luo *et al.* 2012).

Flux extraction. The raw data is obtained by 32 EEV CCD chips (Wei & Stover 1996)(Fig. 1). The fiber position on the CCD is traced in flat-field spectra. The centroid of one fiber in row direction is fitted by a polynomial along column direction. The profile of one fiber in row direction is assumed to be a Gaussian-like profile, which is characterized by $y = ae^{-(x-b)^d/dc^d}$. The coefficients are calculated by flat-field spectra as used for other exposures.

Wavelength calibration. The arc lamp spectra are extracted, and the centroids of the arc emission lines are measured, to which we fit a Legendre polynomial as a function between wavelengths and pixels. The wavelength error of both sides is less than 0.1 Å.

Sky subtraction. More than 20 fibers are used to model the background of a specific region of the sky. A supersky model is constructed by performing iterative fits to the extracted spectra associated with blank sky positions. For each fiber in that region, the supersky background is subtracted. Telluric absorption in four bands is removed.

Flux calibration. About 5 of the fibers are selected to point to flux standard stars. The sky-subtracted flux is divided by the best-fit model at the best-fit redshift. These ratios are called the calibration vectors. The set of flux-calibration vectors for an entire spectrograph are combined using B-splines to compute an average flux-calibration vector per CCD. Finally, for each object in the individual exposures, the spectra from the red and blue sides are combined. The combined spectra are resampled in constant-velocity pixels, with a pixel scale of 69km s^{-1}.

Conclusion. This pipeline has been used for the LAMOST Pilot Survey in which we released about 316,000 spectra. The wavelength calibration is accurate to 10km s^{-1} or better. The sky residual is less than 10%.

Figure 1. Raw data on the CCD.

References

Cui, X., Zhao, Y., Chu, Y., *et al.* 2012, *RAA*, 12, 1197
Stoughton, C., Lupton, R. H., *et al.* 2012, *AJ*, 123, 485
Luo, A., Zhang, H., Zhao, Y., *et al.* 2012, *RAA*, 12, 1243
Wei, M. & Stover R. J. 1996, *Solid State Sensor Arrays and CCD Cameras(San Jose)*, 226

The intriguing life of massive galaxies
Proceedings IAU Symposium No. 295, 2012
D. Thomas, A. Pasquali & I. Ferreras, eds.

© International Astronomical Union 2013
doi:10.1017/S1743921313004778

The evolution of massive galaxies in semi-analytical models of galaxy formation

Carlton M. Baugh

Institute for Computational Cosmology, Department of Physics,
Durham University, Science Laboratories, South Road, Durham, DH1 3LE, UK.
email: `c.m.baugh@durham.ac.uk`

Abstract. Massive galaxies with old stellar populations have been put forwards as a challenge to models in which cosmic structures grow hierarchically through gravitational instability. I will explain how the growth of massive galaxies is helped by features of hierarchical models. I give a brief outline of how the galaxy formation process is modelled in hierarchical cosmologies using semi-analytical models, and illustrate how these models can be refined as our understanding of processes such as star formation improves. I then present a brief survey of the current state of play in the modelling of massive galaxies and list some outstanding challenges.

Keywords. galaxies:formation, galaxies:evolution, galaxies: high-redshift, mehtds:numerical.

---

1. Introduction

The observational evidence in support of the emergence of massive galaxies at early epochs in the history of the Universe is now overwhelming. Different techniques have successfully isolated galaxies at substantial lookback times, which have stellar masses similar to those of the largest galaxies seen today. Often the methods used to identify such galaxies use colour selection, and target galaxies which either have old stellar populations with little recent star formation, or which may have substantial star formation activity, but which appear red because they are enshrouded in dust.

At the same time, the past twenty years have seen a hierarchical cosmology – the cold dark matter model with a cosmological constant (ΛCDM) – cemented in place as the "standard model" of cosmology. In ΛCDM, very small fluctuations in the density of the Universe, as revealed in maps of the cosmic microwave background radiation, are amplified due to gravitational instability. The assembly of self-gravitating structures called dark matter haloes, is hierarchical, proceeding via steady accretion and mergers of more substantial structures. Analytic calculations and N-body simulations have been applied to study the hierarchical clustering of the dark matter to high accuracy. The hierarchical growth of the abundance of dark matter haloes is well established in such calculations.

The apparent dilemma then faced by theoreticians is how do we reconcile the observations of seemingly old, massive galaxies in place at early epochs in the history of the Universe with the steady growth of dark matter haloes, whereby the largest structures are forming now?

In this review I will describe how massive galaxies fit into hierarchical models, explaining how their mass is built up. I will then outline the semi-analytical approach to modelling galaxy formation, setting out the pros and cons of the technique. I will highlight results from different groups running semi-analytical models; due to space restrictions, this will be an unavoidably incomplete overview of the current literature, but will hopefully serve to whet the appetite of the reader to investigate the predictions of this class of model further.

2. How do galaxies acquire their mass?

Two features of hierarchical clustering cosmologies help to build big galaxies at early epochs. These effects are not peculiar to the approach used to model galaxy formation, but instead arise from the process of gravitational instability operating in a model with fluctuations over a broad range of mass scales.

These effects are:

Formation bias. Regions of the Universe which are overdense evolve faster than those that are underdense. The higher density means a faster turnaround from the background expansion and an earlier gravitational collapse, compared with the evolution of a similar sized object in an underdense part of the Universe. Hence within a large overdensity, dark matter haloes which are able to retain gas that has been heated by supernovae will tend to form sooner and so galaxy formation will be advanced compared to an average patch of the Universe. Therefore, if massive galaxies form in massive haloes, then there is a natural tendency for these objects to form earlier, purely due to gravity.

Formation time versus assembly time. In hierarchical models, it is not necessary to form the mass of a galaxy in situ, in one object. The task of accumulating stellar mass can be delegated to a number of progenitor galaxies, which may individually form stars at modest rates, but collectively build stellar mass more prodigiously. Massive galaxies display a distinct shift between the formation time of their stars, e.g. the time when half of the present day mass was in place in all of the progenitors, and the assembly time, when the progenitors merge into one object (Baugh, Cole & Frenk 1996; Kauffmann 1996; De Lucia *et al.* 2006).

3. Semi-analytical modelling of galaxy formation

By itself, the cold dark matter model says nothing directly about galaxy formation. Inferences can be drawn about the sequence of galaxy formation, based on how structures grow in the dark matter. However, without an attempt at a physically motivated calculation of the fate of baryons in a cold dark matter Universe, there is little hope of learning much about galaxy formation or of understanding the implications of observations of high redshift galaxies for the cold dark matter cosmology (see recent reviews, Baugh 2006 and Benson 2010).

White & Rees (1978) argued that galaxy formation is a two-phase process, with the bulk of the mass undergoing a dissipationless collapse which is responsible for building the gravitational potential wells, or halos, in which galaxies form. The baryonic component of the Universe is able to dissipate energy, and therefore to collapse down to smaller scales, forming denser units. This model was able to explain the appearance of clusters of galaxies. However, without an additional process to reduce the efficiency of galaxy formation in shallow gravitational potential wells, the predicted luminosity function is much steeper than is observed at the faint end.

This pioneering work, along with a clutch of papers published around the same time looking at the radiative cooling of gas within gravitational potential wells, laid the groundwork for modern galaxy formation theories. The time taken for all of the gas within a halo to cool radiatively increases with halo mass. This is because cooling is a two-body process (collisionally excited radiative transitions or bremsstrahlung) which depends on the square of the gas density. In hierarchical models, more massive haloes tend to form later when the density of the Universe is lower. It is possible for the cooling time of the gas to exceed the Hubble time, thus limiting the supply of cold gas to form a galaxy (see the review of Fred Hoyle's contributions to galaxy formation theory by Efstathiou 2003).

The first papers to incorporate these ideas fully into the cold dark matter cosmology, introducing the semi-analytical methodology, were published in 1991 (White & Frenk 1991; Cole 1991, Lacey *et al.*1991). This approach tries to follow a wide range of the processes which are thought to be important in determining the fate of the baryons. This is a daunting task. At the time, theories of star formation were rudimentary at best. There has been much progress in this area since 1991, but we are still a long way from having a reliable description of the process which underpins galaxy formation. The regulation of star formation efficieny comes from the stars themselves. Stars above $\approx 5-8$ times the mass of the Sun end their life in a Type II supernova, which injects substantial amounts of energy and momentum into the interstellar medium. This alters the state of the gas in the interstellar medium (ISM), perhaps leading to the ejection of gas from the galactic disk or even the dark matter halo. This process is known as supernova feedback and is critical to the success of any model of galaxy formation.

The absence of a precise description of a key process, such as star formation or supernova feedback, may lead one to consider giving up any hope of ever understanding galaxy formation. Instead, in semi-analytical modelling an attempt is made to write down the differential equation which gives the current best bet model of how the system behaves. As our understanding develops, or when new observations clarify how a process works, then the model can be improved. The differential equation may contain a free parameter. Often there is little guidance as to the appropriate range of values to take for the parameter. In such instances, the only approach is to be pragmatic and see what the model predicts for different parameter values. By comparing the model predictions to observations, the value of the parameter is set as the one which gives the most faithful reproduction of the data. This procedure is exactly what physicists undergo when attempting to describe complex phenomena: start off with a simple model, which can be adjusted or refined to improve the description of observations. I will give an example of this principle in action below.

The semi-analytical framework allows us to model a range of processes together, within the cosmological setting of the formation of structure in the dark matter. The ability to follow the interplay between processes is essential in studying galaxy formation. The models solve the set of differential equations which govern the flow of mass and metals between different reservoirs of baryons: hot gas, cold gas and stars. The output of the models is the full star formation and chemical enrichment histories for a wide range of galaxies, including mergers between galaxies.

Semi-analytical modelling has some features which might be perceived as limitations or drawbacks. One example is the generality of the assumptions which are needed to be able to calculate the fate of the baryonic component (e.g. the assumption of a spherically symmetric hot halo gas profile to compute the cooling rate). Another is the "deterministic" way in which processes such as supernova feedback are modelled. In the semi-analytical model, the mass loading of the supernova driven wind is specified by choosing model parameters, and precisely this amount of gas is ejected from the ISM. In a gas dynamics simulation in which the wind is fully coupled to the hydrodynamics equations, the same number of supernovae could result in a very different mass of gas being ejected. The mass loading could be intricately linked to the resolution of the simulation.

Nevertheless, despite the progress made over the past twenty years, there is still widespread mistrust of semi-analytical modelling. Recently this has led to something akin to grassroots insurgency in the subject, effectively a "tea party" movement for galaxy formation, whereby simplified models have been devised with the aim of elucidating how galaxies form. Examples include the "bathtub" and "reservoir" models. These calculations are inspired by models of supply and demand from economics, and track the inflow (sources) of gas into halos and the "sinks" of cold gas in star formation and supernova feedback. In their simplest form, the models follow one galaxy per halo, and invoke ad-hoc efficiency factors to specify the inflow of gas as a function of halo mass, without any attempt to calculate the rate at which gas can cool or to explain the form of the efficiency factor. Galaxy mergers are ignored. This class of calculation effectively takes one of the equations which has been considered within semi-analytical models for more than a decade and solves it in isolation.

The desire for a better grasp of how galaxy properties are shaped by different processes is understandable, but it is not clear that it can be usefully gained from stripped-down approaches. The perceived "complexity" of semi-analytical modelling is actually the great strength of the technique. The ability to model the interplay between processes is the key to building a realistic model of galaxy formation. By taking a more complete view of galaxy formation rather than a selective one, the consequences of the calculation – the predictions of the model – are more far reaching and therefore more tightly constrained by observations. If the model seems complex, then this is simply a reflection of the nature of the underlying processes, such as star formation and heating by supernovae.

4. An illustration of semi-analytical modelling: the star formation rate in galaxies

An illustration of how semi-analytical models work can be obtained by considering how star formation is modelled within a galaxy.

The bulk of semi-analytical models attempt to predict the global star formation rate within a galaxy. The early modelling of the star formation rate was essentially based on dynamical arguments, with loose motivation coming from a comparison to the Kennicutt-Schmidt law (Bell *et al.* 2003). The star formation rate, ψ, is often parametrised as

$$\psi = \epsilon \frac{M_{\mathrm{cold}}}{\tau},$$

where M_{cold} is the total mass of cold gas, and ϵ is an efficiency factor, which controls the fraction of cold gas which is turned into stars in the timescale τ. The timescale for star formation is generally assumed to scale with the dynamical time within the galaxy:

$$\tau = t_{\mathrm{dyn}} f(v_{\mathrm{disk}}).$$

In some models, $f(v_{\mathrm{disk}}) = 1$; in the Cole *et al.* (2000) model, an explicit scaling of the star formation timescale with the circular velocity of the disk was implemented, to allow the model to produce a better match to the observed gas fraction luminosity relation for spiral galaxies: $f(v_{\mathrm{disk}}) = (v_{\mathrm{disk}}/200\mathrm{kms}^{-1})^{\alpha_*}$. Hence in the most general case, two parameters are required to set the star formation rate: ϵ and α_*. These parameters are set by chosing values, running the model and then comparing the model predictions to observables. The key observables for constraining the values of these star formation parameters are the gas fraction - luminosity relation, the galaxy luminosity function and the colour magnitude relation.

High resolution imaging of galaxies at different wavelengths has revealed that star formation activity correlates better with the molecular hydrogen content of galaxies than with the overall cold gas mass. Lagos *et al.* (2011a) investigated more general star formation models in the GALFORM semi-analytical model, implementing different empirical and theoretically motivated star formation laws (see also Cook *et al.* 2010; Fu *et al.* 2010). The most successful of these was the empirical star formation law proposed by Blitz & Rosolowsky (2006), who suggested that the observational data could be explained if the ratio of molecular to atomic hydrogen is set by the pressure in the mid-plane of galactic disks; gas disks with higher pressure have a higher fraction of H_2.

This work illustrates the modularity of semi-analytical modelling and how it provides a framework in which new and improved descriptions of various processes can be readily implemented. The Blitz & Rosolowsky star formation law involves two observationally determined "parameters". Whereas in the original parameterization of the star formation rate there was little guidance about the range of parameter values which should be considered, there is now a much smaller volume of parameter space to search (at least once the Blitz & Rosolowsky law has been adopted). Furthermore, as the modelling of the star formation becomes more sophisticated, the predictions that can be made by the model expand. Rather than simply outputting the cold gas mass of galaxies, the atomic and molecular hydrogen contents are now predicted, meaning that the model should also be able to reproduce the mass functions of HI and H_2, their evolution and their relation to other galaxy properties (Lagos *et al.* 2011b). By combining GALFORM with the photon dominated region model of Bell *et al.* (2006), it is also possible to predict the different carbon monoxide transitions, and to make contact with observations from ALMA (Lagos *et al.* 2012).

Hence by adopting the improved star formation model, the parameter space open to the model has shrunk in volume and the constraints on the model have increased through the capability to make new predictions which must match the available observations.

5. Massive galaxies in hierarchical models

I now review some of the predictions made by semi-analytical models for the abundance of massive galaxies at different epochs.

5.1. *The massive end of the stellar mass function*

Traditionally semi-analytical models have used observational estimates of the local galaxy luminosity function to set the values of many of the model parameters. The comparison with the luminosity function requires modeling which is already part of the majority of semi-analytical models. This involves taking the star formation and chemical enrichment histories predicted for each galaxy, including bursts of star formation, and combining this with a simple stellar population model, and a model for the attenuation of starlight by dust. The stellar population model depends on the choice of stellar initial mass function. This choice in turn influences the galaxy formation model, by setting the yield of metals and the mass of gas recycled into the ISM during an episode of star formation.

With the advent of large galaxy redshift surveys with multi-band photometry, the present day stellar mass function has now been determined to high accuracy in so far as sampling errors are concerned (Cole *et al.* 2001; Bell *et al.* 2003; Li & White 2009). The Li & White determination of the stellar mass function is based on stellar masses inferred from SDSS photometry and redshifts, assuming a simple model for the star formation history, a universal stellar initial mass function (IMF) and a model for dust extinction.

Hence, the comparison between the model predictions with the observational estimate requires an additional layer of modelling when considering the stellar mass function compared with the luminosity function. A similar set of assumptions to those already made

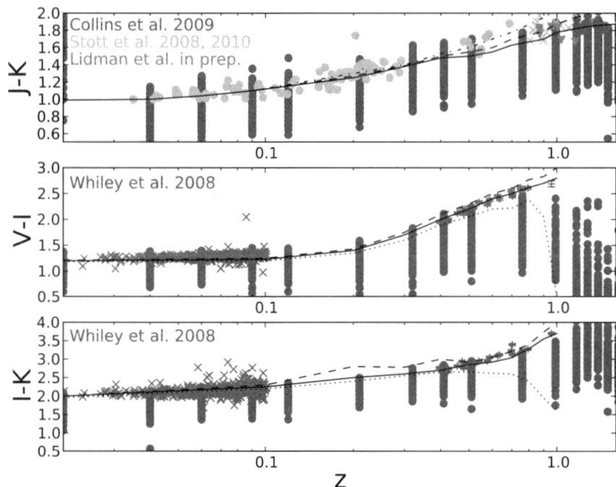

Figure 2. The color evolution of the model BCGs from redshift $z = 0$ to $z \sim 1.6$ (in all panels the model galaxies are represented by *blue points*). *Upper panel*: $J - K$, data by Collins *et al.* (2010; *red points*), Stott *et al.* (2008, 2010; *green points*) and Lidman *et al.* (*orange triangles*). The lines represent toy models discussed in Collins *et al.* (2009) for single stellar populations (synthesized with the BC03 models), in the cases of no luminosity evolution (*dot-dashed line*), with passive evolution with formation redshifts of 5 (*dashed line*) and with formation redshift of 2 (*solid line*). *Middle panel*: $V - I$, data by Whiley *et al.* (2008; *purple symbols*. *Lower panel*: $I - K$, data by Whiley *et al.* (2008; *purple symbols*. In the *middle and lower panels*, the lines represent toy models discussed in Whiley *et al.* (2008) for single stellar populations (synthetised with the BC03 models), in the cases of passive evolution with formation redshifts of 1 (*dotted line*), formation redshift of 2 (*solid line*) and formation redshift of 5 (*dashed line*).

2. The physical properties of BCGs

As described in detail in Tonini *et al.* (2012), the model BCGs evolve following diverse star formation histories, with one feature in common: none of these galaxies is passively evolving, they all have some degree of residual star formation at $z = 0$, and they are actively star forming at $z = 1$. However, the BCG luminosity and colour evolution in a fully hierarchical assembly scenario, with the use of M05 stellar populations, is indistinguishable from a toy-model scenario of passive evolution with other stellar population models. The model BCGs are star forming and assembling a significant fraction of their mass (up to 50%) after $z = 1$ (see Tonini *et al.* 2012). However, how old are these galaxies? Or more accurately, how old are the stars that compose them? Do these galaxies follow a special evolutionary path in the hierarchical galaxy formation scenario?

Figure (3) shows the distribution of ages of the model BCGs as a function of redshift. We define the age of each BCG as the minimum age of a certain fraction of its stars, which corresponds to the lookback time of their formation. The contours represent the number of galaxies that have a certain age at a certain redshift. In the *left panel* we define the mass-weighted age of the BCG as the lookback time by which 50% of its stars were formed, while on the *right panel* we define the same quantity based on 90% of its stars.The *black line* in both plots shows the age of the Universe as a function of redshift. This figure reveals a number of things. First, the evolution of the model BCGs fans out from $z \sim 1$, showing a larger and larger variety in age as time goes by. This behaviour is driven by the differences in mass and environment, and is sustained by the ongoing star formation. This variety causes the colours of the model BCGs to scatter towards the blue. Second, most model BCGs are actually *old*. For most of them, at any given time the majority of their stars are of an age equal at least to half the age of the universe. Third, the BCGs population seems to grow old following the ageing of the universe itself.

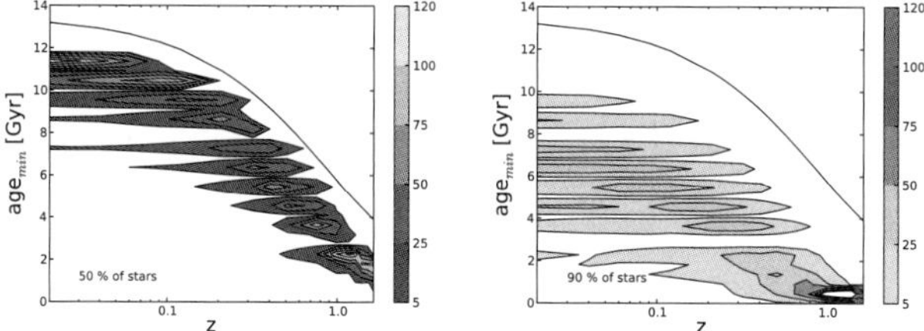

Figure 3. The distribution of ages of the model BCGs, as a function of redshift. We define the age of each BCG as the minimum age of a certain fraction of its stars, which corresponds to the lookback time of their formation. The contours represent the number of galaxies that have a certain age at a certain redshift. In the *left panel* we define the age of the BCG based on the oldest 50% of its stars, while on the *right panel* we define its age based on the oldest 90% of its stars. The *black line* in both plots shows the age of the Universe as a function of redshift.

There is a constant gap between the age of the oldest BCGs and the age of the universe (the black line in the plot), whether one considers the 50% or the 90% mark as an age indicator. Figure (3) means the following: the star formation in the model BCGs is not intense enough to offset their mass-weighted age since $z \sim 1.6$. There is not enough newly formed stellar mass to rejuvenate these objects, whose ageing is always dominated by the ageing of their stellar populations, regardless of where they were formed (main progenitor or satellites). For comparison, a truly actively star forming galaxy (like a spiral) would show a flatter age-redshift relation, which means that the fractional increase of stellar mass is dominated by new stellar populations (and not accretion of old stars), with a rate sufficiently high to lower the mass-weighted age. Since the ageing of the BCG is dominated by the ageing of its stellar populations (1 Gyr every Gyr), implying that the difference between the BCG age and the age of the Universe is constant, this represents an equivalent scenario to that of the ageing of a single stellar population in a passive evolution toy model. The difference is that we are now considering not just one stellar population, but all the populations that end up forming the BCG, i.e. we are looking at the collective ageing of the stellar component in the merger tree. We define this behaviour as 'passive evolution in the hierarchical sense'.

References

Brough, S., Couch, W. J., Collins, C. A., Jarrett, T., Burke, D. J., & Mann, R. G. 2008, *MNRAS*, 385, 103

Collins, C. A., *et al.* 2009, *Nature*, 458, 603

Croton, D. J., *et al.* 2006, *MNRAS*, 365, 11

Daddi, E., *et al.* 2007, *ApJ*, 670, 156

Maraston, C. 2005, *MNRAS*, 362, 799

Stott, J. P., Edge, A. C., Smith, G. P., Swinbank, A. M., & Ebeling, H. 2008, *MNRAS*, 384, 1502

Stott, J. P., *et al.* 2010, *ApJ*, 718, 23

Tonini, C., Maraston, C., Devriendt, J., Thomas, D., & Silk, J. 2009, *MNRAS*, 396, 36

Tonini, C., Maraston, C., Thomas, D., Devriendt, J., & Silk, J. 2010, *MNRAS*, 403, 1749

Tonini, C., Maraston, C., Ziegler, B., Böhm, A., Thomas, D., Devriendt, J., & Silk, J. 2011, *MNRAS*, 415, 811

Tonini, C., Bernyk, M., Croton, D., Maraston, C., & Thomas, D. 2012, *ApJ*, 759, 43

Whiley, I. M., *et al.* 2008, *MNRAS*, 387, 1253

The intriguing life of massive galaxies
Proceedings IAU Symposium No. 295, 2012
D. Thomas, A. Pasquali & I. Ferreras, eds.

© International Astronomical Union 2013
doi:10.1017/S1743921313004791

Structural evolution of
massive early-type galaxies

Ludwig Oser[1], Thorsten Naab[1], Jeremiah P. Ostriker[2],
Peter H. Johansson[3]

[1] Max-Planck-Institut für Astrophysik, Karl-Schwarzschild-Strasse 1, 85741,
Garching, Germany
[2] Department of Astrophysical Sciences, Princeton University, Princeton, NJ 08544, USA
[3] Department of Physics, University of Helsinki, Gustaf Hällströmin katu 2a, FI-00014
Helsinki, Finland
oser@mpa-garching.mpg.de

Abstract. We use a large sample of cosmological re-simulations of individual massive galaxies to investigate the origin of the strong increase in sizes and weak decrease of the stellar velocity dispersions since z=2. At the end of a rapid early phase of star-formation, where stars are created from infalling cold gas, our simulated galaxies are all compact with projected half-mass radii of $\lesssim 1\,\mathrm{kpc}$ and central line-of-sight velocity dispersions of $\approx 262\,\mathrm{km\,s^{-1}}$. At lower redshifts ($z < 2$) those galaxies grow predominantly by the accretion of smaller stellar systems and evolve towards the observed local mass-size and mass-velocity dispersion relations. This loss of compactness is accompanied with an increase of central dark matter fractions. We find that the structural evolution of massive galaxies can be explained by frequent minor stellar mergers, which is the dominant mode of accretion for our simulated galaxies.

Keywords. galaxies: formation, galaxies: evolution, methods: numerical

1. Introduction

When compared to systems at the present day, massive ($\approx 10^{11}\,M_\odot$) quiescent galaxies at redshift 2 are much more compact. Their half-mass radii are typically a factor of three to five times smaller ($\approx 1\,\mathrm{kpc}$) and their effective stellar densities are at least an order of magnitude higher (Bezanson *et al.* 2009). These results are supported by deep observations that cannot find faint, previously missed, stellar envelopes (Carrasco *et al.* 2010) as well as by dynamical mass estimates derived through velocity dispersion measurements. While these massive compact galaxies are common at redshift 2 among the quiescent galaxy population they are extremely rare in the present-day Universe, suggesting that today's massive early-type galaxies underwent significant structural evolution up to the present day.

We show here that the structural evolution of massive early-type galaxies can be understood as a consequence of the two-phase formation scenario (Oser *et al.* 2010, Johansson *et al.* 2012). The compact cores of massive galaxies form during an early rapid phase of dissipational in-situ star formation at $6 \gtrsim z \gtrsim 2$ fed by cold flows and/or gas rich mergers leading to large stellar surface densities. At redshift $z \sim 2$ the observed as well as simulated galaxies are compact ($R_e \sim 1\mathrm{kpc}$), more flattened and disk-like than their low redshift counterparts. The subsequent evolution is dominated by the addition of stars that have formed ex-situ, i.e. outside the galaxy itself. This accretion is dominated by minor mergers and provides an explanation for the structural evolution of massive galaxies. The early domination of in-situ star formation and the subsequent growth by stellar mergers is in agreement with semi-analytical models (De Lucia *et al.* 2006).

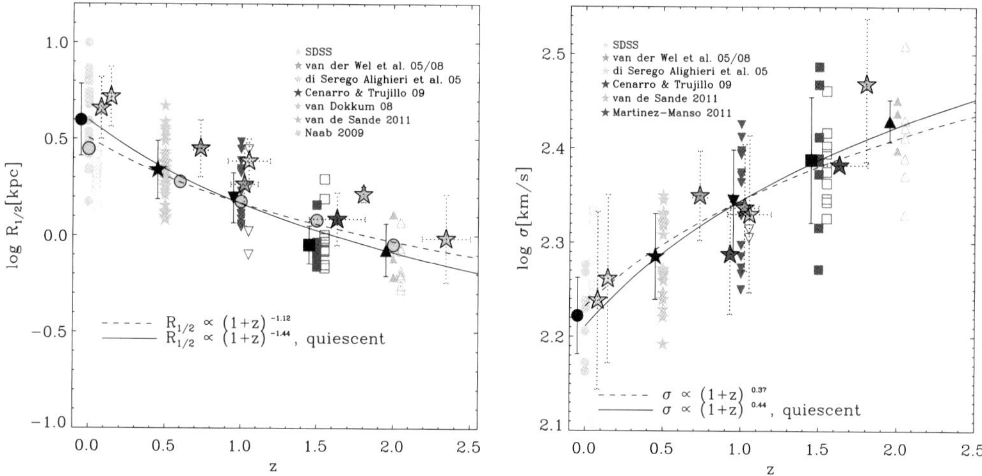

Figure 1. *Left*: Projected stellar half-mass radii of galaxies with stellar masses $M_* > 6.3 \times 10^{10} M_\odot$ as a function of redshift. The black symbols indicate the mean sizes at a given redshift with the error bars showing the standard deviation. The black lines show the result of a power law fit for all (dotted line) and quiescent (solid line and solid symbols) systems, respectively, in good agreement with the observed relations. Observational estimates from different authors are given by the solid star symbols. By z=3 all progenitor galaxies drop below our mass limit. *Right*: Evolution of the central stellar velocity dispersions, with open symbols for star-forming, and closed symbols for quiescent systems. We show the results for a power law fit with dashed (all) and solid (quiescent) lines. The simulated trend follows closely the observed relations (star symbols).

2. Size and velocity dispersion evolution

We present here the results obtained from 40 zoom-in re-simulations described in Oser *et al.* (2010) and Oser *et al.* (2012). These simulations follow the formation and evolution of a massive galaxy at high resolution with the help of a modified version of the TreeSPH code Gadget2 (Springel *et al.* 2005) including radiative cooling, star formation, supernovae feedback and a uniform UV radiation background. In an early dissipational phase, the stellar mass growth is primarily due to in-situ star formation. The baryonic mass is assembled in gaseous form and a significant amount of the potential energy is dissipated (Johansson *et al.* 2009). This leads to compact systems, with our simulated massive galaxies ($\sim 10^{11} M_\odot$) having half-mass radii of about 1 kpc (see left panel of Fig. 1). At later times, the simulated galaxies increase their stellar mass primarily by the accretion of smaller stellar systems, which – since this is a collisionless process – inevitably leads to a strong increase of the half-mass radii (Naab *et al.* 2009, Bezanson *et al.* 2009). The left panel of Fig. 1 shows the evolution of the average sizes of all galaxies with masses between $6.3 \times 10^{10} M_\odot$ and $5 \times 10^{11} M_\odot$ from redshift 2 up to the present day. We define galaxies with a specific star formation rate lower than $0.3/t_{hub}$, where t_{hub} is the Hubble time, as quiescent (solid symbols in Fig. 1). We fit the evolution of galaxy sizes with $R_{1/2} \propto (1+z)^\alpha$. We find $\alpha = -1.12$ for all and $\alpha = -1.44$ for the quiescent galaxies in agreement with the observed values.

At the same time, the central stellar velocity dispersions drop continuously with time from $\sim 260\,\mathrm{km\,s}^{-1}$ at redshift 2 to $\sim 180\,\mathrm{km\,s}^{-1}$ at the present day. We show the evolution of stellar velocity dispersions for our simulated galaxies along with some observations in the right panel of Fig. 1. We use again a power law ($\sigma_{1/2} \propto (1+z)^\beta$) to fit our results

216

O. Gerhard

second galaxy within a $10-20'$ field. With discrete tracers (PNe or GCs) it is possible to eliminate kinematic substructures, which is a prerequisite for a reliable dynamical mass determination in such systems.

4.4. *Dark halo circular velocity curves and densities from stellar kinematics*

Using absorption line kinematics to $1\text{-}2R_e$ for luminous round ETGs, and nonparametric spherical distribution function models in combined luminous and dark halo potentials, Gerhard *et al.* (2001) found that the CVCs of ETGs are approximately flat. A more recent analysis of a sample of ETGs in the Coma cluster with $M_B = [-18.8, -22.6]$ and kinematic data out to $1-3\,R_e$ with axisymmetric Schwarzschild models by Thomas *et al.* (2007) showed somewhat more varied CVC shapes. Massive ETGs tend to have flat CVCs (e.g., NGC 4649, Das *et al.* 2011), while lower-mass ETGs (see §4.5) are also consistent with slightly decreasing CVCs. These studies agree in their derived DM fractions, $\sim 10-40\%$ within R_e, assuming constant M/L for the stellar component, which is also consistent with models of inner 2D kinematics by Cappellari *et al.* (2006).

Both studies also agreed in finding significantly higher central dark matter densities in ETGs than in spiral galaxies of the same luminosity or mass. For the Coma ETGs, Thomas *et al.* (2009) found a factor of $7\times$ higher mean DM density within $2R_e$ at the same stellar mass, and $13\times$ at the same luminosity. Baryonic contraction is not sufficient to explain this difference. One simple explanation is that the cores of ETG halos assembled earlier (at redshifts 1-3) than spirals of same luminosity (Gerhard *et al.* 2001; Thomas *et al.* 2009). Good agreement of the dynamical luminous plus dark matter models was found with SLACS lensing models of Auger *et al.* (2009); see Thomas *et al.* (2011).

4.5. *Mass Distributions of Quasi-Keplerian Ellipticals*

Méndez *et al.* (2001) first showed that the steeply decreasing outer velocity dispersion profile of the intermediate-luminosity elliptical (ILE) galaxy NGC 4697 could be matched well with an isotropic Hernquist model. They thus concluded that no evidence for dark matter had been found out to $3R_e$, but that dark matter could be present if the velocity distribution is anisotropic. Romanowsky *et al.* (2003) drew attention to three further ILEs with steeply decreasing $\sigma_p(R)$, NGC 821, 3379, and 4494, suggesting the presence of little if any dark matter in their halos. However, the well-known mass-anisotropy degeneracy (Binney & Mamon 1982) is much stronger in galaxies such as these ILEs, with de Vaucouleurs type luminosity profiles and steeply decreasing σ_p-profiles, than in galaxies with either more shallow luminosity profiles or radially constant projected velocity dispersions (Gerhard 1993).

With NMAGIC particle models (see de Lorenzi *et al.* 2007) based on a variety of kinematic data, including several hundred PN line-of-sight velocities, de Lorenzi *et al.* (2008, for NGC 4697), de Lorenzi *et al.* (2009, for NGC 3379) and Morganti *et al.* (2013, for NGC 4494) determined the range of quasi-isothermal halo potentials consistent with the data for these galaxies. Fig. 1c shows the 70% confidence boundaries from the PN likelihood for NGC 3379 and 4697, and from a more comprehensive analysis for NGC 4494. While some spread in the allowed CVCs and anisotropies remains, as expected, dark matter is required in all 3 galaxies. Fig. 1c indicates that the most favoured CVCs may be slightly falling, as in the lower-mass systems in Wu *et al.* (2012). Relative to the outer CVCs, the baryonic centers of these galaxies are quite centrally concentrated, as could be expected in gas-rich mergers.

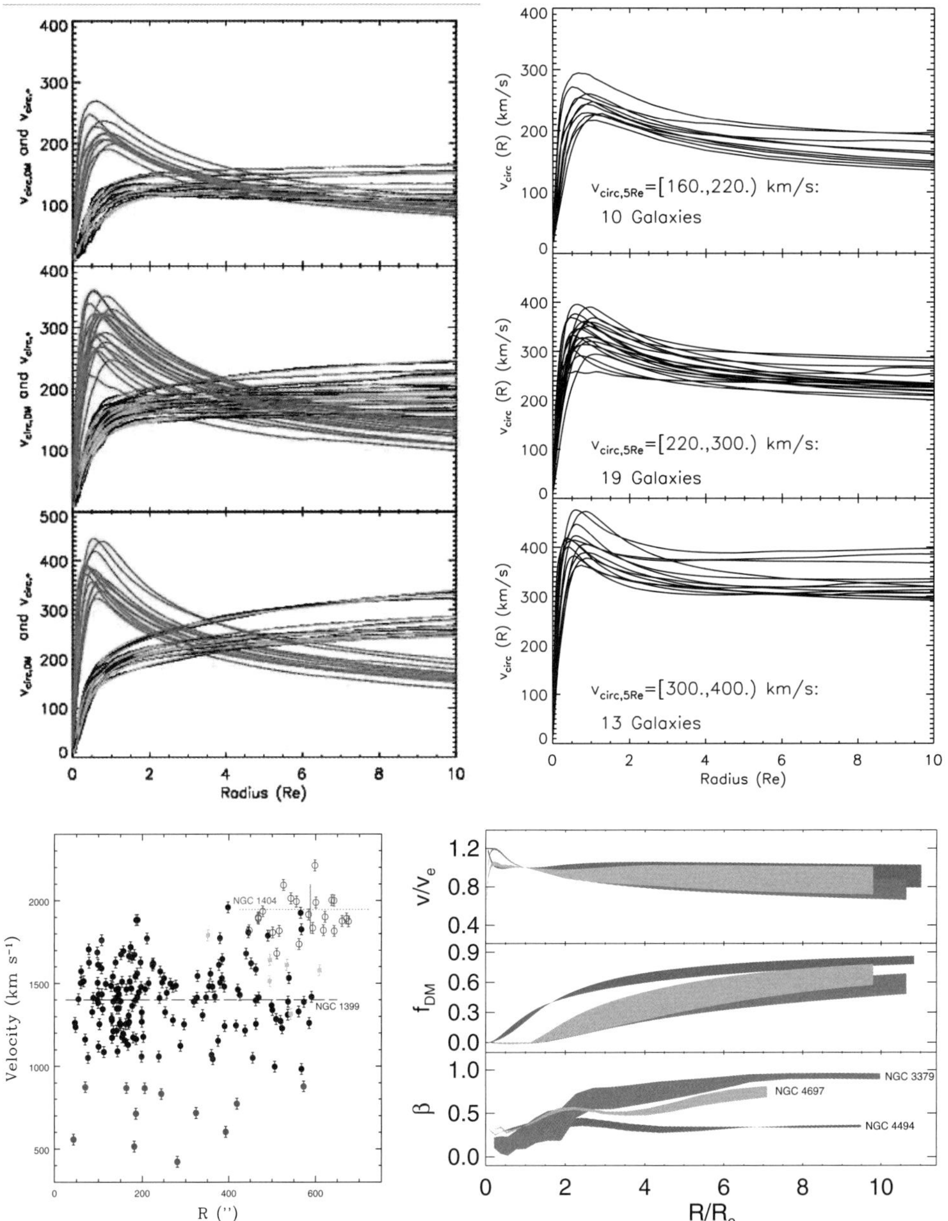

Figure 1. (a) Top: Circular velocity curves from stars (magenta), dark matter (black-blue, both left panel) and total mass distribution (right), for 42 model galaxies from the cosmological simulation of Oser *et al.* (2012), in bins of increasing total circular velocity at $5R_e$, as given on the right panels. From Wu *et al.* (2012). (b) Bottom left: Phase-space distribution of PNe in the nearby BCG galaxy NGC 1399 (black), its neighbour galaxy NGC 1404 (blue), and the low-velocity component (red). From McNeil *et al.* (2010). (c) Bottom right: 70% confidence ranges from dynamical modeling for total circular velocity normalized to value at R_e, DM fraction, and anisotropy parameter, for the three quasi-Keplerian ETGs NGC 3379, 4494, 4697. From Morganti *et al.* (2013).

4.6. *Dark Matter Fraction with Mass*

For radially constant mass-to-light ratio Υ, maximizing the fraction of mass following the luminosity profile results in a lower limit for the DM fraction, $f_{\rm DM,min}$. Dynamical modeling (§4.4, 4.5) gives $f_{\rm DM,min}(R < 1R_e) \simeq 0.1\text{-}0.4$ and $f_{\rm DM,min}(R < 5R_e) \simeq 0.3\text{-}0.8$, with a tendency of larger values for more massive ETGs. The corresponding $\Upsilon_{\rm max}$ also increase with galaxy mass (see Thomas *et al.* 2011). Napolitano *et al.* (2005) estimated gradients of Υ for a sample of ETGs with available dynamical mass modeling and found a correlation between luminosity and Υ-gradient such that the more luminous galaxies were more dark matter dominated. Comparing with DM halo predictions, the underlying reason appeared to be that the stellar bodies of brighter galaxies enclose a larger fraction of their DM halos. Deason *et al.* (2012) used power-law models to estimate dynamical masses at $5R_e$ for a sample of ETGs with extended PN and GC tracer kinematics. From their models they also found an increase of DM fraction inside $5R_e$ with galaxy mass, independent of whether a Chabrier or Salpeter IMF was assumed. The strong lensing results (§3.2) show a similar trend for the DM fraction within R_E.

Because dynamical $\Upsilon_{\rm max}$ increase with galaxy mass faster than stellar population Υ for given IMF, either some of the dark matter in massive ETGs must follow the light distribution, or the IMF must vary systematically with galaxy mass (Thomas *et al.* 2011). The latter interpretation is favoured by the dynamical analysis of Cappellari *et al.* (2012), and by new stellar population modeling of ETG spectra with NIR spectral features sensitive to low mass stars (Conroy & van Dokkum 2012). If the IMF varies between galaxies, it is natural in a hierarchical assembly picture to expect that it would also vary with radius within a galaxy. Reliable mass decomposition in ETGs will then require radially resolved spectral measurements of the IMF.

5. Conclusions

- Pure N-body simulations in ΛCDM cosmology predict nearly universal, prolate-triaxial DM halos. Including baryons and galaxy formation modifies the DM halo shapes, alignment, and inner density profiles. This provides one way to constrain the formation processes of the baryonic components.
- Weak lensing measurements are consistent with the predicted density structure of halos on large scales, and determine the relation between halo mass and galaxy luminosity.
- Mass determinations from strong lensing, X-ray emitting gas, stellar kinematics and PN kinematics out to several R_e find flat, close to isothermal CVCs for massive ETGs, while for lower mass, "quasi-Keplerian" ETGs the CVCs may be slightly falling.
- DM fractions at $5R_e$ inferred from dynamical modeling are in the range $\sim 30\text{-}80\%$. Inner DM densities in ellipticals are ~ 10 times higher than in spiral galaxies of the same stellar mass.
- A suite of recent high-resolution cosmological simulations of massive galaxies grown by minor and major mergers show systematic variations of the CVCs and DM fractions inside $5R_e$ with mass and fitted Sersic index, similar to those observed in ETGs.

References

Abadi, M. G., *et al.* 2010, *MNRAS*, 407, 435
Allgood, B., *et al.* 2006, *MNRAS*, 367, 1781
Arnaboldi, M., *et al.* 1996, *ApJ*, 472, 145
—. 1998, *ApJ*, 507, 759
Auger, M. W., *et al.* 2009, *ApJ*, 705, 1099

—. 2010, *ApJ*, 724, 511

Bailin, J. & Steinmetz, M. 2005, *ApJ*, 627, 647

Bailin, J., *et al.* 2005, *ApJ*, 627, L17

Barnabè, M., *et al.* 2009, *MNRAS*, 399, 21

—. 2011, *MNRAS*, 415, 2215

Berentzen, I. & Shlosman, I. 2006, *ApJ*, 648, 807

Bett, P., *et al.* 2010, *MNRAS*, 404, 1137

Binney, J. & Mamon, G. A. 1982, *MNRAS*, 200, 361

Blumenthal, G. R., Faber, S. M., Flores, R., & Primack, J. R. 1986, *ApJ*, 301, 27

Bolton, A. S., *et al.* 2008, *ApJ*, 682, 964

Brainerd, T. G., Blandford, R. D., & Smail, I. 1996, *ApJ*, 466, 623

Bullock, J. S., *et al.* 2001, *MNRAS*, 321, 559

Buote, D. A. & Humphrey, P. J. 2012, in Astrophysics and Space Science Library, Vol. 378, Astrophysics and Space Science Library, ed. D.-W. Kim & S. Pellegrini, 235

Cappellari, M., *et al.* 2006, *MNRAS*, 366, 1126

—. 2012, *Nature*, 484, 485

Churazov, E., *et al.* 2008, *MNRAS*, 388, 1062

—. 2010, *MNRAS*, 359

Ciotti, L. & Pellegrini, S. 2004, *MNRAS*, 350, 609

Coccato, L., Gerhard, O., & Arnaboldi, M. 2012, *MNRAS*, submitted (arXiv:1209.0356)

Coccato, L., *et al.* 2009, *MNRAS*, 394, 1249

Conroy, C. & van Dokkum, P. G. 2012, *ApJ*, 760, 71

Côté, P., *et al.* 2001, *ApJ*, 559, 828

Das, P., Gerhard, O., Churazov, E., & Zhuravleva, I. 2010, *MNRAS*, 409, 1362

Das, P., *et al.* 2011, *MNRAS*, 415, 1244

de Lorenzi, F., Debattista, V. P., Gerhard, O., & Sambhus, N. 2007, *MNRAS*, 376, 71

de Lorenzi, F., *et al.* 2008, *MNRAS*, 385, 1729

—. 2009, *MNRAS*, 395, 76

Deason, A. J., Belokurov, V., Evans, N. W., & McCarthy, I. G. 2012, *ApJ*, 748, 2

Deason, A. J., *et al.* 2011, *MNRAS*, 415, 2607

Diehl, S. & Statler, T. S. 2007, *ApJ*, 668, 150

Diemand, J. & Moore, B. 2011, Advanced Science Letters, 4, 297

Diemand, J., Moore, B., & Stadel, J. 2004, *MNRAS*, 353, 624

Diemand, J., *et al.* 2008, *Nature*, 454, 735

Doherty, M., *et al.* 2009, *A&A*, 502, 771

Douglas, N. G., *et al.* 2007, *ApJ*, 664, 257

Dubinski, J. & Carlberg, R. G. 1991, *ApJ*, 378, 496

Emsellem, E., *et al.* 2007, *MNRAS*, 379, 401

Ferreras, I., Saha, P., & Williams, L. L. R. 2005, *ApJ*, 623, L5

Fischer, P., *et al.* 2000, *AJ*, 120, 1198

Foster, C., *et al.* 2011, *MNRAS*, 415, 3393

Gao, L., *et al.* 2004, *MNRAS*, 355, 819

Gavazzi, R., *et al.* 2007, *ApJ*, 667, 176

Gerhard, O., Kronawitter, A., Saglia, R. P., & Bender, R. 2001, *AJ*, 121, 1936

Gerhard, O. E. 1993, *MNRAS*, 265, 213

Gnedin, O. Y., Kravtsov, A. V., Klypin, A. A., & Nagai, D. 2004, *ApJ*, 616, 16

Gnedin, O. Y. & Zhao, H. 2002, *MNRAS*, 333, 299

Governato, F., *et al.* 2010, *Nature*, 463, 203

Grillo, C. 2012, *ApJ*, 747, L15

Hahn, O., Teyssier, R., & Carollo, C. M. 2010, *MNRAS*, 405, 274

Hayashi, E., Navarro, J. F., & Springel, V. 2007, *MNRAS*, 377, 50

Hoekstra, H., Yee, H. K. C., & Gladders, M. D. 2004, *ApJ*, 606, 67

Hoekstra, H., *et al.* 2005, *ApJ*, 635, 73

Hui, X., Ford, H. C., Freeman, K. C., & Dopita, M. A. 1995, *ApJ*, 449, 592

Humphrey, P. J., Buote, D. A., O'Sullivan, E., & Ponman, T. J. 2012, *ApJ*, 755, 166
Humphrey, P. J., *et al.* 2006, *ApJ*, 646, 899
Hwang, H. S., *et al.* 2008, *ApJ*, 674, 869
Jing, Y. P. & Suto, Y. 2002, *ApJ*, 574, 538
Kazantzidis, S., *et al.* 2004, *ApJ*, 611, L73
Klypin, A., Kravtsov, A. V., Valenzuela, O., & Prada, F. 1999, *ApJ*, 522, 82
Koopmans, L. V. E., *et al.* 2006, *ApJ*, 649, 599
Kronawitter, A., Saglia, R. P., Gerhard, O., & Bender, R. 2000, *A&AS*, 144, 53
Leier, D., Ferreras, I., Saha, P., & Falco, E. E. 2011, *ApJ*, 740, 97
Macciò, A. V., Dutton, A. A., & van den Bosch, F. C. 2008, *MNRAS*, 391, 1940
Macciò, A. V., *et al.* 2012, *ApJ*, 744, L9
Mandelbaum, R., *et al.* 2006a, *MNRAS*, 370, 1008
—. 2006b, *MNRAS*, 368, 715
Mashchenko, S., Wadsley, J., & Couchman, H. M. P. 2008, *Science*, 319, 174
McNeil, E. K., *et al.* 2010, *A&A*, 518, A44
McNeil-Moylan, E. K., Freeman, K. C., Arnaboldi, M., & Gerhard, O. E. 2012, *A&A*, 539, A11
Méndez, R. H., *et al.* 2001, *ApJ*, 563, 135
Moore, B., *et al.* 1998, *ApJ*, 499, L5
Morganti, L., *et al.* 2013, in preparation.
Murphy, J. D., Gebhardt, K., & Adams, J. J. 2011, *ApJ*, 729, 129
Nagino, R. & Matsushita, K. 2009, *A&A*, 501, 157
Napolitano, N. R., *et al.* 2005, *MNRAS*, 357, 691
—. 2009, *MNRAS*, 393, 329
—. 2011, *MNRAS*, 411, 2035
Navarro, J. F., Frenk, C. S., & White, S. D. M. 1996, *ApJ*, 462, 563
Navarro, J. F., *et al.* 2010, *MNRAS*, 402, 21
Oser, L., Naab, T., Ostriker, J. P., & Johansson, P. H. 2012, *ApJ*, 744, 63
Oser, L., *et al.* 2010, *ApJ*, 725, 2312
Parker, L. C., *et al.* 2007, *ApJ*, 669, 21
Peng, E. W., Ford, H. C., & Freeman, K. C. 2004, *ApJ*, 602, 685
Proctor, R. N., *et al.* 2009, *MNRAS*, 398, 91
Revnivtsev, M., *et al.* 2008, *A&A*, 490, 37
Romanowsky, A. J., *et al.* 2003, *Science*, 301, 1696
—. 2012, *ApJ*, 748, 29
Rusin, D. & Kochanek, C. S. 2005, *ApJ*, 623, 666
Schuberth, Y., *et al.* 2010, *A&A*, 513, A52
Sellwood, J. A. & McGaugh, S. S. 2005, *ApJ*, 634, 70
Springel, V., *et al.* 2008, *MNRAS*, 391, 1685
Strader, J., Brodie, J. P., Spitler, L., & Beasley, M. A. 2006, *AJ*, 132, 2333
Teodorescu, A. M., *et al.* 2005, *ApJ*, 635, 290
—. 2011, *ApJ*, 736, 65
Thomas, J., *et al.* 2007, *MNRAS*, 382, 657
—. 2009, *ApJ*, 691, 770
—. 2011, *MNRAS*, 415, 545
Treu, T. 2010, *ARA&A*, 48, 87
Treu, T. & Koopmans, L. V. E. 2004, *ApJ*, 611, 739
Treu, T., *et al.* 2010, *ApJ*, 709, 1195
Trinchieri, G., *et al.* 2008, *ApJ*, 688, 1000
van Uitert, E., *et al.* 2011, *A&A*, 534, A14
—. 2012, *A&A*, 545, 71
Weijmans, A., *et al.* 2009, *MNRAS*, 398, 561
Woodley, K. A., *et al.* 2010, *AJ*, 139, 1871
Wu, X., *et al.* 2012, *MNRAS*, submitted (arXiv:1209.3741)

The intriguing life of massive galaxies
Proceedings IAU Symposium No. 295, 2012
D. Thomas, A. Pasquali & I. Ferreras, eds.

© International Astronomical Union 2013
doi:10.1017/S1743921313004821

The XLENS Project: Do More Massive Early-Type Galaxies Have More Dark Matter or Different Stellar IMFs?

Chiara Spiniello

Kapteyn Astronomical Institute, University of Groningen,
Postbus 800, 9700 AV Groningen, the Netherlands
email: spiniello@astro.rug.nl

Abstract. The X-shooter Lens Survey (XLENS) aims to study the interplay of dark matter (DM) and stellar content in the inner regions of massive early-type galaxies (ETGs) by combining strong gravitational lensing, dynamical models, and spectroscopic stellar population analysis. XLENS targets a sample of ETGs from the SLACS survey (The Sloan Lens ACS Survey, e.g. Bolton *et al.* 2006) with velocity dispersions $\geqslant 250 \mathrm{km\,s}^{-1}$ using the X-Shooter spectrograph on ESO's *Very Large Telescope*. Recent observations indicate that the internal dark-matter fraction of ETGs increases rapidly with galaxy mass, although some hints for a varying initial mass function (IMF) have also been suggested, where the low-mass end of the stellar IMF steepens with galaxy mass. XLENS first results unambiguously confirm that DM plays an important role already within one effective radius for very massive systems (Spiniello *et al.* 2011). Moreover, studying equivalent widths of certain red spectral features which are indicators of low-mass stars in massive ETGs (e.g. NaI and TiO2) as a function of age and metallicity (i.e. Mgb, Fe, Hβ), and as function of stellar velocity dispersion, has shown that the IMF slope is varying mildly with galaxy mass (Spiniello *et al.* 2012).

Keywords. dark matter, galaxies, IMF, structure, evolution

1. Introduction

The relationship between baryonic matter and dark matter in the internal regions of early-type galaxies (ETGs) is particularly important to comprehend the processes that drive in hierarchical galaxy formation. While stars are supposed to dominate in the innermost region, dark matter – which dominates most of the dynamics during galaxy assembly – is found to play a non-negligible role in the central region, especially for very massive systems. In fact, new observations indicate that the internal DM fraction increases monotonically with the mass of the galaxy (e.g. Auger *et al.* 2010, Barnabé *et al.* 2011), assuming a universal IMF.

When constraining the star formation, metallicity and gas/dust content of galaxies, the initial mass function (IMF) is often assumed to be universal and equal to that of the solar neighbourhood (Kroupa 2001; Chabrier 2003; Bastian, Covey & Meyer 2010). However, evidence has recently emerged that the IMF might evolve (Davé 2008; van Dokkum 2008) or depend on the stellar mass of the system (e.g. Worthey 1992; Trager *et al.* 2000; Graves *et al.* 2009; Treu *et al.* 2010; Auger et al. 2010b; Napolitano 2010; van Dokkum & Conroy 2010; Spiniello et al. 2012). van Dokkum & Conroy (2010; hereafter vDC10) suggested that low-mass stars ($\leqslant 0.3\,M_\odot$) could be more prevalent in more massive ETGs. The increase in the mass-to-light ratio (M/L) of galaxies with galaxy mass may thus be partly due to a changing IMF rather than an increasing dark matter fraction, consistent with previous suggestions (Treu *et al.* 2010, Auger *et al.* 2011, Barnabè *et al.*

2011, Dutton *et al.* 2012, Cappellari *et al.* 2012). To assess whether this result is genuine and model-independent, it is of crucial importance to disentangle stellar and dark matter contributions in the inner regions of galaxies, and to calculate stellar M/L values with an accuracy better than 20% (set by the range in DM fractions). With the XLENS survey we have set out to achieve this ambitious goal in a substantial sample of individual ETGs that have strong gravitational lensing, stellar kinematic and stellar-population information available.

2. The X-Shooter Lens Survey (XLENS)

2.1. *A pilot program: The Cosmic Horseshoe*

The first result from the X-shooter Lens Survey (XLENS) is an analysis of the massive ETG SDSS J1148+1930 at redshift $z = 0.444$ (Spiniello *et al.* 2011). We combine its extended kinematic profile – derived from X-shooter spectra – with strong gravitational lensing and multi-color information derived from SDSS images. We calculated the luminosity-weighted stellar velocity dispersion ($\langle\sigma_*\rangle (\leqslant R_{eff}) = 351 \pm 10\,\mathrm{km\,s^{-1}}$) and we obtain a projected stellar mass fraction ($f_*(< R_\mathrm{E}) = 0.19^{+0.04}_{-0.09}$) from a two component mass model of the lens galaxy. From SDSS colors, we obtain a second independent internal stellar mass fraction for different assumed IMFs (i.e. Chabrier, Salpeter, $x = 3$, and $x = 3.5$, assuming $dN/dM \propto M^{-x}$). We find that the lensing and kinematic constraints on the stellar mass fraction agree well with those independently derived from the SDSS colors for a Salpeter IMF. Dwarf-rich IMFs in the lower mass range of 0.1–0.7 $M_\odot$, with $x \geqslant 3$ are excluded at the $> 90\%$ C.L. and for $x = 3.5$ violate the total lensing-derived mass limit. We conclude that this very massive early-type galaxy is dark-matter dominated inside one effective radius, extending the trend recently found from massive SLACS galaxies (Treu *et al.* 2009, Auger *et al.* 2010) and that a very steep IMF can not fully account for this trend.

2.2. *Evidence for a mild steepening and Bottom-heavy IMF in Massive ETGs*

In Spiniello *et al.* (2012) we investigate possible IMF variations with galaxy mass. We studied equivalent widths (EW) – focussing on two absorption lines (NaI and TiO2) of low-mass stars ($\leqslant 0.3\mathrm{M}_\odot$) – for luminous red galaxy spectra from the Sloan Digital Sky Survey (SDSS) and X-Shooter Lens Survey (XLENS). indicators (Mgb, Fe, Hβ), of NaD, and of stellar velocity dispersion. We compare the NaI and TiO2 EWs to those derived from simple stellar population models computed for different assumed IMFs, ages, [α/Fe], and elemental abundances (Fig. 1). We find that current state-of-the-art SSP models are only able to simultaneously reproduce the observed NaD $\lambda5895$ and NaI $\lambda8190$ features for the lower-mass (around σ_*) ETGs but deviate increasingly for more massive ETGs. In particular, the NaD EWs do not follow the models well (Fig. 1(b)). We conclude that the NaD feature is affected by as-of-yet not understood processes in the more massive ETGs ($\sigma > 250\,\mathrm{km\,s^{-1}}$). Despite this, we find that the TiO2 $\lambda6230$ and the NaI $\lambda8190$ are particularly promising features to decouple the IMF from stellar population, age, metallicity, and abundance pattern, especially when combined with metallicity-dependent indices. We also observe a clear trend of an increasing IMF slope between $\sigma = 200 - 335\,\mathrm{km\,s^{-1}}$ from Salpeter ($x = 2.35$) to $x \approx 3.0$. The XLENS ETG (SDSSJ0912+0029) and one SDSS ETG (SDSSJ0041-0914) appear to require both an extreme dwarf-rich IMF ($x \geqslant 3.0$) and a high sodium enhancement ([Na/Fe] $= +0.4$). However, lensing constraints on the total mass of the XLENS system within its Einstein radius limit a bottom-heavy IMF to a power-law slope of $x \leqslant 3.0$ at the 90% C.L. (Table 1). A full spectral comparison,

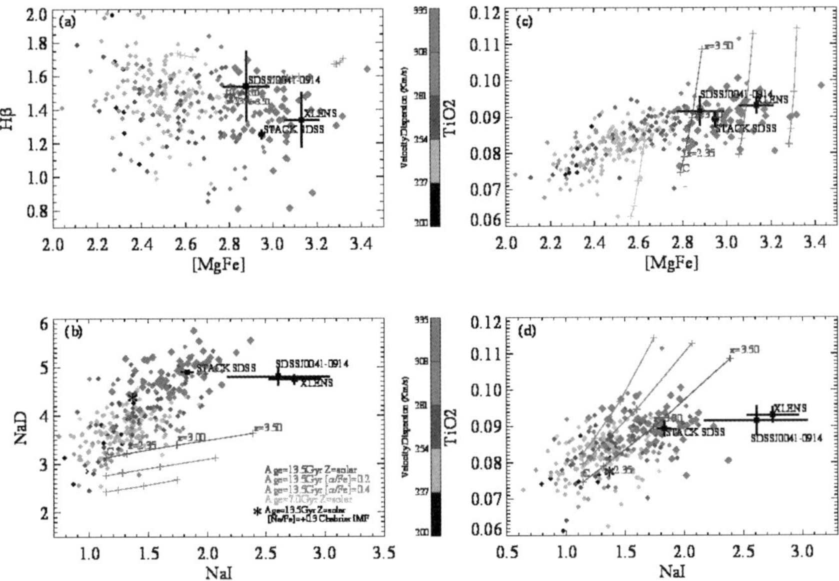

Figure 1. Index-index plots of the main absorption features. Lines and crosses are different SSP models from single population models of Conroy & van Dokkum 2012 with increasing IMF (from Chabrier, to an extremely dwarf-rich IMF with a slope of $x = 3.5$). Points coloured according to their velocity dispersions are individual SDSS galaxies, with index errors similar to SDSS J0041-0914. In the plots showing sodium (Panel (b) and Panel (d)), the XLENS system SDSSJ0912+0029 requires a very steep IMF, violating lensing constraints on its total mass. From Panel (c), it is clear that IMF varies with mass: the most massive ETGs require an IMF slope slightly steeper than Salpeter. A Chabrier-type IMF systematically underestimates the SDSS TiO2 EWs.

Table 1. Variation with IMF of M/L and stellar mass fraction within the Einstein radius.

IMF slope $(dN/dm = M^{-x})$	$(M/L)^*_{DSEP,B}$ $([\alpha/Fe] = 0.0)$	$(M/L)^*_{DSEP,V}$ $([\alpha/Fe] = 0.0)$	f_B^{*} [1]	f_V^{*} [2]
-2.35	10.2 ± 3	7.2 ± 2	0.75 ± 0.2	0.59 ± 0.18
-3.00	22 ± 6	16 ± 5	1.6 ± 0.5	1.4 ± 0.4
-3.50	43 ± 13	29 ± 9	2.4 ± 0.8	2.4 ± 0.7

Notes:
[1]: $f_B^* = (L_{Ein}/M_{Ein}) \times (M/L)^*_B$, [2]: $f_V^* = (L_{Ein}/M_{Ein}) \times (M/L)^*_V$
Constraints on the mass and luminosity within M_{Ein} from Auger *et al.* (2009) and Barnabè *et al.* (2009). All quantities are calculated in the rest-frame V- and B-band.

in combination with more detailed lensing and dynamical constraints is planned to assess whether NaI and NaD (in some instances) are contaminated.

Our results are the first SSP-based indications of a steepening of the low-mass end of the IMF with increasing galaxy mass *within* the class of LRG/ETGs. Our results (i) support a similar trend first found by Treu *et al.* (2010), (ii) extend the evidence based on SSP models that the IMF steepens from spiral to early-type galaxies (vDC10), and (iii) are in agreement with a similar trend found by Cappellari *et al.* (2012). The upper limit of $x \leqslant 3.0$, based on one of the most massive ETG systems in our sample, a gravitational lens, also supports our previous similar finding that extremely bottom-heavy IMFs are excluded (Spiniello *et al.* 2011).

3. Conclusions

I presented the first results from the *X-Shooter Lens Survey (XLENS)* project which aims to study the internal structure and the mass profile of a sample of massive lens ETGs with $\sigma_*^{\mathrm{ETG}} \geqslant 250$ km/s using X-Shooter. The final goal of the survey is to separate the luminous and dark matter in the internal regions of massive red, old galaxies in order to study the interaction between these components and their impact on galaxies assembly and evolution in the hierarchical galaxy formation framework. In this context, characterizing the slope of the IMF and assess whether it is universal and similar to the one of the Milky Way, or if it depends on galaxy mass, on epoch of formation or on local density, is critical. The combination of gravitational lensing, stellar kinematics and spectroscopic tracers of low-mass stars, as a part of the XLENS Survey, has proven to be a powerful tool to constrain the low-mass end of the IMF from galaxy spectra. Our first finding supports the idea of a non-universal IMF. We find that massive ETGs (i.e. $\sigma_*^{\mathrm{ETG}} \geqslant 250$) (i) have an IMF slope most consistent with Salpeter, excluding Chabrier/Kroupa-type IMFs, (ii) show evidence in their optical line-indices (e.g. NaI and TiO2) that their IMF steepens mildly with increasing galaxy mass. By using lensing constraints on the total mass enclosed within the Einstein radius of the lens ETGs, we also find that the IMF slope most likely cannot exceed $x \approx 3$. A full spectral comparison extended to the NIR, where most of the features that show different strenght in M-dwarfs and in cool giants are present, together with new data and more flexible models, are necessary to further strengthen these results and completely break the degeneracy between IMF variations and variations of other SSP parameters such as age and metallicity.

The author would like to thank her supervisors L.V.E. Koopmans and S.C. Trager for their guidance, support, time and new ideas, and her collaborators for useful help and comments.

References

Auger, M. W., Treu, T., Bolton, A. S., *et al.* 2009, *ApJ*, 705, 1099

Auger, M. W., Treu, T., Gavazzi, R., *et al.* 2010, *ApJL*, 721, L163

Auger, M. W., Treu, T., Bolton, A. S., *et al.* 2010, *ApJ*, 724, 511

Barnabè, M., Czoske, O., Koopmans, L. V. E., *et al.* 2011, *MNRAS*, 415, 2215

Bastian, N., Covey, K. R., & Meyer, M. R. 2010, *ARAA*, 48, 339

Bolton, A. S., Burles, S., Koopmans, L. V. E., *et al.* 2006, *ApJ*, 638, 703

Cappellari, M., McDermid, R. M., Alatalo, K., *et al.* 2012, *Nature*, 484, 485

Chabrier, G. 2003, *PASP*, 115, 763

Conroy, C. & van Dokkum, P. 2012, *ApJ*, 747, 69

Davé, R. 2008, *MNRAS*, 385, 147

Dutton, A. A., Mendel, J. T., & Simard, L. 2012, *MNRAS*, 422, L33

Kroupa, P. 2001, *MNRAS*, 322, 231

Graves, G. J., Faber, S. M., & Schiavon, R. P. 2009, *ApJ*, 698, 1590

Napolitano, N. R., Romanowsky, A. J., & Tortora, C. 2010, *MNRAS*, 405, 2351

Spiniello, C., Koopmans, L. V. E., Trager, S. C., Czoske, O., & Treu, T. 2011, *MNRAS*, 417, 3000

Spiniello, C., Trager, S. C., Koopmans, L. V. E., & Chen, Y. P. 2012, *ApJL*, 753, L32

Trager, S. C., Faber, S. M., Worthey, G., & González, J. J. 2000, *AJ*, 120, 165

Treu, T., Auger, M. W., Koopmans, L. V. E., *et al.* 2010, *ApJ*, 709, 1195

van Dokkum, P. G. 2008, *ApJ*, 674, 29

van Dokkum, P. G. & Conroy, C. 2010, *Nature*, 468, 940

Worthey, G. 1992, in: B. Barbuy, A. Renzini (eds.), IAU Symposium 149, *The Stellar Populations of Galaxies*, 149, 507

The intriguing life of massive galaxies
Proceedings IAU Symposium No. 295, 2012
D. Thomas, A. Pasquali & I. Ferreras, eds.

© International Astronomical Union 2013
doi:10.1017/S1743921313004833

Further evidence for large central mass-to-light ratios in massive early-type galaxies

E. M. Corsini[1,2]**, G. A. Wegner**[3]**, J. Thomas**[4]**, R. P. Saglia**[4]**, R. Bender**[4,5] **and S. B. Pu**[6]

[1]Dipartimento di Fisica e Astronomia, Università di Padova, Padova, Italy
email: enricomaria.corsini@unipd.it

[2]INAF–Osservatorio Astronomico di Padova, Padova, Italy

[3]Department of Physics and Astronomy, Dartmouth College, Hanover, NH, USA

[4]Max-Planck-Institut für extraterrestrische Physik, Garching, Germany

[5]Universitäts-Sternwarte München, München, Germany

[6]The Beijing No. 12 High School, Beijing, China

Abstract. We studied the stellar populations, distribution of dark matter, and dynamical structure of a sample of 25 early-type galaxies in the Coma and Abell 262 clusters. We derived dynamical mass-to-light ratios and dark matter densities from orbit-based dynamical models, complemented by the ages, metallicities, and α-element abundances of the galaxies from single stellar population models. Most of the galaxies have a significant detection of dark matter and their halos are about 10 times denser than in spirals of the same stellar mass. Calibrating dark matter densities to cosmological simulations we find assembly redshifts $z_{\rm DM} \approx 1 - 3$. The dynamical mass that follows the light is larger than expected for a Kroupa stellar initial mass function, especially in galaxies with high velocity dispersion $\sigma_{\rm eff}$ inside the effective radius $r_{\rm eff}$. We now have 5 of 25 galaxies where mass follows light to $1 - 3\,r_{\rm eff}$, the dynamical mass-to-light ratio of all the mass that follows the light is large ($\approx 8 - 10$ in the Kron-Cousins R band), the dark matter fraction is negligible to $1 - 3\,r_{\rm eff}$. This could indicate a 'massive' initial mass function in massive early-type galaxies. Alternatively, some of the dark matter in massive galaxies could follow the light very closely suggesting a significant degeneracy between luminous and dark matter.

Keywords. galaxies: abundances, galaxies: elliptical and lenticular, cD, galaxies: formation, galaxies: kinematics and dynamics, galaxies: stellar content.

1. Introduction

In the past years we studied the stellar populations, mass distribution, and orbital structure of a sample of early-type galaxies in the Coma cluster with the aim of constraining the epoch and mechanism of their assembly.

The surface-brightness distribution was obtained from ground-based and *HST* data. The stellar rotation, velocity dispersion, and the H_3 and H_4 coefficients of the line-of-sight velocity distribution were measured along the major axis, minor axis, and an intermediate axis. In addition, the line index profiles of Mg, Fe and Hβ were derived (Mehlert *et al.* 2000; Wegner *et al.* 2002; Corsini *et al.* 2008). Axisymmetric orbit-based dynamical models were used to derive the mass-to-light ratio Υ_* of all the mass that follows the light and the dark matter (DM) halo parameters in 17 galaxies (Thomas *et al.* 2005, 2007a,b, 2009a,b). The comparison with masses derived through strong gravitational lensing for early-type galaxies with similar velocity dispersion and the analysis of the ionized-gas kinematics gave valuable consistency checks for the total mass distribution predicted by dynamical modeling (Thomas *et al.* 2011). The line-strength indices were analyzed by single stellar-population models to derive the age, metallicity, α-element abundance, and mass-to-light ratio $\Upsilon_{\mathrm{Kroupa}}$ (or $\Upsilon_{\mathrm{Salpeter}}$ depending on the adopted initial mass function, IMF) of the galaxies (Mehlert *et al.* 2003).

More recently, we have performed the same dynamical analysis for 8 early-type galaxies of the nearby cluster Abell 262 (Wegner *et al.* 2012). The latter is far less densely populated than the Coma cluster and it is comparable to the Virgo cluster. Moreover, while galaxies in Coma were selected to be mostly flattened, the Abell 262 galaxies we measured appear predominantly round on the sky.

2. Results

2.1. *Evidence for halo mass not associated to the light*

In the Coma galaxy sample, the statistical significance for DM halos is over 95% for 8 (out of 17) galaxies (Thomas *et al.* 2007b), whereas the Abell 262 sample reveals 4 (out of 8) galaxies of this kind (Wegner *et al.* 2012). In Coma, we found only one galaxy (GMP 1990) with $f_{\mathrm{halo}} \approx 0$, i.e., with a negligible halo-mass fraction of the total mass inside r_{eff}. This is also the case of 4 galaxies in Abell 262 (NGC 703, NGC 708, NGC 712, and UGC 1308). The evidence for a DM component in addition to mass that follows light is not directly connected to the spatial extent of the kinematic data, degree of rotation, or flattening of the system. There is no relationship with the age, metallicity, and α-element abundance of the stellar populations.

We cannot discriminate between cuspy and logarithmic halos based on the quality of the kinematic fits, except for NGC 703 where the logarithmic halo fits better. Still, the majority of cluster early-type galaxies have $2 - 10$ times denser halos than local spirals (e.g., Persic *et al.* 1996), implying a $1.3 - 2.2$ times higher $(1 + z_{\mathrm{DM}})$ assuming $\langle \rho_{\mathrm{DM}} \rangle \sim (1 + z_{\mathrm{DM}})^3$, where z_{DM} is the formation redshift of the DM halos. Thus, if spirals typically formed at $z_{\mathrm{DM}} \approx 1$, then cluster early-type galaxies assembled at $z_{\mathrm{DM}} \approx 1.6 - 3.4$.

Averaging over all galaxies, we find that a fraction of $\langle f_{\mathrm{halo}} \rangle = 0.2$ of the total mass inside r_{eff} is in a DM halo distinct from the light. Similar fractions come from other dynamical studies employing spherical models (e.g., Gerhard *et al.* 2001). The Coma and Abell 262 galaxies show an anti-correlation between $\Upsilon_*/\Upsilon_{\mathrm{Kroupa}}$, i.e. the ratio between the dynamical and stellar population mass-to-light ratios, and f_{halo} (Fig. 1, left panel). Galaxies where the dynamical mass following the light significantly exceeds the Kroupa value ($\Upsilon_*/\Upsilon_{\mathrm{Kroupa}} > 3$) seem to lack matter following the halo distribution inside r_{eff} ($\langle f_{\mathrm{halo}} \rangle \approx 0$). In contrast, in galaxies near the Kroupa limit ($\Upsilon_*/\Upsilon_{\mathrm{Kroupa}} < 1.4$), the dark-halo mass fraction is at its maximum ($\langle f_{\mathrm{halo}} \rangle = 0.3$).

2.2. *Mass that follows the light*

As far as the mass-to-light ratios are concerned, the galaxies of Coma and Abell 262 follow a similar trend. While the dynamically determined Υ_* increases strongly with σ_{eff}, i.e., the velocity dispersion averaged within r_{eff} (Fig. 1, right top panel), the stellar population models indicate an almost constant $\Upsilon_{\mathrm{Kroupa}}$ (Fig. 1, right middle panel). This implies that the ratio $\Upsilon_*/\Upsilon_{\mathrm{Kroupa}}$ increases with σ_{eff} (Fig. 1, right bottom panel). Around $\sigma_{\mathrm{eff}} \approx 200$ km s^{-1} the distribution of $\Upsilon_*/\Upsilon_{\mathrm{Kroupa}}$ has a sharp cutoff with almost no galaxies below $\Upsilon_*/\Upsilon_{\mathrm{Kroupa}} = 1$. For $\sigma_{\mathrm{eff}} \gtrsim 250$ km s^{-1} the lower bound of $\Upsilon_*/\Upsilon_{\mathrm{Kroupa}}$ increases to $\Upsilon_*/\Upsilon_{\mathrm{Kroupa}} \gtrsim 2$ at $\sigma_{\mathrm{eff}} \approx 300$ km s^{-1}. Similar trends are also observed in the SAURON sample with dynamical models lacking a separate DM halo (Cappellari

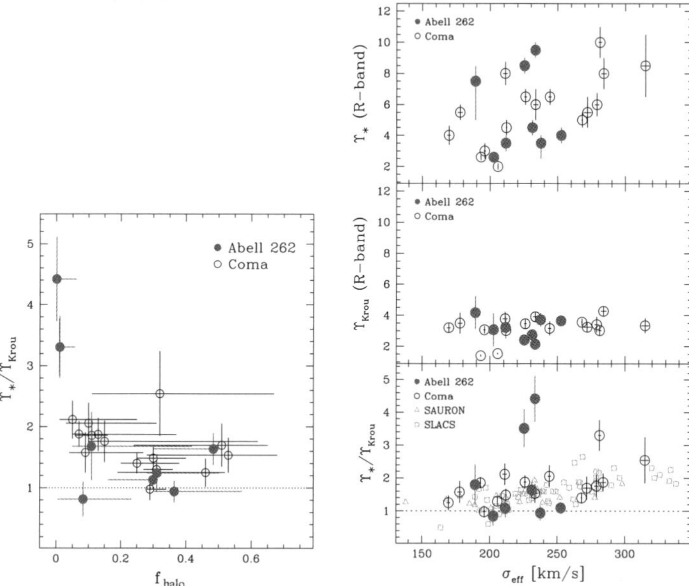

Figure 1. Left: Ratio of dynamical Υ_* to stellar-population $\Upsilon_{\mathrm{Kroupa}}$ as a function of f_{halo}, i.e., the halo-mass fraction of the total mass inside r_{eff}, for galaxies in Coma (open circles, Thomas *et al.* 2011) and Abell 262 (filled circles, Wegner *et al.* 2012). r_{eff}. $\Upsilon_*/\Upsilon_{\mathrm{Kroupa}} = 1.6$ corresponds to a Salpeter IMF. Right: Dynamical Υ_* (upper panel), stellar-population $\Upsilon_{\mathrm{Kroupa}}$ (middle panel), and their ratio (bottom panel) as a function of the effective velocity dispersion, σ_{eff}. In the bottom panel, Coma and Abell 262 galaxies are compared to SLACS (open squares, Treu *et al.* 2010) and SAURON galaxies (open triangles, Cappellari *et al.* 2006).

et al. 2006), in SLACS galaxies with combined dynamical and lensing analysis (Treu *et al.* 2010) and, recently, in the ATLAS3d survey with dynamical models including a DM halo (Cappellari *et al.* 2012).

3. Discussion

Fig. 1 provides strong evidence for large central Υ_* in massive early-type galaxies. However, in all gravity-based methods, there is a fundamental degeneracy concerning the interpretation of mass-to-light ratios. Such methods cannot uniquely discriminate between luminous and dark matter once they follow similar radial distributions. The distinction is always based on the assumption that the mass density profile of the DM differs from that of the luminous matter.

One extreme point of view is the assumption that the stellar masses in early-type galaxies are maximal and correspond to Υ_*. The immediate consequence is that the stellar IMF in early-type galaxies is not universal, varying from Kroupa-like at low velocity dispersions to Salpeter (or steeper) in the most massive galaxies (see Auger *et al.* 2010, Thomas *et al.* 2011, and Cappellari *et al.* 2012, for a detailed discussion). Recent attempts to measure the stellar IMF directly from near-infrared observations point in the same direction (see Conroy & van Dokkum 2012, and references therein). However, we also find the galaxies with the largest $\Upsilon_*/\Upsilon_{\mathrm{Kroupa}}$ have the lowest halo-mass fractions inside r_{eff} and vice versa. A possible explanation for this finding is a DM distribution that follows the light very closely in massive galaxies and contaminates the measured Υ_*, while it is more distinct from the light in lower-mass systems. This has been suggested elsewhere as a signature of violent relaxation.

One option to further constrain the mass-decomposition of gravity-based models is to incorporate predictions from cosmological simulations that confine the maximum amount of DM that can be plausibly attached to a galaxy of a given stellar mass. Since adiabatic contraction increases the amount of DM in the galaxy center, it could be in principle a viable mechanism to lower the required stellar masses towards a Kroupa IMF (Napolitano *et al.* 2010, but see also Cappellari *et al.* 2012). An immediate consequence is that some of the mass that follows the light is actually DM, increasing the DM fraction to about 50% of the total mass inside $r_{\rm eff}$.

Since the (decontracted) average halo density scales with the mean density of the universe at the assembly epoch, we derived the dark-halo assembly redshift $z_{\rm DM}$ for Coma (Thomas et al. 2011) and Abell 262 galaxies (Wegner et al. 2012). We compared the values of $z_{\rm DM}$ to the star-formation redshifts z_* calculated from the stellar-population ages. For the majority of galaxies $z_{\rm DM} \approx z_*$ and their assembly seems to have stopped before $z_{\rm DM} \approx 1$. The stars of some galaxies appear to be younger than the halo, which indicates a secondary star-formation episode after the main halo assembly. The photometric and kinematic properties of the remaining galaxies suggest they are the remnants of gas-poor binary mergers and their progenitors formed close to the $z_{\rm DM} = z_*$ relation. Without trying to overinterpret the result given the assumptions, it seems that Kroupa IMF allows us to explain the formation redshifts of our galaxies. In addition, galaxies in Coma and Abell 262 – where the dynamical mass that follows the light is in excess of a Kroupa stellar population – do not differ in terms of their stellar population ages, metallicities and α-element abundances from galaxies where this is not the case.

Taken at face value, our dynamical mass models are therefore as consistent with a universal IMF, as they are with a variable IMF. If the IMF indeed varies from galaxy to galaxy according to the average star-formation rate (Conroy & van Dokkum 2012) then the assumption of a constant stellar mass-to-light ratio *inside* a galaxy should be relaxed in future dynamical and lensing models.

References

Auger, M. W., Treu, T., Bolton, A. S., *et al.* 2009, *ApJ*, 705, 1099

Cappellari, M., Bacon, R., Bureau, M., *et al.* 2006, *MNRAS*, 366, 1126

Cappellari, M., McDermid, R. M., Alatalo, K., *et al.* 2012, *Nature*, 484, 485

Conroy, C. & van Dokkum, P. 2012, *ApJ*, 760, 71

Corsini, E. M., Wegner, G. A., Saglia, R. P., *et al.* 2008, *ApJS*, 175, 462

Gerhard , O., Kronawitter, A., Saglia, R. P., & Bender, R. 2001, *AJ*, 121, 1936

Mehlert, D., Saglia, R. P., Bender, R., & Wegner, G. A. 2000, *A&AS*, 141, 449

Mehlert, D., Thomas, D., Saglia, R. P., Bender, R., & Wegner, G. A. 2000, *A&A*, 407, 423

Napolitano, N. R., Romanowsky, A. J., & Tortora, C. 2010, *MNRAS*, 405, 2351

Persic, M., Salucci, P., & Stel, F. 1996, *MNRAS*, 283, 1102

Thomas, J., Jesseit, R., Naab, T., *et al.* 2007a, *MNRAS*, 381, 1672

Thomas, J., Jesseit, R., Saglia, R. P., *et al.* 2009a, *MNRAS*, 393, 641

Thomas, J., Saglia, R. P., Bender, R., *et al.* 2005, *MNRAS*, 360, 1355

Thomas, J., Saglia, R. P., Bender, R., *et al.* 2007b, *MNRAS*, 382, 657

Thomas, J., Saglia, R. P., Bender, R., *et al.* 2009b, *ApJ*, 691, 770

Thomas, J., Saglia, R. P., Bender, R., *et al.* 2011, *MNRAS*, 415, 545

Treu, T., Auger, M. W., Koopmans, L. V. E., *et al.* 2010, *ApJ*, 709, 1195

Wegner, G. A., Corsini, E. M., Saglia, R. P., *et al.* 2002, *A&A*, 395, 753

Wegner, G. A., Corsini, E. M., Thomas, J., *et al.* 2012, *AJ*, 144, 78

The intriguing life of massive galaxies
Proceedings IAU Symposium No. 295, 2012
D. Thomas, A. Pasquali & I. Ferreras, eds.

© International Astronomical Union 2013
doi:10.1017/S1743921313004845

The Angular Momentum of Brightest Cluster Galaxies

S. Brough[1], K.-V. Tran[2,3] and A. von der Linden[4]

[1]Australian Astronomical Observatory, PO Box 915, North Ryde, NSW 1670, Australia
[2]George P. & Cynthia W. Mitchell Institute for Fundamental Physics & Astronomy,
Department of Physics & Astronomy, Texas A&M University, College Station, TX 77843, USA
[3]Institute for Theoretical Physics, University of Zurich, CH 8057, Switzerland
[4]Kavli Institute of Particle Astrophysics & Cosmology (KIPAC), Stanford University, 452
Lomita Mall, Stanford, CA 94305, USA

Abstract. Massive Brightest Cluster Galaxies (BCGs) are observed to have a range of angular momenta, suggesting a variety of merging histories.

Keywords. galaxies: elliptical and lenticular, cD, galaxies: kinematics and dynamics

1. Summary

Brightest Cluster Galaxies (BCGs) include the most massive galaxies in the Universe and are predicted to undergo more merging than less massive galaxies. However, the observational evidence for recent BCG growth via merging is contradictory (e.g. Collins *et al.* 2009, Lidman *et al.* 2012). BCGs should also have relatively low angular momentum due to their predicted rich merger histories, but this remains unclear.

We present VLT/VIMOS integral field spectroscopy of 10 BCGs at $z \sim 0.1$ from SDSS (Brough *et al.* 2011, Jimmy *et al.* in prep.). Three have companions within 20 kpc. These galaxies extend the mass range analysed in existing surveys (Fig. 1).

BCG merging activity, as indicated by angular momentum, is diverse - two BCGs (as well as the massive companions) are fast rotators, while others have very low angular momentum (Fig. 1). This may indicate that dry merging is taking place up until today.

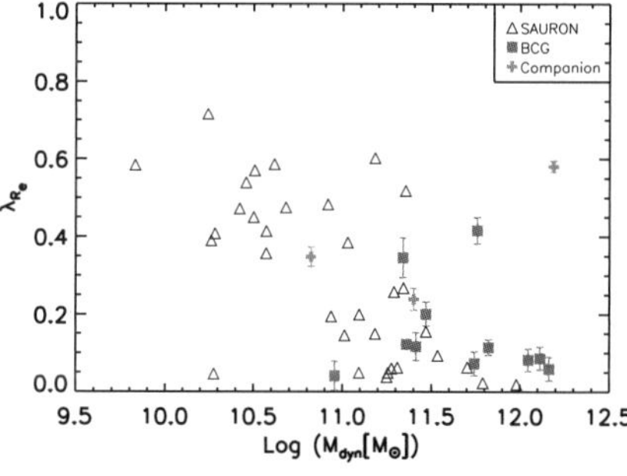
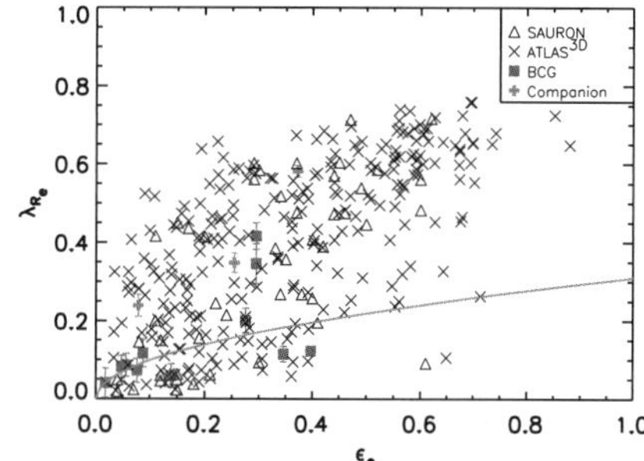

Figure 1. Left: Angular momentum parameter, λ_{Re}, versus dynamical mass. Right: λ_{Re} versus ellipticity. Rotating galaxies lie above the green line. Some of these massive galaxies have unexpectedly high angular momentum.

References

Brough, S., *et al.* 2011, *MNRAS*, 413, 1236
Collins, C. A., *et al.* 2009, *Nature*, 458, 603
Lidman, C., *et al.* 2012, *MNRAS*, 427, 550

The intriguing life of massive galaxies
Proceedings IAU Symposium No. 295, 2012
D. Thomas, A. Pasquali & I. Ferreras, eds.

© International Astronomical Union 2013
doi:10.1017/S1743921313004857

Photometric analysis of Abell 1689

Elena Dalla Bontà[1,2], Roger L. Davies[3], Ryan C. W. Houghton[3], Francesco D'Eugenio[3], Enrico M. Corsini[1,2] and Jairo Méndez-Abreu[4]

[1] Dipartimento di Fisica e Astronomia "G. Galilei", Università di Padova, Padova, Italy
email: elena.dallabonta@unipd.it

[2] INAF Osservatorio Astronomico di Padova, Padova, Italy

[3] Sub-department of Astrophysics, Department of Physics, University of Oxford, Oxford, UK

[4] Instituto Astrofísico de Canarias, La Laguna, Spain

Abstract. We carried out a photometric analysis of a sample of early-type galaxies in Abell 1689 at $z = 0.183$, using *HST*/ACS archive images in the rest-frame V band. We performed a two-dimensional photometric decomposition of each galaxy surface-brightness distribution using the GASP2D fitting algorithm (Méndez-Abreu *et al.* 2008). We adopted both a Sérsic and de Vaucouleurs law. S0 galaxies were analysed also taking into account a disc component described by an exponential law. The derived photometric parameters, together with the ones previously obtained with the curve of growth method (Houghton *et al.* 2012), will be used to analyse the Fundamental Plane of Abell 1689 and quantify how it is affected by the use of different decomposition techniques (Dalla Bontà *et al.* 2013, in preparation). The stellar velocity dispersions of the sample galaxies were derived by using GEMINI-N/GMOS and VLT/FLAMES (D'Eugenio *et al.* 2013) spectroscopic data.

Keywords. galaxies: clusters: individual: Abell 1689, galaxies: elliptical and lenticular, cD

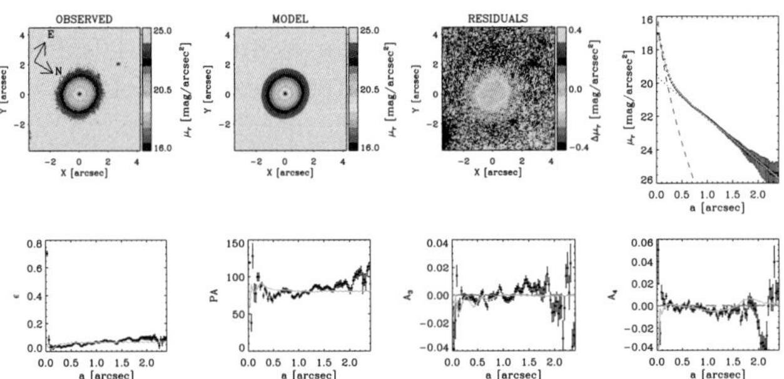

Figure 1. Two-dimensional photometric decomposition of a lenticular galaxy of Abell 1689 at $RA = +13^h 11^m 31^s.26$, $DEC = -1°20'52''.44$. From left to right and top to bottom: map of the observed, modelled, and residual (observed—modelled) surface-brightness distribution of the galaxy; ellipse-averaged radial profile of surface-brightness, ellipticity, position angle, and cosine-harmonic amplitudes A_3 and A_4, measured in the observed (black dots with error-bars) and modelled image (green solid line). The dashed blue and dotted red lines represent the intrinsic surface-brightness radial profiles of the bulge and disc, respectively.

References

D'Eugenio, F., Houghton, R. C. W., Davies, R. L., & Dalla Bontà, E. 2013, *MNRAS*, 429, 1258
Houghton, R. C. W., Davies, R. L., Dalla Bontà, E., & Masters, R. 2012, *MNRAS*, 423, 256
Méndez-Abreu, J., Aguerri, J. A. L., Corsini, E. M., & Simonneau, E. 2008, *A&A*, 478, 353

The intriguing life of massive galaxies
Proceedings IAU Symposium No. 295, 2012
D. Thomas, A. Pasquali & I. Ferreras, eds.

© International Astronomical Union 2013
doi:10.1017/S1743921313004869

The rotation curve and the density model of the Milky Way

Oleksiy Golubov and Andreas Just

Astronomisches Rechen-Institut, Zentrum für Astronomie der Universität Heidelberg,
Mönchhofstraße 12-14, 69120 Heidelberg, Germany,
email: golubov@ari.uni-heidelberg.de

Abstract. We study the asymmetric drift in the Milky Way with the aid of the RAVE data. Then we apply the deduced asymmetric drift correction to the SEGUE data and reconstruct the behaviour of the rotation curve of the Milky Way in the extended solar neighbourhood. The rotation curve appears to be essentially flat. We supplement our data by tangent point measurements of the inner rotation curve and fit it by a density model of the Milky Way.

Keywords. Galaxy: kinematics and dynamics, Galaxy: solar neighborhood

---

We study the asymmetric drift of the Milky Way disc dwarf stars with the sample from the Radial Velocity Experiment (RAVE). We adopt photometric parallaxes by Zwitter *et al.* (2010), bin the stars in colour, and study the mean rotational velocity as function of the squared radial velocity dispersion. We find that the linear Strömberg's relation provides a good fit to the data, with the best fit parameters $R_d = 2.37$ kpc for the radial scale length of the disc and $V_\odot = 4.37$ km s^{-1} for the V-component of the velocity of the Sun with respect to the local standard of rest. When the RAVE data are divided into 2 bins in the Galactic radius and the asymmetric drift correction is performed, we get 2 points in the rotation curve, nearly 300 pc inside and outside the solar radius. There is only a slight hint on the decrease of the circular velocity in the solar neighbourhood.

Then we apply the same Strömberg's relation to the F- and G-dwarfs sample from SDSS Extension for Galactic Understanding and Exploration (SEGUE). Photometric parallaxes of the stars were measured by Lee *et al.* (2011). The stars are at distances up to about 2 kpc from the Sun, and the deduced rotation curve is almost flat over the range of Galactocentric radii 7 to 10 kpc, in contrast to Sofue *et al.* (2009), who assume a strong dip in the rotation curve just outside the solar radius.

We supplement our data for the local rotation curve with the tangent point data for the inner rotation curve from Sofue *et al.* (2009), and fit them by a theoretical rotation curve, corresponding to a 3-component density model of the Galaxy, consisting of a Dehnen bulge, an exponential disc with a hole, and a flattened dark matter halo (Golubov & Just 2013). During this fitting we keep the local surface density and dark matter density in the disc constrained to the values determined by Just & Jahreiß (2010). Thus the density model of the Milky Way is constructed.

References

Golubov, O. & Just, A. 2013, in: T. Wong & J. Ott (eds.), *Molecular Gas, Dust, and Star Formation in Galaxies*, Proceedings of IAU Symposium No. 292 (Cambridge: Cambridge University Press)
Just, A. & Jahreiß, H. 2010, *MNRAS*, 402, 461
Lee, Y. S., Beers, T. C., An, D., Ivesic, Z., Just, A. *et al.* 2011, *ApJ*, 738, 187
Sofue, Y., Honma, M., & Omodaka, T. 2009, *PASJ*, 61, 227
Zwitter, T., Matijevič, G., Breddels, M. A., *et al.* 2010, *A&A*, 522, 54

The intriguing life of massive galaxies
Proceedings IAU Symposium No. 295, 2012
D. Thomas, A. Pasquali & I. Ferreras, eds.

© International Astronomical Union 2013
doi:10.1017/S1743921313004870

Massive bulges are *not* just ellipticals surrounded by disks

Dimitri A. Gadotti

European Southern Observatory
email: dgadotti@eso.org

Abstract. Using results from parametric multi-component multi-band image fitting of 1000 local massive galaxies in the SDSS, I investigate scaling relations of elliptical galaxies and bulges of disk galaxies. I show that ellipticals and bulges occupy different loci in both the edge-on and face-on views of the fundamental plane. In addition, ellipticals and bulges have offset mass-size relations (see Fig. 1). These results imply that massive bulges are not just massive ellipticals with a surrounding disk, a misconception driven by early studies. This is evidence that massive ellipticals and bulges have different formation histories, with important consequences for studies on galaxy formation and evolution. Full details can be seen in Gadotti (2009).

Keywords. galaxies: bulges, galaxies: elliptical and lenticular, cD, galaxies: evolution, galaxies: formation, galaxies: structure

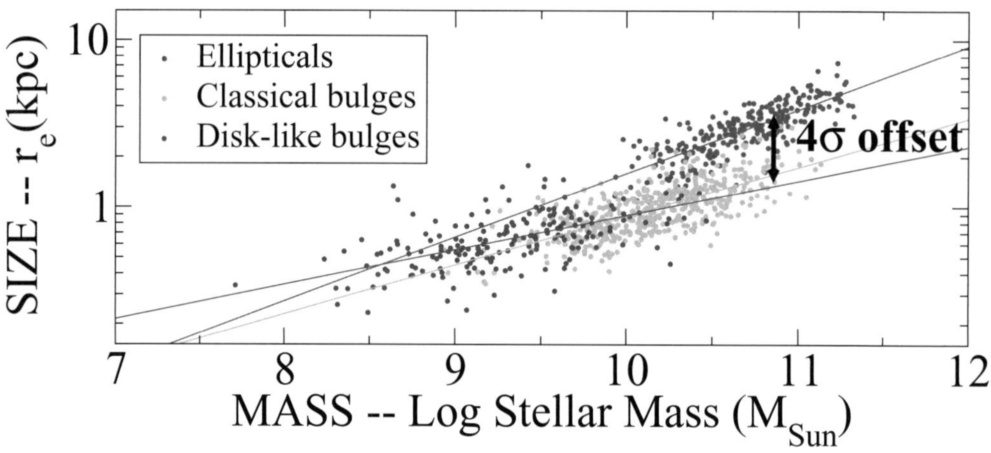

Figure 1. Mass–size relations of elliptical galaxies, classical bulges and disk-like bulges, as indicated. The relations for massive ellipticals and classical bulges in disk galaxies show a clear offset at a 4σ confidence level. If one puts a disk surrounding a massive elliptical galaxy (i.e. with $M_\star \geqslant 10^{10}\ M_\odot$) one does *not* end up with a disk galaxy with a bulge similar to those observed in Nature (adapted from Gadotti 2009).

References

Gadotti, D. A. 2009, *MNRAS*, 393, 1531

The intriguing life of massive galaxies
Proceedings IAU Symposium No. 295, 2012
D. Thomas, A. Pasquali & I. Ferreras, eds.

© International Astronomical Union 2013
doi:10.1017/S1743921313004882

Variations in the Fundamental Plane of massive galaxies with stellar population, morphology and local density

Christina Magoulas[1,2], Christopher Springob[1], Rob Proctor[3], Matthew Colless[1], D. Heath Jones[4], Chiaki Kobayashi[5], Lachlan Campbell[6], John Lucey[7] and Jeremy Mould[8]

[1]AAO, [2]U. Melbourne, email: `c.magoulas@pgrad.unimelb.edu.au`,
[3]U. São Paulo, [4]Monash University, [5]ANU, [6]WKU, [7]Durham University, [8]Swinburne University

We have measured Fundamental Plane (FP) parameters in the 2MASS J, H and K passbands for 10,000 ellipticals, lenticulars and early-type spiral bulges in the 6dF Galaxy Survey (6dFGS) – a spectroscopic survey of the southern sky with $|b| > 10°$ (Jones *et al.* 2009). 6dFGS provides a large near-infrared-selected sample of galaxies for accurately determining the NIR FP and investigating the trends in the FP with stellar population, morphology and environment.

In this study the FP, as defined by $\log R_e = a \log \sigma_0 + b \log I_e + c$, is modelled by a 3D Gaussian distribution and the best-fit parameters are determined using a robust maximum likelihood method (Magoulas *et al.* 2012). The best-fit plane is therefore an unbiased determination of the FP, as the censoring effects and correlated uncertainties present in the data are accounted for with this method.

By fitting the FP to galaxies that span a range of environments, we test the universality of the FP and its consistency across galaxies in the field and galaxies in clusters. We find little variation of the 6dFGS FP slopes with global environment (quantified by group richness) or local environment (quantified by surface density). However there is an offset in r_0 (the $\log R_e$-value of the fitted FP at a fiducial point in the middle of a sample) with environment such that galaxies in the field and low-density regions are on average about 5% larger than those in groups and higher-density regions (Magoulas *et al.* 2012). We also classify the morphological mix of galaxies in our sample and find that ellipticals, lenticulars and the bulges of bulge-dominated spirals lie on the same FP and have similar scatter, although the latter are typically 10% larger at fixed velocity dispersion and surface brightness (Magoulas *et al.* 2012).

Global stellar population trends within the FP were derived by binning individual galaxy measurements of stellar population parameters (age, metallicity) in FP space. A significant trend in r_0 is observed, such that younger galaxies are, in general, 25% larger compared to the oldest galaxies. We find that massive galaxies exhibit clear stellar population trends in FP space: age varies most strongly orthogonal to the FP, while metallicity variation is almost entirely along the intermediate axis of the FP. Remarkably, none of the stellar population parameters vary along the long axis of the plane, which corresponds to luminosity density (Springob *et al.* 2012).

We interpret the lack of any trend in stellar populations along the long axis of the FP as being the effect of differing dry merger histories on the central luminosity density. We speculate that age drives all the trends in stellar population with residuals about the plane through its correlation with environment, morphology and metallicity.

References

Jones D. H., *et al.* 2009, *MNRAS*, 399, 683
Magoulas C., *et al.* 2009, *MNRAS*, 427, 245
Springob C. M., *et al.* 2012, *MNRAS*, 420, 2773

The intriguing life of massive galaxies
Proceedings IAU Symposium No. 295, 2012
D. Thomas, A. Pasquali & I. Ferreras, eds.

© International Astronomical Union 2013
doi:10.1017/S1743921313004894

2D kinematics of the edge-on spiral galaxy ESO 379-006

M. Rosado[1], R. F. Gabbasov[1], P. Repetto[1], I. Fuentes-Carrera[2], P. Amram[3], M. Martos[1] and O. Hernandez[3,4]

[1]Instituto de Astronomía, Universidad Nacional Autónoma de México, México
Apartado Postal 70-264, CP04510, México, D. F.
email: `margarit@astro.unam.mx`

[2]Escuela Superior de Física y Matemáticas, IPN, México
Box 515, SE-75120 Uppsala, Sweden
email: `hoefner@astro.uu.se`

[3]Laboratoire d'Astrophysique de Marseille, France

[4]Université de Montréal, Canada

Abstract. We present a kinematical study of the marginally edge-on galaxy ESO 379-006. With Fabry-Perot spectroscopy at Hα we obtain velocity maps, the radial velocity field, and position-velocity diagrams parallel to the major and to the minor axis of the galaxy. We build the rotation curve of the galaxy and discuss the role of projection effects. The twisting of isovelocities in the radial velocity field of the disk of ESO 379-006 as well as a kinematical asymmetry found in the position-velocity diagrams parallel to the minor axis suggest the existence of non-circular motions that can be modeled by including a radial inflow besides the rotation motion. Extraplanar Diffuse Ionized gas was detected in this galaxy both from the images and from its kinematics. It is possible that the diffuse gas is lagging in rotation.

1. Results

The Fabry-Perot data were obtained at the ESO 3.6m telescope with the CIGALE instrument. These data allowed us to construct velocity maps, the radial velocity field and position-velocity diagrams (PVDs) parallel to the major and minor axis of the galaxy ESO 379-006. Rotation Curves (RCs) were obtained from the intensity peaks of the velocity profiles (barycentric) and by the Envelope-Tracing Method. A comparison between them shows that projection effects are important in the inner regions of the galaxy. We have modeled this galaxy using the GALMOD task of GYPSY. We explored almost all the galactic parameters such as inclination, vertical scale height, velocity dispersion, systemic velocity, center, etc. and the model that fits the best the important features derived from the observations has an inclination of 82.5 degrees and a vertical scale height of 2 arcsec among the main parameters, together with the inclusion of an additional velocity, in the radial direction, corresponding to an inflow (Fig. 1). In addition, the PVD along the major axis shows some low velocity extensions ("the beard") that can be interpreted either as extraplanar Ionized Diffuse Gas (from which we know about its existence and detection in this particular galaxy) or as projection effects due to the high inclination of the galaxy.

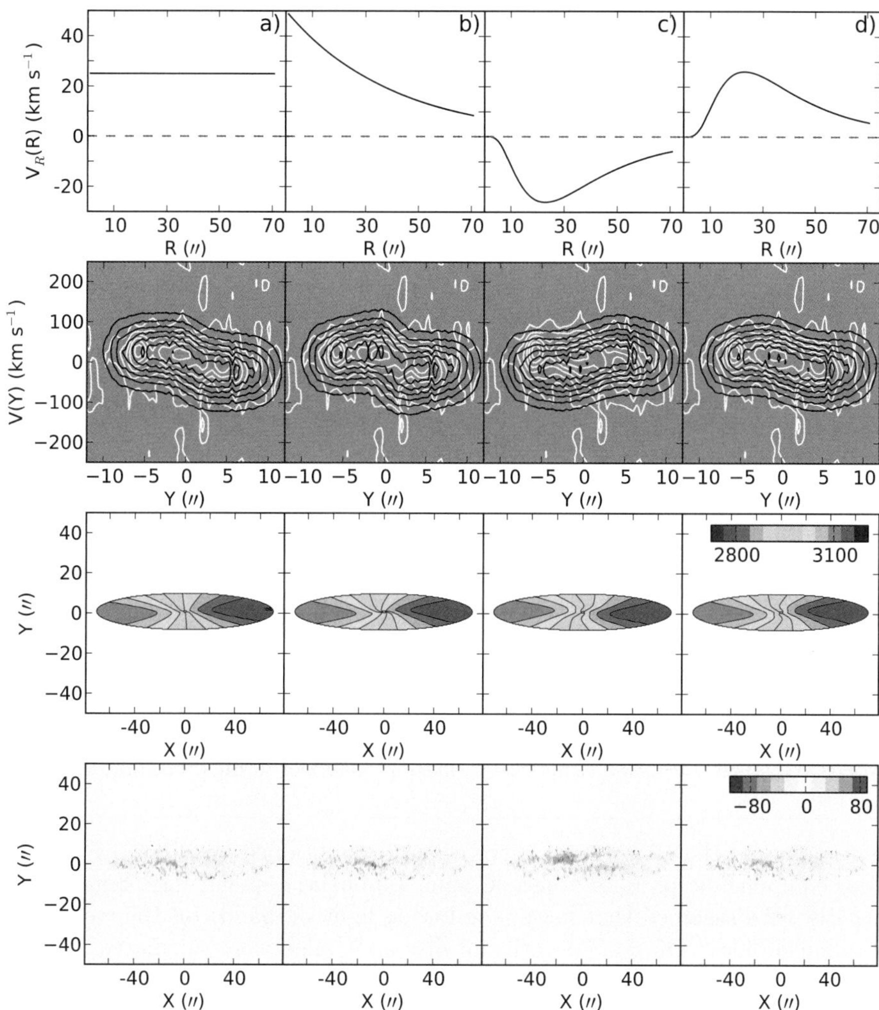

Figure 1. The radial motion distribution (top row), a comparison of modeled (black contours) and observed (white contours) minor axis PVD (second row), synthetic velocity fields (third row) and the residual velocity fields (bottom row) for models with a constant inflow (a), exponential decaying inflow (b), lognormal outflow (c) and the lognormal inflow (d).

Acknowledgements

This paper was done with financial support from grants P-82389 from CONACYT.

The intriguing life of massive galaxies
Proceedings IAU Symposium No. 295, 2012
D. Thomas, A. Pasquali & I. Ferreras, eds.

© International Astronomical Union 2013
doi:10.1017/S1743921313004900

On the general structure of giant low surface brightness galaxy Malin 2

A. S. Saburova[1], A. V. Kasparova[1], I. Yu. Katkov[1], D. V. Bizyaev[2,1] and I. V. Chilingarian[3,1]

[1]Sternberg Astronomical Institute of Moscow State University,
Universitetskii pr. 13, 119992, Moscow, Russia
email: saburovaann@gmail.com

[2]New Mexico State University and Apache Point Observatory, Sunspot, NM 88349, USA

[3]Smithsonian Astrophysical Observatory, Harvard-Smithsonian Center for Astrophysics, 60 Garden Street, Cambridge, MA 02138 USA

Abstract. We carried out the multicolor surface photometry in BVR and griz filters parallel with stellar kinematic measurements from the long slit spectra for the low surface brightness galaxy Malin 2. The use of the multicolor surface photometry as well as the available HI rotation curve allowed us to construct the mass distribution model of the galaxy. Photometrical and dynamical mass estimates agree with the dark halo mass fraction of about 70% within four disc radial scalelengths ($\sim$ 70 kpc). We used our dynamical model to obtain radial profiles of the equilibrium disc volume density and gas pressure in the galaxy midplane based on the available HI and CO data. The observed molecular gas fraction appears to be much higher than in the high surface brightness galaxies for a similar gas pressure.

Keywords. Galaxies: evolution, Galaxies: formation, Galaxies: individual: Malin 2.

We decomposed the HI rotation curve of Malin 2 obtained by Pickering *et al.* (1997) into four components: pseudoisothermal halo, exponential stellar disc, Sersic bulge and gaseous disc. We assumed that mass is following light (R-band) for the stellar disc and bulge radial profiles. The M/L_R ratios of disc and bulge were calculated from the observed color profiles and stellar population models. The obtained photometrical model is close to the "maximum disc" model.

We calculated the equilibrium hydrostatic turbulent gas pressure using an equilibrium self-consistent model (see Kasparova 2012). We find that Malin 2 behaves in a different way than the normal spiral field galaxies. For all points along the radius the values of η (molecular gas fraction) is significantly higher than the expected ones for a given P in field galaxies.

A possible explanation is the long lifetime of molecular clouds in low density regions.

This work was supported by president grant of the Russian Federation MD-3288.2012.2.

References

Das, M., Boone, F., & Viallefond, F. 2010, *A&A*, 523, 63

Pickering, T. E., Impey, C. D., van Gorkom, J. H., & Bothun, G. D. 1997, *AJ*, 114, 1858

Kasparova, A. 2012 *Astron. Lett.*, 38, 63

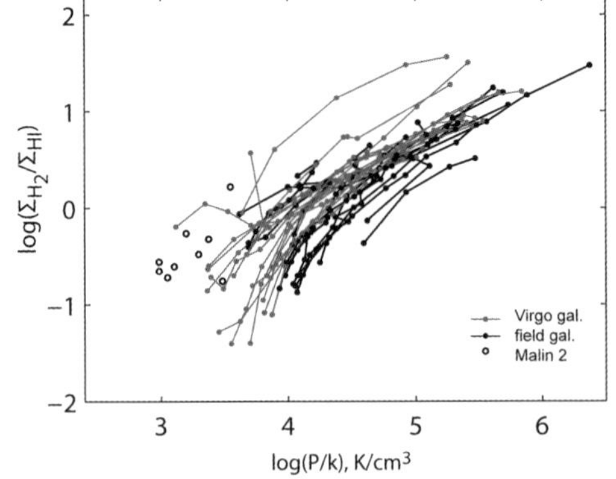

Figure 1. The molecular gas fraction (η) versus the gas pressure.

The intriguing life of massive galaxies
Proceedings IAU Symposium No. 295, 2012
D. Thomas, A. Pasquali & I. Ferreras, eds.

© International Astronomical Union 2013
doi:10.1017/S1743921313004912

Strong Lenses With Single Images

Yiping Shu[1], Adam S. Bolton[1], Joel R. Brownstein[1] and the SLACS Collaborations

[1]Department of Physics and Astronomy, University of Utah,
115 South 1400 East, Salt Lake City, UT 84112, USA
email: ypshu@physics.utah.edu

SLACS for the masses is the extension of the successful Sloan Lens ACS (SLACS) survey (Bolton *et al.* 2006, Treu *et al.* 2006, Koopmans *et al.* 2006, Gavazzi *et al.* 2007 and Bolton *et al.* 2008) but focuses on the lower-mass end of elliptical galaxies (EGs) to yield a more complete strong-lens sample. As to date, 118 out of the 137 proposed candidates have been observed and inspected individually. Among all the targets we have modeled until now, there are:

- 50 grade-A lenses which show clearly lensing features with multiple imaging
- 13 grade-B lenses which have lensing features but no counter-images

The details about image processing can be found in Bolton *et al.* (2006) and Brownstein *et al.* (2012). Here we focus particularly on grade-B lenses which have been barely studied due to the absence of counter-images and the difficulty to construct reliable lens models.

For each grade-B lens, we fix all the other lensing parameters to values suggested by the b-spline fit (Bolton *et al.* 2006) except the Einstein radius $\theta_{\rm Ein}$ which is gradually varied and fit for the lensed image. Eventually we get a chi-square curve as a function of the trial $\theta_{\rm Ein}$ (Figure 1.) from which we can infer its upper limit by looking for a point at which the slope of the chi-square curve changes significantly and the fit goes unreasonable after that point. This set of upper-limit candidates, which are relatively low mass galaxies, extends our understandings of EGs to a wider mass range.

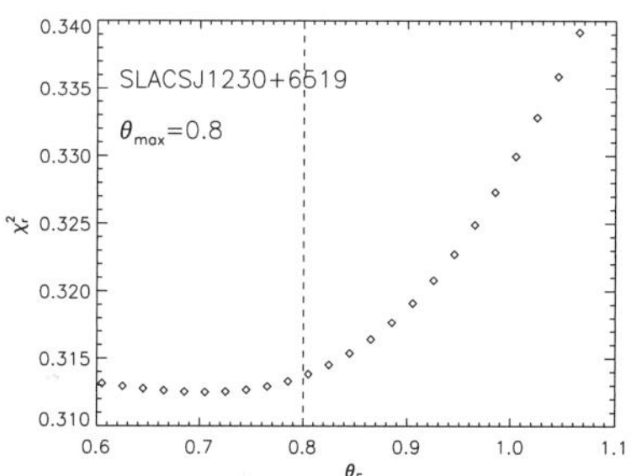

Figure 1. The reduced chi-square curve as a function of the trial Einstein radius for target SLACSJ1230+6519. The dashed line indicates the upper limit of the Einstein radius.

References

Bolton, A. S., Burles S., Koopmans, L. V. E., Treu, T., Moustakas, L. A., & Schlegel, D. J. 2006, *ApJ*, 638, 703

Bolton, A. S., Burles, S., Koopmans, L. V. E., Treu, T., Gavazzi, R., & Moustakas, L. A. 2008, *ApJ*, 682, 964

Brownstein, J. R., Bolton, A. S., Schlegel, D. J., Eisenstein, D. J., Kochanek, C. S., Connolly, N., Maraston, C., Pandey, P., Seitz, S., Wake, D. A., Wood-Vasey, W. M., Brinkmann, J., Schneider, D. P., & Weaver, B. A. 2012, *ApJ*, 744, 41

Gavazzi, R., Treu, T., Rhodes, J. D., Koopmans, L. V. E., Bolton, A. S., Burles, S., Massey, R. J., & Moustakas, L. A. 2007, *ApJ*, 667, 176

Koopmans, L. V. E., Treu, T., Bolton., A. S., Burles, S., & Moustakas, L. A. 2006, *ApJ*, 649, 599

Treu, T., Koopmans, L. V. E., Bolton, A. S., Burles, S., & Moustakas, L. A. 2006, *ApJ*, 640, 662

The intriguing life of massive galaxies
Proceedings IAU Symposium No. 295, 2012
D. Thomas, A. Pasquali & I. Ferreras, eds.

© International Astronomical Union 2013
doi:10.1017/S1743921313004924

Local group analogues – searching for the satellites of the nearest massive galaxies

Ryan Speller and James E. Taylor

Department of Physics and Astronomy, University of Waterloo,
200 University Avenue West, Waterloo, Ontario, N2L 3G1, Canada
email: taylor@uwaterloo.ca

Abstract. We have performed a search for faint companions around the nearest massive galaxies. We see a clear signature of clustering of faint objects, both in projected separation and in velocity offset. The inferred satellite luminosity functions confirm that the abundance of faint satellites seen in the Local Group is typical of other nearby systems.

Keywords. dark matter, galaxies: dwarf, Local Group

The abundance and distribution of dwarf satellites remains a critical test of LCDM on the smallest scales. Complete samples of dwarfs have been limited to the Local Group, however. Beyond 1 (10) Mpc, few galaxies fainter than $M_V \sim -10$ (-15) are known.

Of the seven unobscured massive galaxies within 10 Mpc, Cen A has a much less concentrated satellite distribution than the rest; interestingly it is also the one nearby early-type galaxy. To see if there is a connection between satellite populations and central morphology, we clearly need a larger sample of satellite systems.

To find more faint satellites, we selected isolated primaries from the Atlas3D sample of the brightest galaxies within 42 Mpc (Cappellari *et al.* 2011) and searched for possible satellites within $R_p = 1$ Mpc projected using the Sloan Digital Sky Survey DR8 (http://skyserver.sdss3.org). Fig. 1 shows that an over-density of objects at $R_p \leqslant 500$ kpc is detected at S/N$\sim$9. Our detection corresponds to about 5.4 ± 0.6 satellites per central galaxy, in keeping with expectations from the Local Group (dashed line).

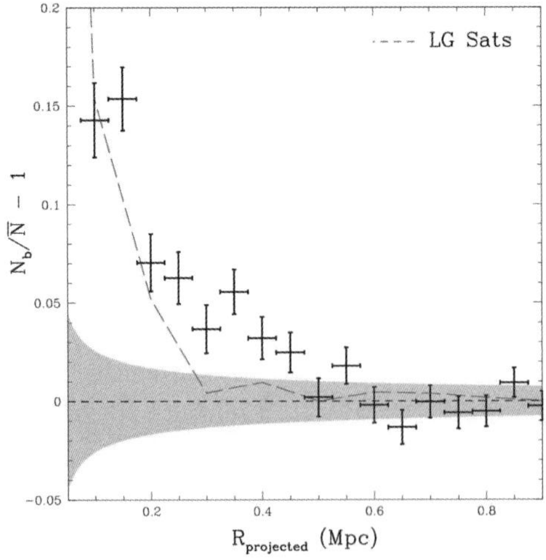

Figure 1. Excess sources detected around the nearby sample. The shading indicates the r.m.s. scatter in the background and the dashed line shows the expectation from the Local Group.

The intriguing life of massive galaxies
Proceedings IAU Symposium No. 295, 2012
D. Thomas, A. Pasquali & I. Ferreras, eds.

© International Astronomical Union 2013
doi:10.1017/S1743921313004936

A Possible Stream of Accreted Globular Clusters in M 31

H.-B. Yuan[1,2]

[1]Kavli Institute for Astronomy and Astrophysics, Peking University, Beijing 100871, China
email: yuanhb4861@pku.edu.cn
[2]LAMOST Fellow

Abstract. Globular clusters (GCs) are excellent tracers for the formation, assembly and evolutionary history of their host galaxies. However, their origin and their role in galaxy evolution are still unclear. There are accumulating evidences that a significant fraction of GCs in massive galaxies (e.g., M 31) are accreted during their assembly history (e.g., Mackey *et al.* 2010). In this contribution, we report the discovery of a possible stream of accreted GCs in M 31 using data from the literature. Unlike previous substructures of GCs identified as clumps in the phase and metallicity space (Ashman, Keith & Christina 1993), the members of this stream are widely spread but tightly correlated in the position-velocity space (see Fig. 1). The tight correlation suggests that they possibly follow very similar orbits. A number of stellar streams have been discovered in the outer halo of M 31 (e.g., Ibata *et al.* 2001; McConnachie *et al.* 2009), one of which may be physically associated with the GC stream. If the association is established, it will not only provide a key evidence for accretion origins of some GCs in M 31, but also place a strong constraint on the mass distribution of M 31.

Keywords. galaxies: individual (M 31), globular clusters: individual (B 077, B 125, B 213, B 399, B 403, B 407, SH 25, EXT 8)

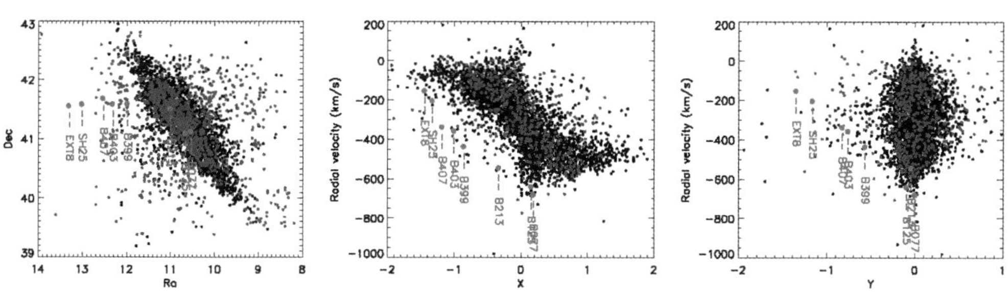

Figure 1. Spatial distribution (left panel) and position-velocity diagrams (middle and right panels) of the accreted GC stream in M 31. Black crosses: planetary nebulae from Merrett *et al.* (2006); Blue crosses: confirmed/candidate GCs from Revised Bologna Catalog V4.0 (Galleti *et al.* 2009); Red dots: members of the GC stream. The IDs of the GC members are also marked.

References

Ashman, K. M. & Bird, C. M. 1993, *AJ*, 106, 2281
Galleti, S., Bellazzini, M., Buzzoni, A., Federici, L., & Fusi Pecci, F. 2009, *A&A*, 508, 1285
Ibata, R., Irwin, M., Lewis, G., Ferguson, A. M. N., & Tanvir, N. 2001, *Nature*, 412, 49
Mackey, A. D., Huxor, A. P., Ferguson, A. M. N., *et al.* 2010, *ApJL*, 717, L11
McConnachie, A. W., Irwin, M. J., Ibata, R. A., *et al.* 2009, *Nature*, 461, 66
Merrett, H. R., Merrifield, M. R., Douglas, N. G., *et al.* 2006, *MNRAS*, 369, 120

The intriguing life of massive galaxies
Proceedings IAU Symposium No. 295, 2012
D. Thomas, A. Pasquali & I. Ferreras, eds.

© International Astronomical Union 2013
doi:10.1017/S1743921313004948

The intriguing properties of local compact massive galaxies: What are they?

A. Ferré-Mateu[1,2], A. Vazdekis[1,2], I. Trujillo[1,2], P. Sánchez-Blázquez[3], E. Ricciardelli[4] and I. G. de la Rosa[1,2]

[1]Instituto de Astrofísica de Canarias, La Laguna, SPAIN
email: aferre@iac.es

[2]Departamento de Astrofísica, Universidad de La Laguna, SPAIN

[3]Departamento de Física Teórica, Univ. Autónoma de Madrid, Madrid, SPAIN

[4]Departament d'Astronomia i Astrofísica, Universitat de València, SPAIN

Abstract. Studying the properties of the few compact massive galaxies that exist in the local Universe (Trujillo *et al.* 2009) might provide a closer look to the nature of their high redshift ($z \geqslant 1.0$) massive counterparts. By this means we have characterized their main kinematics, structural properties, stellar populations and star formation histories with a set of new high quality spectroscopic and imaging data (Ferré-Mateu *et al.* 2012 and Trujillo *et al.* 2012). These galaxies seem to be truly unique, as they do not follow the characteristic kinematics, stellar surface mass density profiles and stellar population patterns of present-day massive ellipticals or spirals of similar mass. They are, instead, more alike their high-z analogs.

Summarizing, local compact massive galaxies are rare, unique and the perfect laboratory to study their high redshift counterparts.

Keywords. galaxies: evolution, galaxies: formation, galaxies: fundamental parameters, galaxies: kinematics and dynamics, galaxies: stellar content

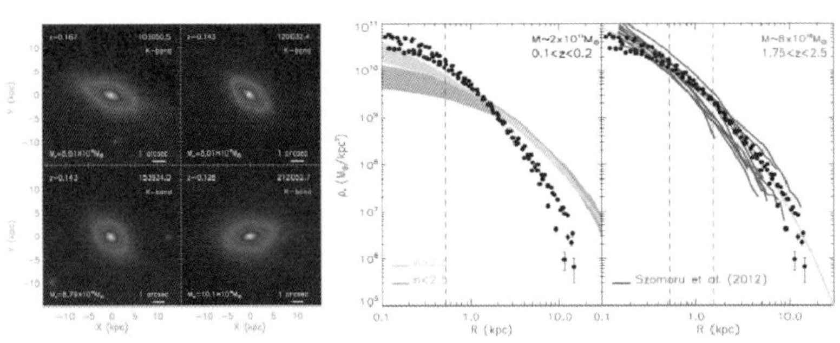

Figure 1. *Left* K-band high resolution imaging of four nearby massive compact galaxies with galaxy stellar mass, spectroscopic redshift, and the equivalent $1''$ angular size (solid line). *Right* Stellar surface mass density profiles of the local compact massive galaxies (black points) *(i)* compared with the stellar mass density profiles of SDSS DR7 $M_* \geqslant 2 \times 10^{11} M_{sun}$ and $0.1 < z < 0.2$ disk-like galaxies (n < 2.5) and spheroid-like (n > 2.5) galaxies. *(ii)* compared with z $\sim$ 2 massive compact galaxies of same stellar mass (Szomoru *et al.* 2012).

References

Ferré-Mateu, A., *et al.* 2012, *MNRAS*, 423, 632

Trujillo, I., *et al.* 2009, *ApJ*, 692, 118

Trujillo, I., Carrasco, R. E., & Ferré-Mateu, A. 2012, *ApJ*, 751,45

Szomoru, D., *et al.* 2012, *ApJ*, 749, 121

The intriguing life of massive galaxies
Proceedings IAU Symposium No. 295, 2012
D. Thomas, A. Pasquali & I. Ferreras, eds.

© International Astronomical Union 2013
doi:10.1017/S174392131300495X

Supermassive black holes: Coevolution (or not) of black holes and host galaxies

John Kormendy

Department of Astronomy, University of Texas at Austin, Austin, TX 78712, USA;
Max-Planck-Institut für Extraterrestrische Physik, Garching bei München, Germany;
Universitäts-Sternwarte, Ludwig-Maximilians-Universität, München, Germany;
email: kormendy@astro.as.utexas.edu

Abstract. Supermassive black holes (BHs) have been found in 75 galaxies by observing spatially resolved dynamics. The *Hubble Space Telescope* (HST) revolutionized BH work by advancing the subject from its 'proof of concept' phase into quantitative studies of BH demographics. Most influential was the discovery of a tight correlation between BH masses $M_\bullet$ and the velocity dispersions σ of stars in the host galaxy bulge components at radii where the stars mostly feel each other and not the BH. Together with correlations between $M_\bullet$ and bulge luminosity, with the 'missing light' that defines galaxy cores, and with numbers of globular clusters, this has led to the conclusion that BHs and bulges coevolve by regulating each other's growth. This simple picture with one set of correlations for all galaxies dominated BH work in the past decade.

New results are now replacing the above, simple story with a richer and more plausible picture in which BHs correlate differently with different kinds of galaxy components. BHs with masses of 10^5—10^6 $M_\odot$ live in some bulgeless galaxies. So classical (merger-built) bulges are not necessary equipment for BH formation. On the other hand, while they live in galaxy disks, BHs do not correlate with galaxy disks or with disk-grown pseudobulges. They also have no special correlation with dark matter halos beyond the fact that halo gravity controls galaxy formation. This leads to the suggestion that there are two modes of BH feeding, (1) local, secular and episodic feeding of small BHs in largely bulgeless galaxies that involves too little energy feedback to drive BH–host-galaxy coevolution and (2) global feeding in major galaxy mergers that rapidly grows giant BHs in short-duration events whose energy feedback does affect galaxy formation. After these quasar-like phases, maintenance-mode BH feedback into hot, X-ray-emitting gas continues to have a primarily negative effect in preventing late-time star formation when cold gas or gas-rich galaxies get accreted. Finally, the highest-mass galaxies inherit coevolution effects from smaller galaxies; the tightness of their BH correlations is caused mainly by averaging during dissipationless major mergers.

Keywords. black hole physics; galaxies: bulges; galaxies: elliptical and lenticular, cD; galaxies: evolution; galaxies: kinematics and dynamics; galaxies: structure; X-rays: galaxies

1. Introduction

This paper summarizes a more detailed *ARAA* review of supermassive black holes (BHs) by Kormendy & Ho (2013). Our understanding of BHs and their coevolution with their host galaxies has begun a period of rapid progress as we have come to realize that BHs correlate differently with different kinds of galaxy components that have different formation histories. This is good news, because it allows us for the first time to distinguish between several different kinds of BH–host-galaxy interactions. But it involves the reader in aspects of the structure and evolution of different kinds of galaxies that deserve full-length reviews of their own. I summarize these subjects as efficiently as I can and provide references for further details. One result that is relevant to this Symposium is evidence that *the main bodies of giant ellipticals form via major mergers.*

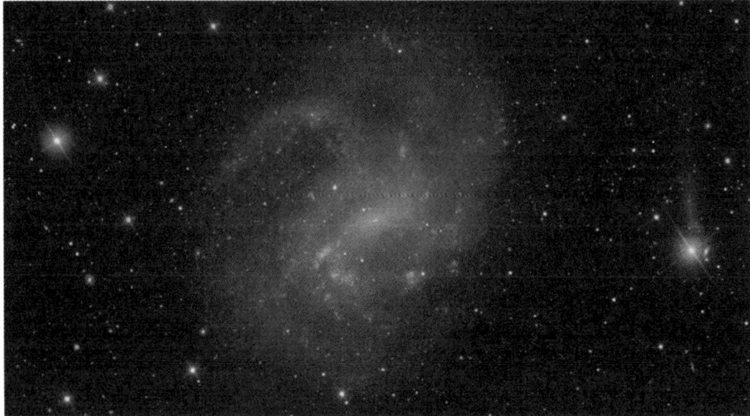

Figure 5. The Sdm galaxy NGC 4395 (SDSS *gri* image). The low surface brightness, the lack of a bulge, and the presence of a nuclear star cluster are characteristic of dwarf, late-type galaxies. M 33 and M 101 are more luminous, higher-surface-brightness analogs (Kormendy *et al.* 2010).

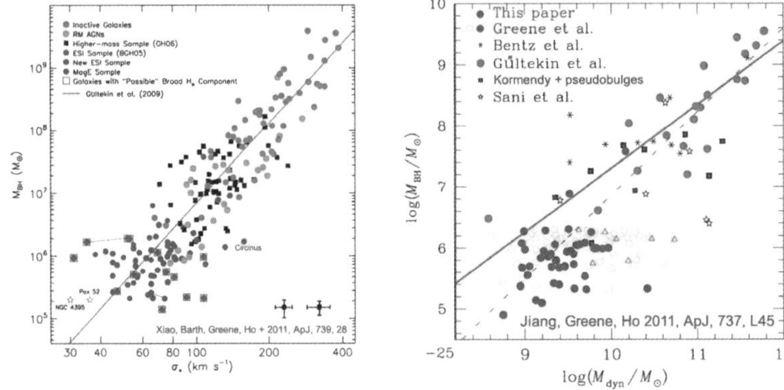

Figure 6. BH – host-galaxy correlations for AGNs with BH masses derived from reverberation mapping (green points) and from broad emission-line widths (blue and black points). Most of these galaxies contain pseudobulges or no bulges. It is important to note that the green, black, and blue points have been zeropointed to the red points for galaxies whose BH masses were measured using observations of spatially resolved kinematics. Shifts in $M_\bullet$ with respect to these points (similar to the offsets of pseudobulge points in Fig. 3) may have been removed.

6. Bulges are not necessary equipment for BH growth

Figure 4 shows the striking difference between M 33 and NGC 4395. M 33 is a bulgeless galaxy with an outer rotation velocity $V_{\rm circ} \simeq 120 \pm 10$ km s^{-1} (Corbelli 2003) and no BH. In contrast, NGC 4395 is a smaller bulgeless galaxy ($V_{\rm circ} \simeq 75 \pm 10$ km s^{-1}; Swaters *et al.* 2009) that contains a BH with $M_\bullet = (3.6 \pm 1.1) \times 10^5\, M_\odot$. The pure-disk nature of NGC 4395 is illustrated in Figure 5. The galaxy does not even contain a pseudobulge, yet it contains a BH and one of the faintest Seyfert 1 active galactic nuclei (AGNs) known (Filippenko *et al.* 1993; Filippenko & Ho 2003). Much work shows that NGC 4395 is no fluke: *Bulges are not necessary equipment for BH formation and growth* (Barth *et al.* 2004; Greene & Ho 2004, 2007b; Thornton *et al.* 2008; Greene *et al.* 2008, 2010; Reines *et al.* 2011; Jiang *et al.* 2011; Xiao *et al.* 2011; Dong *et al.* 2012; see Ho 2008; Kormendy & Ho 2013 for reviews). Most of this evidence comes from Seyfert 1 activity. The large scatter of $M_\bullet$ with host galaxy properties is further illustrated in Figure 6.

7. Feeding the monster with and without BH-galaxy coevolution

The results reviewed in Sections 3–6 led Kormendy *et al.* (2011) to suggest that there are two modes of BH feeding, (1) local, secular and episodic feeding of small BHs in largely bulgeless galaxies that involves too little energy feedback to result in BH-host coevolution and (2) global feeding in galaxy mergers that rapidly grows giant BHs in short-duration events whose energy feedback does affect galaxy formation. Plausibly, the former, smaller BHs are the seeds of the latter, giant BHs.

8. BHs do not correlate directly with dark matter halos

Several authors suggest that $M_\bullet$ correlates at least as well with dark matter (DM) halos as it does with bulges (Ferrarese 2002; Baes *et al.* 2003; Pizzella *et al.* 2005). This is based on proxy parameters: σ for $M_\bullet$ and the asymptotic outer rotation velocity V_{circ} of galaxy disks for DM. The claimed observation is a tight correlation between σ and V_{circ}. The implications would be profound, because the unknown physics of nonbaryonic DM might be necessary to engineer BH-galaxy coevolution. The result would also be attractive for modelers of galaxy formation, because the simplest galaxy property that is provided by hierarchical clustering n-body simulations could then be used to control AGN feedback.

However, Kormendy & Bender (2011) show that BHs do not correlate tightly with DM. Is a BH – DM relation more fundamental than the BH–bulge relation that we know about? A simple test is to ask whether $M_\bullet$ correlates tightly with V_{circ} in the absence of a bulge. Kormendy *et al.* (2010) measure σ in the biggest bulgeless galaxies that are close enough so we can observe their globular-cluster-like nuclear star clusters. Kormendy & Bender (2011) then update the $V_{\mathrm{circ}} - \sigma$ correlation from Ferrarese (2002) as shown in Figure 7. Absent a bulge (color data), the correlation has too much scatter to imply coevolution, even if σ is a valid proxy for $M_\bullet$ (which we do not know). This is true even for giant galaxies ($V_{\mathrm{circ}} \sim 200$ km s^{-1}). Overlapping these in V_{circ}, the black points show galaxies

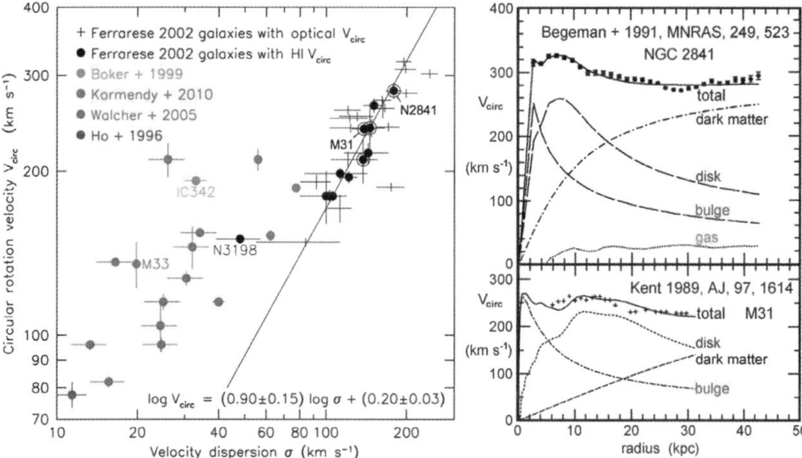

Figure 7. (left) Outer rotation velocities V_{circ} of spiral galaxy disks versus near-central velocity dispersions σ. Black points show the correlation from Ferrarese (2002, circled if the galaxy has a classical bulge) unless the σ measurement was compromised by low wavelength resolution. Added in color are points for galaxies that have no classical bulge and essentially no pseudobulge but that are measured with wavelength resolution (dispersion $\sigma_{\mathrm{instr}} < 10$ km s^{-1}) high enough to resolve the smallest σ in galactic nuclei. The line (equation; velocities are in units of 200 km s^{-1}) is a symmetric least-squares fit. (right) Decomposition of observed rotation curves (data points) into contributions from the bulge, disk, gas, and dark matter for two classical-bulge galaxies labeled in the left panel. From Kormendy & Bender (2011), which lists references in the key.

from Ferrarese (2002) that do show a tight correlation between V_{circ} and σ. But only four of these contain classical bulges (circled points). The rest have pseudobulges, and we already know that σ is not a proxy for $M_\bullet$ for pseudobulges. If pseudo and classical bulges nevertheless show the same, tight V_{circ}–σ correlation, then that correlation has nothing to do with BHs. Instead, Kormendy & Bender (2011) conclude that the correlation of the black points in Figure 7 (left) is a manifestation of the conspiracy between visible and dark matter to have approximately the same maximum V_{circ} (van Albada & Sancisi 1986). This is shown in Figure 7 (right) for two classical bulges in Figure 7 (left). Kormendy & Bender (2011) and Kormendy & Ho (2013) review other arguments against the idea that DM correlates with BHs beyond the bulge correlation.

9. BHs and the fundamental plane correlations of their host galaxies

BHs correlate tightly with classical bulges and ellipticals but not with pseudobulges, disks or dark matter halos. In deciphering coevolution, this focuses our attention on major galaxy mergers (Sections 10, 11). Figure 8 summarizes some additional implications. BHs coevolve only with the galaxy components that define the fundamental plane correlations. Therefore the mass function of the BHs that coevolve with ellipticals is bounded: we know of no ellipticals that are much smaller than M 32. In contrast, the mass function of spiral, irregular, and spheroidal galaxies is not known to be bounded at low masses (green and blue points in Figure 8). However, there are signs that even the BHs that live in these galaxies have mass functions that become bounded at $M_\bullet < 10^6\ M_\odot$ (Greene & Ho 2007a).

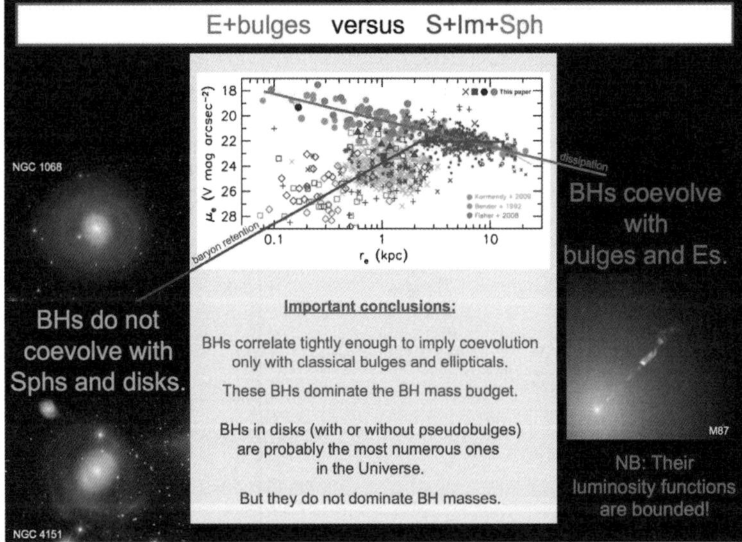

Figure 8. Powerpoint slide showing the relationships of BHs that coevolve with classical bulges and ellipticals (red line and text; pink and brown points; e. g., M 87) and those that do not evolve with disks and Sph galaxies (blue lines and text; blue and green points; Seyfert galaxy examples). Here r_e is the effective radius that contains half of the light of the bulge or disk and μ_e is the surface brightness at r_e. For classical bulges and ellipticals, this correlation (Kormendy 1977: red line) is an almost–edge-on projection of the fundamental plane; it is a sequence of increasing dissipation during galaxy formation at decreasing luminosity (lower-right toward upper-left). For disks (blue), surface brightness is independent of r_e or luminosity L at high L (Freeman 1970). Fainter than absolute magnitude $M_V \simeq -18$ or at $r_e < 2$ kpc, surface brightness decreases with decreasing r_e and L (upper-right toward lower-left); this is a sequence of decreasing baryon retention in shallower gravitational potential wells. Adapted from Kormendy & Bender (2012).

10. A brief introduction to AGN feedback

The idea that BHs and bulges coevolve by regulating each other's growth is popular for many reasons: (1) The tightness of the $M_\bullet - M_{K,\text{bulge}}$, $M_\bullet - \sigma$, $M_\bullet$ – globular-cluster (Burkert & Tremaine 2010; Harris & Harris 2011) and other (Section 11.2) correlations suggests that BH growth and galaxy formation are connected. Particularly compelling is the conclusion by Ferrarese & Merritt (2000) and by Gebhardt *et al.* (2000) that the scatter in the $M_\bullet - \sigma$ correlation is consistent with measurement errors. (2) The binding energy of a BH is much larger than the binding energy of its host bulge; if only a few percent of AGN energy couples to gas in the forming galaxy, then all of the gas can be blown away (e.g., Silk & Rees 1998; Ostriker & Ciotti 2005). Thus BH growth may be self-limiting and AGNs may quench star formation and evolve their hosts from the blue cloud to the red sequence in the color-magnitude correlations. (3) Figure 9 (left) shows that the histories of BH growth and star formation in the universe are closely similar.

AGN feedback has the potential to solve many problems in galaxy formation. E.g., (1) The mass function of visible galaxies drops more steeply at high masses than that of DM halos that is predicted by our standard cosmology. The proposed solution is that AGN feedback is more efficient at higher $M_\bullet$ at keeping galaxies from growing more massive. (2) Episodic AGN feedback is believed to solve the 'cooling flow' problem, i.e. that, in the absence of energy input, X-ray halos in giant galaxies and in clusters of galaxies would cool quickly, but this produces no visible star formation (e.g., Ostriker & Ciotti 2005).

Figure 9 (right) shows the danger of overdoing feedback. The scatter in the $M_\bullet$–host correlations decreases strongly from the smallest to the largest BHs (Fig. 3, 6). Several authors emphasize that such a decrease results naturally from the averaging that is inherent in galaxy mergers. ' "Fine-tuning" through feedback is unnecessary and produces too low a dispersion in [the final] $M_\bullet/L$' (Gaskell 2010). As shown in Figure 9 (right), merging can convert no correlation at all into correlations that are essentially as tight as the ones observed (Jahnke & Macciò 2011; see also Peng 2007; Hirschmann *et al.* 2010).

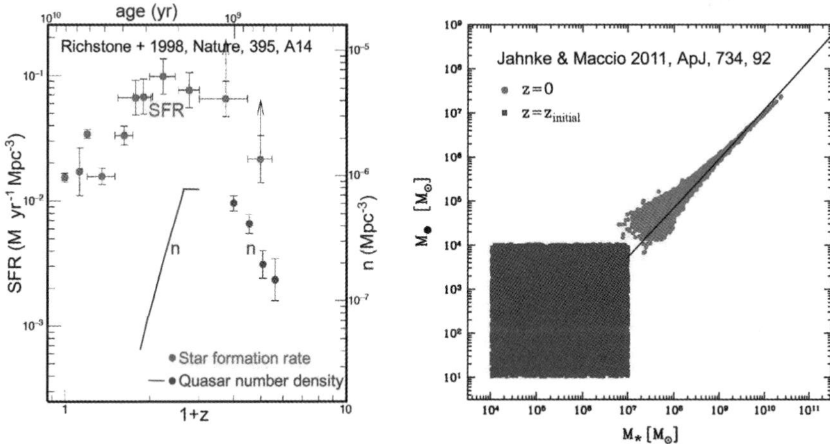

Figure 9. Arguments (left) for and (right) against AGN feedback as a factor in creating tight correlations between $M_\bullet$ and bulge L or σ. The left panel shows that the evolution with redshift (bottom) or age of the Universe (top) of the volume density of bright quasars (blue points and right axis labels) is very similar to the evolution of the star formation rate (red points and left axis labels). The right panel shows how an increasingly tight correlation between $M_\bullet$ and the stellar mass M_* of the host bulge (red points compared with the correlation observed by Häring & Rix 2004) can be manufactured from no correlation (blue square) by averaging in dry mergers.

11. Two kinds of elliptical galaxies have different AGN feedback

The rest of this paper discusses evidence for the two different kinds of AGN feedback that are introduced in the second paragraph of Section 10. They happen in two different kinds of elliptical galaxies summarized here and in Kormendy & Bender (2013) and reviewed in Kormendy *et al.* (2009, hereafter KFCB). These are illustrated in Figure 10.

Giant ellipticals ($M_V \lesssim -21.6$ for $H_0 = 70$ km s^{-1} Mpc^{-1}) *generally* (1) *have cores, i. e., central missing light with respect to an inward extrapolation of the outer Sérsic profile* (Figure 10, left); (2) rotate slowly, so rotation is of little importance dynamically; (3) hence are anisotropic and modestly triaxial; (4) have boxy-distorted isophotes; (5) are made of very old stars that are enhanced in α elements; (6) often contain strong radio sources, and (7) contain X-ray-emitting gas, more of it in more luminous galaxies.

Smaller ellipticals with $M_V \gtrsim -21.5$ *generally* (1) *are coreless – they have central extra light with respect to an inward extrapolation of the outer Sérsic profile* (Figure 10, left); (2) rotate rapidly; (3) are nearly isotropic and oblate spheroidal; (4) have disky-distorted isophotes; (5) are made of (still old but) younger stars with little α-element enhancement; (6) rarely contain strong radio sources, and (7) rarely contain X-ray-emitting gas.

The SAURON/ATLAS^{3D} division of ellipticals into fast and slow rotators (Emsellem *et al.* 2007, 2011; Cappellari *et al.* 2007, 2011) is closely similar.

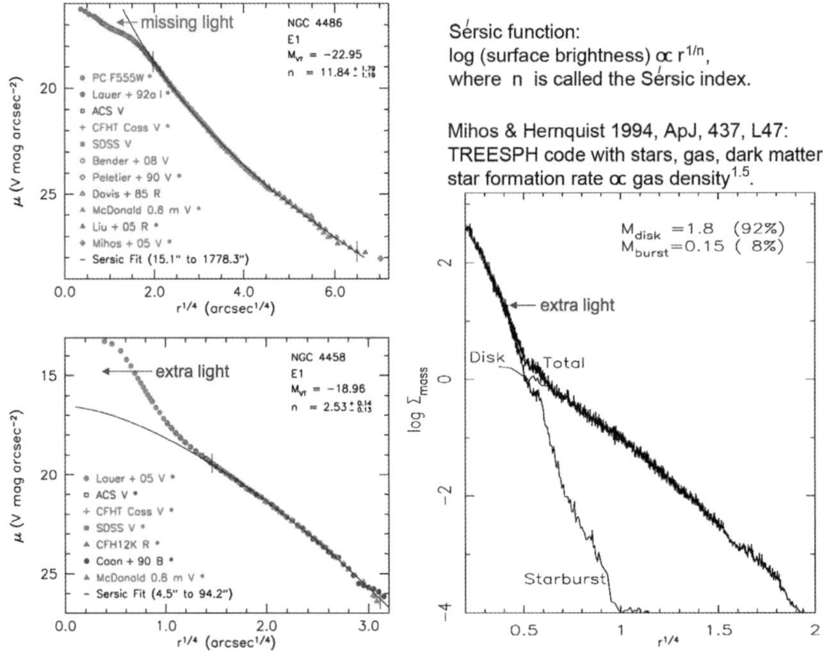

Figure 10. (left) From KFCB, brightness profiles of prototypical elliptical galaxies (top) with and (bottom) without cores. The core is defined to be the region of missing light with respect to the inward extrapolation of the outer Sérsic (1968) function brightness profile. KFCB showed that all Virgo cluster ellipticals with $M_V \leqslant -21.6$ have cores, whereas all Virgo ellipticals with $M_V \geqslant -21.5$ have central extra light above the inward extrapolation of the outer Sérsic profile. This core–no-core difference correlates with many other physical parameters that distinguish the two kinds of ellipticals. (right) Stellar mass density profile of the remnant of a merger of progenitor galaxies that each consisted of a stellar disk containing 92 % of the mass plus cold gas containing 8 % of the mass plus a dark matter halo. During the merger, the gas falls to the center and produces the 'Starburst' profile. Note that the outer profile is well described by a Sérsic function with $n < 4$, exactly as in the extra light elliptical. From Mihos & Hernquist (1994).

11.1. *Quasar-mode AGN feedback in extra light ellipticals*

The extra light seen in coreless/rotating/disky ellipticals closely resembles the compact components that form via central starbursts in wet mergers (Fig. 10, right, from Mihos & Hernquist 1994). This led Kormendy (1999) to suggest that extra light and core ellipticals formed in wet and dry mergers, respectively. Further support is provided by more extensive observations (e. g., Côté *et al.* 2007; KFCB), including mergers in progress (Rothberg & Joseph 2004), and by structural details in some extra light components that point to dissipative formation (disky structure, rapid rotation; KFCB). Extra light components are widely seen in detailed simulations of major mergers with gas (e. g., Springel & Hernquist 2005 [see Fig. 43 in KFCB]; Cox *et al.* 2006; Hopkins *et al.* 2008, 2009). However, if AGN feedback is too strong too early in the merger, the starburst is quenched and the extra light component does not form (e. g., Cox *et al.* 2006). This provides an engineering constraint for AGN feedback in extra light ellipticals.

In this context, an attractive picture that looks consistent with all observations is one in which most BH growth happens in 'quasar' events in the late stages of dissipative major mergers that grow host bulges (e. g., Hopkins *et al.* 2006). This 'quasar-mode feedback' from a bright BH accretion disk is where we need to engineer the magic that produces tight BH–galaxy correlations. Quenching of star formation, evolution of the galaxy from the blue cloud to the red sequence in the color-magnitude diagram, and expulsion of remaining cold gas plausibly happen in this context. AGN feedback may cooperate with starburst-driven outflows (Weiner *et al.* 2009; Newman *et al.* 2012). For both kinds of outflow, the fundamental feature to be exploited is radiation (and possibly jets) fed into a *three-dimensional* surrounding distribution of *very opaque gas*. The devil is in the details.

11.2. *Maintenance-mode AGN feedback in core ellipticals. I.*
Cores as evidence for dry major mergers

In contrast, I will suggest that AGN feedback takes a different form and has different results in core ellipticals. To see why, I need first to provide further evidence for their formation in both *major* and *dry* mergers. I begin with further evidence for dry mergers.

Figure 11 shows that the Faber-Jackson $L - \sigma$ relation is shallower for core ellipticals than it is for coreless ellipticals. This accounts for the steepening of the relation at low L (Tonry 1981; Davies *et al.* 1983). A shallow dependence on stellar mass, $\sigma \propto M_*^{0.091 \pm 0.022}$, is consistent with n-body simulations of *dry major* mergers (Hilz *et al.* 2012). The reason

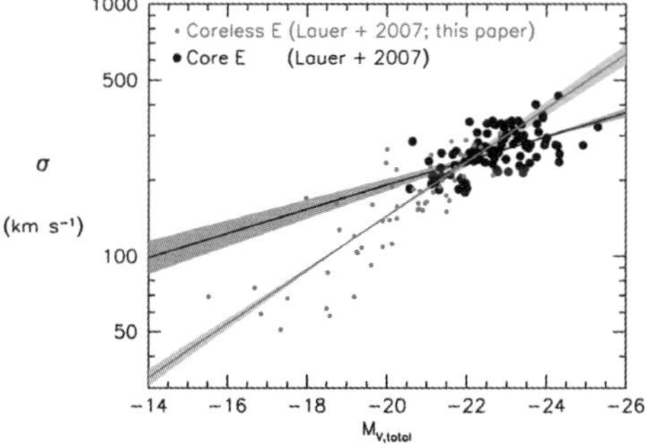

Figure 11. Faber-Jackson (1976) relation between σ and $M_{V,\text{total}}$ for ellipticals without (red) and with (black) cores. The lines are symmetric least-squares fits. Converting $M_{V,\text{total}}$ to stellar mass M_* via $M/L \propto L^{0.32 \pm 0.06}$ (Cappellari *et al.* 2006), $\sigma \propto M_*^{0.20 \pm 0.04}$ for coreless ellipticals and $\sigma \propto M_*^{0.091 \pm 0.022}$ for elliptical galaxies with cores. From Kormendy & Bender (2013).

why σ does not stay constant as in simple virial theorem arguments (Lake & Dressler 1986) is that some exchange in energy between stars and dark matter results in merger remnants that are slightly more compact (smaller size, higher density, higher velocity dispersion) than predicted if no orbital energy is added to the merger remnant.

Note that σ does not decrease with increasing luminosity as predicted for minor mergers or for scenarios in which cores form by the expansion of the center after ejection of large amounts of gas (Naab 2013). This does not mean that minor mergers are unimportant; they may build the large-Sérsic-index *halos* of giant ellipticals (Naab *et al.* 2009; Hopkins *et al.* 2010; van Dokkum *et al.* 2010; Oser *et al.* 2011; Hilz *et al.* 2012; Naab 2013). But the evidence does suggest that the main bodies of ellipticals are made by major mergers.

Another argument for *major* and *dry* mergers follows from Figure 12. The bottom-left panel shows that the mass that is missing in cores with respect to the outer Sérsic profile correlates tightly with $M_\bullet$. The correlation is well known (e. g., Milosavljević *et al.* 2001, 2002) but more accurately measured in KFCB. The RMS scatter (~ 0.20 dex in $\log M_\bullet$) is formally smaller than the scatter (~ 0.30 dex in $\log M_\bullet$) in the $M_\bullet - \sigma$ relation (Tremaine *et al.* 2002). And (Fig. 12, right), although the dispersion in BH mass fractions $M_\bullet/M_*$ around the mean of $0.13\,\%$ (Merritt & Ferrarese 2001; Kormendy & Gebhardt 2001) is large, it is not random: it correlates tightly with core light deficit. Kormendy & Bender (2009) argue that these results are a 'smoking gun' that connects cores with BHs. In fact, we believe that BHs are responsible for cores. Here's why:

Mergers tend to preserve the highest densities in their progenitors. Smaller ellipticals have higher central densities (Fig. 10). So mergers do not naturally make low-density cores (Faber *et al.* 1997). The proposed solution is that cores are scoured by the orbital decay of binary BHs that form in major mergers (e. g., Ebisuzaki *et al.* 1991; Faber *et al.* 1997; Milosavljević *et al.* 2001, 2002). The orbit shrinks as the binary flings stars away. This decreases the brightness and excavates a core. The effect of a series of mergers is cumulative. *Only if core ellipticals are made in several mergers with mass ratios ~ 1 do the resulting BH binaries fling away enough stars to account for the observed cores.*

Why did BH scouring fail in extra light ellipticals? Kormendy & Bender (2009) suggest that starbursts – which are massive (Fig. 12, left) – swamp BH scouring in wet mergers.

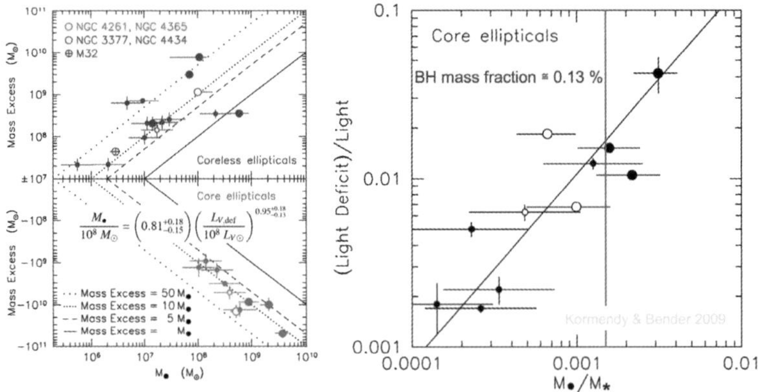

Figure 12. (left) Stellar mass that is 'missing' (in cores, lower panel) or 'extra' (in coreless galaxies, upper panel) as a function of BH mass $M_\bullet$ measured directly using spatially resolved dynamics observations (large symbols) or inferred from σ (small symbols). (right) Fraction of the total V-band luminosity 'missing' in cores versus the ratio of BH mass to the total stellar mass of the galaxy. Filled circles are for ellipticals in the Virgo cluster. From Kormendy & Bender (2009).

11.3. *Maintenance-mode AGN feedback in core ellipticals. II. Importance of X-ray gas*

This brings us to our conclusions about AGN feedback in core ellipticals. If the last (one or several) mergers that make core ellipticals are dry, then quasar-mode feedback presumably cannot operate. A different kind of AGN feedback takes over, as follows.

KFCB suggest that core ellipticals are protected from further gas dissipation and star formation by X-ray-emitting gas that is kept hot by 'maintenance-mode AGN feedback'. Figure 13 (from Bender *et al.* 1989) shows that only core/nonrotating/boxy ellipticals contain large amounts of hot gas. Also, radio ellipticals tend to be core/nonrotating/boxy; coreless/rotating/disky ellipticals are almost always radio-quiet. None of these galaxies have quasar-like BH accretion disks. But radio sources – especially jets – inject substantial energy and momentum into the surrounding hot gas. Evidence reviewed by Cattaneo *et al.* (2009) and by Fabian (2012) supports the idea that this helps to keep the gas hot. Other heating processes exist, too; e. g., continued gas infall from the cosmological hierarchy (Dekel & Birnboim 2006) and recycling of gas lost by dying stars. Further work is required to determine the relative importance of these processes. But this is engineering. The existence of a working surface against which feedback can operate is not in doubt; the Bender *et al.* (2009) results are confirmed by Pellegrini (1999, 2005) and by KFCB.

X-ray gas protection from late star formation is proposed to account for the rapidly quenched star formation implied by α-element enrichment (point 5 in Section 11). This is the '$M_{\rm crit}$ quenching picture' of Cattaneo *et al.* (2006, 2008) and Dekel & Birnboim (2006). A small amount of mass must trickle in to the BH in order to feed AGN activity, but no substantial $M_\bullet$ or M_* growth needs to occur by any means other than dry mergers. So the suggestion is that the role of maintenance-mode AGN feedback is mainly negative, to hold up the completion of galaxy formation by keeping baryons suspended in hot gas.

Meanwhile, the core galaxies at the high-mass end of the BH–galaxy correlations inherit the tightness of the scatter from coreless galaxies that formed by wet mergers (Figure 14). No further magic from AGN feedback is required.

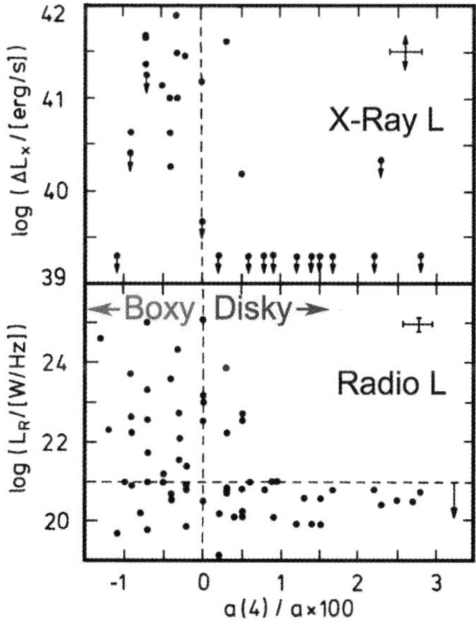

Figure 13. Correlation of (top) X-ray emission from hot gas and (bottom) radio emission with isophote shape parameter a_4 of E galaxies. Boxy isophotes have $a_4 < 0$; disky isophotes have $a_4 > 0$. These correspond to core and coreless galaxies, respectively. From Bender *et al.* (1989).

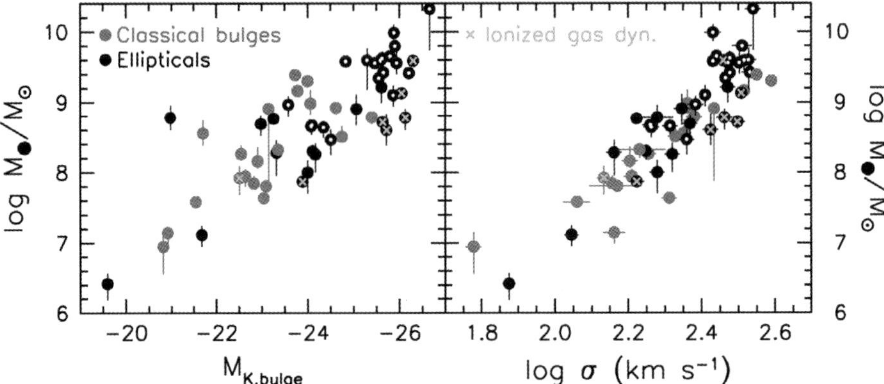

Figure 14. From Figure 2, the $M_\bullet$ – $M_{\rm K,bulge}$ and $M_\bullet$ – σ correlations with core galaxies identified using white dots. Core galaxies inherit the tightness of their correlation scatter from coreless galaxies. From Kormendy & Ho (2013).

Acknowledgements

It is a pleasure to thank the organizers (Daniel, Anna and Ignacio), and the SOC for the invitation to give a plenary talk on BHs at IAU 295. I thank K. Gebhardt, S. Rusli and R. Saglia for allowing me to use their $M_\bullet$ measurements before publication. Many of the ideas in this paper were developed in discussions with Luis Ho; I am most sincerely grateful to Luis for our always productive and enjoyable collaboration. My BH work is supported by the Curtis T. Vaughan, Jr. Centennial Chair in Astronomy at the University of Texas at Austin and by the Max-Planck-Institut für Extraterrestrische Physik (MPE). This review was written during a visit to MPE and to the Universitäts-Sternwarte of the Ludwig-Maximilians-Universität Munich. I am very grateful to MPE Managing Director Ralf Bender and to the staffs of both institutions for their hospitality and support.

References

Baes, M., Buyle, P., Hau, G. K. T., & Dejonghe, H. 2003, *MNRAS*, 341, L44
Barth, A. J., Ho, L. C., Rutledge, R. E., & Sargent, W. L. W. 2004, *ApJ*, 607, 90
Barth, A. J., Sarzi, M., Rix, H.-W., *et al.* 2001, *ApJ*, 555, 685
Bender, R., Surma, P., Döbereiner, S., Möllenhoff, C., & Madejsky, R. 1989, *A&A*, 217, 35
Burkert, A. & Tremaine, S. 2010, *ApJ*, 720, 516
Cappellari, M., Bacon, R., Bureau, M., *et al.* 2006, *MNRAS*, 366, 1126
Cappellari, M., Emsellem, E., Bacon, R., *et al.* 2007, *MNRAS*, 379, 418
Cappellari, M., Emsellem, E., Krajnović, D., *et al.* 2011, *MNRAS*, 416, 1680
Cattaneo, A., Dekel, A., Devriendt, J., Guiderdoni, B., & Blaizot, J. 2006, *MNRAS*, 370, 1651
Cattaneo, A., Dekel, A., Faber, S. M., & Guiderdoni, B. 2008, *MNRAS*, 389, 567
Cattaneo, A., Faber, S. M., Binney, J., *et al.* 2009, *Nature*, 460, 213
Corbelli, E. 2003, *MNRAS*, 342, 199
Côté, P., Ferrarese, L., Jordán, A., *et al.* 2007, *ApJ*, 671, 1456
Cox, T. J., Jonsson, P., Primack, J. R., & Somerville, R. S. 2006, *MNRAS*, 373, 1013
Cretton, N. & van den Bosch, F. C. 1999, *ApJ*, 514, 704
Davies, R. L., Efstathiou, G., Fall, S. M., Illingworth, G., & Schechter, P. L. 1983, *ApJ*, 266, 41
Dekel, A., & Birnboim, Y. 2006, *MNRAS*, 368, 2
Dong, X.-B., Ho, L. C., Yuan, W., *et al.* 2012, *ApJ*, 755, 167
Dressler, A. 1989, in *IAU Symposium 134, Active Galactic Nuclei*, ed. D. E. Osterbrock & J. S. Miller (Dordrecht: Kluwer), p. 217
Ebisuzaki, T., Makino, J., & Okamura, S. K. 1991, *Nature*, 354, 212
Emsellem, E., Cappellari, M., Krajnović, D., *et al.* 2007, *MNRAS*, 379, 401
Emsellem, E., Cappellari, M., Krajnović, D., *et al.* 2011, *MNRAS*, 414, 888

Erwin, P., Vega Beltrán, J. C., Graham, A. W., & Beckman, J. E. 2003, *ApJ*, 597, 929

Faber, S. M. & Jackson, R. E. 1976, *ApJ*, 204, 668

Faber, S. M., Tremaine, S., Ajhar, E. A., *et al.* 1997, *AJ*, 114, 1771

Fabian, A. C. 2012, *ARAA*, 50, 455

Fabian, A. C. 2013, in *IAU Symposium 290, Feeding Compact Objects: Accretion on All Scales*, ed. T. Belloni *et al.* (Cambridge: Cambridge Univ. Press), p. 3

Ferrarese, L. 2002, *ApJ*, 578, 90

Ferrarese, L. & Ford, H. 2005, *Space Sci. Revs*, 116, 523

Ferrarese, L., & Merritt, D. 2000, *ApJ*, 539, L9

Filippenko, A. V. & Ho, L. C. 2003, *ApJ*, 588, L13

Filippenko, A. V., Ho, L. C., & Sargent, W. L. W. 1993, *ApJ*, 410, L75

Freeman, K. C. 1970, *ApJ*, 160, 811

Gaskell, C. M. 2010, in *The First Stars and Galaxies: Challenges for the Next Decade*, ed. D. J. Whalen, V. Bromm, & N. Yoshida (Melville: AIP), p. 261

Gebhardt, K., Bender, R., Bower, G., *et al.* 2000, *ApJ*, 539, L13

Gebhardt, K., Adams, J., Richstone, D., *et al.* 2011, *ApJ*, 729, 119

Gebhardt, K., Lauer, T. R., Kormendy, J., *et al.* 2001, *AJ*, 122, 2469

Gebhardt, K., Richstone, D., Tremaine, S., *et al.* 2003, *ApJ*, 583, 92

Gebhardt, K. & Thomas, J. 2009, *ApJ*, 700, 1690

Genzel, R., Eisenhauer, F., & Gillessen, S. 2010, *Rev. Mod. Phys.*, 82, 3121

Greene, J. E. & Ho, L. C. 2004, *ApJ*, 610, 722

Greene, J. E. & Ho, L. C. 2007a, *ApJ*, 667, 131 ; Erratum 2009, *ApJ*, 704, 1743

Greene, J. E. & Ho, L. C. 2007b, *ApJ*, 670, 92

Greene, J. E., Ho, L. C., & Barth, A. J. 2008, *ApJ*, 688, 159

Greene, J. E., Peng, C. Y., Kim, M., *et al.* 2010, *ApJ*, 721, 26

Gültekin, K., Richstone, D. O., Gebhardt, K., *et al.* 2009, *ApJ*, 698, 198

Häring, N. & Rix, H.-W. 2004, *ApJ*, 604, L89

Harris, G. L. H. & Harris, W. E. 2011, *MNRAS*, 410, 2347

Hilz, M., Naab, T., Ostriker, J. P. 2012, *MNRAS*, 429, 2924

Hirschmann, M., Khochfar, S., Burkert, A., *et al.* 2010, *MNRAS*, 407, 1016

Ho, L. C. 1999, in *Observational Evidence for Black Holes in the Universe*, ed. S. K. Chakrabarti (Dordrecht: Kluwer), p. 157

Ho, L. C. 2008, *ARAA*, 46, 475

Hopkins, P. F., Bundy, K., Hernquist, L., Wuyts, S., & Cox, T. J. 2010, *MNRAS*, 401, 1099

Hopkins, P. F., Cox, T. J., Dutta, S. N., *et al.* 2009, *ApJS*, 181, 135

Hopkins, P. F., Hernquist, L., Cox, T. J., *et al.* 2006, *ApJS*, 163, 1

Hopkins, P. F., Hernquist, L., Cox, T. J., Dutta, S. N., & Rothberg, B. 2008, *ApJ*, 679, 156

Hu, J. 2008, *MNRAS*, 386, 2242

Jahnke, K. & Macciò, A. V. 2011, *ApJ*, 734, 92

Jiang, Y.-F., Greene, J. E., & Ho, L. C. 2011, *ApJ*, 737, L45

Kormendy, J. 1977, *ApJ*, 218, 333

Kormendy, J. 1993, in *The Nearest Active Galaxies*, ed. J. Beckman, L. Colina & H. Netzer (Madrid: Consejo Superior de Investigaciones Científicas), p. 197

Kormendy, J. 1999, in *Galaxy Dynamics: A Rutgers Symposium*, ed. D. Merritt, J. A. Sellwood & M. Valluri (San Francisco: ASP), p. 124

Kormendy, J. 2004, in *Carnegie Observatories Astrophysics Series, Vol. 1: Coevolution of Black Holes and Galaxies*, ed. L. C. Ho (Cambridge: Cambridge Univ. Press), p. 1

Kormendy, J. 2013, in *XXIII Canary Islands Winter School of Astrophysics, Secular Evolution of Galaxies*, ed. J. Falcón-Barroso & J. H. Knapen (Cambridge: Cambridge Univ. Press), p. 1

Kormendy, J. & Bender, R. 2009, *ApJ*, 691, L142

Kormendy, J. & Bender, R. 2011, *Nature*, 469, 377

Kormendy, J. & Bender, R. 2012, *ApJS*, 198, 2

Kormendy, J. & Bender, R. 2013, *ApJ*, 769, L5

Kormendy, J., Bender, R., & Cornell, M. E. 2011, *Nature*, 469, 374

Kormendy, J., Bender, R., Magorrian, J., *et al.* 1997, *ApJ*, 482, L139

Kormendy, J., Drory, N., Bender, R., & Cornell, M. E. 2010, *ApJ*, 723, 54

Kormendy, J., Fisher, D. B., Cornell, M. E., & Bender, R. 2009, *ApJS*, 182, 216 (KFCB)

Kormendy, J. & Gebhardt, K. 2001, in *20th Texas Symposium on Relativistic Astrophysics*, ed. J. C. Wheeler & H. Martel (Melville, NY: AIP), p. 363

Kormendy, J. & Ho, L. C. 2013, *ARAA*, in press

Kormendy, J., & Kennicutt, R. C. 2004, *ARAA*, 42, 603

Kormendy, J. & Richstone, D. 1995, *ARAA*, 33, 581

Lake, G. & Dressler, A. 1986, *ApJ*, 310, 605

Lauer, T. R., Faber, S. M., Richstone, D., *et al.* 2007, *ApJ*, 662, 808

Macchetto, F., Marconi, A., Axon, D. J., *et al.* 1997, *ApJ*, 489, 579

Magorrian, J., Tremaine, S., Richstone, D., *et al.* 1998, *AJ*, 115, 2285

Marconi, A. & Hunt, L. K. 2003, *ApJ*, 589, L21

McConnell, N. J., Ma, C.-P., Gebhardt, K., *et al.* 2011, *Nature*, 480, 215

McConnell, N. J., Ma, C.-P., Murphy, J. D., *et al.* 2012, *ApJ*, 756, 179

McLure, R. J. & Dunlop, J. S. 2002, *MNRAS*, 331, 795

Merritt, D. & Ferrarese, L. 2001, *MNRAS*, 320, L30

Mihos, J. C. & Hernquist, L. 1994, *ApJ*, 437, L47

Milosavljević, M. & Merritt, D. 2001, *ApJ*, 563, 34

Milosavljević, M., Merritt, D., Rest, A., & van den Bosch, F. C. 2002, *MNRAS*, 331, L51

Miyoshi, M., Moran, J., Herrnstein, J., *et al.* 1995, *Nature*, 373, 127

Naab, T. 2013, in *IAU Symposium 295, The Intriguing Life of Massive Galaxies*, ed. D. Thomas, A. Pasquali, & I. Ferreras (Cambridge: Cambridge University Press), arXiv:1211.6892

Naab, T., Johansson, P. H., & Ostriker, J. P. 2009, *ApJ*, 699, L178

Newman, S. F., Genzel, R., Förster-Schreiber, N. M., *et al.* 2012, *ApJ*, 761, 43

Oser, L., Ostriker, J. P., Naab, T., Johansson, P. H., & Burkert, A. 2010, *ApJ*, 725, 2312

Ostriker, J. P. & Ciotti, L. 2005, *Phil. Trans. R. Soc. London*, 363, A667

Pellegrini, S. 1999, *A&A*, 351, 487

Pellegrini, S. 2005, *MNRAS*, 364, 169

Peng, C. 2007, *ApJ*, 671, 1098

Peterson, B. M., Bentz, M. C., Desroches, L.-B., *et al.* 2005, *ApJ*, 632, 799

Pizzella, A., Corsini, E. M., Dalla Bontà, E., *et al.* 2005, *ApJ*, 631, 785

Rees, M. J. 1984, *ARAA*, 22, 471

Reines, A. E., Sivakoff, G. R., Johnson, K. E., & Brogan, C. L. 2011, *Nature*, 470, 66

Renzini, A. 1999, in *The Formation of Galactic Bulges*, ed. C. M. Carollo, H. C. Ferguson & R. F. G. Wyse (Cambridge: Cambridge University Press), p. 9

Richstone, D., Ajhar, E. A., Bender, R., *et al.* 1998, *Nature*, 395, A14

Rothberg, B. & Joseph, R. D. 2004, *AJ*, 128, 2098

Rusli, S., Thomas, J., Saglia, R. P., *et al.* 2013, arXiv:1306.1124

Schulze, A. & Gebhardt, K. 2011, *ApJ*, 729, 21

Sérsic, J. L. 1968, *Atlas de Galaxias Australes* (Córdoba: Obs. Astronómico, Univ. de Córdoba)

Shen, J. & Gebhardt, K. 2010, *ApJ*, 711, 484

Silk, J. & Rees, M. J. 1998, *A&A*, 331, L1

Springel, V. & Hernquist, L. 2005, *ApJ*, 622, L9

Swaters, R. A., Sancisi, R., van Albada, T. S., & van der Hulst, J. M. 2009, *A&A*, 493, 871

Thornton, C. E., Barth, A. J., Ho, L. C., & Rutledge, R. E., Greene J. E. 2008, *ApJ*, 686, 892

Tonry, J. L. 1981, *ApJ*, 251, L1

Tremaine, S., Gebhardt, K., Bender, R., *et al.* 2002, *ApJ*, 574, 740

van Albada, T. S. & Sancisi, R. 1986, *Phil. Trans. R. Soc. London*, 320, A447

van den Bosch, R. C. E. & de Zeeuw, P. T. 2010, *MNRAS*, 401, 1770

van Dokkum, P. G., Whitaker, K. E., Brammer, G., *et al.* 2010, *ApJ*, 709, 1018

Walsh, J. L., Barth, A. J., & Sarzi, M. 2010, *ApJ*, 721, 762

Weiner, B. J., Coil, A. L., Prochaska, J. X., *et al.* 2009, *AJ*, 692, 187

Xiao, T., Barth, A. J., Greene, J. E, *et al.* 2011, *ApJ*, 739, 28

The intriguing life of massive galaxies
Proceedings IAU Symposium No. 295, 2012
D. Thomas, A. Pasquali & I. Ferreras, eds.

© International Astronomical Union 2013
doi:10.1017/S1743921313004961

Hot Gas and AGN Feedback in Galaxies and Nearby Groups

Christine Jones[1], William Forman[1], Akos Bogdan[1], Scott Randall[1], Ralph Kraft[1], and Eugene Churazov[2]

[1]Center for Astrophysics,
60 Garden Street, Cambridge, MA 02138, USA
email: cjones@cfa.harvard.edu, wforman@cfa.harvard.edu, abodgan@cfa.harvard.edu,
srandall@cfa.harvard.edu, rkraft@cfa.harvard.edu

[2]Max Planck Institute for Astrophysik,
Karl-Schwarzchild-Str. 1., Garching, Germany
email: echurazov@mpa-garching.mpg.de

Abstract. Massive galaxies harbor a supermassive black hole at their centers. At high redshifts, these galaxies experienced a very active quasar phase, when, as their black holes grew by accretion, they produced enormous amounts of energy. At the present epoch, these black holes still undergo occasional outbursts, although the mode of their energy release is primarily mechanical rather than radiative. The energy from these outbursts can reheat the cooling gas in the galaxy cores and maintain the red and dead nature of the early-type galaxies. These outbursts also can have dramatic effects on the galaxy-scale hot coronae found in the more massive galaxies. We describe research in three areas related to the hot gas around galaxies and their supermassive black holes. First we present examples of galaxies with AGN outbursts that have been studied in detail. Second, we show that X-ray emitting low-luminosity AGN are present in 80% of the galaxies studied. Third, we discuss the first examples of extensive hot gas and dark matter halos in optically faint galaxies.

Keywords. galaxies: nuclei, galaxies: bulges, galaxies: ISM X-rays: galaxies: clusters

1. Introduction

In both galaxies and clusters, X-ray observations of hot gas have been used to determine the overall morphology and characterize the dynamical histories of these systems, as well as to measure the total system mass and, of particular importance to this discussion, to determine the level of recent AGN activity from the galaxy nucleus. While the hot X-ray emitting gas in clusters is the dominant component of the luminous baryonic mass, in galaxies, the hot gas is only a few percent of the stellar mass. While the hot intracluster medium dominates the X-ray emission in clusters, in galaxies, emission from X-ray binaries, coronally active stars and cataclysmic variables, and the AGN can also be important (e.g. Revnivtsev *et al.* 2006).

2. SMBH Outbursts in Early-type Galaxies

In both galaxies and clusters, the hot X-ray emitting gas provides a fossil record of AGN activity and is often the primary evidence of recent AGN activity. Proof of recent AGN outbursts is commonly found through Chandra and XMM-Newton images, which show cavities in ~30% of galaxies with hot coronae, while 50% of clusters exhibit such cavities (Dunn and Fabian 2008). From these X-ray observations, the total mechanical power of the AGN outburst can be measured directly from the energy required to inflate

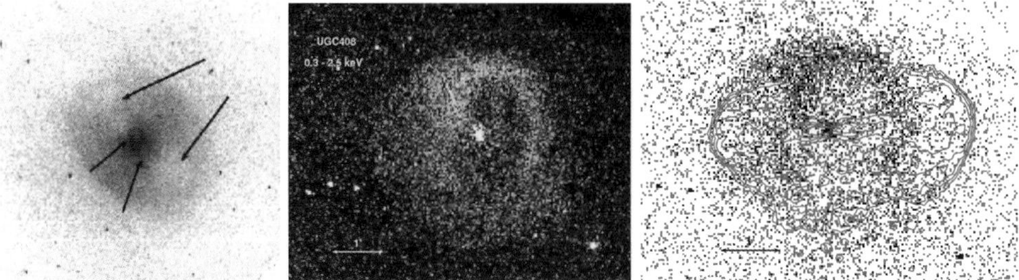

Figure 1. (left) The Chandra image of NGC5813 shows two pairs of cavities (marked with black arrows) indicating multiple episodes of AGN activity. A third pair of cavities lies at slightly larger radii. (center) The Chandra image of the galaxy UGC408 shows a nucleus, jet and bright X-ray rim at 20 kpc radius. (right) UGC408 with 1.4 GHz contours superposed on the X-ray image shows a collimated beam with radio emission extending E-W, while low frequency radio contours (610 Mhz) show the large extent of the relativistic plasma.

the cavities, while the duration and age of the outburst can be measured from the position and rise time of the bubbles or, in a few systems, from the observations of shocks produced by the outburst (see McNamara & Nulsen (2007) and references therein). Computations of the radiative cooling and the energy supplied by the AGN through cavities and shocks show that the AGN feedback in early type galaxies is sufficient to balance the radiative losses of the cooling X-ray gas and thus suppresses star formation. The presence of multiple cavities in the hot gas also allows the recurrence frequency of AGN outbursts to be determined.

Many of the central galaxies in groups exhibit the scars of multiple outbursts. Fig. 1 (left) shows three pairs of X-ray cavities in the hot gas around NGC5813; the two inner pairs of cavities are marked with arrows. In addition, NGC5813 has two confirmed sets of shocks, at the locations of sharp brightness discontinuities. The outer shock has a velocity of $\sim$480 km s^{-1} (Mach 1.15). The outburst that produced the middle set of cavities, which are $\sim$5 kpc in diameter and associated with this shock, had a total energy of 10^{55} ergs and occurred about 2 Myr ago. Additional information on the NGC5813 outbursts can be found in Randall *et al.* (2011).

For the galaxy UGC408 (NGC 193), the combination of radio and X-ray observations suggest two episodes of AGN activity (see Fig. 1 center and right). The X-ray images show a 20 kpc radius ring of enhanced X-ray emission surrounding a large X-ray cavity centered on the nucleus. This cavity is filled with radio plasma emitting at 610 MHz. However the 1.4 GHz radio emission is significantly more collimated and likely arises from a second outburst. An X-ray jet is associated with the eastern radio emission.

3. X-ray Emission from the Supermassive Black Holes in the Nuclei in Early-type Galaxies

While observations of cavities or shocks in the hot gas can measure the energy of an AGN outburst, detection of point-like X-ray emission in the central cores of galaxies identifies the presence and the precise location of the supermassive black hole, as well as measuring the low level of radiative power arising from the black hole at the present epoch. Fig. 2 (left) plots the X-ray emission from galaxy nuclei for a sample of about 200 early-type galaxies observed with Chandra. We detect X-ray emission from the supermassive black hole for approximately 80% of the galaxies with hot corona. The black hole X-ray luminosities range from 10^{38} to 10^{41} ergs s^{-1}. We use each galaxy's velocity dispersion to calculate the mass of the black hole and then compute the ratio of the black

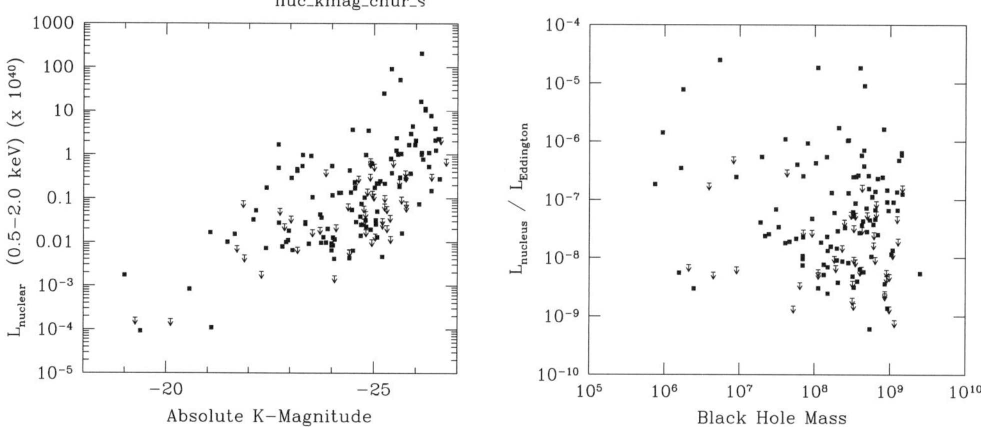

Figure 2. (left) The X-ray luminosities of the nuclear emission in a sample of early-type galaxies plotted against each galaxy's K magnitude. Emission from more than 80% of the supermassive black holes in galaxies with gas coronae are detected in X-rays with Chandra. (right) The Eddington ratio for supermassive black holes, where the mass of the black hole is estimated from the velocity dispersion and the nuclear luminosity from X-ray measurements.

hole's observed luminosity to its possible Eddington luminosity. These Eddington ratios, shown on the right in Fig. 2, in these low luminosity AGN range from $\sim 10^{-5}$ to $\sim 10^{-9}$. The radiative luminosities of these supermassive black holes are much smaller than their mechanical energies, when measured through observations of X-ray cavities. Thus, the present epoch AGN mode is one of mechanical power, not radiative power (see Churazov *et al.* 2005 for a discussion of the AGN transition from radiatively bright to radiatively faint and mechanically powerful.)

4. Dark Matter Haloes, Stellar Bulges and the Growth of Supermassive Black Holes

Although galaxies with large stellar bulges are observed to harbor supermassive black holes (e.g. Gultekin *et al.* 2009), recently, examples of galaxies have been identified which do not have large stellar bulges, but still harbor supermassive black holes. In particular, the lenticular galaxy NGC4342 hosts an unusually massive black hole, compared to its low bulge mass (Cretton & van den Bosch (1999)). As illustrated in Fig. 3, NGC4342 also hosts an X-ray bright halo of hot gas that extends more than 10 kpc, significantly broader than its stellar bulge. NGC4342's diffuse X-ray luminosity is also significantly higher than generally observed for galaxies with bulges of similar stellar mass (Bogdan *et al.* 2012a). For NGC4342, the mass of the hot gas and the dark halo mass, inferred from the X-ray observations, are similar to those found for more optically luminous early-type galaxies.

Similar results also have been found for the elliptical galaxy NGC4291 (Bogdan *et al.* 2012b). The presence of massive dark matter halos, along with supermassive black holes in their cores ($4 \times 10^8 M_\odot$ for NGC4342 and $7 \times 10^8 M_\odot$ for NGC4291; Gultekin *et al.* 2009), but low stellar masses in their bulges suggest that it is the total dark matter halo that plays the fundamental role in governing black hole growth. Bodgan *et al.* also conclude, based on their results for NGC4342 and NGC4291, that supermassive black holes and galaxy stellar bulges may not grow in tandem, but instead the earlier formation

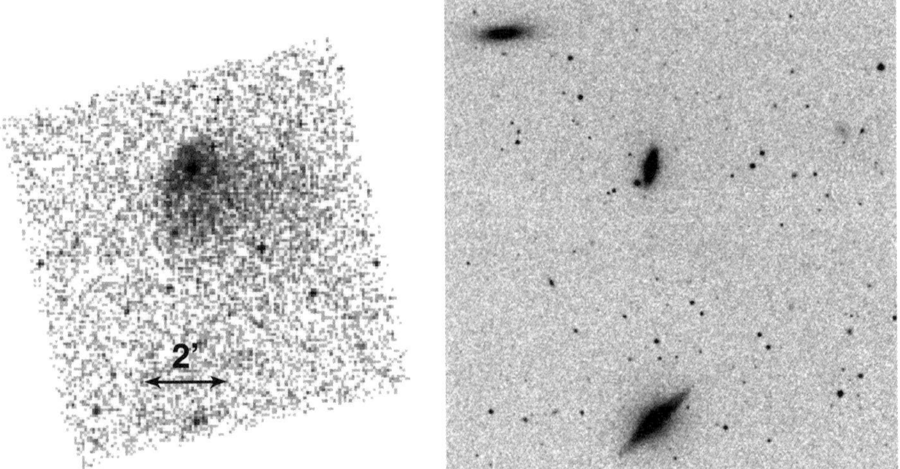

Figure 3. (left) Diffuse X-ray emission from NGC4342 extends beyond 10 kps, much broader than the stellar bulge, shown on the same spatial scale, in the right panel. The sharp edge, seen in the diffuse X-ray emission, northeast of the core, results from the motion of NGC4342 through the surrounding group gas. For details see Bogdan *et al.* (2012a).

and growth of supermassive black holes can suppress star formation, limiting the growth of the stellar bulge.

5. Implications

We reviewed three areas related to hot coronae around early-type galaxies and their supermassive black holes. First, we noted that the cavities and shocks produced in the hot gas by AGN outbursts can be used to measure the mechanical energy and age of the outbursts. Second we showed that low luminosity X-ray AGN are common in apparantly normal early-type galaxies in the local Universe, and third we discussed two galaxies with small stellar bulges that harbor supermassive black holes, along with extended hot coronae and dark matter halos.

We are greatful to many of our colleagues for useful discussions related to this work, particularly Larry David, Paul Nulsen, Marie Machacek and Ming Sun. Support was received from the Smithsonian Institution, the Chandra X-ray Center and the Max Planck Institute for Astrophysik.

References

Bogdan, A., Forman, W., Kraft, R., Jones, C., Blom, C., Randall, S., Zhang, Z., Zhuravleva, I., Churazov, E., Li, Z., Nulsen, P., Vikhlinin, A., & Schindler, S. 2012, *ApJ*, 755, 25

Bogdan, Forman, W., Zhuravleva, I., Mihos, C., Kraft, R., Harding, P., Guo, Q., Li, Z., Churazov, E., Vikhlinin, A., Nulsen, P., Schindler, S., & Jones, C. 2012, *ApJ*, 753, 140

Churazov, E., Sazonov, S., Sunyaev, R., Forman, W., Jones, C., & Bohringer, H. 2005, *MNRAS*, 363, L91

Cretton, N. & van den Bosch, F. C. 1999, *ApJ*, 514, 704

Dunn, R. J. H. & Fabian, A. C. 2008, *MNRAS*, 385, 757

Gultekin, K., *et al.* 2009, *ApJ*, 695, 1577

McNamara, B. R. & Nulsen, P. E. J. 2007, *ARA&A*, 45, 117

Randall, S., *et al.* 2011, *ApJ*, 726, 86

Revnivtsev, M., Sazonov, S., Gilfanov, M., Churazov, E., & Sunyaev, R. 2006, *A&A*, 452, 169

The intriguing life of massive galaxies
Proceedings IAU Symposium No. 295, 2012
D. Thomas, A. Pasquali & I. Ferreras, eds.

© International Astronomical Union 2013
doi:10.1017/S1743921313004973

Chandra and VLA Observations of Supermassive Black Hole Outbursts in M87

William Forman[1], C. Jones[1], and Eugene Churazov[2]

[1]Smithsonian Astrophysical Observatory
60 Garden St., Cambridge, MA USA
email: wrf@cfa.harvard.edu, cjones@cfa.harvard.edu

[2]Max Planck Institute for Astrophysics, Garching Germany
Institute for Space Research, Moscow, Russia
email: churazov@MPA-Garching.MPG.DE

Abstract. We discuss the effects of supermassive black hole (SMBH) outbursts on the hot atmospheres surrounding the central massive galaxies in groups and clusters, as observed with X-ray and radio observations. We focus on a detailed study of the supermassive black hole in M87 at the center of the Virgo cluster using Chandra and VLA observations. We summarize the outburst history and describe the clearly observed energy input from buoyant bubbles of relativistic plasma produced by the central SMBH, uplifted filaments of X-ray emitting gas, and the Mach 1.2 shock together balance the energy lost as gas radiatively cools.

Keywords. galaxies: clusters: individual (M87), galaxies: active, galaxies: evolution

1. Introduction

Outbursts from supermassive black holes have become an important ingredient of galaxy evolution models (e.g., Croton *et al.* 2006, Bower *et al.* 2006, Guo *et al.* 2011). X-ray observations of gas rich systems from clusters to groups to galaxies show that outbursts at the present epoch are very common and serve to replenish the energy lost from radiative cooling in hot gas rich, massive galaxies. M87, the dominant central galaxy of the Virgo cluster, is a classic example of a cooling flow cluster (e.g., Fabian 1994) and is the nearest example of this class. In the absence of any energy input, the inferred cooling rate around M87 implies a cooling flow of ~ 10 M$_\odot$ yr^{-1} (Stewart *et al.* 1984). M87 is the second brightest extragalactic X-ray source (Forman *et al.* 1978) and is 50 times more X-ray luminous than NGC4472, the optically brightest galaxy in Virgo. M87 hosts a 6.6×10^9 M$_\odot$ SMBH (Gebhardt *et al.* 2011) and a well-studied jet, detected from the radio to the X-ray regime (Marshall *et al.* 2002), that is inflating a plasma filled cocoon. With a massive SMBH, an associated jet, and an X-ray luminous atmosphere, M87 is the ideal system for detailed studies of the feedback process between a SMBH and its surrounding gaseous corona (see Churazov *et al.* 2001, Forman *et al.* 2005, Forman *et al.* 2007, Million *et al.* 2010).

2. M87's Environment – Merger Driven Gas Sloshing

The Virgo cluster core, dominated by M87, is a complex environment and is still undergoing mergers. Fig. 1 (left panel) shows the wide field ROSAT image of the cluster core including the "cold front" around M87 and the dramatic merger of M86 with its more than 150 kpc long ram pressure stripped tail (Randall *et al.* 2008). The mergers in Virgo have perturbed the core and are driving "gas sloshing" (Markevitch & Vikhlinin 2007

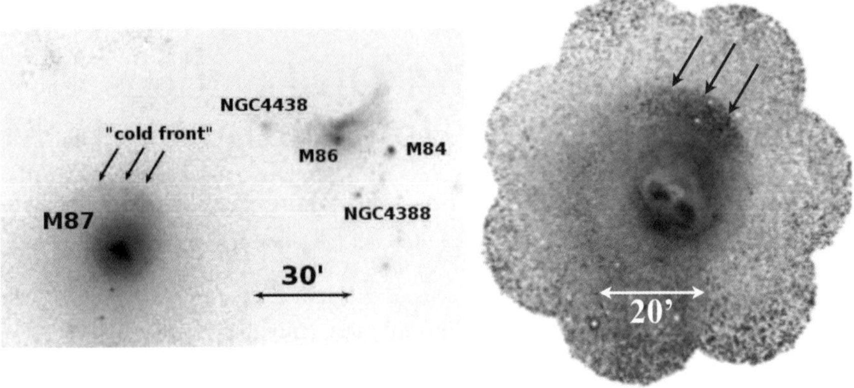

Figure 1. (left)ROSAT view of the X-ray emission surrounding M87. In addition to the very bright emission from M87, we see M86 (NGC4406) and its ram pressure stripped tail, as well as emission from other bright galaxies in the Virgo core, including M84 (NGC4374), NGC4438, and NGC4438. (right) The XMM-Newton observatins of M87 with an average surface brightness profile subtracted shows the sloshing cold front (dark arrows), as well as the cool gas "arms" (extending 5′ to the east and southwest).

for a review; see Simionescu *et al.* 2010 for details of M87). Note that merger-driven gas sloshing is commonly seen in clusters with gas density peaks (14 clusters in a sample of 18 show similar features; Markevitch *et al.* 2003; Johnson 2011). Fig. 1 (right panel) shows the XMM-Newton image of the core region with an average radial profile subtracted to enhance faint features. The northern X-ray edge, a contact discontinuity (see Markevitch & Vikhlinin 2007 for a detailed review of cold fronts) is clearly seen (black arrows) as are the X-ray filamentary structures in the very central region extending to the east and southwest (see Fig. 2 (left panel) for the Chandra high resolution view of the filaments).

3. Shocks and Filaments in M87

The VLA and Chandra observations of M87 show a sequence of outbursts in M87 that are chronicled in Fig. 2 and can be summarized as:

- The outer radio "pancakes" (filamentary circular regions in Fig. 2 right panel; adapted from Owen, Eilek & Kassim 2000). Fig. 1 in Churazov *et al.* (2001), described these as buoyant bubbles flattened into "pancakes" as they rose in M87's atmosphere. Owen *et al.* (2000) estimate the ages of these largest lobes as $100 - 150 \times 10^6$ yr.

- "mushroom cloud" with stem and torus (dark structure in Fig. 2, right panel) extending to the east from the nucleus with an age of a few tens of Myr (the buoyancy time). A radio structure also extends to the southwest, but has been disrupted. These two radio plasma arms are closely related to the X-ray filamentary arms (Fig. 2, left panel) of cool gas that has been uplifted by the buoyant radio-emitting bubbles.

- 13 kpc shock, a nearly complete azimuthal ring (Fig. 2, left center panel, the hard X-ray image of M87) at a radial distance of 13 kpc (2.8′) from the center of M87, driven by an outburst $\sim 12 \times 10^6$ yr old, and its driving piston, the central radio lobes.

- a possible weaker shock at about 5 kpc (about 5×10^6 yr old first suggested as a weak shock in Forman *et al.* 2007; see also Million *et al.* 2010).

- the present outburst with its multi-wavelength jet, flaring knots, and variable gamma-ray emission (Hines *et al.* 1989; Owen *et al.* 2000; Marshall *et al.* 2002; Harris *et al.* 2006; Shi *et al.* 2007; Acciari *et al.* 2010).

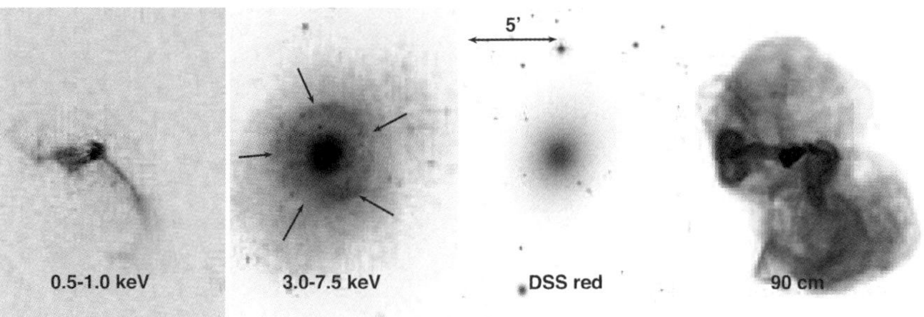

Figure 2. Images of M87 from soft and hard X-ray to optical and radio *at the same angular scale* showing the variety of outbursts. (left) The soft (0.5-1.0 keV) Chandra image showing the thin filamentary arms, uplifted by the buoyant radio bubbles; (left-center) The hard (3.0-7.5 keV) Chandra image showing the main shock. The hard band image is a "pressure map" (integral of pressure squared projected on the sky) and directly shows the 13 kpc shock (dark arrows); (center-right) The DSS (red) image of M87 showing the galaxy scale compared to the X-ray and radio; (right) The 90 cm VLA map of M87 (Owen *et al.* 2000) which shows the large relativistic plasma bubbles as well as the "mushroom cloud" feature buoyantly rising to the east and the disrupted arm extending to the southwest. The radio bright nuclear region, the piston that initiated the 13 kpc outburst, is clearly visible in the 90cm radio image.

Hence, M87 shows a rich sequence of outbursts. We discuss the 13 kpc shock and its associated central "piston" which present the opportunity, through a simple and robust model, to derive the characteristics of a SMBH outburst and the energy partition between the shock and the central buoyant radio lobes.

Fig. 3 (left) shows the azimuthally averaged (northern sector of width 80°) surface brightness profiles in the 1.2-2.5 keV (upper curve) and 3.5-7.5 keV (lower curve) bands (see Forman *et al.* 2007 for details). The profiles show a sharp edge at $\sim 0.6'$ (most distinctly seen in the hard band as a decrease between the fourth and fifth data points), a moderate flattening of the profile at about $1'$, and a strong excess at $2-3'$. This strong excess is the 13 kpc shock. To derive quantitative estimates of the shock jump conditions, we deprojected surface brightness profiles into emissivity profiles (Fig. 3 central panel). The deprojection shows the very pronounced 13 kpc shock at $2'-3'$.

The 1.2-2.5 keV energy band is nearly a direct measure of the gas density. The soft and hard bands combine to yield the temperature profile. The derived values of the gas density and temperature jumps are $\rho_{shock}/\rho_0 = 1.33 \pm 0.02$ and $T_{shock}/T_0 = 1.18 \pm 0.03$ (see Forman *et al.* 2007 for details). These values, combined with the Rankine-Hugoniot shock jump conditions, yield *independent*, consistent values of the Mach number $M \sim 1.2$.

For the 13 kpc shock and its driving outburst, we describe a simple (one dimensional model) that yields the energy balance between shock and "bubble" energy. Constraints on the model include 1) the density and temperature jumps across the shock which yield the present epoch shock Mach number and 2) the size of the driving piston as seen in the radio-filled central cavity. The Mach number yields the outburst energy (almost directly) of about 5×10^{57} ergs. The combined constraints give an estimate of the outburst duration of about 2 Myr. The present radius of the shock, in the context of the model, gives an age of about 12 Myr. The model shows that over the lifetime of the outburst, the energy is partitioned with about half of the energy in the enthalpy of the central piston (cavity/bubble), about a quarter possibly going into shock heating of the gas, and the remaining part is carried to the outskirts of the cluster as the shock weakens to a sound wave. Averaged over the age of the outburst (to the time of today's present outburst), the outburst power matches the radiative cooling of the gas in the cluster core.

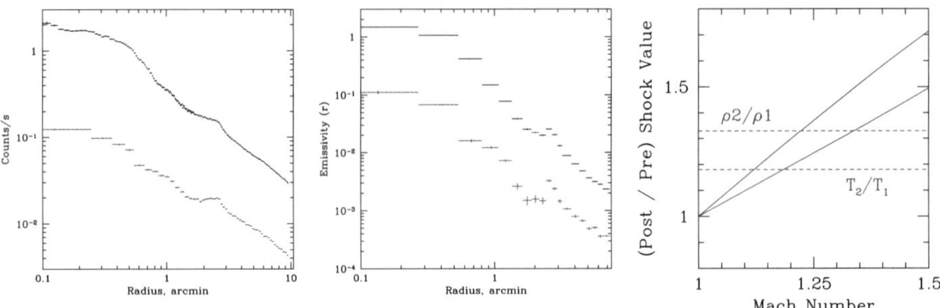

Figure 3. The radial profiles and Mach number of the 13 kpc shock in M87. (left panel) The azimuthally averaged (sector of width 80° centered on north) surface brightness profiles in the 1.2-2.5 keV (upper curve) and 3.5-7.5 keV (lower curve) bands. The 13 kpc shock is seen as the excess at $2-3'$. (central panel) The deprojected emissivity profiles, derived from the curves in the left panel. The deprojection shows the very pronounced 13 kpc shock at $2'-3'$. (right panel) The ratio of the pre-shock and post-shock density (blue solid line) and temperature (red solid line) as a function of shock Mach number as defined by the Rankine-Hugoniot shock jump conditions. The measured ratios of pre/post shock density and temperature are shown as dashed horizontal lines (blue and red for density and temperature, respectively). The intersections of the predicted and measured functions give consistent Mach numbers of ~ 1.2.

4. Conclusions

M87, the nearest active galaxy at the center of a gaseous cluster atmosphere shows multiple outbursts in its radio and X-ray images. We see evidence for *both* shocks and for buoyant relativistic plasma bubbles providing energy to the cluster ICM. A simple model gives details of the partition of energy with roughly 50% in bubbles which will be returned to the hot gas and 50% in the 13 kpc shock of which half is used to heat the gas and half is carried away by the shock as it becomes a weak sound wave.

References

Acciari *et al.* 2010, *ApJ*, 716 819

Bower, R. *et al.* 1993, *MNRAS*, 370, 645

Churazov, E., Brueggen, M. Kaiser, C., Boehringer, H., & Forman, W. 2001, *ApJ*, 554, 261

Croton, D. *et al.* 1995, *MNRAS*, 365, 11

Forman, W. *et al.* 1978 1978, *ApJS*, 38, 357

Gebhardt, K. *et al.* 2011, *ApJ*, 729, 119

Guo, Q. *et al.* 2011, *MNRAS*, 413, 101

Harris, D. *et al.* 2006, *ApJ*, 640, 211

Hines, D., Eilek, J., & Owen, F. 1989, *ApJ*, 347, 713

Johnson, R. 2011, Ph.D. Thesis, Dartmouth College, Publication number: AAT 3477679

Markevitch, M. & Vikhlinin, A. 2007, *Phy. Rev*, 443, 1

Markevitch, M., Vikhlinin, A., & Forman, W. 2003, *ASPC* 301, 37

Marshall, H. *et al.* 2002, *ApJ*, 564, 683

Million, E. *et al.* 2010, *MNRAS*, 407, 2046

Owen, F. Eilek, J., & Kassim, N. 2000, *ApJ*, 543, 611

Randall, S. *et al.* 2008, *ApJ*, 688, 208

Shi, Y., Rieke, G., Hines, D., Gordon, K., & Egami, E. 2007, *ApJ* 655, 781

Simionescu, A. *et al.* 2010, *MNRAS*, 405, 91

Stewart, G., Canizares, C., Fabian, A., & Nulsen, P. 1984, *ApJ*, 278, 536

The intriguing life of massive galaxies
Proceedings IAU Symposium No. 295, 2012
D. Thomas, A. Pasquali & I. Ferreras, eds.

© International Astronomical Union 2013
doi:10.1017/S1743921313004985

Super-massive black hole binaries in gas-rich environments

Jorge Cuadra[1]

[1]Departamento de Astronomía y Astrofísica, Pontificia Universidad Católica de Chile
email: `jcuadra@astro.puc.cl`

Abstract. Black hole binaries form after major galaxy mergers, but their fate is unclear as hardening due to stars gets inefficient at sub-parsec distances. We model an alternative scenario in which the merger is driven by the interaction of the binary with the surrounding gas. †

Keywords. black hole physics

For the expected masses, the disc will be gravitationally unstable and develop spiral structure, transporting out the angular momentum of the binary and making it shrink. Scaling our results analytically, we find that decay due to gas can be faster than that due to stars for separations below 0.01–0.1 pc. The minimum merger time-scale is shorter than the Hubble time for binaries with $M < 10^7 M_\odot$ (Cuadra *et al.* 2009).

The orbit of the binary does not only shrink – it will also become eccentric, tending to $e \approx 0.6$. This is simply due to the instantaneous gravitational interaction of the secondary black hole with the wake it drives in the inner part of the disc. The eccentricity increases the variability of the accretion onto the black holes (Roedig *et al.* 2011).

For massive discs, cooling conditions could produce their fragmentation and star formation. We are currently investigating the fate of binaries within such systems. Preliminary results show a slower decay rate than in the gaseous case, and an interesting enhancement in the rate of stellar tidal disruptions (Amaro-Seoane, Brem & Cuadra, ApJ submitted).

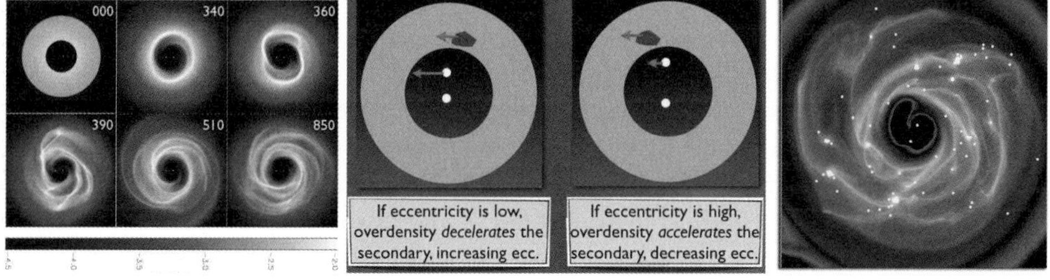

Figure 1. *Left:* Surface density evolution of the disc around the binary. The spiral pattern that develops transfers out the binary angular momentum. *Middle:* Binary eccentricity evolution due to the disc. This process makes the binary approach $e \approx 0.6$. *Right:* Star-forming circumbinary disc. The newly formed stars scatter with the binary, exchanging energy and angular momentum. A relatively large number of stars get tidally disrupted.

References

Cuadra, J., Armitage, P. J., Alexander, R. D., & Begelman, M. 2009, *MNRAS*, 393, 1423
Roedig, C., Dotti, M., Sesana, A., Cuadra, J., & Colpi, M. 2011, *MNRAS*, 415, 3033

† This article is based on work done in collaboration with the co-authors of the cited papers. JC acknowledges support from FONDAP (15010003), FONDECYT (11100240), Basal (PFB0609), VRI-PUC (Inicio 16/2010), the FP7 (LACEGAL) and IAU grant GA1090.

The intriguing life of massive galaxies
Proceedings IAU Symposium No. 295, 2012
D. Thomas, A. Pasquali & I. Ferreras, eds.

© International Astronomical Union 2013
doi:10.1017/S1743921313004997

Fluorescence in the Active Galactic Nuclei NGC 4151

Mattias Eriksson[1] and Hans Veenhuizen[2]

[1]Blekinge Institute of Technology,
SE-371 79 Karlskrona, Sweden
email: meo@bth.se

[2]Linnaeus University, School of Computer Science, Physics and Mathematics
391 82 KALMAR, Sweden
email: hans.veenhuizen@lnu.se

Abstract. The possibility of UV lines from Fe II formed in the BLR of NGC 4151 has been examined. As a result Fe II emission has been shown to play an important role for the topology of the 2000 to 3000 Å wavelength region of the NGC 4151 spectra. The Fe II UV emission originates from two processes, collisional excitation and PAR (photoexcitation by accidental resonance).

Keywords. galaxies: active, galaxies: nuclei

1. The model spectra

NGC 4151 has regions in which emission lines with widths corresponding to several thousand km/s doppler broadening are formed. The Mg II resonance doublet is one of those broad lines (Leech *et al.* 1987). Since the Fe II (^{5}D)4p levels have similar ionization and excitation potentials as the Mg II 3p levels, broad emission lines are likely formed by Fe II too. However, in the case of Fe II the flux would be split in more than hundred individual emission lines which would be difficult to detect. The topology of the resulting emission from all those Fe II emission lines is modeled using the oscillator strengths of Kurucz (1988) and Fuhr *et al.* (1988). In addition to the Fe II collisionally excited lines there is a possibility for PAR excited Fe II lines to be formed in NGC 4151. When Fe^+ ions are subjected to strong emission from H Lyα and C IV λ1550 the Fe II levels y^{4}H$_{11/2}$ and (^{3}F)4p ^{4}G$_{9/2}$ levels can be pumped resulting in a number of fluorescence lines in the 2000 to 3000 Å region. This phenomenon has previously been confirmed in symbiotic stars (Johansson 1983). Since the H Lyα and C IVλ1550 lines are the strongest UV lines in the NGC 4151 spectra PAR is incorporated in the model spectra.

2. Results

The model spectra and observed archive STIS spectra were compared in the 2000 - 3000 Å region. The fractional average difference between the best-fitting model and observations was $\sim$0.08 when excluding the Fe II emission. When incorporating the Fe II emission the fractional average difference decreased to $\sim$0.03.

References

Fuhr, J. R., Martin, G. A., & Wiese, W. L. 1988, *J. Phys. Chem. Ref. Data*, 17, Suppl. 4
Johansson, S. 1983, *MNRAS*, 205, 71
Kurucz, R. L. 1988, *Astrophysics & Space Science*, 138, 41
Leech, K. J., Penston, M. V., Snijders, M. A. J., & Gull, T. R. 1987, *MNRAS*, 225, 837

The intriguing life of massive galaxies
Proceedings IAU Symposium No. 295, 2012
D. Thomas, A. Pasquali & I. Ferreras, eds.

© International Astronomical Union 2013
doi:10.1017/S1743921313005000

An additional production mechanism for the diffuse x-ray background

Allan D. Ernest, Matthew P. Collins and Graeme L. White

Faculty of Science, Charles Sturt University,
Wagga Wagga, NSW, 2678 Australia
email: `aernest@csu.edu.au`

Abstract. We propose a mechanism that contributes energy and particles to the diffuse x-ray halos of galaxies and clusters, based on the dark quantum states of large-scale gravity wells.

Keywords. X-rays: galaxies: clusters, galaxies: halos

X-ray observations have shown that galaxies and clusters of galaxies are surrounded by massive gas halos at temperatures 10^6 to 10^8 K with cooling times $<<$ Hubble time. A source of energy is required to maintain the hot gas and following the recent discovery that particles in gravitational fields form stationary quantum eigenstates (Nesvizhevsky *et al.* 2003) we propose that baryonic particles in eigenspectra composed of dark gravitational eigenstates (Ernest 2009, 2012) could supply energy and visible baryonic matter required for continual star formation. Quantum theory shows that baryonic particles in such states are transparent to photons but Coulomb interactions between protons in dark states might conceivably either (1) shift their eigenspectra to mixtures of more visible quantum states appropriate to classical particles or (2) result in direct Bremsstrahlung production from dark-state electron collisions that take place at a reduced rate than would be expected from laboratory cross-section measurements. The multitude of transfer channels, and the complexity of deep state eigenfunctions makes quantitative ab-initio quantum calculations of dark to visible eigensptral mixtures impossible (at this stage) although we can estimate within a few orders of magnitude, the required interaction rate for comparison with observations. Dark eigenstates are unable to collapse via traditional classical processes but carry with them the equivalent kinetic energy of a viralised orbiting particle. In the halos of clusters of galaxies case these energies correspond to 1-15 keV while for isolated galaxies energies are between 0.3-2 keV.

As an example of process (1), taking typical galactic values of the average temperature T (eV), density of visible baryons n_b, and halo radius r, as 0.3-2 keV, 10^5 m^{-3} and 10^{21} m respectively, gives the x-ray luminosity L as 10^{36} W, using $L = n_b^2 T^{1/2}/(7.7 \times 10^{18})^2$ or an energy loss of 10^{-13} eV s^{-1} per particle. To balance this energy loss via eigenstate collisions, assuming a dark eigenstate number density n about 10 times that of the visible baryons gives a transfer-to-visible-state collision rate constant k $(dn/dt = -k\,n^2)$ of $10^{-27} - 10^{-28}$ m^{3}s^{-1} and corresponding cross section $10^{-33} - 10^{-34}$ m^2 at these equivalent kinetic energies. This corresponds to a visible baryon production rate of 10^8 solar masses over the Hubble time. For clusters the figure is 10^{10} to 10^{11} solar masses.

References

Ernest, A. D. 2009, *J. Phys A: Math. Theor.*, 42, 115207
Ernest, A. D. 2012, in: I. Cotaescu (ed.), *Advances in Quantum Theory* (InTech), p. 221
Nesvizhevsky, V. V., Borner, H. G., & Petukhov, A. K. 2003, *Nature*, 415, 297

The intriguing life of massive galaxies
Proceedings IAU Symposium No. 295, 2012
D. Thomas, A. Pasquali & I. Ferreras, eds.

© International Astronomical Union 2013
doi:10.1017/S1743921313005012

Quasar activity in the neighbor Universe

R. Falomo[1], D. Bettoni[1], K. Karhunen[2], J. Kotilainen[3] and M. Uslenghi[4]

[1] INAF – Osservatorio di Padova, Italy
[2] Tuorla Observatory, University of Turku, Finland
[3] Finnish Center for Astronomy with ESO (FINCA), University of Turku, Finland
[4] INAF – IASF, Milano

Abstract. We investigate the properties of the galaxies hosting quasars in $\sim$400 low redshift ($z < 0.5$) SDSS QSO that are in the "Stripe 82" sky area. For this region deep (r $\sim$ 22.4 mag) u, b,v,r and i images are available and allow us to study both the host galaxies and the Mpc scale environments. This sample outnumbers previous studies of low-z QSOs. We present preliminary results of the properties of quasars activity and in particular we focus on the relationships among host galaxy luminosity, black hole mass, radio emission and the surrounding galaxy environments. We select from the SDSS - QSO Catalogue all the QSOs in the range of redshift $0.1 < z < 0.5$ and in the Stripe82 region. This gives a total of 416 QSO. In this sample we are dominated by radio quiet quasars (about 5% are radio loud). In Fig. 1 we report the distribution of QSO in the plane redshift-M_R ($H_0 = 70$). The mean redshift of the sample is $< z > = 0.39$ and the average absolute magnitude is: $< M_i > = -22.68$. We implemented an automated procedure using AIDA (Uslenghi & Falomo 2011) to decompose the QSO images into nucleus and host galaxy luminosity. After masking of all contaminating sources in the field a 2D fitting is performed using PSF + galaxy model. In Fig. 1 we show an example of a QSO image in the sample and the distribution of the host galaxy absolute magnitude of the resolved objects.

Keywords. galaxies: quasars: general

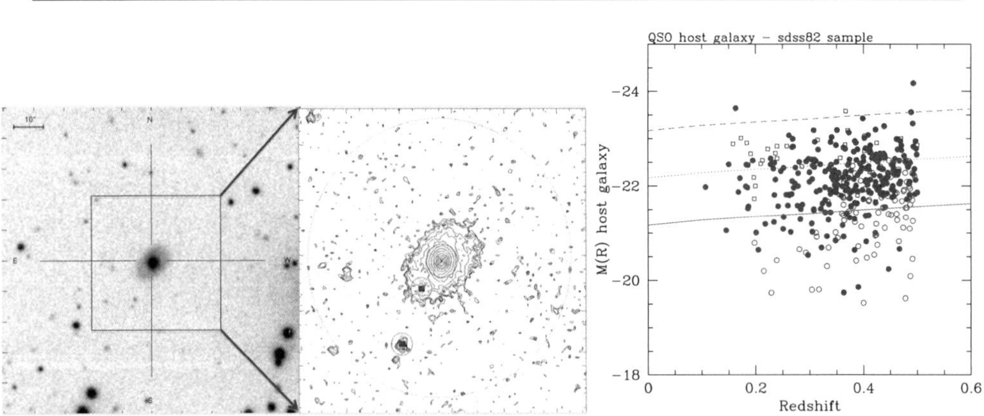

Figure 1. Left: i band image from Stripe82 and the isophotes of a zoom of the central region. Right: M_R-z relation. Filled and open red points are for resolved and marginally resolved QSO respectively. Blue open squared represent QSOs from HST images.

Reference

Uslenghi, M. & Falomo, R. 2011, *Proc. SPIE* 8135, 813524

The intriguing life of massive galaxies
Proceedings IAU Symposium No. 295, 2012
D. Thomas, A. Pasquali & I. Ferreras, eds.

© International Astronomical Union 2013
doi:10.1017/S1743921313005024

Probing the QSO host galaxy evolution through the gas metallicity

B. Husemann[1], L. Wisotzki[1], K. Jahnke[2], S. F. Sánchez[3], and D. Nugroho[2]

[1]Leibniz-Institut für Astrophysik Potsdam, Germany
email: bhusemann@aip.de
[2]Max-Planck-Institut für Astronomie, Heidelberg, Germany
[3]Instituto de Astrofísica de Andalucía (CSIC), Granada, Spain

Abstract. We use the spatially resolved gas-phase metallicity as a new diagnostic for tagging recent interactions in QSO host galaxies. With this technique we also identified a QSO with extremely low gas-phase metallicity as likely evidence for gas accretion from the environment.

Keywords. galaxies: active, quasars: emission lines, HII regions, ISM: abundances

We measured the oxygen abundance distribution of QSO host galaxies at $z < 0.2$ using integral-field spectroscopy. We find that the abundance gradients of disc-dominated QSO hosts are consistent with normal galaxies, but are flat or non-linear in bulge-dominated systems and ongoing mergers. The oxygen abundances at $1R_\mathrm{e}$ are systematically lower in bulge-dominated QSO hosts as shown in Fig. 1. We interpret this as indication for recent (minor) interactions to trigger AGN in bulge-dominated hosts considering the expected metallicity dilution and systematic change in the gradients during merger events.

In Husemann *et al.* (2011), we presented the discovery of an extraordinary low gas-phase metallicity in the NLR of the QSO HE 2158-0107 based on IFS observations. The [OIII] line brightness distribution reveals a 30kpc long tail-like structure of gas. Both facts are likely evidence for the accretion of nearly pristine gas from the environment either through minor merger(s) or smooth accretion of gas.

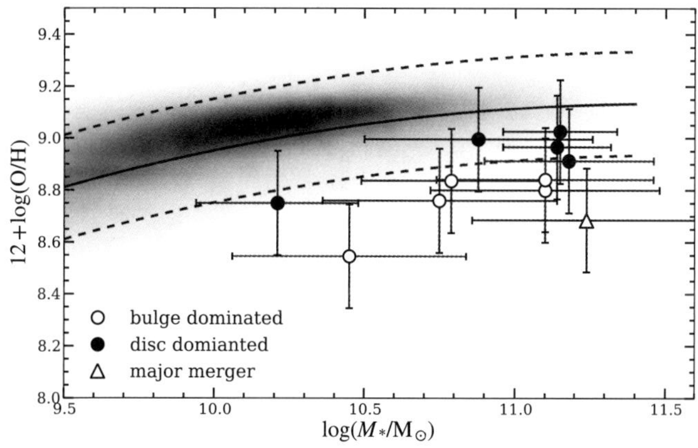

Figure 1. Estimated oxygen abundances at $1R_\mathrm{e}$ for 11 QSO hosts against their total stellar mass. The global SDSS mass-metallicity relation of galaxies is shown for comparison.

Reference

Husemann, B., *et al.* 2011, *A&A*, 535, A72

The intriguing life of massive galaxies
Proceedings IAU Symposium No. 295, 2012
D. Thomas, A. Pasquali & I. Ferreras, eds.

© International Astronomical Union 2013
doi:10.1017/S1743921313005036

Where the active galaxies live: a panchromatic view of radio-AGN in the AKARI-NEP field

Marios Karouzos[1], Myungshin Im[1], Markos Trichas[2], and the AKARI-NEP team

[1] CEOU-Seoul National University, 1 Gwanak-ro, Gwanak-gu, Seoul 151-742, South Korea
email: `mkarouzos@astro.snu.ac.kr`
[2] Harvard-Smithsonian Center for Astrophysics, 60 Garden Street, Cambridge, MA 02138

Abstract. We study the host galaxy properties of radio sources in the AKARI-North Ecliptic Pole (NEP) field, using an ensemble of multi-wavelength datasets. We identify both radio-loud and radio-quiet AGN and study their host galaxy properties by means of SED fitting. We investigate the relative importance of nuclear and star-formation activity in radio-AGN and assess the role of radio-AGN as efficient quenchers of star-formation in their host galaxies.

Keywords. galaxies: active, galaxies: evolution, galaxies: starburst

1. The Project and Results

We construct broad-band SEDs (UV to 24μm; Fig. 1) for 48 radio sources at 1.5GHz with optical spectra in the AKARI-NEP field (Lee *et al.* 2009). Following Trichas *et al.* (2012), we fit an AGN and a starburst component additively to each SED. The fractional contribution and luminosities of both components are derived.

We see a trend for decreasing contribution of active nuclei with increasing radio luminosity (3σ difference between lowest and highest luminosity bins; Fig. 1). The most radio-loud systems show hints for lower star-formation activity than otherwise expected.

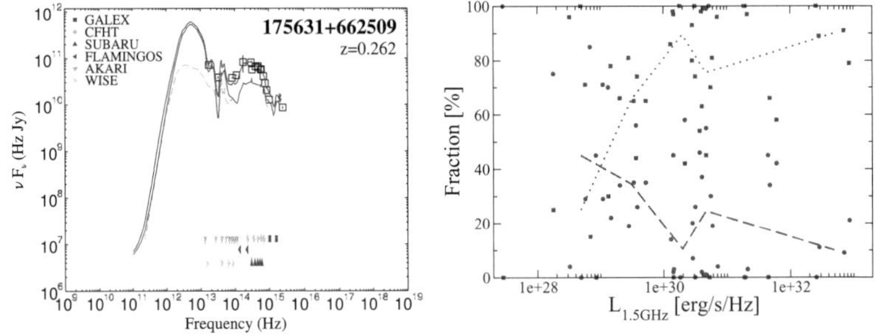

Figure 1. Example SED (left). Fractional contribution of AGN (red) and starburst (blue) components versus radio luminosity (right), for individual (symbols) and average values (lines).

References

Lee, H. M., Kim, S. J., Im, M., *et al.* 2009, *PASJ*, 61, 375
Trichas M., Green P., Silverman J. D., *et al.* 2012, *ApJS*, 200, 17

The intriguing life of massive galaxies
Proceedings IAU Symposium No. 295, 2012
D. Thomas, A. Pasquali & I. Ferreras, eds.

© International Astronomical Union 2013
doi:10.1017/S1743921313005048

All-sky catalog of local radio galaxies

S. van Velzen[1] and H. Falcke[1,2,3]

[1] IMAPP, Radboud University, P.O. Box 9010, 6500 GL Nijmegen, The Netherlands
[2] ASTRON, Dwingeloo, The Netherlands
[3] Max-Planck-Institut für Radioastronomie Bonn, Germany

Abstract. The final episode in the history of black hole accretion and galaxy formation takes place in our cosmic backyard, the local universe. Within this volume must also reside the — until now unknown — sources of observed ultra-high energy cosmic rays (UHECRs). A thorough study of the local universe requires full-sky coverage to obtain a sizable sample and map the matter anisotropy. We recently constructed the first catalog of radio-emitting galaxies that meets this requirement. The sample contains all radio galaxies similar to Centaurus A out to ~ 100 Mpc. Only 3% of the hosts of the powerful radio jets are classified as Spiral galaxies, while for non-radio galaxies of similar mass, this fraction is 34%. The energy injected by radio jets per unit volume indicates that Cen A-like radio galaxies have in principle sufficient power to accelerate cosmic rays to ultra-high energies. A significantly enhanced clustering of radio-loud galaxies compared to normal galaxies of the same luminosity is observed. This indicates a causal relation between galaxy environment and jet power, independent of black hole mass.

Keywords. catalogs, galaxies: jets, galaxies: evolution, accretion, acceleration of particles

Here we highlight two applications of our catalog (van Velzen *et al.* 2012, available at http://ragolu.science.ru.nl), using a volume-limited sample of radio galaxies. (*i*) To find potential UHE proton accelerators, we compare the equipartition B-field and radius (R) of the radio lobes to the minimum value required for containing cosmic rays of this energy. The jet power of these sources, estimated from the total energy in the lobes ($\propto B^2 R^3$) over their dynamical time, exceeds the energy injected into UHECRs. (*ii*) We observe that the projected density of galaxies around radio galaxies ($\rho_{\rm RG}$) is significantly enhanced with respect to the mean density around non-radio galaxies of the same luminosity and Hubble type ($\rho_{\rm matter*}$).

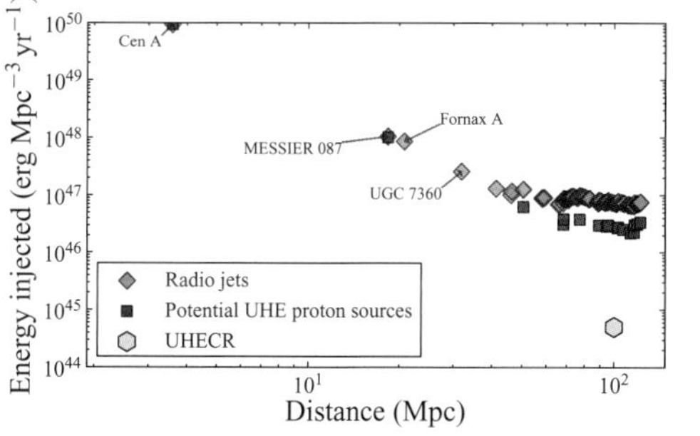

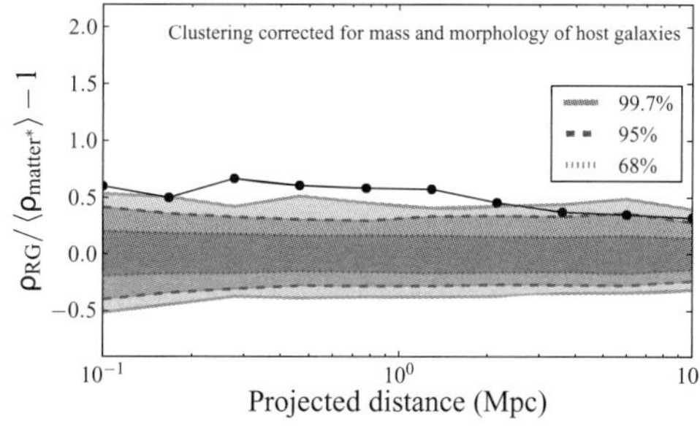

Figure 1. The energy injected in a sphere between us and the galaxy (left) and the clustering (right) of a volume-limited sample of radio galaxies ($L_{\rm 1GHz} > 5 \times 10^{23}$ W Hz^{-1}, $z < 0.03$).

Reference

van Velzen, S., Falcke, H., Schellart, P., & Nierstenhöfer, N., Kampert K. H. 2012, *A&A*, 544, A18

The intriguing life of massive galaxies
Proceedings IAU Symposium No. 295, 2012
D. Thomas, A. Pasquali & I. Ferreras, eds.

© International Astronomical Union 2013
doi:10.1017/S174392131300505X

Stellar Population Models

Claudia Maraston[1]

[1]Institute of Cosmology and Gravitation, University of Portsmouth,
Burnaby Road, PO1 3FX, Portsmouth, United Kingdom
email: `claudia.maraston@port.ac.uk`

Abstract. Modelling stellar populations in galaxies is a key approach to gain knowledge on the still elusive process of galaxy formation as a function of cosmic time. In this review, after a summary of the state-of-art, I discuss three aspects of the modelling, that are particularly relevant to massive galaxies, the focus of this symposium, at low and high-redshift. These are the treatment of the Thermally-Pulsating Asymptotic Giant Branch phase, evidences of an unusual Initial Mass Function, and the effect of modern stellar libraries on the model spectral energy distribution.

Keywords. galaxies: formation; galaxies: evolution; galaxies: high-redshift; galaxies: spiral; galaxies: initial mass function; stars: AGB and post-AGB; stars: atmospheres; stars: carbon.

1. Introduction

Forty years ahead of this Symposium, Beatrice Tinsley (1972) provided the theoretical background for describing the physical properties of populations of stars as functions of relevant parameters, and their evolution with time. The approach she put forward, by exploiting knowledge of stellar evolution, allows the calculation of models which can be evolved with time hence can be used to study the redshift evolution of galaxies. These same models are used to make predictions for galaxy formation (see C. Tonini, this volume).

The stellar populations of galaxies are used as probes of the process of galaxy formation and evolution (see Renzini 2006, and T. Naab and P. Johansson, this volume). Obviously the analysis of stellar populations is based on models, therefore depends on the adopted models and their characteristics. In this review I shall first provide an overview of current models, and highlight results around the most discussed topics, namely the contribution of Asymptotic Giant Branch stars (see also Bruzual, this volume); the latest generation of stellar spectral libraries; the initial mass function of massive galaxies from a population model perspective (see McDermid, Corsini and Renzini, this volume).

2. Overview of stellar population models

Since the pioneering work by Tinsley, much progress has been made on so-called "evolutionary population synthesis" (EPS) models (see Figure 1), thanks to relevant contributions, some of which we list below (see Greggio & Renzini 2011 and Conroy 2013 for reviews). Renzini (1981) introduces analytical relations (the *fuel consumption theorem*) for controlling the energetics of stellar populations. Bruzual (1983) and Bruzual & Charlot (1993, 2003; see G. Bruzual, this volume) propose a convenient approach to population synthesis which exploits isochrones, and made available widely used computer codes. Leitherer *et al.* (1999) focus on the treatment of young massive stars and the effect of their surrounding gas on spectra. Maraston (1998; 2005) extends the *fuel consumption theorem* approach to EPS models and includes a semi-empirical treatment of the

Thermally-Pulsing Asymptotic Giant Branch (TP-AGB) using real Carbon star spectra. Thomas, Maraston & Bender (2003a) calculate comprehensive grids of absorption-lines of EPS models for various element abundance ratios. Conroy, Gunn & White (2009) quantify global uncertainties on EPS models. Conroy & van Dokkum (2012) study the modelling of near-IR lines as a function of the IMF.

The state of the art for the models can be summarised with a few key bullet points.

All major stellar evolutionary phases are now included in the models, but a vivid debate remains on the energetics and spectra of the TP-AGB phase (see Section 3). The latter is relevant to galaxies at high-redshift, and spiral and star forming galaxies at low redshift. It is also important for galaxy formation models (Tonini *et al.* 2009; Monaco & Fontanot 2010; Henriques *et al.* 2011; see Tonini's contribution).

The spectral resolution of models is nowadays optimal in the optical wavelength range (up to 8000 Å) and the ultraviolet, thanks to the advent of several stellar libraries, both empirical (STELIB, Le Borgne *et al.* 2003; ELODIE, Prugniel *et al.* 2007; MILES, Sánchez-Blázquez *et al.* 2006) and theoretical (Rodriguez-Merino *et al.* 2005; Gustfaffson *et al.* 2008), with a resolution as high as 0.1 Å. Several EPS engines have incorporated one or more of these libraries in their modelling, starting with Bruzual & Charlot (2003, based on STELIB); Pegase-HR (based on ELODIE); Vazdekis *et al.* (2010, based on MILES); Maraston & Strömback (2011), which use them all.

Thanks to the same stellar spectral libraries, models of absorption lines with variable abundance-ratios are now flux-calibrated (Thomas, Maraston, Johansson 2011; see Johansson's contribution), which, by being freed of the standard star calibration required to match the Lick index system, allows the use of these models to study galaxies at high-redshift.

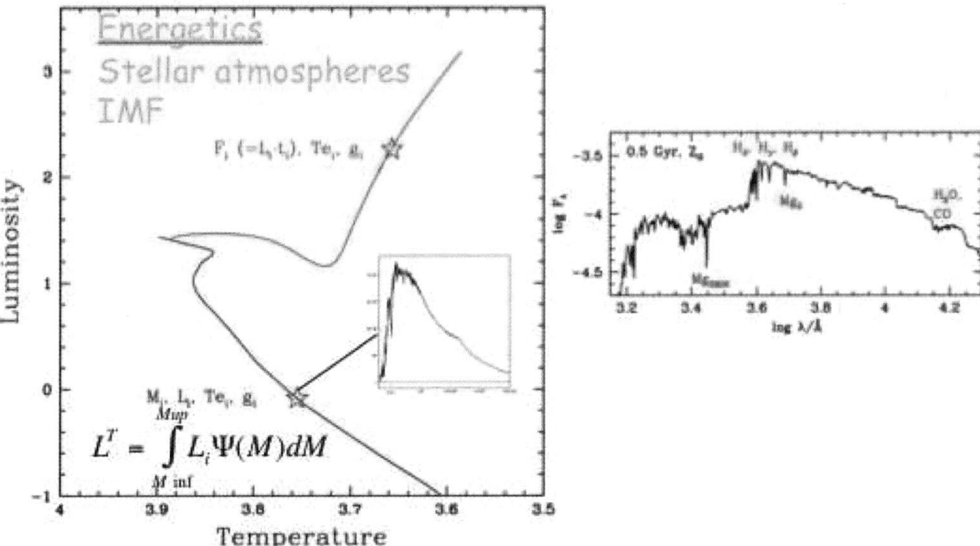

Figure 1. Sketch of evolutionary population synthesis modelling. *Left-hand panel* A theoretical HR diagram reproduced up to the tip of the RGB, with highlighted MS and post-MS stellar phase (blue and red, respectively). In order to obtain the integrated spectral energy distribution (*right-hand panel*) an integral either by mass or using the *fuel consumption theorem* is performed, on the luminosity-weighted stellar spectra (small inset). Models have three main ingredients, the energetics, the stellar spectra and the IMF.

3. The (infamous) TP-AGB

A lively debate arose around the treatment of this evolutionary phase in EPS models since Maraston (1998) pointed out that, if a sizeable TP-AGB contribution is present around $t \sim 1$ Gyr as suggested by studies in Magellanic Clouds star clusters available at the time (Frogel, Mould & Blanco 1990), this would have a strong impact on EPS models, especially in the rest-frame near-IR. Maraston (1998) develops a method to model the contribution of the TP-AGB phase using available data and the fuel consumption theorem. Maraston (2005) extends these models to the description of the full spectral energy distribution, by using real spectra of Carbon and Oxygen-rich TP-AGB stars (Lançon & Mouhcine 2002). When compared to other EPS models (Bruzual & Charlot 2003, Pegase, Starburst99) the Maraston models are substantially redder longward $\sim$ 6000 Å rest-frame and exhibit marked spectral features by Carbon stars.

As we shall discuss below, the comparison with observations is providing a sort of conflicting evidence. But before commenting on the observational evidence, let us explain why this evolutionary phase has gained so much attention and why the modelling is so complicated.

The answer to the first question is that this is a very luminous stellar phase that provides substantial contribution to the bolometric of the model. The answer to the second question is that the theoretical stellar evolution is complicated because these are dynamically unstable stellar configurations in double-shell burning (Iben & Renzini 1983).

There are some facts on which presumably everyone would agree. Firstly, the TP-AGB cannot be fully modelled via first principles because of strong (and poorly known) mass loss affecting stellar evolution during this phase. The theoretical energetic must then be calibrated with data (Maraston 1998, 2005; Marigo & Girardi 2007). Secondly, TP-AGB stars are very cool ($T_{\rm eff} < 3500$ K) and the modelling of their spectral energy distribution, which is an ingredient of integrated models, is complicated, and result in just a few model realisations. Again here, a solution is to use stellar spectra from observations (Maraston 2005; Conroy, Gunn & White 2009).

What is a matter of choice among different authors is how the calibration is accomplished, and which data are used.

Maraston (1998; 2005) put forward the approach of using globular clusters (GCs) in the Magellanic Clouds, assuming their integrated colours are comparable to those of Simple Stellar Population (SSP) models, which are instantaneous bursts with homogeneous chemical composition†. This method is also adopted in Conroy & Gunn (2010). Two are the important assumptions here: i) the adopted ages of the GCs; ii) the observed integrated spectrophotometry.

Marigo & Girardi (2007), Marigo *et al.* (2008) and Bruzual (see Bruzual's talk) use instead resolved AGB stars in the Magellanic Clouds and compare their luminosity function with models with extended star formation histories. Here the crucial assumptions are: i) the star formation history of the Clouds and their metallicity distributions; ii) the photometric data, as before.

Let us now see how the results of the different procedures compare. Figure 2 shows the results of different EPS model calibrations. The left-hand panel shows the calibration of the bolometric contribution of the AGB phase with Magellanic Clouds clusters by Maraston (1998). According to this calibration, the phase onsets at about ~ 0.3 Gyr, peaks at ~ 1 Gyr and then decreases in importance at older ages. This result, and the related EPS models, are obviously completely dependent on the ages assumed for the

---

† The validity of approximating GCs to SSPs has been questioned in recent years

GCs, which in turn depend on the adopted stellar models and method used to derive them (e.g. Main Sequence turnoff fitting, Colour Magnitude Diagram fitting, integrated colour model matching). The results of Maraston (1998) were based on the age calibration by Frogel *et al.* (1990), in turn based on stellar models without inclusion of overshooting. These so-called *classic* tracks were also input to the Maraston's models, hence the procedure was self-consistent, though it may be now obsolete.

Marigo *et al.* (2008) (central panel) perform their calibration using resolved stellar populations. However, the resulting models (their model $V - K$ is the magenta line in the central panel of Figure 2) behave quite similarly to Maraston's models (dashed blue lines in the same figure). A slight shift in age (~ 0.3 Gyr) is evident between the Marigo *et al.* and the Maraston models, which is due to the different adopted age scale. Marigo *et al.* use ages derived with tracks including overshooting, which implies longer timescales on the Main Sequence, hence older ages.

Conroy & Gunn (2010) perform a similar model calibration as Maraston and Marigo, but adopt an age-scale that pushes Magellanic Clouds GCs to even older ages (right-hand panel in Figure 2). Furthermore, they adopt GCs colour data binned in age with a rather coarse age binning. This has the effect of diluting the AGB phase-transition. In the Conroy & Gunn re-calibration of the TP-AGB phase, the Maraston's models appear to be much off the data, but one should notice that the data have been pushed towards the right of the age scale. These and other differences, such as the adopted photometry, explain most part of the offset between the models (see Noell *et al.* 2013, *submitted*).

One may conclude that, in light of more recent GCs age determinations and photometry, the Maraston (2005) models have too a strong contribution from the TP-AGB phase. Zibetti *et al.* (2013) reach the same conclusions as Conroy & Gunn (2010), using post-starburst galaxies, following Lancón *et al.* (1999)'s suggestions that these are the optimal calibrators of the AGB phase.

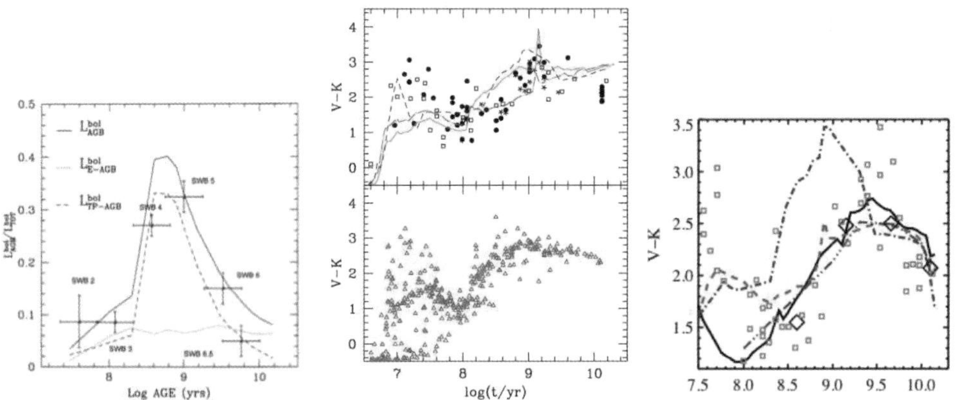

Figure 2. Calibration(s) of the TP-AGB phase for EPS models. *Left-hand panel.* Calibration of the bolometric contribution (split into E-AGB and TP-AGB, short and long-dashed) with Magellanic Clouds clusters (points with errorbars). The observed contribution is an average over clusters with similar ages, performed in order to minimise stochastic fluctuations due to the shortness of the AGB phase ($1 - 4$ Myr). From Maraston (1998). *Central panel.* Calibration of the integrated $V - K$ model colour with Magellanic Clouds clusters (points with errorbars). Models are constrained using resolved stars rather than GCs, but the net result is similar to the one from Maraston. From Marigo *et al.* (2008). *Right-hand panel* As in central panel, but note the shift towards older ages of most data (the x-axis shows ages as $\log t/yr$). From Conroy & Gunn (2010). All calibrations are obviously strongly dependent on the assumed ages for the star clusters.

The situation remains unsettled, however. Bruzual (this volume, see arXiv:1212.5381) comparing the Bruzual & Charlot (2003) models with resolved AGB stars in the Magellanic Clouds conclude, consistently with Marigo *et al.* (2008) that the BC03 models under-predict their LF, in other words these models do not contain enough TP-AGB.

Post-starburst (post-SB) galaxies in the local universe probably have only a small fraction of their stellar mass at the relevant ages, and this mass contribution could scatter between galaxies. Indeed, the integrated near-IR colours of the post-SB galaxies used in Conroy & Gunn (2010) and in Zibetti *et al.* (2013) seem to be quite different. This intrinsic variation, plus uncertainties in their star formation histories do not make post-SB galaxies such ideal calibrators. For example, Zibetti *et al.* succeed at modelling post-SB galaxies with BC03 models but using single-burst rather than composite models (see Figure 3), which appears contrived.

A more stringent test is represented by massive galaxies at high-redshifts ($z \sim 2 - 3$), where a larger fraction of stellar mass could be in the relevant age range. Comparisons using these galaxies suggested models more similar to Maraston (2005) with a substantial TP-AGB contribution. This has been widely discussed in the literature, therefore we focus here on a type of comparison which is relatively new. This uses near-IR data of spiral galaxies in the local universe, in which star formation is ongoing and must have produced at least a fraction of TP-AGB stars. MacArthur *et al.* (2010) can only fit optical and near-IR colours of these galaxies using M05 models, as the colours of other models are too blue.

Riffel *et al.* (2013, *in preparation*) perform spectral fitting in the near-IR of a massive local spiral using different models (see Figure 4 and 5).

The M05 models give a very good fit to the near-IR spectrum (Figure 4, left-hand plot), consistent with a broad age distribution, from $\sim 10^8$ to $\sim 10^{10}$ (Figure 4, right-hand plot, middle panel), which is sensible given the age spread of stars expected to be present in a spiral. When the same fit is performed with M05 models not including any TP-AGB (Figure 5, left-hand plots for the sole solar metallicity), the fits degrade substantially. Moreover, the derived age distribution is now stretched towards old ages, a solution which is obviously an artefact. When fitting with BC03 models and a range in

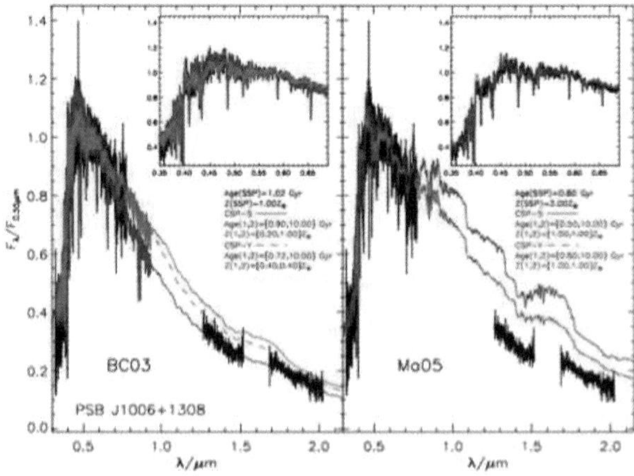

Figure 3. Spectral fitting of post-SB galaxies by Zibetti *et al.* (2013), using BC03 and M05 models (left-hand and right-hand panels, respectively), with single-burst (red) and composite (blue) models. The best fit to the whole optical plus near-IR fluxes is achieved with BC03 SSP models, though the optical fit is not very satisfactory (insets in both panels) and is better with M05 models. From Zibetti *et al.* (2013), see discussion therein.

metallicity (Figure 5, right-hand plots), the fitting improves thanks to the high metallicity components (though it is not as good as with the TP-AGB stars, see Fig. 4). However, the derived ages are still only the old ones.

In conclusion, the question whether the TP-AGB should be as strong as in the M05 or as weak as in the BC03 models has not yet been given a satisfactory answer, with local GCs and post-SB galaxies favouring BC03, and high-redshift and spiral galaxies favouring M05. Work is in progress by all of us now and the question will possibly get a consistent answer in the near future.

4. Modern spectral libraries and high-resolution models

As mentioned in the Introduction, most stellar population synthesis models have now been upgraded to a high spectral resolution, thanks to the advent of a multiplicity of stellar libraries, both empirical and theoretical. The former in particular contain now a sufficient number of stars in the various evolutionary phases which leads to sensible models, though they are still mostly defined in the optical range. Let us just recall the pros and cons of the two approaches. Empirical libraries contain spectra of real stars, but have a pattern of abundance ratios which reflect the chemical history of the Milky Way as a function of total metallicity. Thomas, Maraston & Bender (2003a) correct for such *bias* analytically, for obtaining models which can be applied to distant galaxies, where the chemical pattern is unknown and has to be determined with these models. Another limitation is that TP-AGB Carbon stars are usually not included, nor are Bulge old and $[\alpha/Fe]$-enhanced stars.

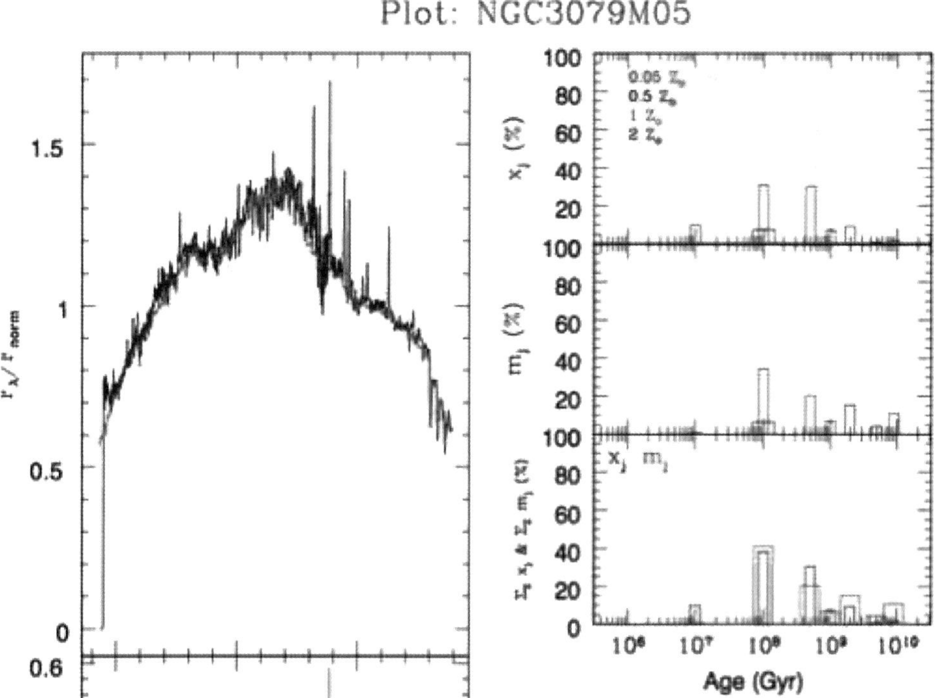

Figure 4. Spectral fitting of the near-IR spectrum of a massive spiral. *Left-hand panel.* Model fitting using M05 models. The spectrum is very well fitted. *Right-hand panel.* Derived age distribution (middle panel), which spans ages from 10^8 yr to 10 Gyr.

Theoretical libraries span an arbitrarily large parameter space in temperature. gravity and metallicity, but suffer from incomplete line list (Korn *et al.* 2005). Maraston & Strömback (2011) calculate stellar population models for a variety of empirical libraries and theoretical libraries, and we refer to this work for comprehensive discussion and comparisons. Here we briefly stress the fact that the theoretical SEDs of population models depend on the adopted library (Figure 6, *left-hand panel*) and in principle the choice of the library, along with the assumed stellar evolution, may affect the resulting galaxy evolution which is deduced from spectral fitting of data. The case is shown in Figure 6 (*right-hand panel*, from Wilkinson *et al.*, *in preparation*), who derive spectral fitting ages of galaxies from the SDSS-III/BOSS survey, using Maraston's models based on different spectral libraries. Older ages are obtained using models based on the MILES library with respect to models based on the STELIB library, which was the backbone of galaxy evolution results obtained with the SDSS-I and II data through the BC03 models. That MILES-based models imply the derivation of older ages was found in Maraston & Strömback (2011) when fitting the Milky Way GC M67.

Again we like to stress how galaxy evolution results depend on the adopted population modes. Works as Wilkinson *et al.* will be important for understanding the effect of population model ingredients on the physical properties of galaxies derived from high-resolution spectral fitting.

5. The evergreen case for a non universal IMF: effect of population models

The Initial Mass Function (IMF) is a mere parameter of population models, and can be included using arbitrary assumptions. On the other hand, the actual IMF is a crucial parameter of our understanding of gas and gravitation physics, telling us how a gas cloud breaks into individual stars and the efficiency of stellar feedback. The IMF assumed in the models is also the important quantity to determine stellar masses and deduce the amount of dark matter (see Cappellari, McDermid and Spiniello *this volume*). The physics around this parameter is complicated, hence we usually rely on empirical determinations of the IMF (or, better said the present mass function) in local fields and star clusters. However, several works have attempted to a *direct* determination of the IMF in distant galaxies. In particular, the past couple of years have witnessed a *revival* of a case for a non-universal IMF, in particular for a bottom-heavy IMF in massive elliptical galaxies. These results

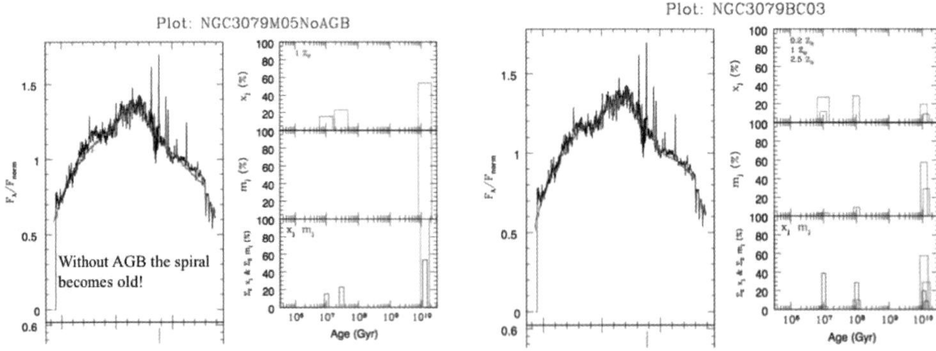

Figure 5. As in Figure 4, but using models with no TP-AGB (M05notp, *left-hand panel*) or with a reduced contribution, but several metallicities (BC03, *right-hand panel*). The spectral fitting degrades and the age distribution shows only old ages, which is clearly a model artefact.

are of two kinds. They can be based on comparing luminous M^*/L to dynamical M/L, or on identifying spectral features sensitive to the IMF and performing spectral fitting (Ferreras *et al.*, this volume). In both cases the results depend on the population models, but more strongly and directly in the second approach. Hence, we briefly summarise recent findings based on spectral fitting and leave to Renzini (this volume) to comment extensively on both cases.

About ten years ago, Saglia *et al.* (2002) (Figure 7, *lower panel*) report the puzzling result that the near-IR calcium triplet (CaT) index observed in massive galaxies cannot be reproduced by population models, unless assuming a substantial steepening of the power-slope exponent of the IMF ($\sim$ 3.5, when the Salpeter's one is 2.35), toward the massive end of the galaxy population. Though Thomas, Maraston, Bender (2003b) discuss possible alternatives such as a depressed Calcium abundance, we are still lacking a robust explanation for the mismatch. Maraston *et al.* (2003) show that in order to match the magnesium vs iron indices of massive galaxies without invoking a substantial $[\alpha/Fe]$-enhancement (which is the accepted solution based on comparison with star clusters, same paper), requires again an exponent as steep as 4 (in the same Salpeter's notation). Recently, van Dokkum & Conroy (2010) resume the near-IR indices introduced in the '70s and other near-IR features, including the CaT, to argue for a dwarf-dominated IMF in massive local ellipticals (Figure 7, *upper panel*). Spectral features in the near-IR are sensitive to gravity, hence are sensitive to the stellar dwarf-to-giant ratio, which depends on the assumed IMF. On the other hand, these same features are also sensitive to other population parameters, such as the element abundance ratios, the total metallicity and also the age composition of the stellar populations. These intricate model dependencies are discussed at length in Conroy & van Dokkum (2012), and more work in this direction will be beneficial to isolate the actual usable features and assess whether there is a steepening of the IMF with increasing galaxy mass.

In fact, the Wing-Ford band which was supposed to be highly sensitive to the dwarf end of the IMF, does not appear to depend much on the galaxy velocity dispersion (Figure 8, upper plot, central panel) within the same galaxies (black points) from which Cappellari *et al.* detect a dwarf-dominated IMF from dynamical considerations.

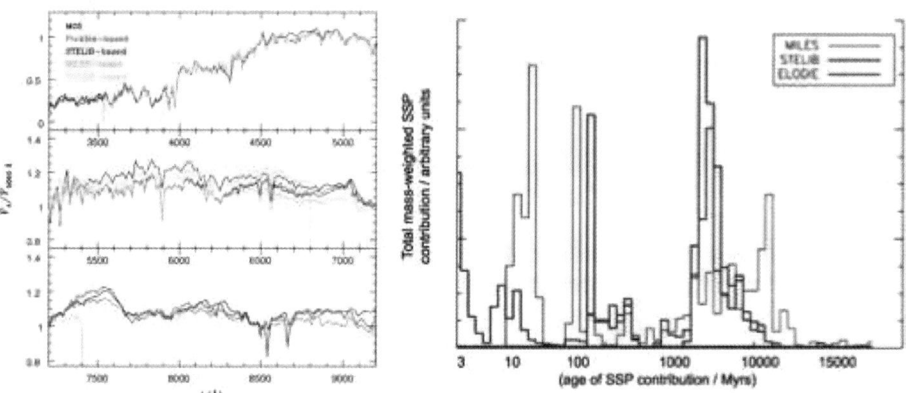

Figure 6. *Left-hand panel* Spectrum of a 12 Gyr old stellar population model obtained using different spectral libraries as input. From Maraston & Strömback 2011. *Right-hand panel* Galaxy ages derived using Maraston models based on different spectral libraries, from Wilkinson *et al.* (2013, *in preparation*).

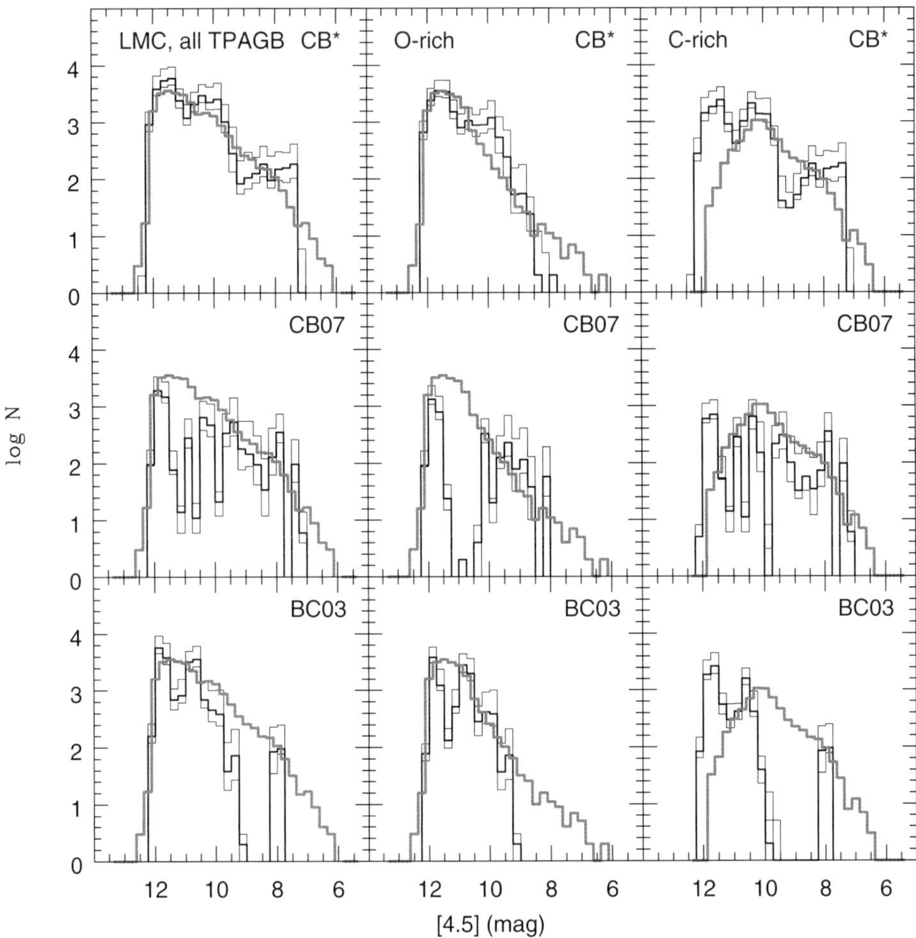

Figure 2. LF derived from our Monte Carlo simulations of the star formation history of the LMC using the CB*, CB07, and BC03 models, and from the observed *SAGE* data set (Srinivasan *et al.* 2009) in the *IRAC* $[4.5]\mu m$ band. The left column corresponds to all the TP-AGB stars. In the central and right column only the O-rich and C-rich TP-AGB stars are shown, respectively. For the simulations we assumed the Salpeter IMF, $Z = 0.008$ isochrones, and we kept all the stars with apparent $K \leqslant 12$ mag. The *heavy gray-line* corresponds to the *SAGE* LF. The *heavy black-line* corresponds to the simulation LF using the central value of the LMC SFH (Harris & Zaritsky 2009). The bracketing *light black-lines* correspond to the upper and lower limit of the SFH derived from the error bars given by these authors.

the case of the SMC the chemical evolution indicated by Harris & Zaritsky (2004) is included in our simulations. Inspection of Figs. 2-4 shows that the LFs computed with the CB* models are in closer agreement with the observations that those computed with the BC03 and CB07 models. These results are consistent with the findings by Kriek *et al.* (2010), Melbourne *et al.* (2012), and Zibetti *et al.* (2013) (see caption to Fig. 1), and support our choice for the treatment of TP-AGB stars in the CB* models. We do not have at hand enough information (isochrones) to perform the same kind of comparison with the M05 models. Details of this work will be published in a coming paper.

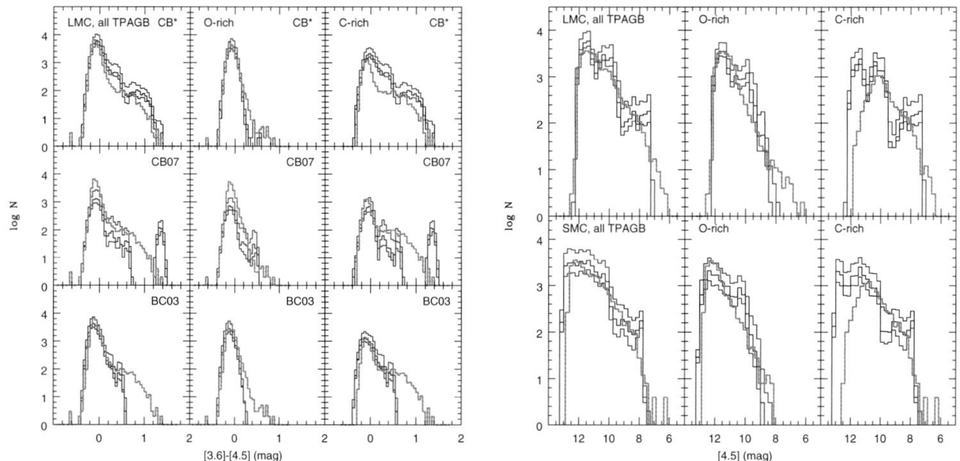

Figure 3. Same as Fig. 2 but for the $IRAC$ $[3.6]-[4.5]\mu m$ color distribution (left) and including the results of the simulations for the SMC using the SFH of Harris & Zaritsky (2004) (right).

G. Bruzual acknowledges support from the National Autonomous University of México through grants IA102311 and IB102212.

References

Aringer, B. *et al.* 2009, *A&A*, 503, 913

Bertelli, G. *et al.* 2008, *A&A*, 484, 815

Bruzual A., G. 2002, in Extragalactic Star Clusters, IAU Symposium Ser., Vol. 207, eds. D. Geisler, E. K. Grebel, & D. Minniti, Provo:ASP, 616

Bruzual, G. 2010, *RSPTA*, 368, 783

Bruzual, G. 2007, in Proceedings of the IAU Symposium No. 241 "Stellar populations as building blocks of galaxies", eds. A. Vazdekis and R. Peletier, Cambridge: Cambridge University Press, 125 (arXiv:astro-ph 0703052)

Bruzual, G. 2011, in Proceedings of the XIII LARIM, eds. W. J. Henney and S. Torres-Peimbert, *Rev. Mex. Astron. Astrofis., Conf. Ser.*, 40, 36-41

Bruzual, G. & Charlot, S. 2003, *MNRAS*, 344, 1000 (BC03)

Charlot, S. & Bruzual, G. 2007, unpublished, models distributed on demand (CB07)

Girardi, L. *et al.* 2010, *ApJ*, 724, 1030

González-Lópezlira, R. *et al.* 2010, *MNRAS*, 403, 1213

Harris, J. & Zaritsky, D. 2004, *AJ*, 127,153

Harris, J. & Zaritsky, D. 2009, *AJ*, 138,1243

Kriek M. *et al.* 2010, *ApJ*, 722, L64

Lançon, A. & Mouhcine, M. 2002, *A&A*, 393, 167

Le Borgne, J.-F. *et al.* 2003, *A&A*, 402, 433

Maraston C. 2005, *MNRAS*, 362, 799 (M05)

Marigo, P. & Girardi, L. 2007, *A&A*, 469, 239

Marigo, P. *et al.* 2008, *A&A*, 482, 883

Melbourne, J. *et al.* 2012, *ApJ*, 748, 47

Rayner, J. T. *et al.* 2009, *ApJS*, 185, 289

Sánchez-Blázquez, P. *et al.* 2006, *MNRAS*, 371, 703

Srinivasan, S. *et al.* 2009, *AJ*, 137, 4810

Valdes, F. *et al.* 2004, *ApJS*, 152, 251

Westera, P. *et al.* 2002, *A&A*, 381, 524

Zibetti, S. *et al.* 2013, *MNRAS*, submitted (arXiv:1205.4717)

The intriguing life of massive galaxies
Proceedings IAU Symposium No. 295, 2012
D. Thomas, A. Pasquali & I. Ferreras, eds.

© International Astronomical Union 2013
doi:10.1017/S1743921313005073

Galaxy spectra from the UV to the mid-IR

Michael J. I. Brown[1], John Moustakas[2], J.-D. T. Smith[3], Elisabete da Cunha[4], Masatoshi Imanishi[5,6], Lee Armus[7], and Bernhard R. Brandl[8]

[1] School of Physics, Monash University, Clayton, Victoria 3800, Australia
email: `Michael.Brown@monash.edu`
[2] Department of Physics and Astronomy, Siena College, 515 Loudon Road, Loudonville, NY 12211-1462, USA
[3] Department of Physics and Astronomy, University of Toledo, Ritter Obs., MS #113, Toledo, OH 43606, USA
[4] Max Planck Institute for Astronomy, Königstuhl 17, 69117 Heidelberg, Germany
[5] Subaru Telescope, 650 North A'ohoku Place, Hilo, HI 96720, USA
[6] Department of Astronomy, School of Science, Graduate University for Advanced Studies (Sokendai), Mitaka, Tokyo 181-8588, Japan
[7] Spitzer Science Center, California Institute of Technology, Pasadena, CA 91125, USA
[8] Leiden Observatory, Leiden University, P.O. Box 9513, 2300 RA Leiden, The Netherlands

Abstract. Templates and models of galaxy spectra are often essential for determining the physical properties of galaxies. Many commonly used templates have large systematic errors, which significantly impact photometric redshifts and k-corrections. We present a new library of 110 galaxy template spectra spanning from the ultraviolet to the mid-infrared. The templates combine optical, *Spitzer* and *Akari* spectra with MAGPHYS models, all normalised and verified with matched aperture photometry. Our library contains more galaxies, spans a broader range of colours and has smaller systematic errors than previous libraries of galaxy spectra.

Keywords. galaxies: general, galaxies: fundamental parameters, galaxies: distances and redshifts, infrared: galaxies

1. Introduction

Libraries and models of galaxy spectra are often essential for determining rest-frame luminosities and photometric redshifts. The Coleman *et al.* (1980) and Kinney *et al.* (1996) templates have proved exceptionally useful over recent decades, but have limited wavelength coverage. With the advent of the *Spitzer* Space Telescope, *Akari* (Astro-F) and the Wide-field Infrared Space Explorer (WISE), it became both important and possible to extend galaxy spectral templates into the mid-infrared. Templates from Polletta *et al.* (2008) and others now include the emission from dust and polycyclic aromatic hydrocarbons (PAHs), but systematic errors remain that can overwhelm random errors when determining k-corrections and photometric redshifts.

Galaxy spectral energy distributions can be modelled, using models of stellar populations, nebular emission lines, dust obscuration and dust emission. The models have improved considerably over the decades, and in some instances do a remarkably good job of reproducing galaxy spectra. However, the models can have a large number of free parameters and cannot be expected to reproduce the spectral energy distribution of a galaxy with a complex star formation history when only limited photometry is available. This being the case, there is still a role for spectral templates derived from real galaxies.

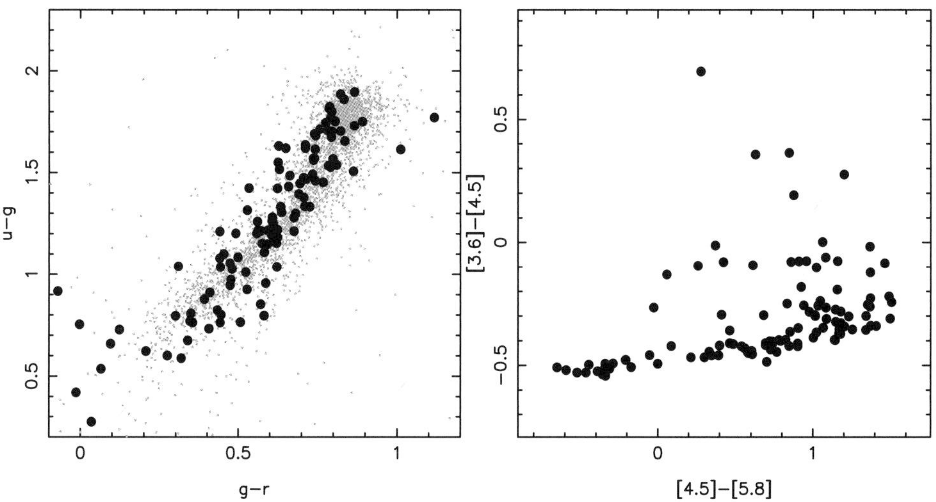

Figure 1. The optical and Spitzer infrared colours of the sample galaxies. While many galaxies do fall on a narrow galaxy locus, there are exceptions including star forming galaxies with strong nebular emission lines and some luminous infrared galaxies.

2. A New Library of Galaxy Spectra

We present a new library template spectra of galaxies spanning from the ultraviolet to the mid-infrared. Our sample consists of local galaxies that have drift-scan optical spectroscopy from Moustakas & Kennicutt (2006), Moustakas *et al.* (2010), Kennicutt (1992) or Gavazzi *et al.* (2004). With the exception of a few elliptical galaxies, almost all of our galaxies also have spectra from *Spitzer's* Infrared Spectrograph (IRS). We make use of International Ultraviolet Explorer (IUE) and *Akari* Infrared Camera (IRC) spectroscopy when they capture a high fraction of the total flux and have high signal-to-noise. We used published spectra when possible (e.g., Smith *et al.* 2007, Imanishi *et al.* 2010), but for some galaxies we reduced data from the *Spitzer* and *Akari* archives.

Matched aperture photometry is critical for normalising and verifying our spectra. We make use of imaging from GALEX Data Release 6 (GR6), XMM-Newton optical/UV monitor telescope, Sloan Digital Sky Survey III (SDSS-III), Two Micron All Sky Survey (2MASS), Wide-field Infrared Survey Explorer (WISE) and *Spitzer*. The aperture is typically matched to the aperture used for the optical spectroscopy. Most galaxies are bright, so we expect systematic errors to dominate over random errors, but for the faintest galaxies we measure uncertainties using 24 sky apertures placed around the target galaxy.

We plot optical and infrared colour-colour diagrams for our galaxies in Figure 1. While many galaxies lie on a tight stellar locus, there are notable exceptions. Galaxies with strong nebular emission lines produce a wedge at the bottom-left of the optical colour-colour diagram. While the Rayleigh-Jeans tail dominates the spectra of most galaxies at $\sim 4\ \mu$m, some galaxies have strong absorption features and hot dust emission. Libraries with a limited number of spectra may not capture the true diversity of local galaxies.

We normalise and verify our spectra with the matched aperture photometry. For compact galaxies with stare mode spectra from *Spitzer's* IRS, we use the photometry to correct for the varying fraction of light that passes through the spectrograph slit as a function of wavelength. To fill the gaps in our spectral coverage, we use MAGPHYS (da Cunha, Charlot & Elbaz 2008) models constrained with our photometry. As MAGPHYS does not include nebular emission lines, we correct our photometry for the contribution of emission lines (using our spectra) before fitting models to the data. In Figure 2 we

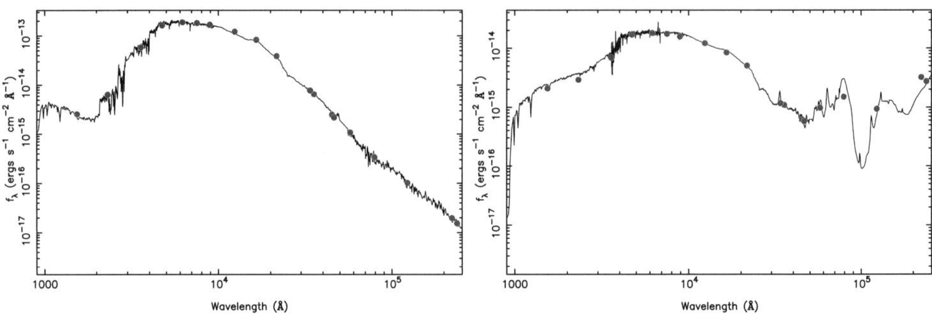

Figure 2. The spectral energy distributions of NGC 584 (left) and IC 4553 (Arp 220; right). The photometry used to normalise and verify the spectra are shown with the blue dots.

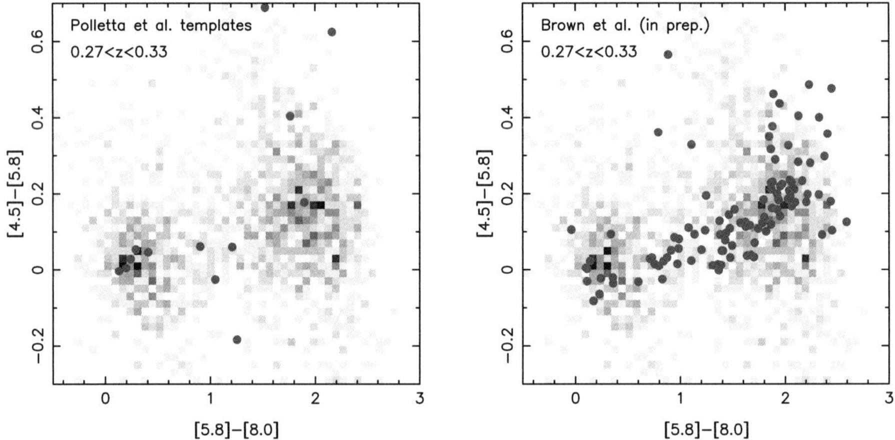

Figure 3. A comparison of infrared colours of $z_{spec} \sim 0.3$ galaxies with the templates of Polletta *et al.* (2008) (left panel) and our templates (right panel; Brown *et al.* in prep.). The loci of passive and star forming galaxies are evident to the left and right of each panel respectively. Our 110 template spectra clearly follow the observed galaxy locus and there are few outliers.

plot example spectra for NGC 584 and IC 4553 (Arp 220). While NGC 584 falls on the galaxy locus in the optical and infrared, IC 4553 has a redder than normal $[3.6] - [4.5]$ colour due to ice absorption and hot dust emission (Imanishi *et al.* 2010).

In Figure 3 we compare our templates and the Polletta *et al.* (2008) templates with the observed infrared colours of $z_{spec} \sim 0.3$ galaxies in Bootes (Jannuzi & Dey 1999, Kochanek *et al.* 2012). While several Polletta *et al.* (2008) templates populate the passive galaxy locus, only one template falls in the centre of the star forming galaxy locus. With over 100 templates, we define a very tight galaxy locus with only a small number of outliers. As our outliers have had their spectral energy distributions cross checked with matched aperture photometry, we are confident they represent real galaxies rather than errors in the spectra.

Our primary goal is to provide improved template spectral energy distributions for photometric redshifts and k-corrections. In Figure 4 we plot the rest-frame optical colour-magnitude diagrams for $z_{spec} \sim 0.5$ galaxies in Bootes determined using the Polletta *et al.* (2008) templates and our templates. To determine the rest-frame luminosities we used the method of Rudnick *et al.* (2003), where the spectral energy distributions are used to interpolate between two observed filters that bracket the relevant rest-frame filters. While this method reduces the impact of systematic errors on the rest-frame colours, we

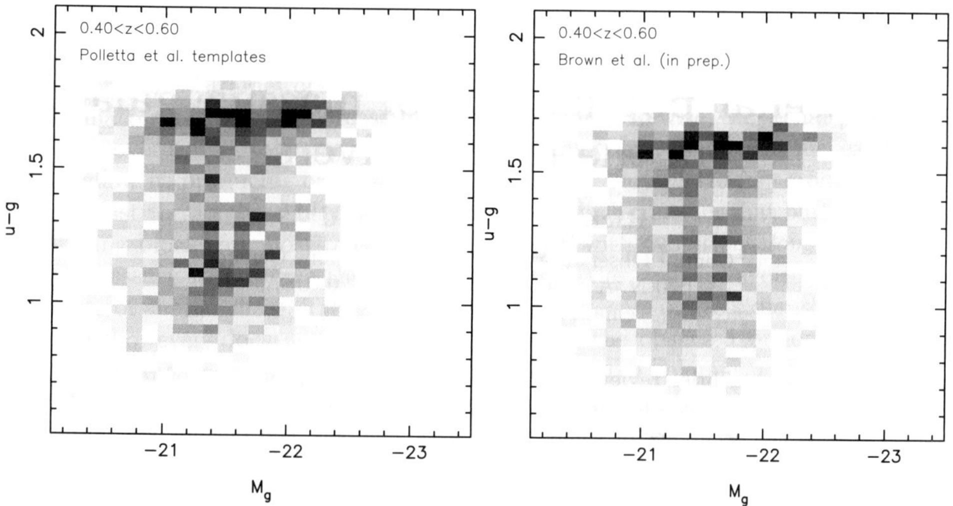

Figure 4. The rest-frame optical colour-magnitude diagrams of $z_{spec} \sim 0.5$ galaxies in the Bootes field. In the left panel the Polletta *et al.* (2008) templates were used to derive the rest-frame colours while in the right panel our templates were used. Despite using the method of Rudnick *et al.* (2003), which reduces the impact of systematic errors on k-corrections, we do observe significant differences when using different libraries of template spectra.

still see significant differences between the k-corrections derived from the two template libraries. Relative to the Polletta *et al.* (2008) templates, our templates shift the red sequence slightly blueward and produce a more dispersed locus of star forming galaxies.

3. Summary

Motivated by systematic errors in existing libraries of galaxy template spectra, we have developed a new library of 110 template spectra. The library contains more spectra, spans a broader wavelength range, has smaller uncertainties and has smaller systematic errors than comparable libraries. Preliminary tests indicate that the new library produces improved photometric redshifts and k-corrections. We find that galaxy k-corrections have a significant dependence on the libraries of template spectra used, even when applying methods that should mitigate the impact of systematic errors on the k-corrections.

References

Coleman, G. D., Wu, C.-C., & Weedman, D. W. 1980, *ApJS*, 43, 393
da Cunha, E., Charlot, S., & Elbaz, D. 2008, *MNRAS*, 388, 1595
Gavazzi, G., Zaccardo, A., Sanvito, G., Boselli, A., & Bonfanti, C. 2004, *A&A*, 417, 499
Imanishi, M., Nakagawa, T., Shirahata, M., Ohyama, Y., & Onaka, T. 2010, *ApJ*, 721, 1233
Jannuzi, B. T. & Dey, A. 1999, Photometric Redshifts and the Detection of High Redshift Galaxies, *ASP Conference Series*, 191, 111
Kennicutt, R. C., Jr. 1992, *ApJS*, 79, 255
Kinney, A. L., Calzetti, D., Bohlin, R. C., *et al.* 1996, *ApJ*, 467, 38
Kochanek, C. S., Eisenstein, D. J., Cool, R. J., *et al.* 2012, *ApJS*, 200, 8
Moustakas, J. & Kennicutt, R. C., Jr. 2006, *ApJS*, 164, 81
Moustakas, J., Kennicutt, R. C., Jr., Tremonti, C. A., *et al.* 2010, *ApJS*, 190, 233
Polletta, M., Omont, A., Berta, S., *et al.* 2008, *A&A*, 492, 81
Rudnick, G., Rix, H.-W., Franx, M., *et al.* 2003, *ApJ*, 599, 847
Smith, J. D. T., Draine, B. T., Dale, D. A., *et al.* 2007, *ApJ*, 656, 770

The intriguing life of massive galaxies
Proceedings IAU Symposium No. 295, 2012
D. Thomas, A. Pasquali & I. Ferreras, eds.

© International Astronomical Union 2013
doi:10.1017/S1743921313005103

Stellar population gradients in GASS

Jonas Johansson[1], Guinevere Kauffmann[1] and Sean Moran[2]

[1] Max-Planck Institut für Astrophysik, Karl-Schwarzschild-Str. 1,
D-85741 Garching, Germany
[2] Department of Physics and Astronomy, The Johns Hopkins University,
3400 N. Charles Street, Baltimore, MD 21218, USA

Abstract. We study relationships between the stellar populations and interstellar medium in massive galaxies using the Galex Arecibo SDSS Survey (GASS). The sample consists of HI-observations ($\sim$1000 galaxies) and complementary H_2-observations (330 galaxies) and long-slit spectroscopy (230 galaxies). Luminosity-weighted stellar population ages, metallicitites and element abundance ratios, are derived by fitting stellar population models of absorption line indices. We find that the ages correlate more strongly with molecular gas fraction (M_{H_2}/M_*) than with neutral Hydrogen fraction (M_{HI}/M_*). This result strengthens the theory that H_2 is a better tracer of star-formation than HI. The sample is dominated by negative metallicity-gradients and flat Mg/Fe-gradients. Galaxies with high M_{H_2}/M_*-ratios show in general flat or weakly negative age-gradients. For low M_{H_2}/M_*-ratios the age-gradients are overall negative. These results are in agreement with the inside-out galaxy formation scenario. For galaxies with high r_{90}/r_{50}-ratios, a sub-population show positive age-gradients indicating additional formation channels. Furthermore, for galaxies with high M_{H_2}/M_*-ratios more massive systems have older stellar populations in their centers, suggesting downsizing within the inside-out formation scenario.

Keywords. galaxies: evolution – galaxies: elliptical and lenticular, cD – galaxies: spiral – galaxies: stellar content

1. Introduction

The sizes of galaxies, i.e., the effective radii, have been found to increase with decreasing redshift, suggesting an inside-out formation scenario (van Dokkum *et al.* 2010). Trujillo *et al.* (2007) show that this occurs for both disc galaxies as well as spheroids for redshifts up to $z \sim 2$. A possible mechanism for size growth is minor merging. Szomoru *et al.* (2012) found that this mechanism is responsible for size evolution of elliptical galaxies since $z \sim 2$ through the accretion of satellite galaxies, adding stellar content to the outskirts of the galaxies. Marmol-Queralto *et al.* (2013) find that satellites typically have younger stellar ages than the central galaxies. Hence, the accretion of such systems would produce negative stellar age-gradients. On the other hand, several authors have found flat or positive age-gradients in elliptical galaxies (e.g. La Barbera *et al.* 2012).

Cold gas accretion is an additional way of driving galaxy growth, e.g. Dekel & Birnboim (2006) predict disc growth through cold streams. Kauffmann *et al.* (2006) found star-formation in regions of high stellar density, fed by cold gas accretion. Wuyts *et al.* (2012) also found enhanced star-formation in regions of high stellar density, located at off center radii and consequently older stellar ages in the centers of star-forming galaxies. These results indicate negative age-gradients in the cold gas accretion scenario.

In this work we derive stellar population gradients, including age, metallicity and element abundance ratios, for a sample of 182 galaxies drawn from the GALEX Arecibo SDSS Survey (GASS, Catinella *et al.* 2010). This allows us to study the stellar population gradients in relation to neutral and molecular Hydrogen mass fractions for an unbiased sample of massive galaxies.

2. Data

2.1. *GASS, COLD GASS and long-slit spectroscopy*

GASS comprises observations of the 21cm line of neutral Hydrogen, obtained with the Arecibo radio telescope. In the final stage, observations for an unbiased sample of $\sim$1000 galaxies, in the redshift range $0.025 < z < 0.05$ and with stellar masses greater than 10^{10} $M_\odot$, will be gathered. The sample was selected using the Sloan Digital Sky Survey (SDSS) spectroscopic and Galaxy Evolution Explorer (GALEX) imaging surveys.

In addition to GASS, COLD GASS (Saintonge *et al.* 2011) constitutes molecular gas masses determined through observations of the CO (J = 1 - 0) line using the IRAM 30m telescope. COLD GASS is a sub-sample of GASS and consists of 330 galaxies. Furthermore, Moran *et al.* (2012) obtained long-slit spectroscopy for a sub-sample of 182 galaxies with the Blue Channel Spectrograph on the 6.5m MMT telescope on Mt. Hopkins, AZ, and 51 galaxies at the Apache Point Observatory (APO). In this work we focus on the former observations that were obtained through a 1.25" wide slit in the wavelength range 3900-7000 Å and at a spectral resolution of $\sim$4 Å FWHM. Each two-dimensional, rotation-curve corrected spectrum has been binned adaptively in the spatial direction until a signal-to-noise (S/N) of 15 is typically reached for a maximum bin size of 3". These limits were set to change with radius to ensure a balance between S/N and bin size for each co-added spectrum.

2.2. *Stellar population parameters*

We utilise the SED-fitting code `GANDALF` (Sarzi *et al.* 2006), which is based on the penalised pixel-fitting (`PPXF`) method of Cappellari & Emsellem (2004), to obtain clean absorption spectra. These are used for deriving the Lick absorption line indices (Worthey & Ottaviani 1997; Trager *et al.* 1998), defined for 25 prominent absorption features in the optical. The resolution of the spectra are downgraded to the Lick/IDS resolution ($\sim$8-10 Å, Worthey & Ottaviani 1997) prior to measuring the indices. The index measurements are corrected for velocity dispersion broadening.

From the derived Lick indices we determine the stellar population parameters age and total metallicity and the element abundance ratios accessible through integrated light spectroscopy of galaxies (O/Fe, Mg/Fe, C/Fe, N/Fe, Ca/Fe and Ti/Fe). Using the method described in detail in Johansson *et al.* (2012), we fit the stellar population models of absorption line indices from Thomas *et al.* (2011). These are single stellar population models with variable element abundance ratios for the 25 Lick indices.

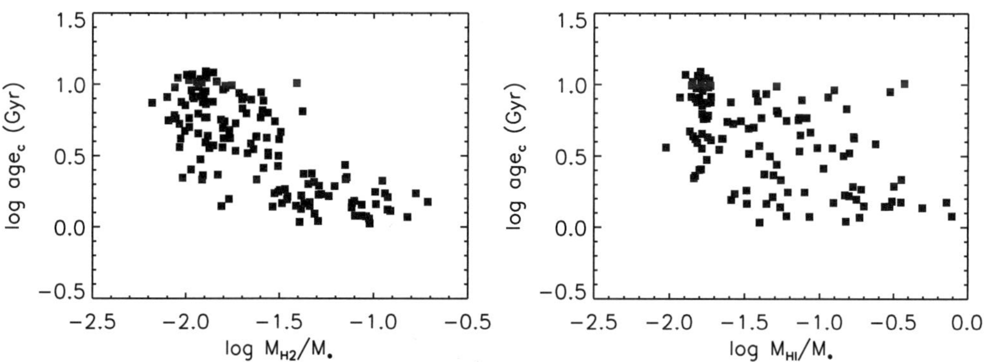

Figure 1. Stellar age within r<0.4R$_{50}$ versus M_{H_2}/M_* (left panel) and $M_{H\,I}/M_*$ (right panel).

3. Results and Discussion

We calibrate the Lick indices derived from the long-slit spectra to those derived from the corresponding SDSS spectra and find in general a very good agreement. Derived central stellar population ages from stacked spectra within $r < 0.4r_{50}$ for each galaxy (age_c) are compared with molecular gas fraction (left panel) and neutral Hydrogen fraction (right panel) in Fig. 1. It is clear that the ages correlate stronger with molecular gas fraction than with neutral Hydrogen fraction. This result strengthens the theory that molecular gas fraction is a better tracer of star formation than neutral hydrogen.

We compute the residuals between the stellar population parameters derived from central stacks ($r < 0.4r_{50}$) and those derived from stacked spectra of outer radial bins ($r > 0.4r_{50}$). The residuals are compared to the molecular gas fractions and we find in general higher metallicities in the central stacks, suggesting negative metallicity gradients. The [Mg/Fe] residuals instead suggest overall flat gradients. Fig. 2 shows the residuals in stellar population age as a function of molecular gas fraction. We find that galaxies with high molecular gas fractions show slightly older stellar ages in the central parts. As we go towards lower gas fractions the spread becomes larger, but the majority of the galaxies show residuals suggesting negative age-gradients. Galaxies with $r_{90}/r_{50} > 3$ (filled squares in Fig. 2), i.e., the most bulge-dominated objects, have in general low molecular gas fractions, as expected, and show the largest spread in the age-residuals. This result indicates a wide range of age-gradients and possibly different formation histories.

We further stack the galaxies in three bins of molecular gas fraction together with six radial bins (r/r_{50}). The derived stellar population parameters for these stacks are shown in Fig. 3. It is clear that the overall stellar ages, metallicities and Mg/Fe ratios increase with decreasing molecular gas fraction. It should be noted that the average stellar mass also increase slightly with decreasing molecular gas fraction.

Fig. 3 shows negative luminosity-weighted age-gradients that increase in strength for lower molecular gas fractions. This result is consistent with the inside-out galaxy formation scenario (van Dokkum *et al.* 2010). We can further see that all bins of molecular gas fraction show steep metallicity gradients, but flat Mg/Fe-gradients. In the inside-out formation scenario we can understand this if stars keep forming from recycled, in-falling gas with high Mg/Fe ratios and lower total metallicities and/or accretion of stellar systems with these characteristics. The bottom right panel of Fig. 3 shows the results from stacking objects in two bins of molecular gas fraction together with two mass-bins. These results indicate that more massive, high molecular gas fraction galaxies have ceased

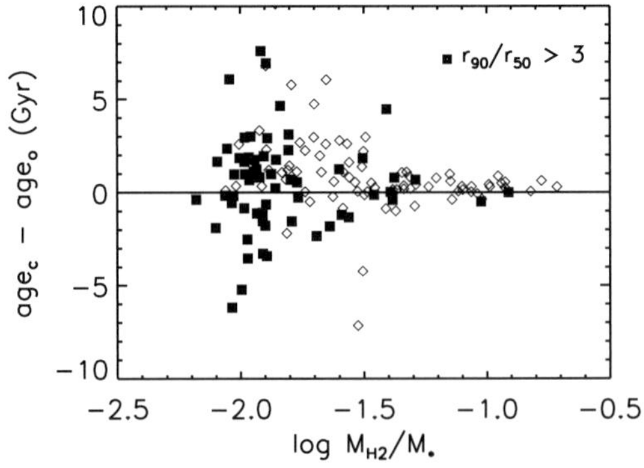

Figure 2. Central stellar age (age_c, $r < 0.4r_{50}$) minus outer stellar age (age_o, $r > 0.4r_{50}$) as a function of M_{H_2}/M_*. Solid square are galaxies with $r_{90}/r_{50} > 2.5$.

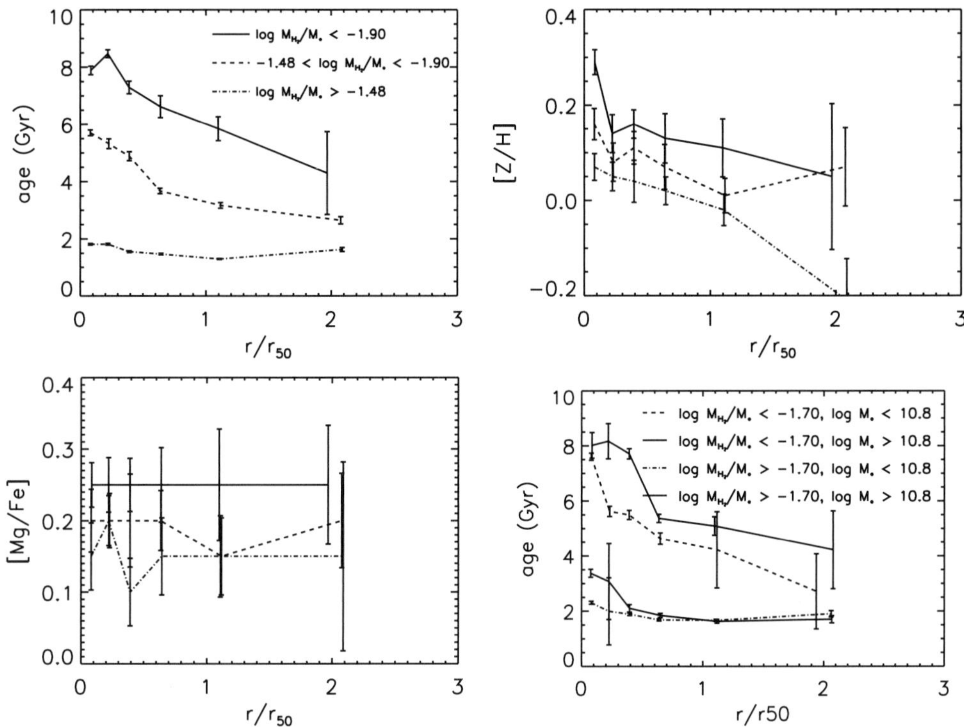

Figure 3. Age- (top left), Z/H- (top right) and Mg/Fe-gradients (bottom left) of stacked galaxies in three bins of M_{H_2}/M_* given by the labels in the top left panel. Bottom right: age-gradients in bins of M_{H_2}/M_* and stellar mass as given by the labels.

forming stars in the central parts, suggesting downsizing within the inside-out formation scenario.

References

Cappellari, M. & Emsellem, E., 2004, *PASP*, 116, 138

Dekel, A. & Birnboim, Y., 2006, *MNRAS*, 368, 2

Johansson, J., Thomas, D., & Maraston, C., 2012, *MNRAS*, 421, 1908

Kauffmann, G., *et al.*, 2006, *MNRAS*, 367, 1394

La Barbera, *et al.*, 2012, *MNRAS*, 426, 2300

Marmol-Queralto, E., Trujillo, I., Villar, V., Barro, G., & Perez-Gonzalez, P., 2013, *MNRAS*, 429, 792

Moran, S., *et al.*, 2012, *ApJ*, 745, 66

Catinella, B., 2010, *MNRAS*, 403, 683

Saintonge, A., Kauffmann, G., Wang, J., *et al.* 2011, *MNRAS*, 415, 61

Sarzi, M., *et al.*, 2006, *MNRAS*, 366, 1151

Szomoru, D., Franx, M., & van Dokkum, P., 2012, *ApJ*, 749, 121

Thomas, D., Maraston, C., & Johansson, J., 2011, *MNRAS*, 412, 2183

Trager, S. C., Worthey, G., Faber, S. M., Burstein, D., & Gonzalez, J. J., 1998, *ApJS*, 116, 1

Trujillo, I., *et al.*, 2007, *MNRAS*, 382, 109

van Dokkum, P. G., *et al.*, 2010, *ApJ*, 709, 1018

Worthey, G. & Ottaviani, D. L., 1997, *ApJS*, 111, 377

Wuyts, S., 2012, *ApJ*, 753, 114

The intriguing life of massive galaxies
Proceedings IAU Symposium No. 295, 2012
D. Thomas, A. Pasquali & I. Ferreras, eds.

© International Astronomical Union 2013
doi:10.1017/S1743921313005115

Optical and near-infrared color distributions of the NGC 4874 globular cluster system

Hyejeon Cho[1], John P. Blakeslee[2], Eric W. Peng[3,4] and Young-Wook Lee[1]

[1]Department of Astronomy & Center for Galaxy Evolution Research, Yonsei University, Seoul 120-749, Korea; email: hyejeon@yonsei.ac.kr

[2]Dominion Astrophysical Observatory, Herzberg Institute of Astrophysics, National Research Council of Canada, Victoria, BC V9E 2E7, Canada

[3]Department of Astronomy, Peking University, Beijing 100871, China

[4]Kavli Institute for Astronomy and Astrophysics, Peking University, Beijing 100871, China

Abstract. Examining both optical and optical-infrared color distributions of the globular cluster (GC) systems in large elliptical galaxies is the key to study how non-linearities in the color-metallicity relations of their GC systems are linked to bimodal optical color distributions. In order to do this for the core of the Coma cluster of galaxies (Abell 1656), centered on the giant elliptical galaxy NGC 4874, we have combined F160W (H_{160}) near-infrared (NIR) imaging data acquired with the Wide Field Camera 3 IR Channel (WFC3/IR), installed on *Hubble Space Telescope* (*HST*) in 2009, with F475W (g_{475}) and F814W (I_{814}) optical imaging data from the *HST* Advanced Camera for Surveys (ACS). Since optical-NIR color distributions of extragalactic GC systems reflect the underlying features of the metallicity distributions, we have probed not only optical $g_{475}-I_{814}$ and optical-NIR $I_{814}-H_{160}$ color distributions but also the color-color relation for this GC system. The features of these color distributions have been quantitatively analyzed using the Gaussian Mixture Modeling code. We find that brighter GCs have a much redder mean color than fainter ones. The optical color distribution of the GC system in the Coma cluster core shows the typical bimodality, while the evidence for bimodality is significantly weaker in the optical-NIR color distribution.

Keywords. galaxies: elliptical and lenticular, cD, galaxies: individual (NGC 4874), galaxies: star clusters

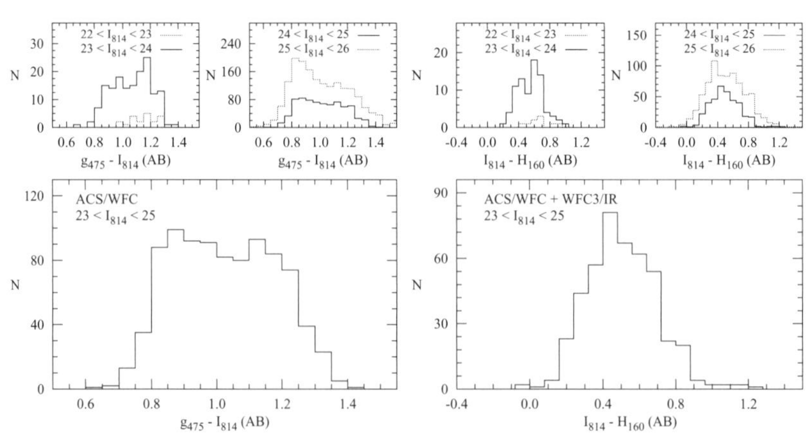

Figure 1. Top panels: The ACS $g_{475}-I_{814}$ and ACS+WFC3/IR $I_{814}-H_{160}$ color distributions for different I_{814} magnitude ranges for matched GC candidates across the bands. Bottom panels: Same as top panels but in the broader magnitude range and with color errors less than 0.2 mag.

The intriguing life of massive galaxies
Proceedings IAU Symposium No. 295, 2012
D. Thomas, A. Pasquali & I. Ferreras, eds.

© International Astronomical Union 2013
doi:10.1017/S1743921313005127

Galactic structure studies with SCUSS and SDSS surveys

Cuihua Du[1] and Xiyan Peng[1,2]

[1] College of Physical Sciences, University of the Chinese Academy of Sciences,
Beijing 100049, China
email: ducuihua@ucas.ac.cn

[2] National Astronomical Observatories, Chinese Academy of Sciences, Beijing 100012, China

Abstract. Based on the South Galactic Cap U-band Sky Survey (SCUSS) and SDSS observations, we adopted the star-count method to analyze the stellar distribution in different directions of the Galaxy. We find that the scale height of the disk may be variable with the observed direction, which cannot simply be attributed to statistical errors. The main reason can be possibly attributed to the disk (mainly the thick disk) being flared, with a scale height increasing with radius. The axis ratio of the Galactic halo is in the range $0.4 - 0.6$. This finding supports Galactic models with a flattened inner halo, partly formed through a merger early in the Galaxy's history.

Keywords. Galaxy: fundamental parameters, Galaxy: halo, Galaxy: stellar component, Galaxy: structure

The star-count analysis has provided a picture of the basic structural and stellar populations of the Galaxy. Up to now, the basic stellar components of the Milky Way are the thin disk, thick disk, stellar halo and central bulge, albeit that the inter-relationships and distinction amongst different components remain subject to some debate. In this study, we use the South Galactic Cap U-band Sky Survey (SCUSS) and SDSS observations to study the Galactic structure. We will report on our investigation of star counts, extending our analysis to include different directions within the Galaxy. The fields used in this paper are towards the Galactic centre and the anticentre direction ($0° < l < 180°$) at the South Galactic Cap, $|b| > 45°$. Since star counts at high Galactic latitudes are not strongly related to the radial distribution, they are well suited to study the vertical distribution of the Galaxy. The SCUSS magnitude u and SDSS magnitudes (Juric *et al.* 2008) g, r, i, and z were used for a total number of 4 039 908 stars in 18 fields (320 deg ^{2} in total). Here, our work is restricted to the magnitude range $15 < g < 23$ for the study of the Galactic model parameters. The adopted model of stellar density distribution in this report includes only two disks (thin disk and thick disk) and a halo. We have chosen to use an exponential functional form for the disk and a de Vaucouleurs law for the halo. We try to derive the structural parameters (e.g. scale height) of the thin and thick disk populations using our data set. For this we calculate the stellar space density as a function of distance from the Galactic plane.

From the χ^2 fit to the direct measurement of the stellar density distribution, we determine that the range of scale height for the thin disk varies from 180 to 250 pc. Although 180 pc seems an extreme value, it is close to the lower limit in the literature. The range of scale height for the thick disk is from 400 to 800 pc, and the corresponding space number density normalization is $14.0 - 5.0\%$ of the thin disk. Our results show that the scale height is variable with the observation direction, which cannot be attributed to statistical errors. The main reason can possibly be attributed to the disk (mainly the thick disk) being flared, with a scale height that increases with radius. This is consistent with the merger origin for the thick disk formation (Du *et al.* 2003, 2006, 2008). For the halo, we derive an estimate of its large-scale flattening and find that the axial ratio towards the Galactic center is more flatter (~ 0.4), while the shape of the halo in the anticentre direction is rounder with $c/a > 0.4$. These results prove that the axial ratio of the halo may vary with distance and observation direction. It reflects the shape of the inner halo. In a word, the star counts for different lines of sight can be used directly to obtain a rough estimate of the shape of the stellar halo. Our solutions support the Galactic models with a flattened inner halo, possibly formed in part through a merger early in the Galaxy's history.

References

Du, C. H., Zhou, X., Ma, J., *et al.* 2003, *A&A*, 407, 541
Du, C. H., Ma, J., Wu, Z. Y., & Zhou, X. 2006, *MNRAS*, 72, 1304
Du, C. H., Wu, Z. Y., Ma, J., & Zhou, X. 2008, *ChJAA*, 5, 566
Juric, M., Ivezic, Z., Brooks, A., *et al.* 2008, *ApJ*, 673, 864

The intriguing life of massive galaxies
Proceedings IAU Symposium No. 295, 2012
D. Thomas, A. Pasquali & I. Ferreras, eds.

© International Astronomical Union 2013
doi:10.1017/S1743921313005139

Probing the Build-Up of Stellar Mass in the Center of IR Luminous Major Mergers with HST

Sebastian Haan[1], Jason Surace[2], Lee Armus[2] and Aaron Evans[3,4]

[1] CSIRO Astronomy & Space Science, Marsfield NSW 2122, Australia, email:
sebastian.haan@csiro.au
[2] Spitzer Science Center, California Institute of Technology, Pasadena, CA 91125, USA
[3] National Radio Astronomy Observatory, Charlottesville, VA 22903, USA
[4] Department of Astronomy, University of Virginia, Charlottesville, VA 22904, USA

Abstract. Interactions and mergers are important drivers of galaxy evolution, transform spiral galaxies into massive ellipticals, and fuel both powerful starbursts and massive nuclear black holes. In particular one galaxy population, namely Luminous Infrared Galaxies (LIRGs), are believed to be responsible for most of the star formation that happened in the history of the universe (see e.g. Le Floch *et al.* 2005, Caputi *et al.* 2007, Magnelli *et al.* 2009), and hence represent a critical phase in the evolution of galaxies where most of the galaxies mass is building up. During a merger process, violent relaxation acts on stars present in gas-rich progenitor disks, while the centers are structured by the relics of dissipational, compact starbursts, imprinting a central "extra light" component or "cusp" into the surface brightness profiles of merger remnants. Our *HST* NICMOS/WFC3 imaging program of the 88 most luminous LIRGs in the Great Observatories Allsky LIRG Survey (GOALS, see Armus *et al.* 2009) shows that the central luminosity surface density in nearby LIRGs increases significantly along the merger sequence, indicating that the gas inflow fuels a central starburst and subsequently builds a compact stellar cusp (Haan *et al.* 2011). A large fraction of all galaxies in our sample possess double or multiple nuclei ($\sim$63%). Half of these double nuclei are not visible in the *HST* B-band images due to dust obscuration, which implies strong limitations on the ability to detect the true nuclear structures of luminous infrared galaxies at high-redshift (z > 2) and may explain some of the apparent discrepancy of the LIRG population and merger ratio between local and high-redshift galaxies. We find that ULIRGs ($\log[L_{IR}/L_\odot] > 12.0$) have significantly smaller nuclear separations than LIRGs ($\log[L_{IR}/L_\odot] = 11.4 - 12.0$) with a median value of 1.2 kpc and 6.7 kpc, respectively. In our sample, merger (regardless of whether LIRG or ULIRG) seem to be prevalent at two time scales (based on the projected nuclear separation and mass ratio of the nuclei): First, at a remaining merger time scale of $0.3 < [t - t_{merg}] < 1.3$ Gyr (53% of mergers in our sample), and second, at $[t - t_{merg}] \sim 0$ (26%), likely representing the first passage of interacting galaxies and the final nuclear coalescence, respectively, with a post-merger time (starburst phase after the nuclei merged) of roughly 300 Myrs.

Keywords. galaxies: interactions, galaxies: nuclei, galaxies: starburst, galaxies: structure, cosmology: observations

---

References

Armus, L., *et al.* 2009, *PASP*, 121, 559
Caputi, K. I., *et al.* 2007, *ApJ*, 660, 97
Haan, S., Surace, J. A., Armus, L., *et al.* 2011, *AJ*, 141, 100
Le Floch, E., *et al.* 2005, *ApJ*, 632, 169
Magnelli, B., Elbaz, D., Chary, R. R., Dickinson, M., Le Borgne, D., Frayer, D. T., & Willmer, C. N. A. 2009, *AAP*, 496, 57

The intriguing life of massive galaxies
Proceedings IAU Symposium No. 295, 2012
D. Thomas, A. Pasquali & I. Ferreras, eds.

© International Astronomical Union 2013
doi:10.1017/S1743921313005140

Bayesian analysis of galaxy spectral energy distributions with BayeSED

Yunkun Han[1,2] and Zhanwen Han[1,2]

[1]National Astronomical Observatories / Yunnan Observatory, Chinese Academy of Sciences, 100012, Beijing, China
email: `hanyk@ynao.ac.cn`

[2]Key Laboratory for the Structure and Evolution of Celestial Objects, Chinese Academy of Sciences, 650011, Kunming, China
email: `zhanwenhan@ynao.ac.cn`

Abstract. In Han & Han (2012), we have preliminarily built BayeSED and applied it to a sample of hyperluminous infrared galaxies. The physically reasonable results obtained from Bayesian model comparison and parameter estimation show that BayeSED could be a useful tool for understanding the nature of complex systems, such as dust obscured starburst-AGN composite galaxies, from decoding their complex SEDs. In this contribution, we present a more rigorous test of BayeSED by making a mock catalog from model SEDs with the value of all parameters to be known in advance.

Keywords. galaxies: active, galaxies: ISM, methods: data analysis, methods: statistical

---

1. Introduction

As a first simple example, we have chosen a black body as the model SED, since its shape only depends on the temperature. For such a simple model, the ANN and PCA methods that are used in BayeSED are completely unnecessary. However, we can use this simple model to test the reliability of these methods.

Firstly, we have built a SED library of black body spectra with temperatures (T [K]) in the range of [100:300] and the logarithm of luminosity (L [erg/s]) in the range of [40:48]. Then, as a common procedure of the application of BayeSED, we use the PCA method to reduce this library and train a 1:20:16 ANN with it. On the other hand, we have built a mock catalog by using SEDs randomly selected from the original library. These model SEDs are convolved with the transmission function of some selected filters to generate the fluxes in different bands. Then, noise with a Gaussian distribution is added to these fluxes to simulate the real observations. Finally, we use BayeSED to analyze the SEDs of this mock catalog to estimate the parameters values. We found that when the noise was selected from a Gaussian distribution with a standard deviation of 10% of the fluxes, the temperatures and logarithm of luminosities can be recovered with a R.M.S error of only 0.85 and 0.029, respectively.

Based on the current state-of-the-art Bayesian inference tool, we have built BayeSED for performing Bayesian analysis of galaxy multi-wavelength SEDs. The preliminary application of this code gives physically reasonable results. The more rigorous test by using a mock catalog presented here shows that BayeSED could be a reliable tool for understanding the nature of objects by decoding their complex SEDs. BayeSED will soon be publicly available at http://www.ynao.ac.cn/jgsz/kybm/dybhxyhyjtz/jj.

Reference

Han, Y. & Han, Z. 2012, *ApJ*, 749, 123

The intriguing life of massive galaxies
Proceedings IAU Symposium No. 295, 2012
D. Thomas, A. Pasquali & I. Ferreras, eds.

© International Astronomical Union 2013
doi:10.1017/S1743921313005152

The intriguing life of cD galaxies

S. N. Kemp[1], Víctor Hugo Ramírez-Siordia[2], and Ernesto Pérez-Hernández[1]

[1]Instituto de Astronomía y Meteorología , Universidad de Guadalajara
Av. Vallarta 2602, Col. Arcos Vallarta, 44130, Guadalajara, Jalisco, México
email: snk@astro.iam.udg.mx

[2]Centro de Radioastronomía y Astrofísica, Universidad Nacional Autónoma de México
Campus Morelia, Apartado Postal 3-72, 58090, Morelia, Michoacán, México

cD galaxies are supergiant elliptical galaxies found generally in the central parts of rich clusters, which have an extended halo-like component (envelope) in addition to the underlying de Vaucouleurs-Sérsic elliptical galaxy-like component. This envelope can extend to radial distances of > 500 kpc (Oemler 1976, Schombert 1988). There have been many theories to explain the formation of these envelopes. These include tidal stripping, where material is stripped from neighbouring galaxies; mergers and fusions, where the envelope is built up hierarchically by successive mergers with large and small galaxies; primordial origin, where the envelope is formed at the same time as the rest of the elliptical galaxy (which appears to be related to theories of early formation of the largest galaxies); and cooling flows: in clusters with X-ray emission there is often a minimum temperature in the centre interpreted as a flow of cooling gas towards the centre of the cluster, where the gas can cool sufficiently, forming stars. The colours of the stars in the envelopes will be affected by their process of formation and subsequent evolution.

We carried out a programme of deep surface photometry on a sample of 10 cD envelopes using the 2.1m and 1.5m telescopes at San Pedro Mártir, Baja California, México, obtaining 30-90 min total exposure per filter (Kemp *et al.* 2009). For the majority of the galaxies we obtained generally flat colour profiles within the errors, varying by no more than ± 0.1 mag in $B - V$. The profiles of 6 cDs do not show appreciable gradients. About 30% of the galaxies do have gradients, generally with bluer centres, implying some star formation activity may result from mergers (A2199) or cooling flows (A426). The flat profiles imply stellar populations with similar ages in the underlying galaxy and the envelope, which could favour the primordial origin or downsizing hypothesis, suggesting that any subsequent evolution that the galaxy experiences does not usually dominate its global colours. Variations of age and metallicity of the stellar populations with position could result in similar colours. To investigate this possibility, we are carrying out a programme of medium-resolution spectroscopy of cDs at the 2.1m in San Pedro Mártir and with OSIRIS/GTC to obtain age and metallicity of stellar populations. We are also carrying out a 2D analysis of surface brightness profiles with GALFIT to fit components to the underlying galaxy and envelopes. The main result so far is that the extended halo component frequently has a Sérsic index $n \sim 1$, i.e. an exponential fall-off with radius.

References

Kemp, S. N., Guzmán Jiménez, V., Ramírez Beraud, P., *et al.* 2009, *New Quests in Stellar Astrophysics II: Ultraviolet Properties of Evolved Stellar Populations, Springer*, 91
Oemler, A. Jr. 1976, *ApJ*, 209, 693
Schombert, J. M. 1988, *ApJ*, 328, 475

The intriguing life of massive galaxies
Proceedings IAU Symposium No. 295, 2012
D. Thomas, A. Pasquali & I. Ferreras, eds.

© International Astronomical Union 2013
doi:10.1017/S1743921313005164

Stellar discs in massive galaxies

D. Krajnović,[1] K. Alatalo,[2] L. Blitz,[2] M. Bois,[3] F. Bournaud,[4]
M. Bureau,[5] M. Cappellari,[5] R. L. Davies,[5] T. A. Davis,[1]
P. T. de Zeeuw,[1,6] E. Emsellem,[1,8] S. Khochfar,[7] H. Kuntschner,[1]
R. M. McDermid,[9] R. Morganti,[10] T. Naab,[11] M. Sarzi,[12] N. Scott,[13]
P. Serra,[10] A. Weijmans[14] and L. M. Young[15]

[1]ESO, Garching, Germany; [2]University of California, Berkeley, USA; [3]Observatoire de Paris,
France; [4]Université Paris Diderot, France; [5]University of Oxford, UK; [6]Leiden University, The
Netherlands; [7]MPE, Garching, Germany; [8]Université de Lyon, France; [9]Gemini Observatory,
Hilo, USA; [10]ASTRON, Dwingeloo, The Netherlands; [11]MPA, Garching, Germany;
[12]University of Hertfordshire, Hatfield, UK; [13]CAS, Swinburne University of Technology,
Australia; [14]University of Toronto Canada; [15]New Mexico Tech, Socorro, USA

Summary. Excluding those unsettled systems undergoing mergers, bright galaxies come
in two flavours: with and without discs. In this work we look for photometric evidence for
presence of discs and compare it with kinematic results of the ATLAS^{3D} survey (Cappellari *et al.* 2011). We fit a Sérsic (1968) function to azimuthally averaged light profiles of
ATLAS^{3D} galaxies to derive single component fits and, subsequently, we fit a combination
of the Sérsic function (free index n) and an exponential function ($n=1$) with the purpose
of decomposing the light profiles into "bulge" and "disc" components (B+D model) of all
non-barred sample galaxies. We compare the residuals of the B+D models with those of
the single Sérsic fits and select the B+D model as preferred only when the improvement is
substantial and there are no correlations within residuals. We find that the high angular
momentum objects (fast rotators) are disc dominated systems with bulges of typically
low n (when their light profiles can be decomposed) or are best represented with a single
Sérsic function with a low Sérsic index ($n < 3$). Single component systems with large
Sérsic indices are characteristic of low angular momentum objects (slow rotators).

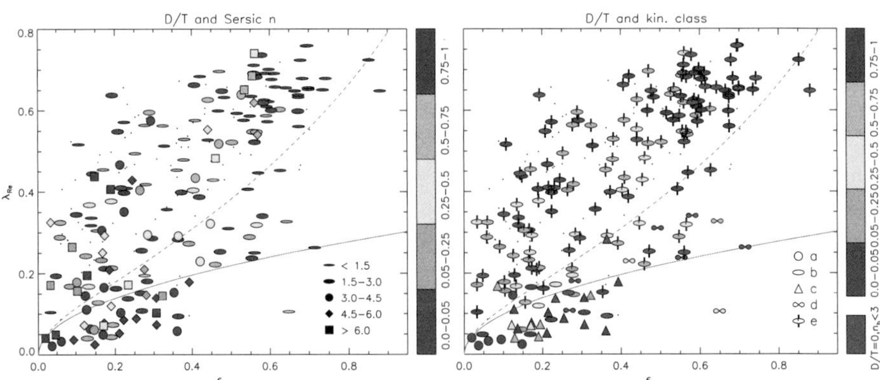

Figure 1. Angular momentum λ_R versus ellipticity ϵ of ATLAS^{3D} galaxies. *Left:* Symbols
represent Sérsic indices as shown on the legend. *Right:* Symbols show different types of kinematics
from Krajnović *et al.* (2011) and are described in the legend: a - non rotating galaxies, b -
featureless non-regular rotators, c - KDC, d - counter-rotating discs, and e - regular rotators.
On both plots colours quantify D/T ratios, as shown on the respective colour bars.

References

Cappellari, M., *et al.* 2011,*MNRAS*,413,813
Sérsic, J. L.,1968, Atlas de galaxias australes, Observatorio Astrónomico, Córdoba, Argentina

The intriguing life of massive galaxies
Proceedings IAU Symposium No. 295, 2012
D. Thomas, A. Pasquali & I. Ferreras, eds.

© International Astronomical Union 2013
doi:10.1017/S1743921313005176

A star cluster view of the life of massive galaxies

Myung Gyoon Lee[1] and Hong Soo Park[1]

[1] Seoul National University, Seoul, Korea
email: mglee@astro.snu.ac.kr

Abstract. We present an overview of recent findings on the kinematics, age, and metallicity of globular cluster systems in nearby giant elliptical galaxies and their implications for understanding how giant elliptical galaxies formed and evolved.

Keywords. galaxies: star clusters, galaxies: evolution, galaxies:elliptical and lenticular, cD

We derived the kinematics, metallicity, age, and $[\alpha/\mathrm{Fe}]$ of globular clusters in several giant elliptical galaxies (gEs) (NGC 4636, M87, M49, M60, NGC 5128, NGC 1399, and NGC 1407) from the spectroscopic line indices in the literature (Park *et al.* 2012). We made a combined sample of gE globular clusters by combining all the globular cluster data in gEs. The globular clusters in the combined sample of gEs show a bimodal metallicity distribution with peaks at $[\mathrm{Fe/H}] = -1.25(\sigma = 0.32)$ and $-0.42(\sigma = 0.25)$, consistent with color distributions. The metal-poor globular clusters are on average $\sim 3\,\mathrm{Gyr}$ older than the metal-rich ones. They show a large range in age from 2 to $15\,\mathrm{Gyr}$, including a significant number of young globular clusters, in contrast to the Milky Way (MW) globular clusters, which are all older than $7\,\mathrm{Gyr}$. The metal-rich globular clusters show a broader age distribution than the metal-poor ones, showing that young ones are mostly metal-rich. The mean $[\alpha/\mathrm{Fe}]$ value of the combined gE globular clusters is smaller than that of the MW globular clusters. In the case of the old group, the mean ages for the metal-rich and metal-poor globular clusters are about $13\,\mathrm{Gyr}$, and there is little difference in the mean ages between the two populations. ones. In the case of the young group, the mean age for the metal-rich globular clusters is about $5\,\mathrm{Gyr}$, which is about $2\,\mathrm{Gyr}$ younger than that for the metal-poor ones. Considering observational clues on photometry, metallicity, age, kinematics and spatial distribution of globular clusters in gEs lead to the mixture (Bibimbap) model for formation of globular clusters in gEs (Lee *et al.* 2010a, Lee *et al.* 2010b). (1) Metal-poor globular clusters are formed mostly in low-mass dwarf galaxies very early, and preferentially in dwarf galaxies located in a high density environment like a galaxy cluster. (2) Metal-rich globular clusters are formed together with stars in massive galaxies or dissipational merging galaxies later than metal-poor globular clusters, but not much later than metal-poor globular clusters. (3) New metal-rich globular clusters are formed during dissipational merging. A significant fraction of metal-poor globular clusters in gEs we see today are from dissipationless merging or accretion. In summary, galaxies grow via dissipationless and dissipational merging of galaxies of various types and via accretion of dwarf galaxies.

References

Park, H. S., *et al.* 2012, *ApJ*, in press
Lee, M. G., Park, H. S., & Hwang, H. S. 2010a, *Science*, 328, 334
Lee, M. G., *et al.* 2010b, *ApJ*, 709, 1983

The intriguing life of massive galaxies
Proceedings IAU Symposium No. 295, 2012
D. Thomas, A. Pasquali & I. Ferreras, eds.

© International Astronomical Union 2013
doi:10.1017/S1743921313005188

Stellar population gradients in brightest cluster galaxies

S. I. Loubser[1] and P. Sánchez-Blázquez[2]

[1] Centre for Space Research, North-West University, Potchefstroom 2520, South Africa
email: `Ilani.Loubser@nwu.ac.za`

[2] Departamento de Física Teórica, Universidad Autónoma de Madrid, E28049, Spain
email: `psanchezblazquez@googlemail.com`

Abstract. We present the stellar population and velocity dispersion gradients for a sample of 24 brightest cluster galaxies (BCGs) in the nearby Universe for which we have obtained high quality long-slit spectra at the Gemini telescopes. With the aim of studying the possible connection between the formation of the BCGs and their host clusters, we explore the relations between the stellar population gradients and properties of the host clusters, as well as the possible connections between the stellar population gradients and other properties of the galaxies. We find mean stellar population gradients (negative $\Delta[Z/H]/\log r$ gradient of -0.285 ± 0.064; small positive $\Delta\log(\text{age})/\log r$ gradient of $+0.069 \pm 0.049$; and null $\Delta[E/Fe]/\log r$ gradient of -0.008 ± 0.032), that are consistent with those of normal massive elliptical galaxies. However, we find a trend between metallicity gradients and velocity dispersion (with a negative slope of -1.616 ± 0.539), that is not found for the most massive ellipticals. Furthermore, we find trends between the metallicity gradients and K-band luminosities (with a slope of 0.173 ± 0.081) as well as the distance from the BCG to the X-ray peak of the host cluster (with a slope of -7.546 ± 2.752). The latter indicates a possible relation between the formation of the cluster and that of the central galaxy.

Keywords. galaxies: evolution – galaxies: elliptical and lenticular, cD – galaxies: stellar content

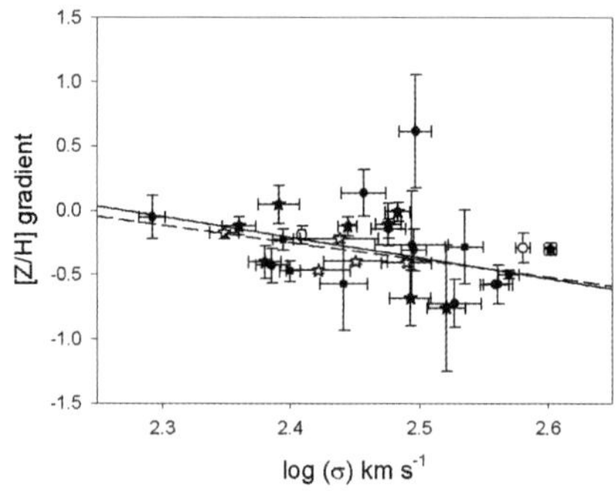

Figure 1. Metallicity gradients plotted against central velocity dispersion. The BCGs in our sample are shown with black symbols (those observed within 15 degrees of the MA are indicated with stars, and the others with circles. The BCGs from other studies are shown with empty symbols (see Loubser & Sánchez-Blázquez 2012). The fitted metallicity – mass correlation for our sample is shown with a solid line, and the correlation fitted to all the data with a dashed line.

Reference

Loubser, S. I. & Sánchez-Blázquez, P. 2012, *MNRAS*, 425, 841

The intriguing life of massive galaxies
Proceedings IAU Symposium No. 295, 2012
D. Thomas, A. Pasquali & I. Ferreras, eds.

© International Astronomical Union 2013
doi:10.1017/S174392131300519X

On the Recovery of Galaxy Properties from Spectral Fits

G. Magris C.[1], C. Mateu[1], G. Bruzual A.[2], and I. Cabrera[1,3]

[1]Centro de Investigaciones de Astronomía (CIDA), A.P. 264, Mérida 5101A, Venezuela
email: `magris,mateu,cabrera@cida.ve`

[2]Centro de Radioastronomía y Astrofísica, UNAM, A.P. 372, 58090, Morelia, México.
email:`g.bruzual@crya.unam.mx`

[3]Postgrado de Física Fundamental, Universidad de Los Andes, Mérida, Venezuela

Abstract. We show the results of a non-parametric, fully bayesian implementation of a spectral fitting algorithm, designed to calculate the main physical parameters that govern the galaxy assembly process. In this work, we present results from a statistical treatment of SED fitting that allows for easy recovery and visualization of the galaxy physical parameters .

Keywords. methods: data analysis; methods: statistical; galaxies: stellar content

1. Method Overview and Results

METHOD. Our fitting method is a new implementation of *gaspex* (Mateu 2001), which consists in modeling the target galaxy SED as a linear combination of a few SEDs, corresponding to simple stellar populations of arbitrary age and metallicity from the Charlot & Bruzual (2013, hereafter CB*) stellar population synthesis models. We use the Markov Chain Monte Carlo sampler *emcee* (Foreman-Mackey *et al.* 2012), which allows for easy fine-tunning of the chain parameters, to construct the posterior probability distribution function (pdf). To test the reliability of our fits, we choose an arbitrary subset of spectra generated with the CB* library . We add to the flux of the synthetic model SED random gaussian noise with a constant S/N ratio equal to 20. We assume uniform prior probability distribution functions in the model parameters: age t_i, mass weight a_i, and metallicity Z_i, of each simple stellar population in the posterior pdf $p(a, t, Z | Fobs) = \mathrm{Prior}(a, t, Z) \exp(-\chi^2/2)$. For each galaxy we compute the marginalized 1-D and 2-D posterior pdf, expectation value and confidence intervals for the explored physical parameters: total mass, mean mass-weighted-log(age), and mean $\log(Z)$. We also compute a bayesian estimate of the SFH, which we define as the mass expectation value as a function of log(age).

RESULTS. Our non-parametric model is capable of successfully recovering the star formation rate and total mass of galaxies with different histories of star formation. When metallicity distribution is correctly recovered, the bayesian spectrum yields residuals comparable to those obtained with the minimum χ^2 solution. Otherwise, the solution spectrum deviates in the UV.

GBA acknowledges support from UNAM through grants IA102311 and IB102212.

References

Charlot, S., Bruzual A., G. 2013, in preparation
Foreman-Mackey, D. *et al.* 2013, *PASP*, 125, 306
Mateu, J. 2001, *ASPC*, 215, 120

The intriguing life of massive galaxies
Proceedings IAU Symposium No. 295, 2012
D. Thomas, A. Pasquali & I. Ferreras, eds.

© International Astronomical Union 2013
doi:10.1017/S1743921313005206

Undressing M87 by Exposing its Most Private Globulars

Mireia Montes[1,2]†, José A. Acosta-Pulido[1,2], M. Almudena Prieto[1,2], and Juan A. Fernández-Ontiveros[3]

[1] Instituto de Astrofísica de Canarias (IAC), Vía Láctea s/n, La Laguna, E-38200, Spain
[2] Departamento de Astrofísica, Universidad de La Laguna, La Laguna, E-38207, Spain
[3] Max-Planck-Institut für Radioastronomie (MPIfR), Bonn, D-53121, Germany

Abstract. We present a multiwavelength photometric analysis of the innermost (3×3 kpc^2) Globular Clusters (GCs) of M87. Their Spectral Energy Distributions (SEDs) were built with J and K_s imaging obtained with NaCo at the VLT, along with HST UV-optical archival data. Using both Galatic GC templates and stellar population models, we derived ages (> 10 Gyr) and metallicities ([Fe/H]~ -0.7) for these clusters (e.g: Cohen *et al.* 1998). These GCs have lower metallicities than its host galaxy. This agrees with the idea that the GC population formed earlier than the bulk of the stars.

Keywords. galaxies: star clusters, galaxies: infrared

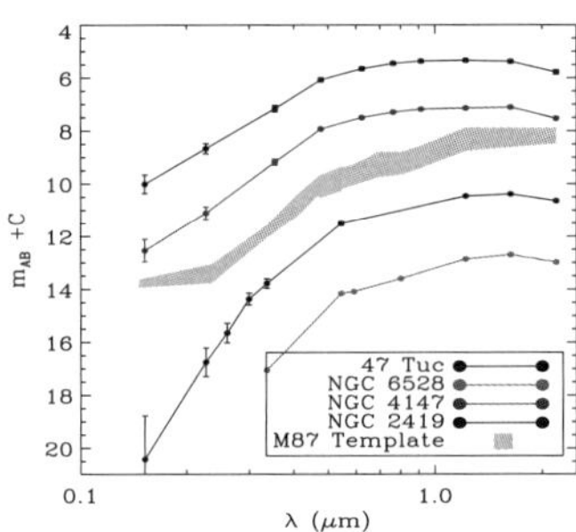

Figure 1. SEDs for MW globulars. The orange polygon is the average SED for M87 GCs.

Our SEDs cover a spectral range from 1400 Å to 2.2 μm. All the globular clusters have very similar properties. We derived SEDs of Galactic GCs in order to make a direct comparison with our GCs. M87 GCs and 47 Tuc are similar in the optical-IR ranges. SSP model fitting (Charlot & Bruzual 2007) lead to the same conclusions. UV fluxes show the presence of Extreme Horizontal Branch stars (Sohn *et al.* 2006). The mean derived metallicity for the GC system (Z~ 0.004) is lower than its host galaxy (Z~ 0.03, Kuntschner *et al.* 2010).

More information at www.eso.org/sci/meetings/2012/ESOat50/videos/20120905_mmontes.html

References

Charlot, S., & Bruzual, G. 2007 models (See 2003, *MNRAS*, 344, 1000)
Cohen, J. G., Blakeslee, J. P., & Ryzhov, A. 1998, *ApJ*, 496, 808
Kuntschner, H. *et al.* 2010, *MNRAS*, 408, 97-132
Sohn, S. T., *et al.* 2006, *AJ*, 131, 866-888

† E-mail: mmontes@iac.es (MM)

The intriguing life of massive galaxies
Proceedings IAU Symposium No. 295, 2012
D. Thomas, A. Pasquali & I. Ferreras, eds.

© International Astronomical Union 2013
doi:10.1017/S1743921313005218

The stellar metallicity distribution in intermediate-latitude fields

Xiyan Peng[1], Cuihua Du[2] and Zhenyu Wu[1]

[1] National Astronomical Observatories, Chinese Academy of Sciences
100012, Beijing, China
email: pengxiyan12@gmail.com

[2] College of Physical Sciences, University of Chinese Academy of Sciences
100049, Beijing, China

Abstract. Based on BATC and SDSS photometric data, we adopt the spectral energy distribution (SED) fitting method to evaluate stellar metallicities in the Galaxy. We find that the mean metallicity shifts from metal-rich to metal-poor with the increase of distance from the Galactic Centre.

Keywords. Star: abundances, Galaxy: disc, Galaxy: formation, Galaxy: halo, Galaxy: structure.

1. Stellar Atmospheric Parameter Estimation

We combine 15 BATC colors with 5 SDSS colors for the sample stars. We use a theoretical library of synthetic stellar spectra from Lejeune, Cuisinier & Buser (1997) to derive synthetic colors of the BATC and SDSS photometric system. The sample SEDs simulation with template SEDs can be used to derive the parameter of sample stars (Du *et al.* 2004).

2. Results

Metallicity variation with Galactic longitude. The mean metallicity is about -1.5 ± 0.2 dex in the interval $10 < r \leqslant 20\,\mathrm{kpc}$ and -1.3 ± 0.1 dex for $8 < r \leqslant 10\,\mathrm{kpc}$. The mean metallicity smoothly decreases from -0.4 to -0.8 at $0 < r \leqslant 5\,\mathrm{kpc}$. For $4 < r \leqslant 8\,\mathrm{kpc}$, the overall distribution of mean metallicity has a maximum at $l \sim 200°$. This feature may reflect a fluctuation from streams (such as the Monoceros stream) that are accreted from nearby galaxies (Peng *et al.* 2012).

The vertical metallicity gradient. We find that the vertical abundance gradient for the thin disk ($0 < z < 2\,\mathrm{kpc}$) is $\mathrm{d[Fe/H]/dz} \sim -0.21 \pm 0.05$ dex kpc^{-1}, and the vertical gradient -0.16 ± 0.06 dex kpc^{-1} at z distance interval $2 < z \leqslant 5\,\mathrm{kpc}$ where the thick disk stars are dominant. The vertical gradient $\mathrm{d[Fe/H]/dz} \sim -0.05 \pm 0.04$ dex kpc^{-1} is found over the z distance interval $5 < z \leqslant 15\,\mathrm{kpc}$. Therefore, there is little or no gradient in the halo (Peng *et al.* 2012).

References

Du, C.-H., Zhou, X., Ma, J., Shi, Y. R., Chen, A. B. C., Jiang, Z. J., & Chen, J. S. 2004, *AJ*, 128, 2265
Lejeune, T., Cuisinier, F., & Buser, R. 1997, *AAS*, 125, 229
Peng, X. H., Du, C.-H., & Wu, Z. H. 2012, *MNRAS*, 422, 2756

The intriguing life of massive galaxies
Proceedings IAU Symposium No. 295, 2012
D. Thomas, A. Pasquali & I. Ferreras, eds.

© International Astronomical Union 2013
doi:10.1017/S174392131300522X

Did brightest cluster galaxies experience more than one star formation epoch?

D. N. Viljoen[1] and S. I. Loubser[1]

[1] Centre for Space Research, North-West University, Potchefstroom, 2520, South Africa
email: 20569513@nwu.ac.za

Abstract. We use the full spectrum fitting ability of *ULySS*, with the Pegase.HR stellar population model to fit the observed spectra of 40 brightest cluster galaxies in order to determine whether a single or a composite stellar population provided the most probable representation of the star formation history (SFH). We find that some galaxies in the sample have more complex SFHs.

Keywords. galaxies: evolution, galaxies: general – galaxies: stellar content.

1. Introduction

Recent literature, i.e. Loubser (2009) refer to brightest cluster galaxies (BCGs) as the central, dominant galaxy in a cluster with a typical mass of $\sim 10^{13}$ M$_\odot$ (Katayama *et al.* 2003). It has been widely excepted that BCGs are dormant elliptical galaxies with a red photometric color, implying that old stellar populations are present. However, Liu, Mao & Meng (2012) have found that some BCGs have the presence of blue cores and UV excess which implies that star formation recently took place.

2. Results and Conclusions

We use the full spectrum fitting software package *ULySS* with the Pegase.HR stellar population model to fit this model against the observed spectra of 40 galaxies from Loubser (2009), to determine whether a simple or composite stellar population were a more probable representation of the star formation histories (SFHs). We found that 22 galaxies could be represented by a single stellar population (SSP) and the remaining 18 by composite stellar populations.

Our findings suggest that although 55% of the sample could be represented by an SSP, the remaining 18 galaxies experienced more than one star formation event. Hence, some BCGs have a more complex SFH than first assumed.

Acknowledgments

We thank the Square Kilometre Array project for the financial support.

References

Katayama, H., Hayashida, K. I., Takahara, F., & Fujita, Y. 2003, *ApJ*, 585, 687
Liu, F. S., Mao, S., & Meng, X. M. 2012, *MNRAS*, 423, 422
Loubser, S. I. 2009, *Kinematics and stellar population in brightest cluster galaxies*, Ph.D. thesis, University of Central Lancashire

The intriguing life of massive galaxies
Proceedings IAU Symposium No. 295, 2012
D. Thomas, A. Pasquali & I. Ferreras, eds.

© International Astronomical Union 2013
doi:10.1017/S1743921313005231

The stellar populations of host galaxies of supernovae

X. Shao[1,2,3,4], Y. C. Liang[1,3], M. Dennefeld[5], X. Y. Chen[1,3], G. H. Zhong[1,3], F. Hammer[6], L. C. Deng[1,3], and B. Zhang[1,4]

[1] National Astronomical Observatories, CAS, 20A Datun Road, 100012, Beijing, PR China
email: xshao@bao.ac.cn
[2] University of Chinese Academy of Sciences, 19A Yuquan Road, 100049, Beijing, PR China
[3] Key Laboratory of Optical Astronomy, NAOC, 20A Datun Rd. 100012, Beijing, China
[4] Department of Physicals, Hebei Normal University, Shijianzhuang 050016, China
[5] Institut d'Astrophysique de Paris, CNRS, 98bis Bd Arago, F-75014 Paris, France
[6] GEPI,Observatoire de Paris-Meudon, 92195 Meudon, France

Abstract. We study and compare the stellar populations of host galaxies of different types of supernovae (SNe): SN Ia and core collapse SN (SN II and SN Ibc) at the same time. The 234 sample galaxies are selected by cross-matching the Asiago Supernova Catalogue (ASC) and the SDSS-DR7 main galaxy sample (MGS). The STARLIGHT software is used to analyze their stellar populations by fitting the continua and absorption lines of the hosts.

Keywords. galaxies: evolution, galaxies: star formation, galaxies: starburst

We performed cross-matching on the ASC and the SDSS-DR7 MGS with 30 arcsec radius to select supernova host galaxies. We select galaxies for which the light-fraction (see details in Liang *et al.* 2010) of their SDSS spectral observations are > 0.15 to ensure that the 3 arcsec fiber can cover most of their global light. In total 234 SN host galaxies are selected, which are divided into two subsamples: emission-line galaxies and absorption-line galaxies. We fit the stellar continua and absorption lines of the hosts using Starlight (Cid Fernandes *et al.* 2005, Chen *et al.* 2009). The results are shown in Table 1. Among the 137 emission-line galaxies, the fraction of young stellar populations is higher in hosts of SN II than in hosts of SN Ia and Ibc. Mots of the 97 absorption-line galaxies host a SN Ia, and they have a large fraction of old stellar populations. The 137 hosts with emission lines contain much younger stellar populations.

Table 1. The contributed light fraction of stellar populations in age-bins for SN host galaxies.

	emission-line galaxies			absorption-lines galaxies		
hosts of	SN Ia	SN II	SN Ibc	SN Ia	SN II	SN Ibc
Young (< 0.2 Gyr)	30.2	56.5	22.2	12.5	26.8	25.8
Intermediate (0.2-2 Gyr)	42.2	31.5	51.6	28.0	30.1	39.2
Old (> 2 Gyr)	27.6	12.0	26.2	59.5	43.1	35.0

Acknowledgements. The authors thank the symposium organizers for their invitation to this poster. This work was supported by the Natural Science Foundation of China (NSFC) Foundation under Nos.10933001, 11273026.

References

Chen, X. Y., Liang, Y. C., Hammer, F. *et al.* 2009, *A&A*, 495, 457
Cid Fernandes, R., Mateus, A., Sodre, L. *et al.* 2005, *MNRAS*, 358, 363
Liang, Y. C., Zhong, G. H., Hammer, F. *et al.* 2010, *MNRAS*, 409, 213

The intriguing life of massive galaxies
Proceedings IAU Symposium No. 295, 2012
D. Thomas, A. Pasquali & I. Ferreras, eds.

© International Astronomical Union 2013
doi:10.1017/S1743921313005243

The stellar initial mass function in the early universe revealed from old stellar populations in our neighbourhood

Yutaka Komiya[1], Shimako Yamada[2], Takuma Suda[1], and Masayuki Y. Fujimoto[2]

[1]National Astronomical Observatory of Japan,
Osawa, Mitaka, Tokyo 181-8588, Japan

[2]Dept. of Cosmophysics, Hokkaido University,
Sapporo, Hokkaido 060-0810, Japan

Abstract. We present a new method to investigate the IMF in the early universe from observations of extremely metal-poor (EMP) stars. EMP stars are the low-mass survivors of stars which are formed in the early universe. We can give constraints on the IMF from statistics of the elemental abundances of the EMP stars in the Galactic halo.

Keywords. stars: abundances, stars: Population II, binaries, Galaxy: halo, Galaxy: formation

1. Constraints on the IMF in the early universe

It is known that a large fraction ($10 - 25\%$) of metal-poor stars show carbon enhancement and many of them show enhancement of s-process elements. The abundance anomaly of these stars is thought to be due to the binary mass transfer from intermediate-mass stars. Abundances of s-process elements on their surface depend on the mass of formerly primary stars. Therefore, from the statistics of observed carbon-enhanced metal-poor stars, we can give constraints on the distribution of primary stars of binaries.

Another constraint is given by the total number of EMP stars which survive to date. We can estimate the averaged yield of iron per massive stars as a function of the IMF from the number density of observed EMP stars. Only a top-heavy IMF peaked around $10\,M_\odot$ predicts iron yields in agreement with theoretical studies of supernova nucleosynthesis (Komiya *et al.* 2009).

2. When and where did the IMF change?

We search for the signatures of IMF change on the abundance distribution of metal-poor stars. The fraction of carbon rich stars depends on the IMF. Observationally, the fraction of carbon-rich stars drops around [Fe/H]$= -2$.

Another signature is hypernova origin elements. Stars with $m > 20\,M_\odot$ are thought to explode as hypernovae with huge explosion energies and the production of large amounts of Zn (Umeda & Nomoto 2002). Observationally, the relative abundance of Zn decreases at [Fe/H]$\gtrsim -2.2$. These can be signatures of an IMF change at [Fe/H]~ -2.2.

References

Komiya, Y. *et al.* 2007, *ApJ*, 658, 367
Komiya, Y., Suda, T., & Fujimoto, Y. M. 2009, *ApJ*, 694, 1577
Umeda, H. & Nomoto K. 2002, *ApJ*, 565, 385

The intriguing life of massive galaxies
Proceedings IAU Symposium No. 295, 2012
D. Thomas, A. Pasquali & I. Ferreras, eds.

© International Astronomical Union 2013
doi:10.1017/S1743921313005255

Multi-wavelength study of star formation properties in barred galaxies

Zhi-Min Zhou[1], Chen Cao[2], and Hong Wu[1]

[1]National Astronomical Observatories, Chinese Academy of Sciences, Beijing 100012, China
email: zmzhou@bao.ac.cn

[2]Shandong University at Weihai, Weihai, Shandong 264209, China

Abstract. Stellar bars are important internal drivers of the secular evolution of disk galaxies. Using a sample of nearby barred galaxies with weak and strong bars, we evaluate the correlations between star formation properties in different galactic structures and their associated bars, and try to interpret the complex process of bar-driven secular evolution. We find that weaker bars tend to associate with lower concentrical star formation activities, while stronger bars appear to have large scatter in the distribution of the global star formation activities. In general, the star formation activities in early- and late-type galaxies have different behavior, with similar star formation rate density distributions. In addition, there are only weak trends toward increased star formation activities in bulges and galaxies with stronger bars, which is consistent with previous works. Our results suggest that the different stages of the evolutionary sequence and many factors besides bars may contribute to the complexity of this process. Furthermore, significant correlations are found between the star formation activities in different galactic structures, in which barred galaxies with intense star formation in bulges tend to also have active star formation in their bars and disks. Most bulges have higher star formation densities than their associated bars and disks, indicating the presence of bar-driven evolution. Therefore, we derived a possible criterion (Figure 1) to quantify the different stages of a bar-driven evolutionary sequence. Future work is needed to improve on the uncertainties of this study.

Keywords. Secular Evolution, Barred Galaxies, Star Formation

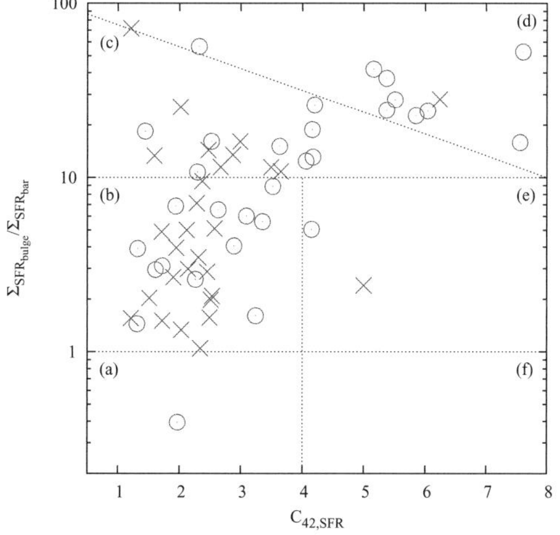

Figure 1. The possible criterion to quantify the different stages of bar-driven secular evolution. From (a) to (d), the evolutionary stages are from early to late, while stage (e) and (f) may be the results of other effects. Red circles are for early-type galaxies and blue crosses for late-type ones.

The intriguing life of massive galaxies
Proceedings IAU Symposium No. 295, 2012
D. Thomas, A. Pasquali & I. Ferreras, eds.

© International Astronomical Union 2013
doi:10.1017/S1743921313005267

Revealing the origin of the cold ISM in massive early-type galaxies

T. A. Davis,[1] K. Alatalo,[2] M. Bureau,[3] L. Young,[4] L. Blitz,[2] A. Crocker,[5] E. Bayet,[3] M. Bois,[6] F. Bournaud,[7] M. Cappellari,[3] R. L. Davies,[3] P-A. Duc,[7] P. T. de Zeeuw,[1,8] E. Emsellem,[1,9] J. Falcon-Barroso,[10] S. Khochfar,[11] D. Krajnovic,[1] H. Kuntschner,[1] P.-Y. Lablanche,[1] R. M. McDermid,[12] R. Morganti,[13] T. Naab,[14] M. Sarzi,[15] N. Scott,[16] P. Serra,[13] and A. Weijmans[17]

[1]ESO, Germany; [2]UC Berkeley, USA; [3]Univ. Oxford, UK; [4]New Mexico Tech, USA; [5]Univ. Toledo, USA; [6]Observatoire de Paris, France; [7]Univ. Paris Diderot, France; [8]Leiden Univ., The Netherlands; [9]Univ. de Lyon, France; [10]IAC, Spain; [11]MPE, Germany; [12]Gemini Observatory, USA; [13]ASTRON, The Netherlands; [14]MPA, Germany; [15]Univ. Hertfordshire, UK; [16]Swinburne Univ., Australia; [17]Univ. Toronto, Canada;

Abstract. Recently, massive early-type galaxies have shed their red-and-dead moniker, thanks to the discovery that many host residual star formation. As part of the ATLAS-3D project, we have conducted a complete, volume-limited survey of the molecular gas in 260 local early-type galaxies with the IRAM-30m telescope and the CARMA interferometer, in an attempt to understand the fuel powering this star formation. We find that around 22% of early-type galaxies in the local volume host molecular gas reservoirs. This detection rate is independent of galaxy luminosity and environment. Here we focus on how kinematic misalignment measurements and gas-to-dust ratios can be used to put constraints on the origin of the cold ISM in these systems. The origin of the cold ISM seems to depend strongly on environment, with misaligned, dust poor gas (indicative of externally acquired material) being common in the field but completely absent in rich groups and in the Virgo cluster. Very massive galaxies also appear to be devoid of accreted gas. This suggests that in the field mergers and/or cold gas accretion dominate the gas supply, while in clusters internal secular processes become more important. This implies that environment has a strong impact on the cold gas properties of ETGs.

Keywords. galaxies: elliptical and lenticular, cD – galaxies: evolution – galaxies: ISM – ISM: molecules – ISM: evolution – stars: mass-loss

1. Introduction

In order for early-type galaxies (ETGs) to fade and join this tight red sequence, it is thought that the fuel for star formation must be consumed, destroyed or removed on a reasonably short timescale (Faber *et al.* 2007). This should leave galaxies on the red sequence with little or no cold ISM, and thus no star formation. Evidence is mounting, however, that a reasonable fraction of ETGs do have cold gas reservoirs (e.g. Serra *et al.* 2012) and residual star-formation (e.g. Yi *et al.* 2005) . Young *et al.* (2011) have presented an unbiased census of the molecular gas content of the ATLAS^{3D} sample of nearby ETGs (Cappellari *et al.* 2011), and report that 22% of optically-selected, morphologically-classified ETGs have substantial molecular gas reservoirs (10^7–10^9 $M_\odot$ of H_2). If galaxies form in a hierarchical manner, then these observations pose a challenge to the standard view that ETGs join and remain on the red-sequence due to a lack of cold gas. One must either demonstrate that it is possible to create a tight red sequence without removing all of the cold ISM, or that galaxies can regenerate or acquire cold gas after becoming red.

Various lines of enquiry suggest that the gas we detect in these galaxies is not left over from the progenitors that formed the ETG (e.g. Kuntschner *et al.* 2010). Unless the remnant gas can be made stable against star-formation this thus suggests that we are seeing regenerated or newly accreted gas. *Internal* gas return from stellar populations (dominated by mass loss from red giant branch, [post-]asymptotic giant branch stars and planetary nebulae; Parriott & Bregman 2008, Bregman & Parriott 2009) must be occurring in all ETGs at all times, at a rate depending on the number of stars present in each galaxy and its star formation history. One would thus expect many ETGs to have detectable molecular gas (assuming that some fraction of the hot gas reservoir does cool). Galaxies can also regain a cold ISM via *external* processes, such as (major and minor) mergers and/or cold mode accretion from the intergalactic medium.

Determining the dominant source of the cold ISM in ETGs is vital in order to understand their evolution. If stellar mass loss can build up molecular reservoirs, then galaxies can transform in isolation from spheroidal to disky systems, and perhaps even evolve back into the blue cloud (e.g. Kannappan *et al.* 2009). If however mergers are the dominant source of the gas, then star-formation episodes are likely to be short-lived.

Observationally the origin of the gas may be addressed by comparing the angular momentum of the ISM with that of the underlying stellar population. Because of angular momentum conservation stellar mass loss must produce gas that is kinematically aligned with the bulk of the stars which produced it, while material from external sources can enter a galaxy with any angular momentum. In this proceedings we report the result of our attempt to use the ATLAS^{3D} sample of ETGs to constrain the importance of externally acquired gas in this way (as presented in Davis *et al.* 2011). Here we show results comparing the kinematic misalignment of stars and ionised gas (from SAURON data; Cappellari *et al.* 2011). Identical results are found when one compares the stars and molecular gas (from CO interferometry with CARMA; Alatalo, *et al.* 2012). Ionised, atomic and molecular gas in local ETGs seem to be linked, always having similar kinematics and thus presumably sharing a common origin.

2. Results and Discussion

2.1. *Gas misalignments in the field and in clusters*

Figure 1 shows the measured kinematic misalignment between the stars and the ionised (and molecular) gas in the fast-rotating ATLAS^{3D} ETGs, split by environment. In the field $\approx 42\%$ of galaxies have kinematically misaligned ionised gas. In many cases, fast-rotating galaxies with kinematically aligned molecular and ionised gas have HI distributions that again suggest an external origin for the gas. This suggests that mergers and accretion play a dominant role in supplying gas to field ETGs. However, in the Virgo cluster, the molecular and ionised material in fast-rotators is nearly always kinematically aligned with the bulk of the stars, pointing to gas supplied by purely internal processes.

We tentatively suggest that preprocessing in groups (see Section 2.2) may help to explain this environmental dichotomy. Merger-induced starbursts in groups and cluster outskirts could consume any kinematically misaligned molecular gas that is present, and leftover ionised material would be ram pressure-stripped. Once galaxies settle into a cluster or HI-poor group, external gas accretion and mergers are suppressed, allowing stellar mass loss to regenerate a kinematically aligned gas reservoir. It is possible that features such as bars and rings could funnel dust and gas lost by stars to the centre of the galaxy, or collect them together, and could explain the greater efficiency with which galaxies in dense environments must recreate their dense gas reservoirs. Alternatively, we could be detecting the remnant gas left over from the morphological transformation of

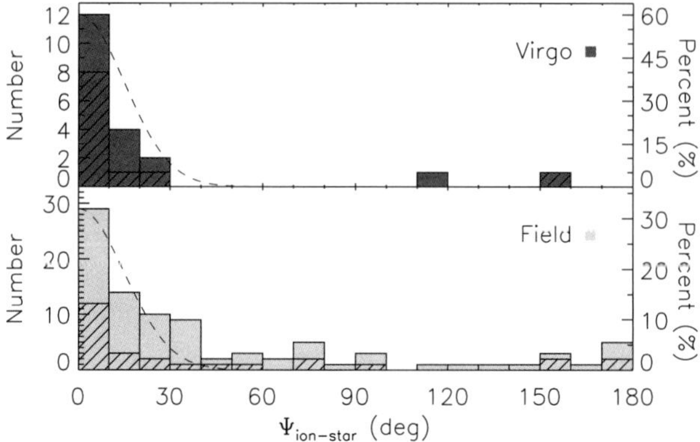

Figure 1. *Top:* Histogram showing the kinematic misalignment angle between the ionised gas and the stars for all fast-rotators in Virgo (from Davis *et al.* 2011). The hatched area indicates the number of galaxies in each bin that were also mapped in molecular gas. *Bottom:* As above, but for all field fast-rotators.

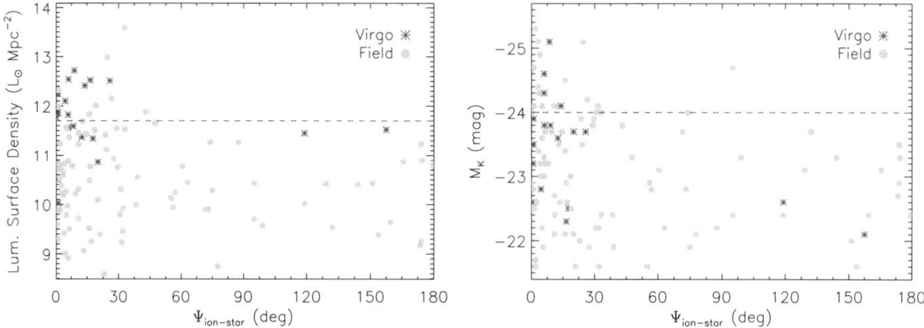

Figure 2. The kinematic misalignment angle between the ionised gas and the stars for fast-rotating ETGs, plotted against the local luminosity surface density (left), and the total absolute K_s-band magnitude (right; both from Davis *et al.* 2011). Virgo galaxies are plotted with red stars, and field/group galaxies with solid blue circles. The dashed line is a guide to the eye, at the suggested critical density/mass where essentially every galaxy becomes aligned. The error on each kinematic misalignment angle measurement is $\approx 15°$.

spiral galaxies into ETGs as they enter the cluster. Both of these possibilities, however, fail to explain why the detection rate of molecular gas (and the molecular gas mass fractions) are similar inside and outside of the Virgo cluster.

2.2. *The effect of group environments and mass*

Figure 2 shows that fast-rotators in dense groups, and with high masses also appear to always have aligned gas kinematics. Analysis (in Davis *et al.* 2011) suggests that these two effects are independent of each other. Thus both environmental and galaxy scale processes (e.g. AGN feedback, the ability for a galaxy to host a hot X-ray gas halo, and/or a halo mass threshold) must be at work, reducing the probability that cold, kinematically misaligned gas can be accreted onto these galaxies.

2.3. *Constraints from dust*

Figure 3 shows the dust to total gas (HI+H_2) ratio (from Crocker *et al.*, in prep), which seems to be systematically lower for ETGs with misaligned molecular gas. This supports a picture where the externally accreted gas comes from *minor* (rather than major) mergers (e.g. Kaviraj *et al.* 2011). Mergers with lower mass systems should result in the accretion

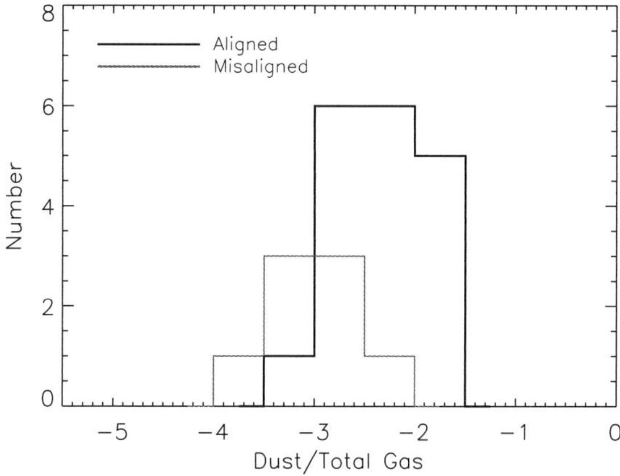

Figure 3. Histogram of the dust to total gas (HI+H$_2$) ratio (from Crocker *et al.*, in prep) for aligned and misaligned ETGs from the ATLAS^{3D} sample.

of lower metallicity gas, with a smaller dust content. Direct investigations of the gas phase metallicity itself are underway (Davis *et al.*, in prep), and should reveal if the dust-to-gas fraction in ETGs scales with gas-phase metallicity in the same way as in spiral galaxies.

2.4. *Future Prospects*

More work is clearly required to unambiguously determine the full importance of internal and external gas sources, and the role of environment and mass in the regeneration of gas reservoirs in ETGs. Equally important is understanding where stellar mass loss material goes in the 78% of ETGs that do not have cold molecular gas. It would be highly beneficial to extend the sort of kinematic analysis presented here to other clusters and groups, to see if the results reported here hold true in yet denser environments. The Fornax cluster in the southern hemisphere and the Coma cluster in the north are obvious nearby targets, which should become accessible with next generation facilities; e.g. the Large Millimeter Telescope (LMT) and the Atacama Large Millimeter/sub-millimeter Array (ALMA) in the millimetre, and instruments like KMOS/MUSE in the optical and infrared.

References

Bregman J. N. Parriott J. R., 2009, *ApJ*, 699, 923
Cappellari M., *et al.*, 2011, *MNRAS*, 413, 813
Davis, T. A. *et al.* 2011, *MNRAS*, 417, 882
Faber, S. *et al.* 2007, *ApJ*, 665, 265
Kannappan, S. J., Guie, J. M., & Baker, A. J. 2009, *AJ*, 138, 579
Kaviraj, S., Tan, K.-M., Ellis, R. S., & Silk, J. 2011, *MNRAS*, 411, 2148
Kuntschner H., *et al.*, 2010, *MNRAS*, 408, 97
Parriott J. R. Bregman J. N., 2008, *ApJ*, 681, 1215
Serra, P. *et al.* 2012, *MNRAS*, 422, 1835
Yi S. K., *et al.*, 2005, *ApJ*, 619, L111
Young L. M., *et al.*, 2011, *MNRAS*, 414, 940

The intriguing life of massive galaxies
Proceedings IAU Symposium No. 295, 2012
D. Thomas, A. Pasquali & I. Ferreras, eds.

© International Astronomical Union 2013
doi:10.1017/S1743921313005279

Origin and Ionization of the Warm Ionized Gas in Massive Early-type Galaxies

Renbin Yan[1,2] and Michael R. Blanton[2]

[1]Department of Physics and Astronomy, University of Kentucky, Lexington, KY, 40506, USA
email: renbin@pa.uky.edu

[2]Center for Cosmology and Particle Physics, Department of Physics, New York University,
New York, NY, 10003, USA
email: michael.blanton@nyu.edu

Abstract. Most early-type galaxies are not devoid of cold and warm gas. The origin and ionization of this gas reveal the intriguing ongoing evolution of these galaxies. In most cases, the warm ionized gas shows emission-line spectra similar to low-ionization nuclear emission-line regions (LINERs). Their ionization mechanism has been hotly debated. We will present evidence from line ratio gradient that rules out AGN and shocks as the dominant ionization mechanism, and suggests the ionizing sources follow the stellar density profile. Hot evolved stars are the favorite candidates but bring new puzzles.

This finding allows us to obtain a gas-phase metallicity calibration in these early-type galaxies, using the line emission. We will show how the metallicity of the warm gas depends on stellar mass and stellar age, and what it tells us about the origin of the warm gas in these galaxies.

Keywords. galaxies: elliptical and lenticular, cD, galaxies:ISM, stars: AGB and post-AGB

1. Introduction

Massive early-type galaxies have old stellar populations and are not actively forming stars. The most massive ones among them have stopped growing over the past 7 billion years. The shape difference between galaxy luminosity function and the halo mass function also requires a lower star formation efficiency in massive haloes. However, simulations suggest the dark matter haloes should still accrete more gas and this gas should cool and turn into stars if there were no additional heating. Also, the recycled gas from stellar evolution could also provide fuel for star formation. What is the fate of the accreted gas and the recycled gas? For the interstellar medium of these massive early-type galaxies, one important probe that has been under-utilized is the warm ionized gas.

The majority of massive early-type galaxies contain warm ionized gas which display optical line emission in their spectra. This has been known since the 1980s with long slit spectroscopy surveys (Phillips *et al.* 1986, Kim 1989, Buson *et al.* 1993, Goudfrooij *et al.* 1994, Macchetto *et al.* 1996, Zeilinger *et al.* 1996). It has been confirmed with integral field spectroscopy surveys, such as SAURON (Sarzi *et al.* 2006) and ATLAS3D (Davis *et al.* 2011). From these observations, we know that the warm ionized gas is spatially extended and can extend to kpc scales (Sarzi *et al.* 2006). Where does this gas come from? Does it come from stellar mass loss? Is it newly accreted gas? Or is it cooled from the hot X-ray-emitting gas? We would be able to tell its origin if we could measure the metallicity of the gas. However, such a metallicity calibration is not yet available, because the ionization mechanism is unsettled. In this contribution, we describe a new constraint on the ionizing source of the gas and explore a gas-phase metallicity calibration.

2. Ionization Sources

The line ratios displayed by the warm gas in most early-type galaxies satisfy the criteria of Low-ionization Nuclear Emission-line Regions (LINERs, Heckman 1980) on all major line-ratio diagnostic diagrams. It has been hotly debated what source ionized the gas. There are multiple ionization mechanisms that can produce the same kind of line ratios. These include photoionization by a central AGN (Ferland & Netzer 1983, Halpern & Steiner 1983, Groves *et al.* 2004), photoionization by hot evolved stars, such as post-AGB stars (di Serego Alighieri *et al.* 1990, Binette *et al.* 1994), collisional ionization by fast shocks (Dopita & Sutherland 1995), photoionization by hot X-ray emitting gas (Voit & Donahue 1990, Donahue & Voit 1991), conductive heating or turbulent mixing (Sparks *et al.* 1989). Therefore, determining the ionization mechanism requires other information.

What is going to distinguish AGN and stellar sources is the gradient in certain line ratios that are sensitive to the ionization parameter. AGN and distributed stellar sources will produce different flux density profiles. Given a fixed density profile, they will produce different ionization parameter profiles. By measuring the spatial gradient in a line ratio that is sensitive to ionization parameter, we can distinguish the AGN photoionization scenario and the distributed stellar ionizing sources convincingly.

Currently available IFU surveys or long-slit spectroscopy either cover too small a wavelength range or do not have sensitivity on the outskirts of a galaxy. We therefore turn to the SDSS survey and use the aperture effect to measure the spatial gradient. SDSS used an angular aperture of 3". The same angular aperture corresponds to a larger physical scale at larger distances. By selecting the same population of galaxies at all distances, we can statistically study the spatial distribution line emission and line ratio gradient. We select only red-sequence galaxies on a color-magnitude diagram and build a volume limited sample with $0 < z < 0.1$. We remove dusty-star-forming galaxies by rejecting 30% of galaxies that have the lowest D4000 in each redshift bin. This way we sample the same population of galaxies at all redshifts. With such a sample, we found that the median [O3]/[S2] ratio among the 25% brightest Hα emitter in each redshift bin is fairly flat with scale, which rules out AGN as the ionizing source and strongly favors a distributed ionizing source that follows the stellar density profile (Yan & Blanton 2012). See Figure. 1.

Additionally, the stellar photoionizing model has a prediction about luminosity dependence. Since bright early-types have slightly shallower profiles than faint early-types, they should produce different ionizing flux profiles and different line ratio gradients. By separating the sample according to broadband luminosity, we verified the prediction using the real data. This strongly suggests that the ionizing source is spatially distributed like the stars.

To evaluate the likelihood of the shock ionization and turbulent mixing models, we measured the temperature of the gas in coadded spectra. We selected galaxies from the above mentioned sample, but limited in redshift $0.06 < z < 0.15$ for which the median line ratio varies very slightly with the redshift (i.e. aperture). We excluded the 3% strongest line emitter among them so that the line ratio is not dominated by any galaxies hosting AGNs. We separated the resulting sample according to [N2]/[O2] ratio which is a metallicity indicator of the gas. We removed the stellar light in the resulting stacked spectra and were able to measure the $[N\ II]\ \lambda5755$ line. The $[N\ II]\ 6584/5755$ ratio yields a temperature measurement of $15,000K$ for the low-metallicity sample and $8,000K$ for the high-metallicity sample. These temperatures are consistent with the photoionization but are inconsistent with shock models or turbulent mixing models.

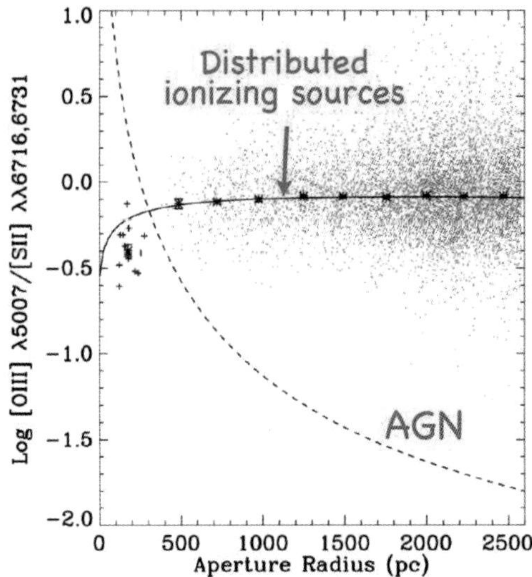

Figure 1. [O III]/[S II] ratio as a function of aperture size for a volume-limited sample of non-star-forming line-emitting red galaxies, most of which have LINER-like emission line ratios. Asterisks indicate median values in each bin. Crosses are for the Palomar sample of Ho, Filippenko & Sargent (1997). Grey points are from SDSS. The curves show the model predictions for photoionization by an AGN (dashed) or by distributed ionizing sources following the stellar density profile (solid). Assuming the line ratio profile in all of these galaxies are similar to each other and the warm gas density profile follows that of the hot X-ray gas, then the median trend suggests the gas is ionized by a population of hot evolved stars. However, this urgently needs to be verified in individual galaxies and with direct warm gas density measurements.

3. Gas-phase Metallicity in Early-type Galaxies

Given that the gas is ionized by stars, we can now derive a gas-phase metallicity calibration using CLOUDY spectral synthesis code (Ferland *et al.* 1998). For the input ionizing spectrum we use a 13 Gyr-old simple stellar population from Bruzual & Charlot (2003), assuming solar metallicity and a Chabrier initial mass function. We assume a gas density of 200 cm^{-3} and the default solar abundance pattern stored in CLOUDY except for Nitrogen. Because Nitrogen is a secondary element in the high metallicity regime, the N/O abundance ratio increases with Oxygen abundance. We adopt the scaling provided by Vila Costas & Edmunds (1993). We vary the ionization parameter from $10^{-4.5}$ to 10^{-2} and the metallicity (Z) from $0.0625Z_\odot$ to $4Z_\odot$. Fig. 2 left panel shows the resulting grid in [O III]/[O II] vs. [N II]/[O II], overplotted with a sample of dust-free, non-star-forming galaxies from SDSS. We can see that the [O III]/[O II] ratio provides a good proxy for ionization parameter and the [N II]/[O II] ratio provides a good proxy for metallicity.

Using the calibration derived, we measure the Oxygen abundance for these massive galaxies and plot them as a function of their stellar mass in the right panel of Fig. 2. We compare them with the star-forming galaxies sample from Tremonti *et al.* (2004). The metallicies in these galaxies are slightly lower than those in star-forming galaxies with the same mass. This indicates that the gas cannot simply be residual gas left over from past star formation. At least it has to be diluted by infalling gas. We also see a hint of a mass-metallicity relation among early-type galaxies, the origin of which will be investigated further.

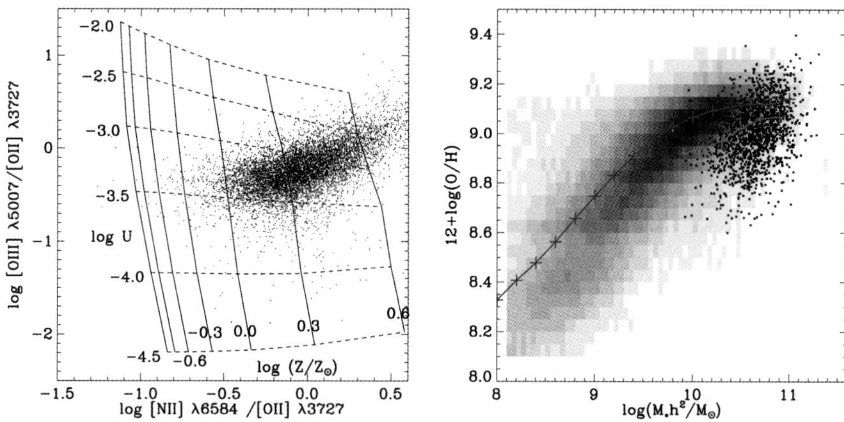

Figure 2. Left: [O III]/[O II] vs. [N II]/[O II] for a grid of CLOUDY models with different ionization parameter and metallicity. Overplotted are a sample of passive red galaxies from SDSS that have a Balmer decrement consistent with zero extinction. Right: Oxygen abundance in non-star-forming red sequence galaxies as a function of stellar mass. The red galaxies are denoted by dark points. The background grey scale indicate star-forming galaxy from Tremonti *et al.* (2004). The large colored crosses indicate the median in each stellar mass bin for passive red galaxies (red crosses) and star-forming galaxies (blue crosses).

References

Binette, L., Magris, C. G., Stasinska, G., & Bruzual, A. G. 1994, *Astro. & Astrophys.*, 292, 13

Bruzual, G. & Charlot, S. 2003, *MNRAS*, 344, 1000

Buson, L. M., *et al.* 1993, *Astro. & Astrophys.*, 280, 409

Davis, T. A., *et al.* 2011, *MNRAS*, 417, 882

di Serego Alighieri, S., Trinchieri, G., & Brocato, E. 1990, in Windows on Galaxies, eds. G. Fabbiano, J. S. Gallagher, & A. Renzini, Astrophysics and Space Science Library, (Dordrecht:Kluwer), Vol. 160., p. 301

Donahue, M. & Voit, G. M. 1991, *Astrophys. J.*, 381, 361

Dopita, M. A. & Sutherland, R. S. 1995, *Astrophys. J.*, 455, 468

Ferland, G. J., Korista, K. T., Verner, D. A., Ferguson, J. W., Kingdon, J. B., & Verner, E. M. 1998, *PASP*, 110, 761

Ferland, G. J. & Netzer, H. 1983, *Astrophys. J.*, 264, 105

Goudfrooij, P., Hansen, L., Jorgensen, H. E., & Norgaard-Nielsen, H. U. 1994, *Astro. & Astrophys. Supp.s*, 105, 341

Groves, B. A., Dopita, M. A., & Sutherland, R. S. 2004, *Astrophys. J. Supp.*, 153, 75

Halpern, J. P. & Steiner, J. E. 1983, *Astrophys. J. Lett*, 269, L37

Heckman, T. M., 1980, *Astro. & Astrophys.*, 87, 152

Kim, D.-W. 1989, *Astrophys. J.*, 346, 653

Macchetto, F., Pastoriza, M., Caon, N., Sparks, W. B., Giavalisco, M., Bender, R., & Capaccioli, M. 1996, *Astro. & Astrophys. Supp.*, 120, 463

Phillips, M. M., Jenkins, C. R., Dopita, M. A., Sadler, E. M., & Binette, L. 1986, *Astro. J.*, 91, 1062

Sarzi, M., *et al.* 2006, *MNRAS*, 366, 1151

—. 2010, *MNRAS*, 402, 2187

Sparks, W. B., Macchetto, F., & Golombek, D. 1989, *Astrophys. J.*, 345, 153

Tremonti, C. A. *et al.* 2004, *Astrophys. J.*, 613, 898

Vila Costas, M. B. & Edmunds, M. G. 1993, *MNRAS*, 265, 199

Voit, G. M. & Donahue, M. 1990, *Astrophys. J. Lett*, 360, L15

Yan, Renbin & Blanton, Michael R., 2012 *Astrophys. J.*, 747, 61

Zeilinger, W. W., *et al.* 1996, *Astro. & Astrophys. Supp.*, 120, 257

The intriguing life of massive galaxies
Proceedings IAU Symposium No. 295, 2012
D. Thomas, A. Pasquali & I. Ferreras, eds.

© International Astronomical Union 2013
doi:10.1017/S1743921313005280

Dust Emission in Early-Type Galaxies with the Herschel Virgo Cluster Survey

Sperello di Serego Alighieri[1] and members of the HeViCS team

[1] INAF - Osservatorio Astrofisico di Arcetri,
Largo E. Fermi 5, 50125 Firenze, Italy
email: sperello@arcetri.astro.it

Abstract. We have searched for dust in an optical sample of 910 Early-Type Galaxies (ETGs) in the Virgo cluster (447 of which are optically complete at $m_{pg} \leqslant 18.0$), extending also to the dwarf ETGs, using *Herschel* images at 100, 160, 250, 350 and 500 μm. Dust was found in 52 ETGs (46 are in the optically complete sample), including M87 and another 3 ETGs with strong synchrotron emisssion. Dust is detected in 17% of ellipticals, 41% of lenticulars, and in about 4% of dwarf ETGs. The dust-to-stars mass ratio increases with decreasing optical luminosity, and for some dwarf ETGs reaches values similar to those of the dusty late-type galaxies. Slowly rotating ETGs are more likely to contain dust than fast rotating ones. Only 8 ETGs have both dust and HI, while 39 have only dust and 8 have only HI, surprisingly showing that only rarely dust and HI survive together. ETGs with dust appear to be concentrated in the densest regions of the cluster, while those with HI tend to be at the periphery. ETGs with an X-ray active SMBH are more likely to have dust and vice versa the dusty ETGs are more likely to have an active SMBH.

Keywords. galaxies: ISM; galaxies: elliptical and lenticular, cD; ISM: dust, extinction

---

1. Introduction

Most of the baryons in a cluster of galaxies are in the hot intracluster medium, some of which is associated with the most massive galaxies. The hot gas interacts in many ways with the cold phases of the interstellar medium of these galaxies, and these interactions have a fundamental effect on the evolution of the galaxies themselves. In order to understand the evolution of massive galaxies (the topic of this Symposium), particularly in clusters, it is therefore important to study also the coldest phases of their interstellar medium. The most massive galaxies are Early-Type Galaxies (ETGs), and, since they may have formed by merging or accretion of smaller ones, it is useful to include in the study also the dwarf ETGs. In di Serego Alighieri *et al.* (2007) we have systematically studied the neutral atomic gas (HI) content of a large and complete sample of ETGs in the Virgo cluster, using the Arecibo Legacy Fast ALFA 21-cm survey (Giovanelli *et al.* 2007). HI is found in very few massive ETGs, where the cold gas could have a recent external origin, and in a few peculiar dwarf galaxies at the edge of the ETG classification. The *Herschel Space Observatory* (Pilbratt *et al.* 2010) is giving us the opportunity to study the dust content of the same sample of Virgo ETGs. We have done so within the Herschel Virgo Cluster Survey (HeViCS, Davies *et al.* 2010), an open-time Key Programme for a confusion-limited imaging survey of a large fraction of the Virgo cluster in 5 bands: at 250, 350 and 500 μm with SPIRE (Griffin *et al.* 2010) and at 100 and 160 μm with PACS (Poglitsch *et al.* 2010). We describe here the main results of this work. A more detailed and complete account has been submitted to A&A (hereafter dSA12).

2. Input sample, analysis and results

We start with a sample of Virgo ETGs selected in the optical from the GOLDMine compilation (Gavazzi *et al.* 2003), mostly based on the Virgo Cluster Catalogue (VCC, Binggeli *et al.* 1985), to be ETG (i.e. equal to or earlier than S0a) and excluding those with $v_{hel} < 3000$ km/s. With these selection criteria 925 ETGs are within the 4 HeViCS fields and constitute our input sample. Out of these, 447 are brighter than the VCC completeness limit ($m_{pg} \leqslant 18.0$) and form the optically complete part of our input sample. Out of the input sample, 287 ETGs have inaccurate positions in the literature, based only on the original work of Binggeli *et al.* (1985), insufficient to find reliable counterparts in the HeViCS images. Using r-band SDSS images, we then remeasured the position for these ETGs, except for 15 (all with $m_{pg} > 19.0$), for which the identification is unsure. We have looked for a reliable far-IR counterpart in the HeViCS 250 μm mosaic image for all the 910 Virgo ETGs with accurate coordinates, and found one for 52 of them at $S/N > 6$. For these sources we measured the flux in each of the 5 HeViCS bands using an aperture of 30 arcsec radius, large enough to contain the PSF also at 500 μm. For 12 sources, which have far-IR emission exceeding this aperture, we used larger apertures, up to 78 arcsec radius. 46 far-IR counterparts have $F_{250} \geqslant 25.4$ mJy, which is our completeness limit at 250 μm. We detect dust above the synchrotron component in the 4 ETGs with radio emission, including M87. Given the large number of background sources present in the 250 μm images, following the methods of Smith *et al.* (2011), we estimate that on average 1.5 ETGs (most likely dwarfs), out of our input sample of 910, have a far-IR counterpart which is a background source. We have used the distance given in GOLDMine, which distinguishes various components in the Virgo cluster at 17, 23 and 32 Mpc (Gavazzi *et al.* 1999).

Dust appears to be very concentrated, much more than stars. The only ETG with a considerable amout of off-nuclear dust is M86, where it appears to be mostly in a filament at 2 arcmin (about 10 kpc in projection) to the South-East (Gomez *et al.* 2010). Dust masses and temperatures have been estimated for the 52 ETGs with a far-IR counterpart by fitting a modified black-body to the measured far-IR fluxes, assuming a spectral index $\beta = 2$ and a MW emissivity, and taking into account colour and aperture corrections. We have also estimated stellar masses with the methods of Zibetti *et al.* (2009), using the available optical/IR broad-band photometry. The dust temperature ranges between 15 and 30 K, and correlates with the stellar mass and with the B-band average surface brigthness within the effective radius (dSA12). The latter correlation suggests that most of the dust heating is due to radiation produced by stellar sources.

3. Discussion

Dust detection rates for the complete samples (i.e. 43 far-IR counterparts with $F_{250} \geqslant 25.4$ mJy out of the 447 input ETGs with $m_{pg} \leqslant 18.0$ and accurate position) are 9.6% for all ETGs, 17.1% for ellipticals, 41.4% for lenticulars and 3.7% for dwarf ETGs. The latter rate becomes 3.6%, if we take into account that 1.5 of the assumed far-IR counterpart of dwarf ETGs are in fact counterparts of background sources (see the previous section), and that about 8 of the dwarf ETGs of the input sample without a measured radial velocity are likely background galaxies. These rates are smaller than those previously measured on samples of bright ETGs (Knapp *et al.* 1989, Temi *et al.* 2004, Smith *et al.* 2012), as can be expected since our sample extends to faint galaxies and the dust detection rate correlates strongly with optical luminosity (dSA12). The dust-to-star mass ratio varies over almost 6 orders of magnitude, anticorrelates with the optical luminosity, and for

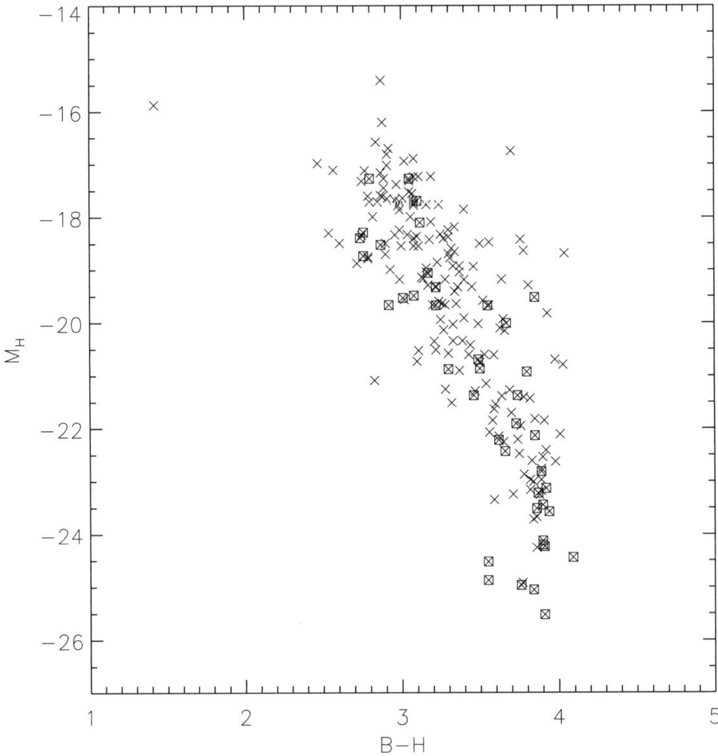

Figure 1. Optical/IR CMD for the Virgo ETGs with accurate photometry. The dusty ones (squares) are not bluer, i.e. not more star-forming, than the other galaxies (crosses).

some dwarf ETGs reaches very high values (around a few 10^{-2}), as high as for the dusty late-type galaxies. This is surprising, also given that the dusty ETGs do not show signs of star formation. In fact the colours of the dusty ETGs are not bluer than those of the non-dusty ones (Fig. 1).

The distinction between fast and slow rotators appears to be an important one for ETGs (Emsellem *et al.* 2011, and references on the ATLAS^{3D} project). For the ETGs of our input sample the detailed kinematical information necessary for this distinction is available only for the 49 ones, which are in common with the ATLAS^{3D} sample. Since we detect dust in $69\pm23\%$ of the slow rotators (in 9 out of 13) and in $28\pm9\%$ of the fast ones (in 10 out of 36), it appears that the former are considerably more likely to have dust. This is the opposite to what is seen for molecular gas in the whole of the 260 ETGs of the ATLAS^{3D} sample. In fact Young *et al.* (2011) find that the CO detection rate is $6\pm4\%$ in slow rotators and $24\pm3\%$ in fast ones. This is surprising, since dust and molecular gas are thought to be closely associated (Draine *et al.* 2007, Corbelli *et al.* 2012); in fact, for the dust-detected ETGs of our sample which have information on the molecular gas content, the dust-to-molecular-gas mass ratio is always 2×10^{-2} within a factor of two, and lower limits are consistent with this range. We suggest that a possible explanation of this difference could be an environmental effect, since most of the ATLAS^{3D} galaxies are outside of the Virgo cluster. In fact, of the 19 dust-detected ETGs, which we have in common with the ATLAS^{3D} sample, molecular gas is detected in 3 slow rotators and in 5 fast ones, a much more balanced situation than found by Young *et al.* (2011) in the whole ATLAS^{3D} sample, and all slow rotators with molecular gas in this whole sample are actually in the Virgo cluster. The difference we find could be due to the presence of kinematically peculiar objects among the dusty slow rotators in the Virgo cluster, like galaxies with counter-rotating components mimicking slow rotation. We can exclude this possibility, since the brightest and most regular ellipticals and lenticulars in the Virgo cluster like M49, M84, M86, M87, M89, NGC 4261 and NGC 4526 are among the dusty slow rotators, reinforcing our suggestion about an environmental effect.

We have also looked at the relationship between dust and HI for the Virgo ETGs, updating the work done by di Serego Alighieri *et al.* (2007) on the HI content of Virgo ETGs to include the 4-8 degrees declination strip, which has become available in the ALFALFA HI survey since their work (Haynes *et al.* 2011). We find an intriguing incompatibility between dust and HI in Virgo ETGs: we detect both dust and HI in only 8 ETGs, while 39 ETGs have dust but no HI, and 8 have HI but no dust. This dichotomy between dust and HI is reinforced by the position of the parent galaxies in the cluster. Dusty ETGs appear to concentrate in the densest regions of the Virgo cluster, while HI-rich ETGs tend to be at the periphery. While the presence of ETGs with dust but no HI can be explained by the longer dust survival times and by the stronger effects of ram pressure stripping on HI, more difficult to understand is that there are ETGs with HI but no dust, also given that these include two rather bright S0 galaxies: NGC 4262 and NGC 4270.

Concerning the relationship between the presence of dust and that of an AGN in Virgo ETGs, our input sample has 71 ETGs in common with the sample of Virgo ETGs for which Gallo *et al.* (2010) have looked for the presence of a supermassive black-hole (SMBH) by observing with *Chandra* the nuclear X-ray luminosity down to a few 10^{38} erg/s. Out of the 71 common ETGs, 25 (35%) have X-rays, most likely from a SMBH, and 14 (20%) have dust. Of the X-ray luminous ETGs 36% have dust, and out of the dusty ETGs 64% are X-ray luminous. It appears that ETGs with an X-ray active SMBH are more likely to have dust, and that ETGs with dust are more likely to have an X-ray active SMBH.

References

Binggeli, B., Sandage, A., & Tammann, G. A., 1985, *AJ*, 90, 1681
Corbelli, E., Bianchi, S., Cortese, L., *et al.*, 2012, *A&A*, 542, A32
Davies, J. I., Baes, M., Bendo, G. J., *et al.*, 2010, *A&A*, 518, L48
di Serego Alighieri, S., Gavazzi, G., Giovanardi, C., *et al.*, 2007, *A&A*, 474, 851
Draine, B. T., Dale, D. A., Bendo, G., *et al.*, 2007, *ApJ*, 663, 866
Emsellem, E., Cappellari, M., Krajnović, D., *et al.*, 2011, *MNRAS*, 414, 888
Gallo, E., Treu, T., Marshall, P. J., *et al.*, 2010, *ApJ*, 714, 25
Gavazzi, G., Boselli, A., Scodeggio, M., Pierini, D., & Belsole, E., 1999, *MNRAS*, 304, 595
Gavazzi, G., Boselli, A., Donati, A., Franzetti, P., & Scodeggio, M., 2003, *A&A*, 400, 451
Giovanelli, R., Haynes, M. P., Kent, B. R., *et al.*, 2007, *AJ*, 133, 2569
Gomez, H. L., Baes, M., Cortese, L., *et al.*, 2010, *A&A*, 518, L45
Haynes, M. P., Giovanelli, R., Martin, A. M., *et al.*, 2011, *AJ*, 142, 170
Knapp, G. R., Guhathakurta, P., Kim, D.-W., & Jura, M., 1989, *ApJSS*, 70, 329
Pilbratt, G. L., Riedinger, J. R., Passvogel, T., *et al.*, 2010, *A&A*, 518, L1
Smith, D. J. B., Dunne, L., Maddox, S. J., *et al.*, 2011, *MNRAS*, 416, 857
Smith, M. W. L., Gomez, H. L., Eales, S. A., *et al.*, 2012, *ApJ*, 748, 123
Temi, P., Brighenti, F., Mathews, W. G., & Bregman, J. D., 2004, *ApJSS*, 151, 237
Young, L. M., Bureau, M., Davis, T. A., *et al.*, 2011, *MNRAS*, 414, 940
Zibetti, S., Charlot, S., & Rix, H.-W., 2009, *MNRAS*, 400, 181

The intriguing life of massive galaxies
Proceedings IAU Symposium No. 295, 2012
D. Thomas, A. Pasquali & I. Ferreras, eds.

© International Astronomical Union 2013
doi:10.1017/S1743921313005292

The role of cold gas on the stellar mass - metallicity relation of nearby galaxies

Thomas M. Hughes

Kavli Institute for Astronomy & Astrophysics, Peking University, China
email: tmhughes@pku.edu.cn

Abstract. We investigate the relationship between stellar mass, metallicity and gas content for a magnitude- and volume-limited sample of 260 nearby late-type galaxies in different environments. Combining new oxygen abundance measurements with ultraviolet to near-infrared photometry and H I 21 cm line observations, we observe the relationship between stellar mass and metallicity. We also find that, at fixed stellar mass, galaxies with lower gas fractions typically possess higher oxygen abundances. Gas-poor galaxies are typically more metal-rich. Our results indicate that internal evolutionary processes, rather than environmental effects, play a key role in shaping the stellar mass-metallicity relation.

Keywords. cosmology: observations, galaxies: evolution, galaxies: spiral

1. What governs the scatter of the M-Z relation?

We derive gas-phase oxygen abundance estimates using new integrated, drift-scan optical spectroscopy (Boselli *et al.* 2012) of galaxies in the Herschel Reference Survey (Boselli *et al.* 2010), and the base metallicity calibrations of Kewley & Ellison (2008). Details of our analysis and results can be found in Hughes *et al.* (2012). To briefly summarize, the observed correlations between metallicity and gas content (see Fig. 1) suggest that the gas content governs the shape and scatter of the stellar mass - metallicity relation.

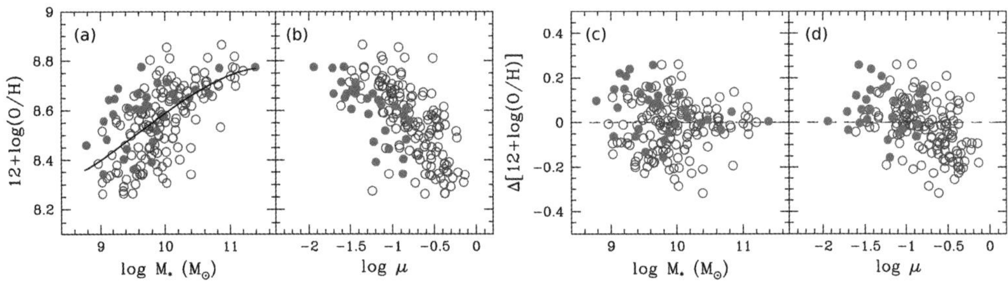

Figure 1. The relationships between oxygen abundance and (a) stellar mass, and (b) gas fraction, where $\mu = M_{HI}/(M_* + M_{HI})$. Dividing the sample by H I content demonstrates that gas content is anti-correlated with metallicity for H I normal galaxies (blue open circles). Furthermore H I deficient systems (red solid circles) are preferentially found at higher metallicities. Using a best fit polynomial (solid black line) to define the (c) residual oxygen abundance as the scatter in the M-Z relation, we find (d) a correlation between residual metallicity and the gas fraction.

References

Boselli, A., Eales, S., Cortese, L., Bendo, G., Chanial, P., *et al.* 2003, *PASP*, 122, 261
Boselli, A., Hughes, T. M., Cortese, L., Gavazzi, G., & Buat, V. 2013, *A&A*, 550, 114
Hughes, T. M., Cortese, L., Boselli, A., Gavazzi, G., & Davies, J. I. 2013, *A&A*, 550, 115
Kewley, L. & Ellison, S. 2008, *ApJ*, 681, 1183

The intriguing life of massive galaxies
Proceedings IAU Symposium No. 295, 2012
D. Thomas, A. Pasquali & I. Ferreras, eds.

© International Astronomical Union 2013
doi:10.1017/S1743921313005309

The [NII]/Hα calibration of the metallicity of galaxies from T_e-based abundances

Yanchun Liang[1,2], **Xu Shao**[1,2], **Francois Hammer**[3] and **Shaoying Yin**[4]

[1] National Astronomical Observatories, Chinese Academy of Sciences, Beijing 100012, China
email: `ycliang@bao.ac.cn`

[2] Key Laboratory of Optical Astronomy, NAOC, 20A Datun Rd. Beijing, 100012, China
[3] GEPI,Observatoire de Paris-Meudon, 92195 Meudon, France
[4] Department of Physics, College of Technology, Harbin University, Haerbin 150086, China

Abstract. We gather a sample of both metal-rich and metal-poor galaxies. Both samples have oxygen abundances estimated from electron temperature (T_e). The calibration of the emission-line ratio, N2($\equiv \log$([NII]6583/Hα), to oxygen abundances is then re-derived from this combined sample, finding good agreement for the wide metallicity and line-ratio ranges considered.

Keywords. galaxies: abundances, galaxies: evolution, galaxies: spiral, galaxies: starburst

---

The index of N2($\equiv \log$([NII]6583/Hα) is very useful for metallicity estimates. Fig. 1 shows the re-derived calibrations of N2 to O/H from a sample of both metal-rich and metal-poor galaxies, both of them have T_e-based abundances.

This work was supported by the Chinese NSFC Foundation under Nos.10933001, 11273026.

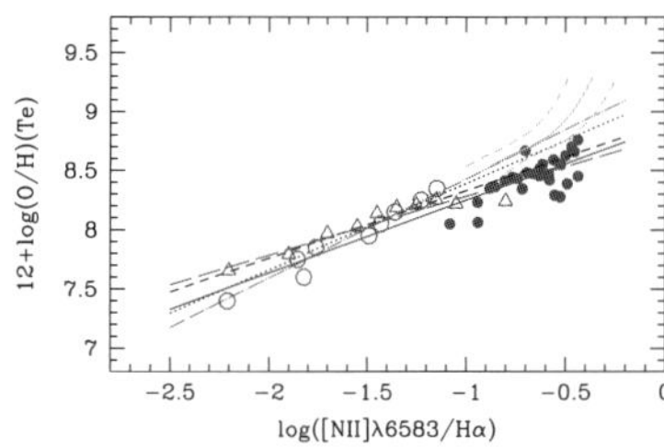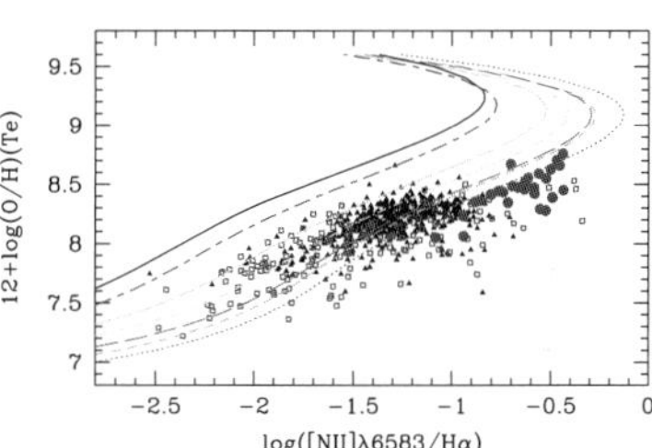

Figure 1. New N2 indicator for metallicity calibration of galaxies. **Left**: the filled blue circles are from Liang *et al.* (2007) for metal-rich ones, where the O/H abundances are derived from T_e by measuring [OII]7320,7330 on stacked spectra from thousand galaxies in each mass bin; The small data points are from Yin *et al.* (2007, Y07) with T_e-derived from [OIII]4363 (filled triangles for SDSS galaxies, and open squares from the literature). The model lines are from Kewley & Dopita (2002). **Right**: both the open circles and triangles are median values of the data from Y07, binning in O/H and N2, respectively. The solid line (12+log(O/H)=8.78 + 0.50×N2) refers to the least-square fitting for the combined data of filled circles plus open circles; the long dashed-line (12+log(O/H)=8.86 + 0.61×N2) is for the combined data of filled circles plus open triangles. Other lines are from the literature: dashed from Pettini & Pagel (2004), dotted from Denicolo *et al.* (2002), dot-dashed from Y07; 3 green curves are from Liang *et al.* (2006) for metal-rich galaxies with O/H from the strong-line method.

References

Denicolo, G., Terlevich, R., & Terlevich, E. 2002, *MNRAS*, 330, 69
Kewley, L. J. & Dopita, M. A. 2002, *ApJS*, 142, 35
Liang, Y. C., Yin, S. Y., Hammer, F., *et al.* 2006, *ApJ*, 652, 257
Liang, Y. C., Hammer, F., Yin, S. Y., Flores, H., *et al.* 2007, *A&A*, 473, 411
Pettini, M. & Pagel, B. E. J. 2004, *MNRAS*, 348, L59
Yin, S. Y., Liang, Y. C., Hammer, F., Brinchmann, J., *et al.* 2007, *A&A*, 462, 535 (Y07)

The intriguing life of massive galaxies
Proceedings IAU Symposium No. 295, 2012
D. Thomas, A. Pasquali & I. Ferreras, eds.

© International Astronomical Union 2013
doi:10.1017/S1743921313005310

Gas, dust and star formation in nearby galaxies as seen with the JCMT

José Ramón Sánchez-Gallego[1,2] and Johan H. Knapen[1,2]

[1]Instituto de Astrofísica de Canarias,
E-38205, La Laguna, Tenerife, Spain

[2]Departamento de Astrofísica, Universidad de La Laguna,
E-38200 La Laguna, Tenerife, Spain
email: jrsg@iac.es

Keywords. galaxies: ISM, ISM: molecules, stars: formation.

We presented results using the Nearby Galaxies Legacy Survey (NGLS), which is being carried out with the James Clerk Maxwell Telescope (JCMT) on Mauna Kea in Hawai'i. We have obtained CO $J = 3 - 2$ data for 155 nearby galaxies to trace the dense molecular gas in which stars are believed to be born. The sample of this survey covers a wide morphological range and has been selected to include galaxies that have been thoroughly studied in the literature, and for which abundant observational data are available. In parallel, we have observed the same sample of galaxies using the Hα spectral line, which traces massive star formation.

Using these data and infrared (IR), far-ultraviolet (FUV) and atomic hydrogen data from the literature, we analysed and discussed the distribution of star formation and gas in the nearby galaxy M 81, an object showing very little CO content (limited to some regions in the spiral arms) but enhanced star formation. We find a strong correlation between the different tracers of star formation both at a global scale and in the small detail. A correlation between CO and star formation is also found, but is much weaker.

We also presented a study of the massive star formation across the NGLS sample by using our Hα imaging and deriving star formation rates (SFRs) and equivalent widths (EWs). We find an overall SFR of $0.25\,M_\odot\,\mathrm{yr}^{-1}$ for the whole sample, which is slightly lower than other values found in comparable previous studies. Average SFR values for different subsamples show strong variations, from $0.1\,M_\odot\,\mathrm{yr}^{-1}$ for Virgo galaxies to $0.6\,M_\odot\,\mathrm{yr}^{-1}$ for large spirals. EWs are in the range of $1 - 880\,\text{Å}$ with a median value of $27\,\text{Å}$. We find a significant correlation between SFR and B-band luminosities, although our different subsamples show disparate behaviours. There is no correlation between EW and luminosity. A weak trend is found with morphology, with late spiral and irregulars showing higher EWs.

We combined CO $J = 3 - 2$ luminosities and Hα SFRs for the whole sample and find a good correlation, similar to previous studies that used lower transitions in the CO line. Several groups of galaxies, including M 81, show a peculiar behaviour in the $\mathrm{SFR}(\mathrm{H}\alpha) - L_{\mathrm{CO}\ J=3-2}$ plot, which disappears when using IR data. We analysed these regions and propose several hypotheses to explain this behaviour, including group interactions and very recent ($\lesssim 20\,\mathrm{Myr}$) bursts of star formation only traced by Hα emission. While atomic hydrogen is hardly related with SFR, molecular and total gas (H I +H$_2$) show similar correlations with Hα luminosities. No correlation is found between CO $J = 3 - 2$ luminosity and metallicity, although galaxies with low metallicity are more likely to remain undetected. Although CO $J = 3 - 2$ does not seem to be a significantly better tracer of the star-forming gas, its independence on metallicity may be used to constrain the conversion factor between CO and H$_2$.

Finally, we presented an ongoing project aimed to obtain continuum-subtracted Hα imaging for a large area of the Hercules Cluster (Abell 2151) using the Isaac Newton Telescope on La Palma. Additional data in several optical bands (u, g, r, i, z) are currently being acquired using the VLT Survey Telescope (VST). Using this variety of data, we aim to study the morphology, SFR and EW of the galaxies in the cluster across different environments.

The intriguing life of massive galaxies
Proceedings IAU Symposium No. 295, 2012
D. Thomas, A. Pasquali & I. Ferreras, eds.

© International Astronomical Union 2013
doi:10.1017/S1743921313005322

Plasma diagnostics of emission-line galaxies in SDSS

Zhitai Zhang[1,2]**, Yanchun Liang**[1]** and François Hammer**[3]

[1]National Astronomical Observatories, CAS, 20A Datun Road, 100012, Beijing, PR China
email: ztzhang@nao.cas.cn

[2]University of Chinese Academy of Sciences, 19A Yuquan Road, 100049, Beijing, PR China

[3]GEPI, Observatoire de Paris, CNRS, University Paris Diderot,
5 place Jules Janssen, 92195 Meudon, France

Abstract. We analyse a sample of 15,019 narrow emission-line galaxies, i.e. Seyferts, LINERs, composites and star-forming galaxies, from SDSS DR7 to study the differences between the different emission-line classes. We report two clear sequences of electron temperature (T_e) and density (n_e): $T_{e-LINER} \gtrsim T_{e-composite} > T_{e-Seyfert} > T_{e-star-forming}$ and $n_{e-Seyfert} \gtrsim n_{e-LINER} > n_{e-composite} > n_{e-star-forming}$. General transitions of n_e and T_e from central regions to disks are quantitatively confirmed.

Keywords. galaxies: active, galaxies: ISM, galaxies: Seyfert, galaxies: starburst, surveys

Brief Overview & Quick Results

We select an emission-line galaxy sample with S/N > 5, and divide into four classes by applying the galaxy classification schemes of Kewley *et al.* (2006). Plasma diagnostics are obtained through $I[\text{S II}]\lambda6716 /\lambda6731$ and $I[\text{O III}] \lambda5007/\lambda4363$ with simultaneous determination for $n_e[\text{S II}]$ and $T_e[\text{O III}]$ in 15,019 galaxies. We further identify three groups according to physical aperture size of the SDSS 3-arcsec diameter fibers ϕ (kpc) and $FWHM$ (km s^{-1}) of Hα (Bennert *et al.* 2006; Kollatschny & Wang 2006): $FWHM >$ 300 for "NLR-dominated" (labeled *ND*; ln$\phi < 1$) and "disk-contaminated NLR" (labeled *DC*; ln$\phi > 1$) objects; $FWHM < 300$ for "non-NLR" objects (labeled *NN*). See Table 1.

Table 1. Summary of the mean values of n_e and T_e.

	Seyfert			LINER			composite			Star-forming		
	ND	DC	NN	ND	DC	NN	ND	DC	NN	ND	DC	NN
Number	89	1,052	701	7	75	13	35	348	955	3	165	11,576
n_e [cm^{-3}]	415	332	160	230	201	113	208	150	77	166	152	57
T_e [10^4 K]	1.40	1.32	1.29	2.13	1.37	2.73	1.76	1.68	1.61	1.78	1.77	1.37

Notes. ND: "NLR-dominated" objects; DC: "disk-contaminated NLR" objects; NN: "non-NLR" ones.

The authors thank the symposium organizers for their invitation to this poster. This work was supported by the Natural Science Foundation of China (NSFC) Foundation under Nos.10933001, 11273026.

References

Bennert N., Jungwiert B., Komossa S., Haas M., & Chini R. 2006, *A&A*, 456, 953
Kewley L. J., Groves B., Kauffmann G., & Heckman T. 2006, *MNRAS*, 372, 961
Kollatschny W. & Wang T. G. 2008, *Ap&SS*, 303, 123

The intriguing life of massive galaxies
Proceedings IAU Symposium No. 295, 2012
D. Thomas, A. Pasquali & I. Ferreras, eds.

© International Astronomical Union 2013
doi:10.1017/S1743921313005334

Modelling the formation of today's massive ellipticals

Thorsten Naab[1]

[1]Max-Planck-Institute for Astrophysics, Karl-Schwarzschild-Str. 1, 85741 Garching, Germany
email: naab@mpa-garching.mpg.de

Abstract. The discovery of a population of massive, compact and quiescent early-type galaxies has changed the view on plausible formation scenarios for the present day population of elliptical galaxies. Traditionally assumed formation histories dominated by 'single events' like early collapse or major mergers appear to be incomplete and have to be embedded in the context of hierarchical cosmological models with continuous gas accretion and the merging of small stellar systems (minor mergers). Once these processes are consistently taken into account the hierarchical models favor a two-phase assembly process and are in much better shape to capture the observed trends. We review some aspects of recent progress in the field.

Keywords. galaxies: elliptical and lenticular, galaxies: formation, galaxies: evolution

1. Introduction

During the formation and assembly of massive galaxies merging is a natural process in modern hierarchical cosmological models. It is expected to play a significant role for the structural and morphological evolution (e.g. Kauffmann *et al.* 1996; Kauffmann 1996; De Lucia *et al.* 2006; Khochfar & Silk 2006; De Lucia & Blaizot 2007; Guo & White 2008; Kormendy *et al.* 2009; Hopkins *et al.* 2010a). In the light of these theoretical expectations and direct observations of 'dry' mergers of gas poor elliptical galaxies up to high redshift (van Dokkum 2005; Tran *et al.* 2005; Bell *et al.* 2006a,b; Lotz 2008; Jogee 2009; Newman *et al.* 2012; Man *et al.* 2012) simulations of idealized collisionless mergers have again received attention and new studies were triggered. Merger simulations of already existing spheroidal galaxies have focused in detail on the evolution of abundance gradients, shapes and kinematics, scaling relations, sizes and dark matter fractions (White 1978, 1979; Makino & Hut 1997; Boylan-Kolchin *et al.* 2005; Naab *et al.* 2006b; Boylan-Kolchin *et al.* 2006, 2008; Di Matteo *et al.* 2009; Nipoti *et al.* 2009b, 2012a). If the progenitors were two disk galaxies (which then can include a gaseous component) the aim was to investigate the morphological transformation, i.e. the formation of new dynamically hot spheroidal elliptical galaxies from two dynamically cold progenitor spiral galaxies (Gerhard 1981; Farouki & Shapiro 1982; Negroponte & White 1983; Barnes 1988; Barnes & Hernquist 1992; Hernquist 1992). Apart from studies of the effect of the merger mass-ratio (Barnes 1998; Bekki 1998; Bendo & Barnes 2000; Naab & Burkert 2003; Bournaud *et al.* 2004, 2005; González-García & Balcells 2005) the tidal torquing of gas, its inflow to the central regions, the impact on the stellar orbits (Barnes & Hernquist 1996; Naab *et al.* 2006a; Hoffman *et al.* 2010), subsequent starbursts (Mihos & Hernquist 1994, 1996; Barnes 2004; Di Matteo *et al.* 2008) and the potential growth of black holes (Hernquist 1989; Springel *et al.* 2005; Di Matteo *et al.* 2005; Johansson *et al.* 2009a; Younger *et al.* 2009) was investigated in numerous studies together with influential studies on the origin of early-type galaxy scaling relations (Robertson *et al.* 2006; Dekel & Cox 2006; Cox *et al.* 2006; Hopkins *et al.* 2008, 2009c,b,d; Debuhr *et al.* 2010; Moster *et al.* 2011). However, despite the detailed insights on the stellar and gas dynamical processes in simulated galaxy mergers, the 'binary merger' approach is limited in scope and seems not to be able to naturally explain all properties of present day massive elliptical galaxies (Naab & Ostriker 2009).

The most massive elliptical galaxies (or their progenitors) are considered to start forming their stars at high redshift ($z \sim 6$, or higher) in a dissipative environment, rapidly become very massive ($\sim 10^{11} M_\odot$) by $z = 2$ (Keres *et al.* 2005; Khochfar & Silk 2006; De Lucia *et al.* 2006; Kriek *et al.* 2006; Naab *et al.* 2007, 2009; Joung *et al.* 2009; Dekel *et al.* 2009; Keres *et al.* 2009; Oser *et al.* 2010; Feldmann *et al.* 2010; Domínguez Sánchez 2011; Feldmann *et al.* 2011; Oser *et al.* 2012). A significant fraction of this high redshift population is observed to be already quiescent at $z \sim 2$, on average 4-5 times more compact (part of this apparent evolution might driven by selection effects, see e.g. Poggianti *et al.* 2012), and typically a factor of two less massive than their low redshift descendants (Daddi *et al.* 2005; van der Wel *et al.* 2005; di Serego Alighieri *et al.* 2005; Trujillo 2006; Longhetti *et al.* 2007; Toft *et al.* 2007; Buitrago *et al.* 2008; van Dokkum *et al.* 2008; van der Wel *et al.* 2008; Cimatti *et al.* 2008; Franx *et al.* 2008; Damjanov *et al.* 2009; Cenarro & Trujillo 2009; Bezanson *et al.* 2009; van Dokkum *et al.* 2010; van de Sande *et al.* 2011; Whitaker *et al.* 2012). It is reasonable to assume that the high-redshift population forms the cores of at least some, if not all, present day massive ellipticals. This rapid structural evolution is supposed to happen in an inside-out fashion, mainly by adding stellar mass to the outer parts of the galaxies over time, however, without the formation of a significant fraction of new stars (Hopkins *et al.* 2009a; van Dokkum *et al.* 2010; Szomoru *et al.* 2012; Saracco *et al.* 2012). In this respect the growth of massive quiescent high-redshift galaxies is markedly different to the star formation driven inside-out growth of disk galaxies

The implications of these observational findings for the formation and evolution of massive elliptical galaxies are many-fold. They are unlikely to have formed by an initial 'monolithic collapse' followed by passive evolution as their present day counterparts would be too small and too red (van Dokkum *et al.* 2008; Kriek *et al.* 2008; Bezanson *et al.* 2009; Ferré-Mateu *et al.* 2012). In addition the evolution of these systems cannot be explained by just a single 'binary merger of disk galaxies'. The compact high-redshift systems might have formed in such a process (Wuyts *et al.* 2010; Bournaud *et al.* 2011), if it were gas-rich, but the subsequent structural evolution requires additional processes which are not driven by the formation of new stars. Observational results that almost none of these massive compact galaxies were able to survive to the present day (Trujillo *et al.* 2009; Taylor *et al.* 2010) indicate that a general and common physical mechanism must be at work. Spectacular events alone, like major early-type galaxy mergers, might be too rare.

2. Minor mergers vs. major mergers

Minor merges, however, are expected to happen frequently in the lifetime of a massive galaxy and have received particular attention as they provide a natural way to increase the size of a galaxy. With only a few assumptions the virial theorem provides a simple estimate of how a one-component system evolves during major and minor mergers (Cole *et al.* 2000; Naab *et al.* 2009; Bezanson *et al.* 2009). Following Naab *et al.* (2009) we assume that a compact initial stellar system has formed (e.g. involving gas dissipation) with a total energy E_i, a mass M_i, a gravitational radius $r_{g,i}$, and the mean square speed of the stars is $\langle v_i^2 \rangle$. According to the virial theorem (Binney & Tremaine 2008) the total energy of the system is

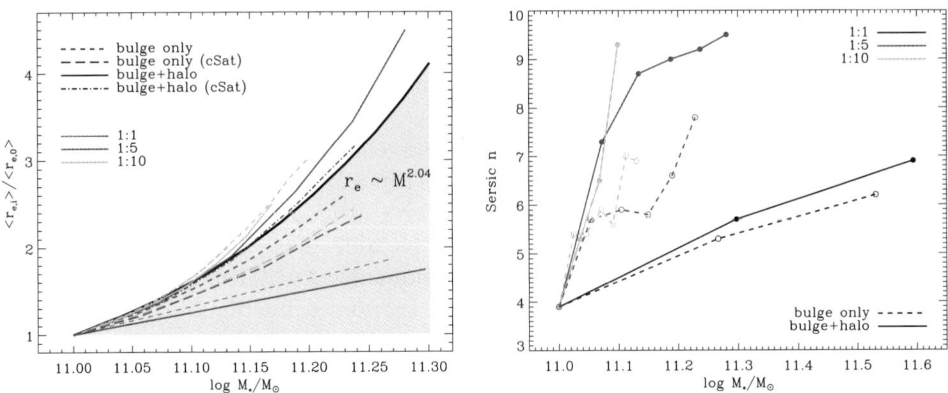

Figure 1. *Left:* Simulated size evolution as a function of bound stellar mass for mergers with mass-ratios 1:1 (blue), 5:1 (red), and 10:1 (green). The observationally expected relation is indicated by the black line (van Dokkum *et al.* 2010). The presence of dark matter significantly boosts the size evolution of 5:1 and 10:1 mergers. *Right:* This also leads to a significantly stronger evolution of the Sersic index (figures taken from Hilz *et al.* 2012a)

$$E_i \;=\; K_i + W_i = -K_i = \frac{1}{2}W_i$$
$$\;=\; -\frac{1}{2}M_i\langle v_i^2\rangle = -\frac{1}{2}\frac{GM_i^2}{r_{g,i}}.$$

This system then merges (on zero energy orbits) with other systems of a total energy E_a, total mass M_a, gravitational radii r_a and mean square speeds averaging $\langle v_a^2\rangle$. The fractional mass increase from all the merged galaxies is $\eta = M_a/M_i$ and the total kinetic energy of the material is $K_a = (1/2)M_a\langle v_a^2\rangle$, further defining $\epsilon = \langle v_a^2\rangle/\langle v_i^2\rangle$. Under the assumption of energy conservation (results from Khochfar & Burkert (2006) indicate that most halos merge on parabolic orbits) the ratio of initial to final mean square speeds, gravitational radii and densities can be then written as (Naab *et al.* 2009)

$$\frac{\langle v_f^2\rangle}{\langle v_i^2\rangle} = \frac{(1+\eta\epsilon)}{1+\eta}, \; \frac{r_{g,f}}{r_{g,i}} = \frac{(1+\eta)^2}{(1+\eta\epsilon)}, \; \frac{\rho_f}{\rho_i} = \frac{(1+\eta\epsilon)^3}{(1+\eta)^5}.$$

For mergers of two identical systems, $\eta = 1$, the mean square speed would remain unchanged, the size increases by a factor of two and the densities drop by a factor of four. In the limit that the mass is accreted in the form of very weakly bound stellar systems with $\langle v_a^2\rangle << \langle v_i^2\rangle$ or $\epsilon << 1$, the mean square speed is reduced by a factor two, the size increases by a factor four and the density drops by a factor of 32. These estimates are, however, idealized assuming one-component systems, no violent relaxation and zero-energy orbits with fixed angular momentum.

Hilz *et al.* (2012b) have recently re-investigated in detail the collisionless dynamics of major and minor mergers of systems including concentrated stellar spheroidal components embedded in extended dark matter halos. They present more accurate versions of the above equations including the effect of escapers and the interaction of the stellar baryonic with dark matter and describe in detail how the presence of a massive dark matter halos alter the evolution of the merging systems. One result of this study was that both minor and major mergers lead to size growth and an increase of the dark matter fraction. The physical processes are, however, different. Violent relaxation in major mergers mixes

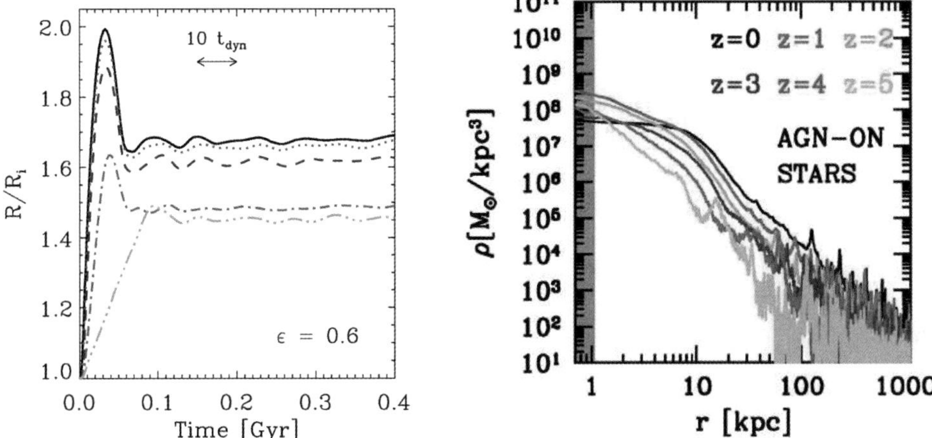

Figure 2. *Left:* Size evolution driven by rapid mass-loss from the idealized simulations of isolated galaxies. If 40 per cent of the mass is lost ($\epsilon = 0.6$) the sizes can rapidly increase by 60 per cent. From the black to the green line the ejection varies from immediate ejection to an ejection time of 80 Myrs (taken from Ragone-Figueroa & Granato 2011). *Right:* Evolution of the stellar surface density profiles of a cosmological zoom-simulation of a brightest cluster galaxy in a model with strong AGN feedback. Due to the gas explusion from the AGN the system is significantly more extended than in the no-AGN case and even develops a central core (taken from Martizzi *et al.* 2012).

dark matter to the central regions. Escaping, unbound, particles limit the expected size growth to values below the ones expected from the idealized equations above. In minor mergers (mass-ratios of 1:5 and 1:10), the stellar satellites are stripped at large radii where the host galaxies dominated by dark matter and the stellar effective radii and the dark matter fractions grow more rapidly than expected from the simple virial equations (see also Laporte *et al.* 2012). Due to the addition of stellar satellite material at large radii (Villumsen 1983), the stellar mass distribution changes significantly resulting in a significant increase of the Sersic index (see Fig. 1 and Hilz *et al.* 2012a). The general results on size evolution are in agreement with similar studies by e.g. Oogi & Habe (2012). However, there is an ongoing debate of whether the size growth by minor mergers is sufficient to explain the observed cosmological size evolution of elliptical galaxies. Whereas Oogi & Habe (2012) argue that the size growth by minor mergers alone might be sufficient, studies by Nipoti *et al.* (2012b), Cimatti *et al.* (2012), and Newman *et al.* (2012) have combined idealized numerical simulations embedded in a cosmological context and new observational constraints. They come to the conclusion that minor mergers might be able to explain the observed size growth from redshift $z \sim 1$ to the present. However, at higher redshift minor and major mergers might not be frequent enough to explain the rapid size evolution observed at $z \gtrsim 1$ and therefore an additional physical mechanism might be required.

A potential candidate for such an additional process is AGN driven outflow of gas from a massive high-redshift gas-rich and compact galaxy (Fan *et al.* 2008; Hopkins *et al.* 2010b; Fan *et al.* 2010). In general, stellar systems suffering from central mass-loss $\epsilon_{loss} = M_{final}/M_{initial}$ will expand (Hills 1980) and for rapid and slow mass-loss simple relations for the ratio of the final to the initial radius can be derived:

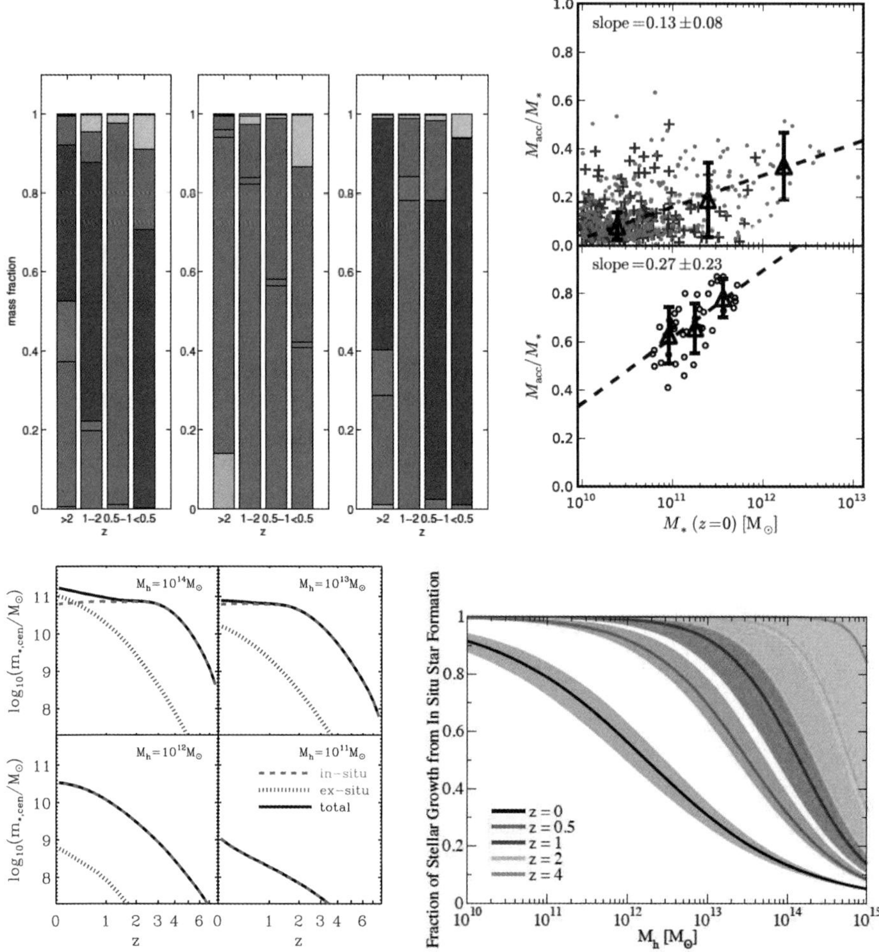

Figure 3. *Upper left:* Examples for the assembly history (stellar origin) of three massive galaxies in high-resolution cosmological zoom simulations. At high redshift the formation is dominated by in-situ star formation (red colors). The low redshift assembly is dominated by merging of stellar systems (blue colors, taken from Feldmann *et al.* 2010). *Upper right:* Ratio of the accreted over final stellar mass versus final stellar mass for galaxies in a cosmological simulation box (void: blue, cluster: red) including strong supernova feedback (upper panel). The fraction of accreted stars is about a factor 2 -3 lower than in the high-resolution zoom simulations of Oser *et al.* (2010) without strong supernova feedback (lower panel); the trend with mass is similar but less strong (taken from Lackner *et al.* 2012). *Lower left:* Independent estimate of the ratio of accreted to in-situ formed stars as a function of halo mass from abundance matching studies (Moster *et al.* 2012). *Lower right:* Similar estimates from a study by Behroozi *et al.* (2012). Both studies find a strong trend that the assembly of galaxies in more massive halos is more dominated by the accretion of stars rather than in-situ star formation.

$$\frac{R_{final,rapid}}{R_{initial}} = \frac{\epsilon_{loss}}{2\epsilon_{loss} - 1},$$
$$\frac{R_{final,slow}}{R_{initial}} = \frac{1}{\epsilon_{loss}}.$$

It is worth noting that rapid mass-loss of more than half the total mass can unbind the whole system. This process is well known and has been studied for star clusters (Hills

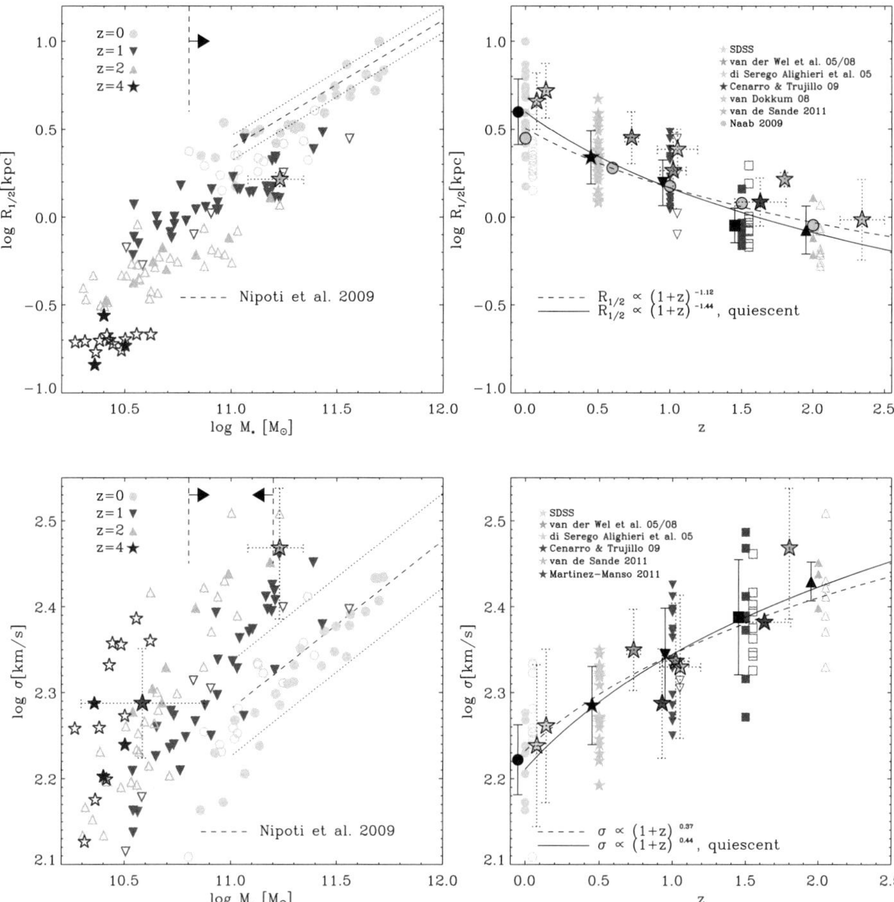

Figure 4. *Upper panels:* The present day mass-size relation (left) for a sample of high-resolution zoom simulations (blue points, full symbols are quiescent galaxies) compared to observations. The evolution of the relation is driven by accretion of stars and is indicated by the location of the most massive progenitor galaxies at different redshifts. For all galaxies more massive than the mass limit indicated on the left plot ($\log(M_*) = 10.8$) the average size evolution agrees well with observations. *Lower panels:* Similar plot for the evolution of the mass-dispersion relation (left). In a fixed mass range galaxies have higher dispersions at higher redshifts (right). Again, the simulated evolution is very similar to the observed one (figures are taken from Oser *et al.* 2012).

1980), galaxies (Hills 1980; Hopkins *et al.* 2010b; Ragone-Figueroa & Granato 2011; Pontzen & Governato 2012), as well as cores of galaxy clusters (see Fig. 2 and Martizzi *et al.* 2012).

3. The cosmological two-phase assembly

The assembly histories of massive galaxies in currently favored hierarchical cosmological models are significantly more complex than a single binary merger. They grow - in particular at high redshift - by smooth accretion of gas, major mergers but also numerous minor mergers covering a large range of mass-ratios which can dominate the amount of assembled stars. The picture that is emerging from semi-analytical models and high-resolution cosmological simulations of massive galaxies bears a two-phase characteristic

(De Lucia & Blaizot 2007; Guo & White 2008; Genel *et al.* 2008; Feldmann *et al.* 2010; Oser *et al.* 2010; Feldmann *et al.* 2011; Hirschmann *et al.* 2012).

At high redshifts the formation is dominated by dissipative processes (i.e. significant radiative energy losses) and in-situ star formation leading to compact progenitors with high phase space densities. In a second phase massive galaxies are growing by the addition of stars at large radii that have formed early outside the main galaxies in other galaxies that were accreted later-on. This assembly phase is dominated by collisionless dynamics and radiative energy losses are of minor importance (see e.g. Johansson *et al.* 2009b; Lackner & Ostriker 2010; Laporte *et al.* 2012).

Independent studies using cosmological simulations based on different numerical methods come to similar conclusions that - on average - the mass assembly of massive galaxies is dominated by minor mergers with mass-ratios $\sim 1 : 5$ (Oser *et al.* 2012; Lackner *et al.* 2012; Gabor & Davé 2012). The relative importance of accreted versus in-situ formed stars increases with galaxy mass, a result that was already predicted by semi-analytical models (De Lucia *et al.* 2006; De Lucia & Blaizot 2007; Guo & White 2008) and has been confirmed by independent estimates from abundance matching techniques (Moster *et al.* 2012; Behroozi *et al.* 2012). The absolute fractions are model dependent and can vary e.g. by $\sim 50\%$ for different feedback models (see Fig. 3). Studies based on cosmological zoom simulations make a plausible point that the present day scaling relations might be set by the stellar accretion history of massive galaxies, i.e. the above mentioned fraction of in-situ to accreted stars (Oser *et al.* 2012). In addition, based on still small samples, in high-resolution cosmological simulations the evolution of the scaling relations appears to be in accordance with observations (Feldmann *et al.* 2011; Oser *et al.* 2012; Johansson *et al.* 2012). However, in general the cosmological simulations of massive galaxies still fail to reproduce all observational constraints at the same time and are still limited with respect to either resolution and statistics as well as the algorithmic implementation of relevant feedback processes. In particular feedback from super-massive black holes might help to finally meet observational constraints for massive ellipticals (McCarthy *et al.* 2010, 2011; Puchwein & Springel 2012).

TN acknowledges support by and valuable disucssions with Peter Johansson, Ludwig Oser and Jeremiah P. Ostriker.

References

Barnes, J. E. 1988, *ApJ*, 331, 699

Barnes, J. E. 1998, in Saas-Fee Advanced Course 26: Galaxies: Interactions and Induced Star Formation, 275

Barnes, J. E. 2004, *MNRAS*, 350, 798

Barnes, J. E. & Hernquist, L. 1992, *ARA&A*, 30, 705

Barnes, J. E. & Hernquist, L. 1996, *ApJ*, 471, 115

Behroozi, P. S., Wechsler, R. H., & Conroy, C. 2012, *ApJ*, submitted, arXiv:1207.6105

Bekki, K. 1998, *ApJL*, 502, L133

Bell, E. F., *et al.* 2006a, *ApJ*, 640, 241

Bell, E. F., Phleps, S., Somerville, R. S., Wolf, C., Borch, A., & Meisenheimer, K. 2006b, *ApJ*, 652, 270

Bendo, G. J. & Barnes, J. E. 2000, *MNRAS*, 316, 315

Bezanson, R., van Dokkum, P. G., Tal, T., Marchesini, D., Kriek, M., Franx, M., & Coppi, P. 2009, *ApJ*, 697, 1290

Binney, J. & Tremaine, S. 2008, Galactic Dynamics: Second Edition

Bournaud, F., Chapon, D., Teyssier, R., Powell, L. C., Elmegreen, B. G., Elmegreen, D. M., Duc, P.-A., Contini, T., Epinat, B., & Shapiro, K. L. 2011, *ApJ*, 730, 4

Bournaud, F., Combes, F., & Jog, C. J. 2004, *A&A*, 418, L27

Bournaud, F., Jog, C. J., & Combes, F. 2005, *A&A*, 437, 69

Boylan-Kolchin, M., Ma, C., & Quataert, E. 2005, *MNRAS*, 362, 184

Boylan-Kolchin, M., Ma, C.-P., & Quataert, E. 2006, *MNRAS*, 369, 1081

Boylan-Kolchin, M., Ma, C.-P., & Quataert, E. 2008, *MNRAS*, 383, 93

Buitrago, F., Trujillo, I., Conselice, C. J., Bouwens, R. J., Dickinson, M., & Yan, H. 2008, *ApJL*, 687, L61

Cenarro, A. J. & Trujillo, I. 2009, *ApJL*, 696, L43

Cimatti, A., *et al.* 2008, *A&A*, 482, 21

Cimatti, A., Nipoti, C., & Cassata, P. 2012, *MNRAS*, 422, L62

Cole, S., Lacey, C. G., Baugh, C. M., & Frenk, C. S. 2000, *MNRAS*, 319, 168

Cox, T. J., Dutta, S. N., Di Matteo, T., Hernquist, L., Hopkins, P. F., Robertson, B., & Springel, V. 2006, *ApJ*, 650, 791

Daddi, E., *et al.* 2005, *ApJ*, 626, 680

Damjanov, I., *et al.* 2009, *ApJ*, 695, 101

De Lucia, G. & Blaizot, J. 2007, *MNRAS*, 375, 2

De Lucia, G., Springel, V., White, S. D. M., Croton, D., & Kauffmann, G. 2006, *MNRAS*, 366, 499

Debuhr, J., Quataert, E., Ma, C.-P., & Hopkins, P. 2010, *MNRAS*, 406, L55

Dekel, A. & Cox, T. J. 2006, *MNRAS*, 370, 1445

Dekel, A., Sari, R., & Ceverino, D. 2009, *ApJ*, 703, 785

Di Matteo, P., Bournaud, F., Martig, M., Combes, F., Melchior, A.-L., & Semelin, B. 2008, *A&A*, 492, 31

Di Matteo, P., Jog, C. J., Lehnert, M. D., Combes, F., & Semelin, B. 2009, *A&A*, 501, L9

Di Matteo, T., Springel, V., & Hernquist, L. 2005, *Nature*, 433, 604

di Serego Alighieri, S., *et al.* 2005, *A&A*, 442, 125

Domínguez Sánchez, H. *et al.* 2011, *MNRAS*, 417, 900

Fan, L., Lapi, A., Bressan, A., Bernardi, M., De Zotti, G., & Danese, L. 2010, *ApJ*, 718, 1460

Fan, L., Lapi, A., De Zotti, G., & Danese, L. 2008, *ApJL*, 689, L101

Farouki, R. T. & Shapiro, S. L. 1982, *ApJ*, 259, 103

Feldmann, R., Carollo, C. M., & Mayer, L. 2011, *ApJ*, 736, 88

Feldmann, R., Carollo, C. M., Mayer, L., Renzini, A., Lake, G., Quinn, T., Stinson, G. S., & Yepes, G. 2010, *ApJ*, 709, 218

Ferré-Mateu, A., Vazdekis, A., Trujillo, I., Sánchez-Blázquez, P., Ricciardelli, E., & de la Rosa, I. G. 2012, *MNRAS*, 2790

Franx, M., van Dokkum, P. G., Schreiber, N. M. F., Wuyts, S., Labbé, I., & Toft, S. 2008, *ApJ*, 688, 770

Gabor, J. M. & Davé, R. 2012, *MNRAS*, 427, 1816

Genel, S., *et al.* 2008, *ApJ*, 688, 789

Gerhard, O. E. 1981, *MNRAS*, 197, 179

González-García, A. C. & Balcells, M. 2005, *MNRAS*, 357, 753

Guo, Q. & White, S. D. M. 2008, *MNRAS*, 384, 2

Hernquist, L. 1989, *Nature*, 340, 687

Hernquist, L. 1992, *ApJ*, 400, 460

Hills, J. G. 1980, *ApJ*, 235, 986

Hilz, M., Naab, T., & Ostriker, J. P. 2012a, ArXiv e-prints

Hilz, M., Naab, T., Ostriker, J. P., Thomas, J., Burkert, A., & Jesseit, R. 2012b, *MNRAS*, 425, 3119

Hirschmann, M., Naab, T., Somerville, R. S., Burkert, A., & Oser, L. 2012, *MNRAS*, 419, 3200

Hoffman, L., Cox, T. J., Dutta, S., & Hernquist, L. 2010, *ApJ*, 723, 818

Hopkins, P. F., Bundy, K., Croton, D., Hernquist, L., Keres, D., Khochfar, S., Stewart, K., Wetzel, A., & Younger, J. D. 2010a, *ApJ*, 715, 202

Hopkins, P. F., Bundy, K., Hernquist, L., Wuyts, S., & Cox, T. J. 2010b, *MNRAS*, 401, 1099

Hopkins, P. F., Bundy, K., Murray, N., Quataert, E., Lauer, T. R., & Ma, C.-P. 2009a, *MNRAS*, 398, 898

Hopkins, P. F., Cox, T. J., Dutta, S. N., Hernquist, L., Kormendy, J., & Lauer, T. R. 2009b, *ApJS*, 181, 135

Hopkins, P. F., Hernquist, L., Cox, T. J., Dutta, S. N., & Rothberg, B. 2008, *ApJ*, 679, 156

Hopkins, P. F., Hernquist, L., Cox, T. J., Keres, D., & Wuyts, S. 2009c, *ApJ*, 691, 1424

Hopkins, P. F., Lauer, T. R., Cox, T. J., Hernquist, L., & Kormendy, J. 2009d, *ApJS*, 181, 486

Jogee, S. *et al.* 2009, *ApJ*, 697, 1971

Johansson, P. H., Naab, T., & Burkert, A. 2009a, *ApJ*, 690, 802

Johansson, P. H., Naab, T., & Ostriker, J. P. 2009b, *ApJL*, 697, L38

Johansson, P. H., Naab, T., & Ostriker, J. P. 2012, *ApJ*, 754, 115

Joung, M. R., Cen, R., & Bryan, G. L. 2009, *ApJL*, 692, L1

Kauffmann, G. 1996, *MNRAS*, 281, 487

Kauffmann, G., Charlot, S., & White, S. D. M. 1996, *MNRAS*, 283, L117

Kereš, D., Katz, N., Fardal, M., Davé, R., & Weinberg, D. H. 2009, *MNRAS*, 395, 160

Kereš, D., Katz, N., Weinberg, D. H., & Davé, R. 2005, *MNRAS*, 363, 2

Khochfar, S. & Burkert, A. 2006, *A&A*, 445, 403

Khochfar, S. & Silk, J. 2006, *ApJL*, 648, L21

Kormendy, J., Fisher, D. B., Cornell, M. E., & Bender, R. 2009, *ApJS*, 182, 216

Kriek, M., van der Wel, A., van Dokkum, P. G., Franx, M., & Illingworth, G. D. 2008, *ApJ*, 682, 896

Kriek, M., *et al.* 2006, *ApJL*, 649, L71

Lackner, C. N., Cen, R., Ostriker, J. P., & Joung, M. R. 2012, *MNRAS*, 425, 641

Lackner, C. N. & Ostriker, J. P. 2010, *ApJ*, 712, 88

Laporte, C. F. P., White, S. D. M., Naab, T., Ruszkowski, M., & Springel, V. 2012, *MNRAS*, 424, 747

Longhetti, M., *et al.* 2007, *MNRAS*, 374, 614

Lotz, J. M. *et al.* 2008, *ApJ*, 672, 177

Makino, J. & Hut, P. 1997, *ApJ*, 481, 83

Man, A. W. S., Toft, S., Zirm, A. W., Wuyts, S., & van der Wel, A. 2012, *ApJ*, 744, 85

Martizzi, D., Teyssier, R., Moore, B., & Wentz, T. 2012, *MNRAS*, 422, 3081

McCarthy, I. G., Schaye, J., Bower, R. G., Ponman, T. J., Booth, C. M., Dalla Vecchia, C., & Springel, V. 2011, *MNRAS*, 412, 1965

McCarthy, I. G., *et al.* 2010, *MNRAS*, 406, 822

Mihos, J. C. & Hernquist, L. 1994, *ApJL*, 431, L9

Mihos, J. C. & Hernquist, L. 1996, *ApJ*, 464, 641

Moster, B. P., Macciò, A. V., Somerville, R. S., Naab, T., & Cox, T. J. 2011, *MNRAS*, 415, 3750

Moster, B. P., Naab, T., & White, S. D. M. 2012, arXiv:1205.5807

Naab, T. & Burkert, A. 2003, *ApJ*, 597, 893

Naab, T., Jesseit, R., & Burkert, A. 2006a, *MNRAS*, 372, 839

Naab, T., Johansson, P. H., & Ostriker, J. P. 2009, *ApJL*, 699, L178

Naab, T., Johansson, P. H., Ostriker, J. P., & Efstathiou, G. 2007, *ApJ*, 658, 710

Naab, T., Khochfar, S., & Burkert, A. 2006b, *ApJL*, 636, L81

Naab, T. & Ostriker, J. P. 2009, *ApJ*, 690, 1452

Negroponte, J. & White, S. D. M. 1983, *MNRAS*, 205, 1009

Newman, A. B., Ellis, R. S., Bundy, K., & Treu, T. 2012, *ApJ*, 746, 162

Nipoti, C., Treu, T., & Bolton, A. S. 2009b, *ApJ*, 703, 1531

Nipoti, C., Treu, T., Leauthaud, A., Bundy, K., Newman, A. B., & Auger, M. W. 2012a, *MNRAS*, 422, 1714

Nipoti, C., Treu, T., Leauthaud, A., Bundy, K., Newman, A. B., & Auger, M. W. 2012b, *MNRAS*, 422, 1714

Oogi, T. & Habe, A. 2012, *MNRAS*, 42

Oser, L., Naab, T., Ostriker, J. P., & Johansson, P. H. 2012, *ApJ*, 744, 63

Oser, L., Ostriker, J. P., Naab, T., Johansson, P. H., & Burkert, A. 2010, *ApJ*, 725, 2312

Poggianti, B., Calvi, R., Bindoni, D., *et al.* 2012, arXiv:1211.1005

Pontzen, A. & Governato, F. 2012, *MNRAS*, 421, 3464

Puchwein, E. & Springel, V. 2012 arXiv e-prints

Ragone-Figueroa, C. & Granato, G. L. 2011, *MNRAS*, 414, 3690

Robertson, B., Cox, T. J., Hernquist, L., Franx, M., Hopkins, P. F., Martini, P., & Springel, V. 2006, *ApJ*, 641, 21

Saracco, P., Gargiulo, A., & Longhetti, M. 2012, *MNRAS*, 422, 3107

Springel, V., Di Matteo, T., & Hernquist, L. 2005, *MNRAS*, 361, 776

Szomoru, D., Franx, M., & van Dokkum, P. G. 2012, *ApJ*, 749, 121

Taylor, E. N., Franx, M., Glazebrook, K., Brinchmann, J., van der Wel, A., & van Dokkum, P. G. 2010, *ApJ*, 720, 723

Toft, S., *et al.* 2007, *ApJ*, 671, 285

Tran, K.-V. H., van Dokkum, P., Franx, M., Illingworth, G. D., Kelson, D. D., & Schreiber, N. M. F. 2005, *ApJL*, 627, L25

Trujillo, I., Cenarro, A. J., de Lorenzo-Cáceres, A., Vazdekis, A., de la Rosa, I. G., & Cava, A. 2009, *ApJL*, 692, L118

Trujillo, I. *et al.* 2006, *ApJ*, 650, 18

van de Sande, J., *et al.* 2011, *ApJL*, 736, L9

van der Wel, A., Franx, M., van Dokkum, P. G., Rix, H.-W., Illingworth, G. D., & Rosati, P. 2005, *ApJ*, 631, 145

van der Wel, A., Holden, B. P., Zirm, A. W., Franx, M., Rettura, A., Illingworth, G. D., & Ford, H. C. 2008, *ApJ*, 688, 48

van Dokkum, P. G. 2005, *AJ*, 130, 2647

van Dokkum, P. G., *et al.* 2008, *ApJL*, 677, L5

van Dokkum, P. G., *et al.* 2010, *ApJ*, 709, 1018

Villumsen, J. V. 1983, *MNRAS*, 204, 219

Whitaker, K. E., Kriek, M., van Dokkum, P. G., Bezanson, R., Brammer, G., Franx, M., & Labbé, I. 2012, *ApJ*, 745, 179

White, S. D. M. 1978, *MNRAS*, 184, 185

White, S. D. M. 1979, *MNRAS*, 189, 831

Wuyts, S., Cox, T. J., Hayward, C. C., Franx, M., Hernquist, L., Hopkins, P. F., Jonsson, P., & van Dokkum, P. G. 2010, *ApJ*, 722, 1666

Younger, J. D., Hayward, C. C., Narayanan, D., Cox, T. J., Hernquist, L., & Jonsson, P. 2009, *MNRAS*, 396, L66

The intriguing life of massive galaxies
Proceedings IAU Symposium No. 295, 2012
D. Thomas, A. Pasquali & I. Ferreras, eds.

© International Astronomical Union 2013
doi:10.1017/S1743921313005346

Red Galaxies from Hot Halos in Cosmological Hydro Simulations

Jared Gabor[1]

[1]CEA Saclay
Bat. 709, Gif-sur-Yvette, 91191 France
email: `jared.gabor@cea.fr`

Abstract. I highlight three results from cosmological hydrodynamic simulations that yield a realistic red sequence of galaxies: 1) Major galaxy mergers are not responsible for shutting off star-formation and forming the red sequence. Starvation in hot halos is. 2) Massive galaxies grow substantially ($\sim \times 2$ in mass) after being quenched, primarily via minor (1:5) mergers. 3) Hot halo quenching naturally explains why galaxies are red when they either (a) are massive or (b) live in dense environments.

Keywords. galaxies: evolution, galaxies: interactions

1. Major mergers are not responsible for quenching

The physical mechanism(s) responsible for shutting down star-formation and leading to the red sequence are still debated (Bower *et al.* 2006, Croton *et al.* 2006, Hopkins *et al.* 2008, Somerville *et al.* 2008). Two leading pictures have emerged: a) major mergers trigger starbursts and possibly AGN, whose combined gas consumption and feedback can rid the galaxy of gas to fuel star-formation; and b) hot gas coronae form in massive halos, shock-heating any infalling gas, and some heating process (such as a radio AGN) prevents that gas from cooling. Often, these are respectively called "quasar" mode and "radio" mode AGN feedback, alluding to the possible importance of AGN.

I have independently tested simplistic versions of these two mechanisms in cosmological hydrodynamic simulations. For merger quenching, I identify mergers on-the-fly during the simulation and eject all the gas from remnants in a 1000 km s^{-1} wind. For hot halo quenching, I identify galaxies whose halos are dominated by hot gas ($T > 10^{5.4}$ K; Kereš *et al.* 2005), and continuously add thermal energy to the circum-galactic gas around them. The results of these simulations are shown in Figure 1 (cf. Gabor *et al.* 2011) as color-magnitude diagrams and luminosity functions. Merger quenching fails to yield a significant red sequence at $z = 0$, whereas hot halo quenching forms red galaxies in numbers consistent with observations.

In the context of cosmological models, major mergers are neither necessary nor sufficient to explain the red sequence. They are not *sufficient* because galaxies are constantly accreting new gas from the cosmic web, even after mergers. Since the accreted gas provides fuel for star-formation, this accretion must be stopped to ensure that galaxies become red and stay red. Mergers are not *necessary* because an alternative quenching mechanism – hot halo quenching – appears to produce enough red galaxies. Note that there are enough major mergers to explain the numbers of red galaxies, if only the galaxies stopped accreting after the merger (Gabor *et al.* 2010).

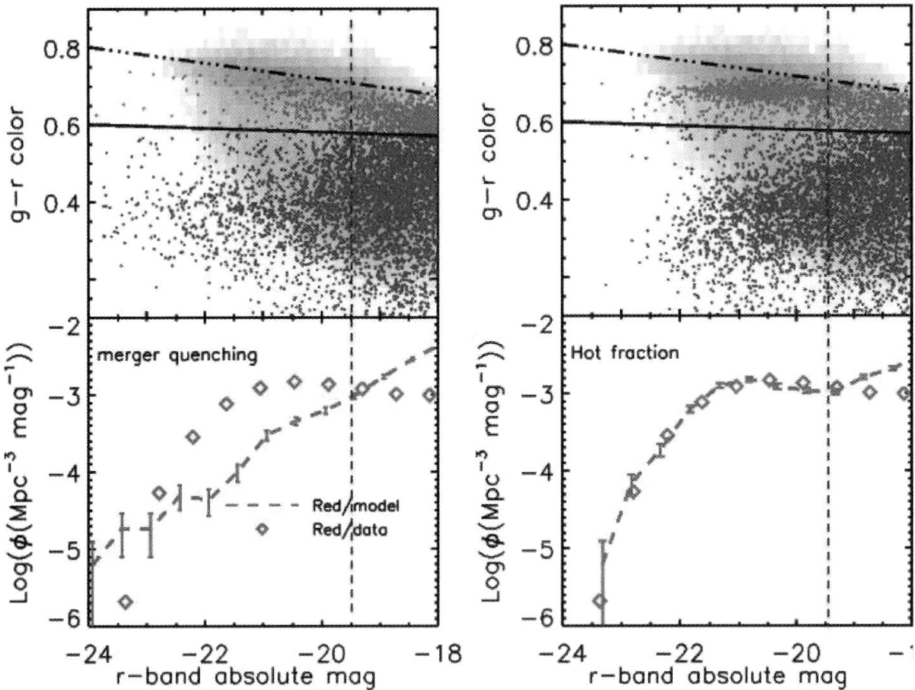

Figure 1. Merger quenching (left) does not produce enough red galaxies, but hot gas quenching (right) does. In color-magnitude diagrams (top panels) colored points represent simulated galaxies, and the grayscale represents SDSS galaxies. In the luminosity functions (bottom panels), dashed lines with error bars show the number densities of simulated red galaxies, and diamonds those from SDSS. Only hot halo quenching matches the $z \sim 0$ number densities of red galaxies. Based on Gabor *et al.* (2011).

2. Massive galaxies grow substantially after quenching

In the hot halo quenching simulation, massive galaxies typically turn red at $z > 0.5$, and grow significantly in mass after being quenched. The mass growth is shown in the left panel of Figure 2. Since star-formation is negligible in these galaxies, there are only two ways to change in stellar mass: mass loss from stellar evolution, and galaxy mergers. By number, most galaxies do not change much in mass – these are mostly recently-quenched satellites which are unlikely to merge with other satellites. Massive galaxies, on the other hand, grow by factors up to 3, with a large scatter driven by variations in merger history.

These massive galaxies are typically central galaxies accreting their small satellites in minor mergers. The right-hand panel of Figure 2 shows the mean merger mass ratio (for those galaxies with at least one merger) as a function $z = 0$ stellar mass. Here I have weighted each merger event by the mass of the smaller galaxy. Thus the plot shows the mass ratio which has been most important for adding mass to galaxies in each bin of stellar mass. In all cases, the characteristic merger is below the typical 1:3 major merger threshold, with a typical value of 1:5. Minor mergers dominate the mass growth of quenched galaxies. This result implies strong growth in the *sizes* of quenched galaxies at high redshift (Gabor *et al.* 2012, Oser *et al.* 2012).

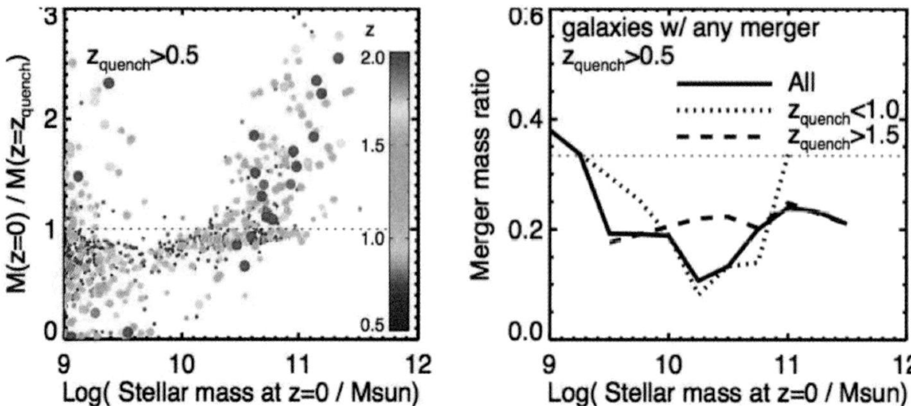

Figure 2. After becoming red, galaxies grow significantly via minor mergers. **Left:** Mass growth of galaxies quenched at $z > 0.5$, expressed as the $z = 0$ stellar mass divided by the mass at the time of quenching. Galaxies are color-coded by the redshift at which they were quenched. Massive galaxies have typically grown by factors ~ 2. **Right:** Mass-weighted mean merger mass-ratio for galaxies quenched at $z > 0.5$. Dashed line at 0.33 shows the typical division between major and minor mergers (1:3). Quenched galaxies grow mostly via minor (~ 1:5) mergers. Taken from Gabor & Davé (2012).

3. Hot gas quenching explains both "mass quenching" and "environment quenching"

In the hot halo quenching model, any galaxies that live in a hot corona are starved of incoming fuel for star-formation. A galaxy has two alternative paths to live inside a hot halo: a) it is the central galaxy in a halo of $> 10^{12} M_\odot$, where a hot coronae is likely to form; or b) it is a satellite galaxy in such a halo. Case (a) can be thought of as "mass quenching" or "central quenching", and case (b) can be thought of as "environment quenching" or "satellite quenching."

These two modes of quenching are apparent in Figure 3, inspired by Peng *et al.* (2010) and taken from Gabor & Davé (2012). The fraction of red galaxies increases with overdensity (i.e. in denser environments) and with stellar mass. Moreover, the "boxy" shape of the contours suggests that these modes are independent. In our simulation, they both result from hot gas cutting off the fuel supply for star-formation.

"Central quenching" and "satellite quenching" can be explained naturally in this model. In hydrodynamic simulations, halos above $\sim 10^{12} M_\odot$ are all dominated by hot gas (Birnboim & Dekel 2003, Kereš *et al.* 2005, Gabor *et al.* 2010). Furthermore, in the absence of quenching, stellar mass closely tracks halo mass. So a star-forming galaxy will increase its stellar mass as its halo mass increases due to accretion. Then, when the halo reaches $\sim 10^{12} M_\odot$ (i.e. stellar mass reaches $\sim 10^{10.5} M_\odot$), a hot corona will form which quenches star-formation. This manifests as a strong increase in red galaxy fraction at stellar masses $\gtrsim 10^{10.5} M_\odot$ – mass quenching.

Once a massive galaxy is quenched in this way, its satellite galaxies will also be quenched since they live in the same hot halo. The dark matter halo will continue to grow as additional galaxies fall in. Such infalling galaxies typically start out as star-forming centrals, but after becoming satellites they will be quenched by the hot gas halo. This is satellite quenching. Satellites become quenched regardless of their masses, as long as they live in sufficiently dense environments where hot gas dominates.

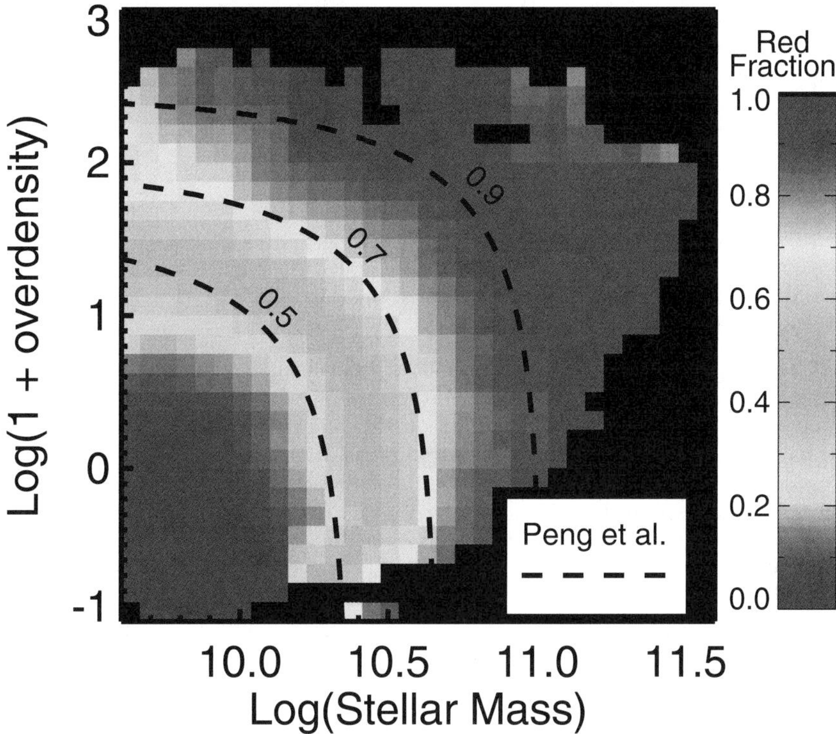

Figure 3. Galaxies are quenched at high stellar masses and high overdensities. I show the red fraction (color-coding) of galaxies as a function of stellar mass and local overdensity, a measure of environment. Dashed lines are from the model of Peng *et al.* 2010. The boxy contour shape suggests that "environment quenching" and "mass quenching" are independent, when in fact they both result from the presence of hot gas. Taken from Gabor & Davé (2012).

4. Summary

Despite its simplicity, the hot halo quenching model does a remarkable job of matching basic observables. Although hot halos do not tell the whole story, they appear to be the dominant factor in forming the red sequence. Mergers must play some role in the formation of today's red ellipticals, but halting the inflow of new gas is crucial.

References

Birnboim, Y. & Dekel, A. 2003, *MNRAS*, 345, 349

Bower, R. G. *et al.* 2006, *MNRAS*, 370, 645

Croton, D. *et al.* 2006, *MNRAS*, 365, 11

Gabor, J. M., Davé, R., Finlator, K., & Oppenheimer, B. D. 2010, *MNRAS*, 407, 749

Gabor, J. M., Davé, R., Oppenheimer, B. D., & Finlator, K. 2011, *MNRAS*, 417, 2676

Gabor, J. M. & Davé, R. 2012, *MNRAS*, 427, 1816

Hopkins, P. F., Cox, T. J., Kereš, D., & Hernquist, L. 2008, *ApJS*, 175, 390

Kereš, D., Katz, N., Weinberg, D. H., & Davé, R. 2005, *MNRAS*, 363, 2

Oser, L., Naab, T., Ostriker, J. P., & Johansson, P. H. 2012, *ApJ*, 744, 63

Peng, Y.-j. *et al.* 2010 *ApJ*, 721, 193

Somerville, R. S., Hopkins, P. F., Cox, T. J., Robertson, B. E., & Hernquist, L. 2008, *MNRAS*, 391, 481

The intriguing life of massive galaxies
Proceedings IAU Symposium No. 295, 2012
D. Thomas, A. Pasquali & I. Ferreras, eds.

© International Astronomical Union 2013
doi:10.1017/S1743921313005358

Assembly Histories and Observational Properties of Simulated Early-type Galaxies

Peter H. Johansson

Department of Physics, University of Helsinki,
Gustaf Hällströmin katu 2a, FI-00014 Helsinki, Finland
email: `Peter.Johansson@helsinki.fi`

Abstract. We demonstrate that massive simulated galaxies assemble in two phases, with the initial growth dominated by compact in situ star formation, whereas the late growth is dominated by accretion of old stars formed in subunits outside the main galaxy. We also show that 1) gravitational feedback strongly suppresses late star formation in massive galaxies contributing to the observed galaxy colour bimodality that 2) the observed galaxy downsizing can be explained naturally in the two-phased model and finally that 3) the details of the assembly histories of massive galaxies are directly connected to their observed kinematic properties.

Keywords. galaxies: elliptical and lenticular, galaxies: formation, galaxies: evolution

---

1. The two phased formation of early-type galaxies

High-resolution numerical zoom-in simulations of massive early-type galaxies have shown that there are two distinct phases in their formation histories (Naab *et al.* 2007; Oser *et al.* 2010; Lackner *et al.* 2012). At high redshifts of $z \sim 3-6$ the galaxies assemble rapidly through compact ($r < r_{\rm eff}$) in situ star formation. The later growth of the galaxies proceeds predominantly through the accretion of stars formed in subunits outside the main galaxy. The majority of the accreted stars are added to the outskirts of the galaxies at larger radii ($r > r_{\rm eff}$) providing a satisfactory explanation for the observed size growth of massive early-type galaxies (ETGs) since $z \sim 3$ until the present-day (Naab *et al.* 2009; Bezanson *et al.* 2009; Oser *et al.* 2012).

Here we show that in addition to the size growth of ETGs, the two phased formation picture can also shed light on other observational results of ETGs, such as the observed galaxy bimodality, the downsizing of massive galaxies and the observed dichotomy of the kinematic properties of ETGs. The simulation sample we use to address these questions contains 9 galaxies simulated at high-resolution (Johansson *et al.* 2012) using the Gadget-2 code (Springel 2005). The simulations include cooling for a primordial composition, star formation and feedback from type II supernovae, but exclude supernova driven winds and AGN feedback. We run three models (A2,C2,E2) at very high spatial ($\epsilon_\star = 0.125$ kpc) and mass resolution ($m_\star \sim 10^5 M_\odot$) with the remaining six simulations (U,Y,T,Q,M,L) simulated at a somewhat lower resolution ($\epsilon_\star = 0.25$ kpc, $m_\star \sim 10^6 M_\odot$). We organize the galaxies in our simulation sample into three groups of three galaxies each depending on whether their late assembly history ($z \lesssim 2$) is dominated primarily by dissipationless minor merging (mostly accreted stars: galaxies C2,U,Y), a mixed dissipationless/dissipational (mostly accreted and some in situ stars: galaxies A2,Q,T) or a primarily dissipational formation history (significant amount of in situ stars: galaxies E2,M,L). We find that this classification of the assembly histories of ETGs is useful as it broadly separates more massive ETGs forming dissipationlessly from less massive ETGs with a more dissipational formation history.

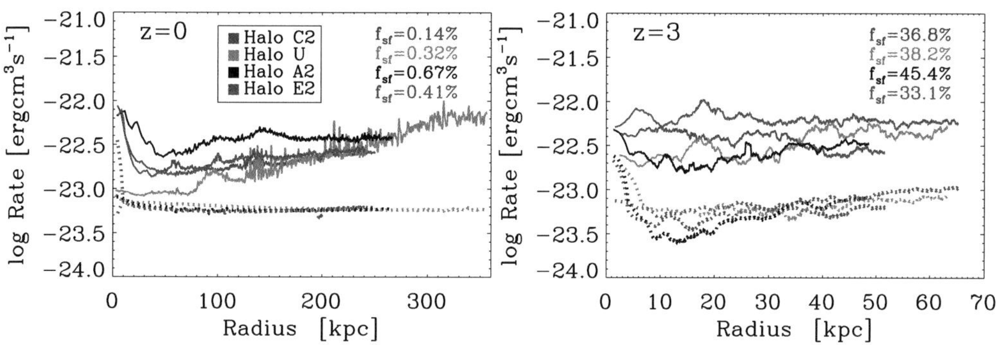

Figure 1. The net heating (solid line) and net cooling rates (dashed lines) for virial ($r < r_{\mathrm{vir}}$) non-starforming diffuse gas, for which the density is below the star formation threshold of $n_{\mathrm{th}} < 0.205$ cm^{-3} shown at redshifts of $z = 0$ (left) and $z = 3$ (right). The fraction f_{sf} of dense starforming gas ($n > n_{\mathrm{th}}$) is also given. Typically, the heating rate dominates over the cooling rate at all redshifts for the low-density non-starforming gas.

2. The bimodality of the local galaxy population

Recent large-field sky surveys have unequivocally demonstrated that local galaxies show a bimodal distribution. Local galaxies with stellar masses above a critical mass of $M_{\mathrm{crit}} \simeq 3 \times 10^{10} M_\odot$ are typically red spheroidal galaxies with old stellar populations, whereas galaxies below this critical mass are typically blue, star-forming galaxies with somewhat younger stellar populations. The observed bimodality is usually explained theoretically using models in which the star formation is efficiently quenched in massive haloes above $M \sim 10^{12} M_\odot$ (Dekel & Birnboim 2006; Gabor *et al.* 2011).

The late assembly of our simulated galaxies is dominated by dry minor merging building up the accreted stellar component. The infalling satellite galaxies captured through dynamical friction will cause gaseous wakes from which energy is transferred to the surrounding gas. Collectively together with the heating caused by supersonic collisions and shocks caused by infalling cold gas this process is deemed gravitational heating. In Fig. 1 we compare the shock-induced heating rates of the diffuse non-starforming gas with the corresponding cooling rates. The heating rates dominate over the cooling rates at all redshifts with the more massive galaxies (U,A2) showing higher heating rates than the slightly lower mass galaxies (C2,E2), as expected by the scaling of the gravitational feedback energy, $(\Delta E)_{\mathrm{grav}} \propto v_c^2$, where v_c is the circular velocity of the galaxy (Johansson *et al.* 2009b). The inclusion of gravitational heating helps in maintaining a hot gaseous halo and thus inhibits star formation contributing to the observed galaxy bimodality. However, gravitational heating is predominantly important in the outer parts of galaxies and some form of additional feedback, most probably AGN feedback (e.g. Johansson *et al.* 2009a), is required to stop late central star formation in very massive galaxies (e.g. galaxy U at $z = 0$).

3. Downsizing of massive galaxies

Several recent observations have shown that old, massive red metal-rich galaxies were already in place at high redshifts of $z \sim 2 - 3$. This observational result can be seen as a manifestation of cosmic downsizing, in which galaxies seem to form anti-hierarchically in the sense that the most massive galaxies formed a significant proportion of their stars at high redshifts, compared to lower mass systems that exhibit a more continuous star formation history throughout the cosmic epoch (e.g. Glazebrook *et al.* 2004).

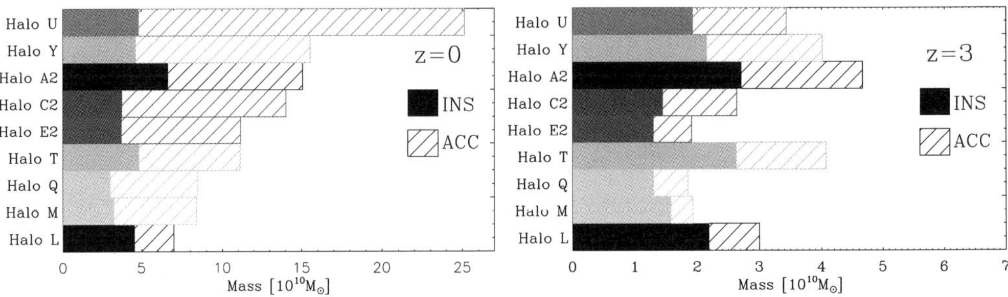

Figure 2. The stellar mass assembly histories of our simulated galaxy sample, with the solid colour bars showing the contribution of in situ formed stellar mass and the dashed colour bars representing the contribution of accreted stellar mass shown at redshifts of $z = 0$ (left) and $z = 3$ (right). At high redshifts ($z \gtrsim 3$) the galaxies assemble rapidly through in situ star formation, whereas the late ($z \lesssim 3$) assembly history is dominated by accreted stars, with the more massive galaxies ending up with a proportionally larger fraction of accreted stars.

In Fig. 2 we show the assembly histories of all our simulated galaxies depicting separately the masses in the in-situ formed and accreted stellar components. At $z = 3$ the stellar components in all galaxies have been assembled rapidly through mainly in situ star formation, fueled by cold gas flows and hierarchical mergers of multiple star-bursting subunits. At lower redshifts ($z \lesssim 3$) the subsequent growth of the stellar component proceeds predominantly through the accretion of existing stellar clumps. The galaxies in Fig. 2 are ordered in decreasing final stellar mass from top to bottom and we can immediately see that the fraction of accreted stars at $z = 0$ increases as a function of galaxy mass. The accreted stellar component forms on average in low mass galaxies very early in the Universe. However, the accreted stellar component is added to the massive galaxies much later and at lower redshifts than the in situ formed stars. In addition, the metallicity of the central in situ formed stars is on average higher than the metallicity of the accreted stars that are formed in lower mass galaxies and added later to the outskirts of galaxies, resulting in a negative metallicity gradient, in agreement with the observations. Thus, the counter-intuitive concept of downsizing can be explained in the two phased formation mechanism. Massive galaxies form their central stellar mass in situ and then accrete substantial amounts of stars that were formed even earlier in smaller subsystems. Hence by $z \sim 2 - 3$ the most massive galaxies have the oldest stellar populations compared to lower mass galaxies that are still forming in situ stars.

4. Kinematic properties of massive galaxies

Observations in the late 1980s using slit spectroscopy found that the ETG population can be broadly separated into a population of massive slowly-rotating systems ($v/\sigma < 0.1$) with boxy isophotes and a population of fast-rotating ($v/\sigma \sim 1$) ETGs with more disky isophotes found typically at somewhat lower masses. Recent results from the volume limited ATLAS^{3D} survey utilizing a modern integral-field-unit (IFU) confirmed this dichotomy showing that about $\sim 15\%$ of local ETGs rotate slowly with no indications of an embedded disk component, whereas the majority ($\sim 85\%$) of the local ETGs show significant disk-like rotation (Cappellari et al. 2011).

We derive the kinematic properties of our simulated galaxies using 500 random projections in order to assess the mean properties of the simulated galaxies averaged over all sightlines. In Fig. 3 we plot the 95% probability of finding a simulated galaxy in the $\epsilon_{\rm eff} - (v_{\rm maj}/\sigma_0)$ plane, where $\epsilon_{\rm eff}$ measures the effective ellipticity of the galaxies and

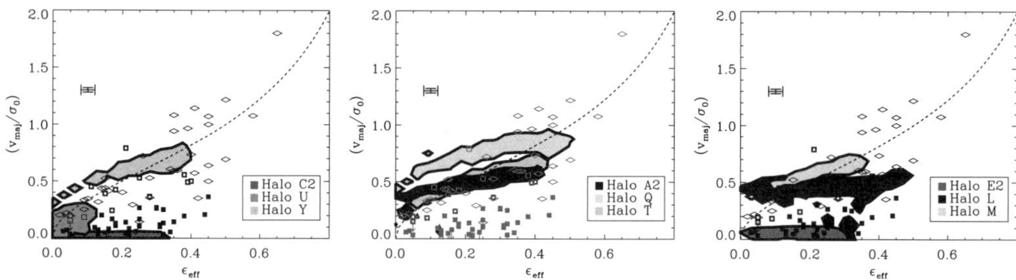

Figure 3. The effective ellipticity (ϵ_{eff}) is plotted against the ratio of the major axis rotation and central velocity dispersion ($v_{\mathrm{maj}}/\sigma_0$). The contours show the 95% probability location of the galaxies in the ($\epsilon_{\mathrm{eff}} - v_{\mathrm{maj}}/\sigma_0$)-plane derived from 500 random viewing angles of our simulation data. The overplotted symbols are observational data from (Bender *et al.* 1994) demonstrating that our simulated galaxies seem to be largely consistent with the observations. The dashed line shows the theoretical value for an oblate isotropic rotator.

the ratio of the major axis rotation and velocity dispersion ($v_{\mathrm{maj}}/\sigma_0$) is a measure of the rotational support of the galaxies. Galaxies U,C,E2 show projections with very low rotational support, whereas the other galaxies (Y,A2,Q,T,L,M) show significantly more rotational support in very good agreement with the theoretical prediction for an oblate isotropic rotator shown by the dashed line in Fig. 3. The most massive galaxy U shows mostly round projections with very low ϵ_{eff}, whereas all the other galaxies show projections extending in ϵ_{eff} from zero up to 0.4. We see a correlation between the rotational support and the in situ/accreted fraction, both galaxies U,C that have a large accreted fraction are slowly rotating and galaxy Y would most probably also been slowly rotating if it had not experienced a late ($z < 0.5$) major merger. Thus, within our rather narrow mass range the majority of the simulated galaxies are consistent with being rotationally supported disky ellipticals, whereas the most massive galaxy in our sample is consistent with being a roundish slow-rotator.

References

Bender, R., Saglia, R. P., & Gerhard, O. E. 1994, *MNRAS*, 269, 785

Bezanson, R., van Dokkum, P. G., Tal, T., *et al.* 2009, *ApJ*, 697, 1290

Cappellari, M., Emsellem, E., Krajnović, D., *et al.*, 2011, *MNRAS*, 413, 813

Dekel, A. & Birnboim, Y. 2006, *MNRAS*, 368, 2

Gabor, J. M., Davé, R., Oppenheimer, B. D., & Finlator, K., 2011, *MNRAS*, 417, 2676

Glazebrook, K., Abraham, R. G., McCarthy, P. J., *et al.* 2004, *Nature*, 430, 181

Johansson, P. H., Naab, T., & Burkert, A. 2009a, *ApJ*, 690, 802

Johansson, P. H., Naab, T., & Ostriker, J. P. 2009b, *ApJL*, 697, L38

Johansson, P. H., Naab, T., & Ostriker, J. P. 2012, *ApJ*, 754, 115

Lackner, C. N., Cen, R., Ostriker, J. P., & Joung, M. R., 2012 *MNRAS*, 425, 641

Naab, T., Johansson, P. H., Ostriker, J. P., & Efstathiou, G. 2007, *ApJ*, 658, 710

Naab, T., Johansson, P. H., & Ostriker, J. P. 2009, *ApJL*, 699, L178

Oser, L., Ostriker, J. P., Naab, T., Johansson, P. H., & Burkert, A. 2010, *ApJ*, 725, 2312

Oser, L., Naab, T., Ostriker, J. P., & Johansson, P. H. 2012, *ApJ*, 744, 63

Springel, V. 2005, *MNRAS*, 364, 1105

The intriguing life of massive galaxies
Proceedings IAU Symposium No. 295, 2012
D. Thomas, A. Pasquali & I. Ferreras, eds.

© International Astronomical Union 2013
doi:10.1017/S174392131300536X

Probing the mass assembly of massive nearby galaxies with deep imaging

P.–A. Duc,[1] J.-C. Cuillandre,[2] K. Alatalo,[3] L. Blitz,[3] M. Bois,[4]
F. Bournaud,[1] M. Bureau,[5] M. Cappellari,[5] P. Côté,[6] R. L. Davies,[5]
T. A. Davis,[7] P. T. de Zeeuw,[7,8] E. Emsellem,[7,4] L. Ferrarese,[6]
E. Ferriere,[1] S. Gwyn,[6] S. Khochfar,[9] D. Krajnovic,[7] H. Kuntschner,[7]
P.-Y. Lablanche,[4] R. M. McDermid,[10] L. Michel-Dansac,[4]
R. Morganti,[11] T. Naab,[12] T. Oosterloo,[11] M. Sarzi,[13] N. Scott,[5]
P. Serra,[11] A. Weijmans[14] and L. M. Young[15]

[1] AIM Paris-Saclay, France; [2] CFHT, USA; [3] University of California, Berkeley, USA;
[4] Observatoire de Lyon, France; [5] University of Oxford, UK; [6] Herzberg Institute of
Astrophysics, Victoria, Canada; [7] ESO, Garching, Germany; [8] Leiden University, The
Netherlands; [9] MPE, Garching, Germany; [10] Gemini Observatory, Hilo, USA; [11] ASTRON,
Dwingeloo, The Netherlands; [12] MPI, Garching, Germany; [13] University of Hertfordshire,
Hatfield, UK; [14] Dunlap Institute for Astronomy & Astrophysics, University of Toronto,
Canada; [15] New Mexico Tech, Socorro, USA

Abstract. According to a popular scenario supported by numerical models, the mass assembly and growth of massive galaxies, in particular the Early-Type Galaxies (ETGs), is, below a redshift of 1, mainly due to the accretion of multiple gas–poor satellites. In order to get observational evidence of the role played by minor dry mergers, we are obtaining extremely deep optical images of a complete volume limited sample of nearby ETGs. These observations, done with the CFHT as part of the ATLAS^{3D}, NGVS and MATLAS projects, reach a stunning 28.5 – 29 mag.arcsec^{-2} surface brightness limit in the g' band. They allow us to detect the relics of past collisions such as faint stellar tidal tails as well as the very extended stellar halos which keep the memory of the last episodes of galactic accretion. Images and preliminary results from this on-going survey are presented, in particular a possible correlation between the fine structure index (which parametrizes the amount of tidal perturbation) of the ETGs, their stellar mass, effective radius and gas content.

Keywords. galaxies: evolution, galaxies: interactions, galaxies: elliptical and lenticular, cD

1. Introduction

Early-Type Galaxies play a key role in modern cosmology: according to the standard hierarchical cosmological model, galaxies build up from successive mergers, associated by a series of morphological transformations. The massive ETGs we see today are the end-product of this process. *At high redshifts*, few ETGs are observed, but surprisingly they have not all disappeared, raising questions on how they formed so quickly in the traditional merging scenario. They appear to be also very compact (e.g. Buitrago *et al.*, 2008). *At low redshift*, the early-type galaxies are observed to be larger but also to be much more complex and lively that once believed. As shown in the presentations of the ATLAS^{3D} results in this volume, a large fraction of them contain in particular gas, some with non-regular kinematics, and thus may still be involved in active transformation processes.

Simulations indicate that while the global light profile of ETGs can be rather easily obtained with various processes, getting their total mass and large radius is much more challenging and may require multiple collisions in the recent past. Several recent papers have highlighted the role of minor mergers in the growth of galaxies (e.g., Johansson *et al.*, 2012; Newman *et al.*, 2012). These late mergers impact the properties of the stellar populations mostly at large galactocentric radii. In particular minor mergers bring low metallicity stars from the infalling dwarf satellites, and create radial color gradients. On the other hand, a major merger induces large radial mixing, leading to a washing up of metallicity/age and thus color gradients.

Furthermore, the mass assembly of galaxies leaves various imprints in their surroundings, such as shells, streams and tidal tails. The frequency, shape and properties of these fine structures depend on the mechanism driving the mass assembly (see the review by Duc & Renaud, 2011). Any analysis of the fine structures around galaxies, however, should take into account (1) that the resulting stellar debris have a very low surface brightness, (2) that such debris fade with age, or may be dispersed by the local environment, and (3) that galaxies may have formed by multiple processes.

Deep imaging surveys coupled with numerical simulations done in cosmological context can now address these issues. Several studies have quantified the importance of fine structures around massive galaxies and their variation with environment and redshift (Tal *et al.*, 2009; Bridge *et al.*, 2010; Adams *et al.*, 2012) but were restricted to the census of luminous tidal features. So far, the surface brightness limit required to apply a genuine galactic archeology technique has been reached for only local galaxies for which stellar counts may be done. As however shown, among others, by Martinez-Delgado *et al.* (2010) and Duc *et al.* (2011), deep optical images obtained under specific conditions can also reveal the diffuse light associated with very low-surface brightness structures.

2. Observations and data reduction

Our targets are the 260 nearby ETGs located at distances below 42 Mpc from the ATLAS^{3D} survey (Cappellari *et al.*, 2011). Very deep optical images in the g',r' and i' bands are currently being obtained with the large field of view camera MegaCam installed on the CFHT. The observations are carried out as part of several projects: ATLAS^{3D}, the Next Generation Virgo Cluster Survey (NGVS, covering the full Virgo Cluster area; Ferrarese *et al.*, 2012), and the recently accepted CFHT MATLAS Large Programme. The typical integration time is about one hour per band. The limiting sensitivity is outstanding with respect to previous generations of optical images: about $28.5 - 29$ mag.arcsec^{-2} in the g'–band. This could be achieved using dedicated observing strategies and data reduction softwares. On traditional images obtained by MegaCam, the presence of scattered light patterns masks extended features below surface brightness of 27 mag.arcsec^{-2}. Investigations motivated by the NGVS have shown that this problem can be overcome carrying out a sequence of observations with large offsets between the images, as it is usually done with infrared observations. A sky is then computed and subtracted from the individual images, before they are stacked.

The majority of the fine structures disclosed by the survey are very extended and directly show up on surface brightness or color maps. To disclose those located more towards the inner regions, several methods were used: unsharp masking, best at revealing sharp-edged structures such as shells or narrow filaments; subtraction of the ETGs modeled by an ellipse fitting or by a multi-component GALFIT model, which helps to detect extended asymmetric features. As a first step to quantify the amount of tidal perturbations, a fine structure index was determined, counting by eye the number of structures, and weighting them according to their nature.

Figure 1. Deep optical images of a sub-sample of nearby Early-Type Galaxies obtained with MegaCam on the CFHT as part of the ATLAS^{3D} and Next Generation Virgo Cluster Survey. The figure illustrates the variety of low surface brightness structures that show up around these galaxies: long tidal tails and shells, telling us about past major mergers (a,d); narrow stellar filaments associated with disrupted dwarf satellites, revealing future minor mergers (b); regular low surface brightness star–forming disks (c); extended featureless stellar halos (e).

3. Preliminary results

Figure 1 illustrates the variety of low surface brightness structures detected around the ETGs: 200 kpc long tidal tails revealing past 2-3 Gyr old major merger (for instance around NGC 5557; Duc *et al.*, 2011), narrow filaments around a disrupted dwarf, with their typical S-shape and wrapping around the host, diffuse halos, some remaining regular even up to large radii.

At time of writing, more than half of the ATLAS^{3D} ETGs benefit from deep Mega-Cam images. This initial sub-sample is however somehow biased towards the most massive ETGs, those that are slow rotating and/or gas–rich. In such conditions, providing the percentage of disturbed galaxies is premature. Nevertheless some initial trends were found.

First of all, statistically, the galaxies that contain atomic hydrogen, in particular in their outskirts, have a higher fine structure index. This is a strong indication that the HI clouds are associated with collisional debris. A rather large fraction of ATLAS^{3D} ETGs contain detectable molecular gas, as traced by CO. Those for which the CO is kinematically misaligned with respect to the stellar component and for which an external origin of the molecular gas had been proposed (Davis *et al.* 2011), exhibit collisional debris, just as expected. Finally one of the most promising results is the trend with mass and size, two fundamental parameters in the scaling laws of galaxies. As shown on Figure 2, the more massive and the more extended† the galaxy is, the higher its fine structure index. Conversely, many, usually fast rotating low–mass, ETGs do not show

† As mentioned a number of times during the symposium, the measure of the effective radius, the radius containing half of the stellar light, is tricky at high redshift. Our study indicates that this is even the case at low redshift. For the ATLAS^{3D} ETGs showing on MegaCam images a

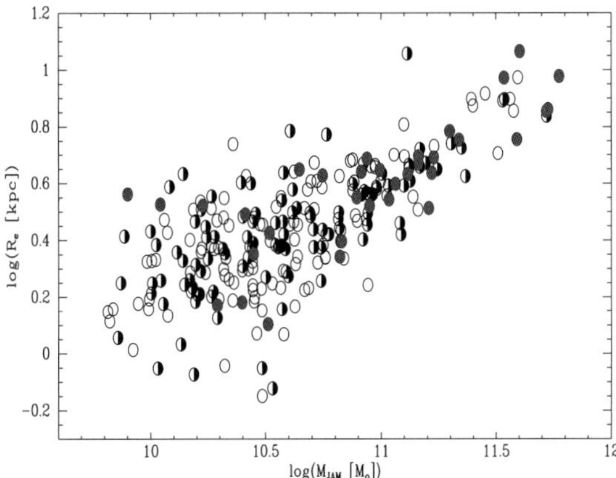

Figure 2. Effective radius, R_e^{max}, vs M_{JAM}, a proxy of the stellar mass, for the full sample of ATLAS^{3D} ETGs (open circles, Cappellari *et al.*, 2012). The galaxies shown with the half filled circles have already deep MegaCam images available from either the ATLAS^{3D} or NGVS surveys. The images for the other galaxies will be obtained as part of the MATLAS CFHT Large Programme. Those that have a high fine structure index, i.e. galaxies that are either strongly tidally perturbed, or have multiple stellar streams in their vicinity, are shown with the red filled circles.

any sign of external perturbation. This tells that the most massive galaxies in the local Universe have had a rich recent mass accretion history. At which level it accounts for their mass/size growth remains to be determined. Since the optical images only reveal structures brighter than 29 mag.arcsec^{-2}, simulations will be key to extrapolate from the observations the quantity of accreted material since $z = 1$. Another crucial parameter will be the survival time of tidal features, which likely depends on the environment. The study of the fine structure index as a function of the local density – our volume limited sample covers a large range of environments: field, groups and the Virgo Cluster – will give interesting constrains on that matter.

References

Adams, S. M., Zaritsky, D., Sand, D. J., *et al.* 2012, *AJ*, 144, 128
Bridge, C. R., Carlberg, R. G., & Sullivan, M. 2010, *ApJ*, 709, 1067
Buitrago, F., Trujillo, I., Conselice, C. J., *et al.* 2008, *ApJ*, 687, L61
Cappellari, M., Emsellem, E., Krajnovic;, D, *et al.* 2011, *MNRAS*, 413, 813
Cappellari, M., McDermid, R. M., Alatalo, *et al.*2012, *MNRAS*, in press (arXiv:1208.3523)
Davis, T. A., Alatalo, K., Sarzi, M., Bureau, M., Young, L. M., *et al.* 2011, *MNRAS*, 417, 882
Duc, P.-A., Cuillandre, J.-C., Serra, P., *et al.* 2011, *MNRAS*, 417, 863
Duc, P.-A. & Renaud, F. 2013, in Tides in Astronomy and Astrophysics, Lecture Notes in Physics, (Springer-Verlag Berlin Heidelberg), Vol. 861, p. 327
Ferrarese, L., Côté;, P., Cuillandre, J.-C., *et al.* 2012, *ApJS*, 200, 4
Johansson, P. H., Naab, T., & Ostriker, J. P. 2012, *ApJ*, 754, 115
Martinez-Delgado, D., Gabany, R. J., Crawford, K., Zibetti, S. *et al.* 2010, *AJ*, 140, 962
Newman, A. B., Ellis, R. S., Bundy, K. *et al.* 2012, *ApJ*, 746, 162
Tal, T., van Dokkum, P. G., Nelan, J., & Bezanson, R. 2009, *AJ*, 138, 1417

very extended low surface brightness halo, the value had to be revised, with differences with earlier published values reaching factors up to 2.

The intriguing life of massive galaxies
Proceedings IAU Symposium No. 295, 2012
D. Thomas, A. Pasquali & I. Ferreras, eds.

© International Astronomical Union 2013
doi:10.1017/S1743921313005371

The role of Active Galactic Nuclei feedback in the formation of the brightest cluster galaxies

Davide Martizzi[1], Romain Teyssier[1,2] and Ben Moore[1]

[1]Institute for Theoretical Physics, University of Zurich, CH-8057 Zürich, Switzerland
email: martdav@physik.uzh.ch, teyssier@physik.uzh.ch, moore@physik.uzh.ch
[2]CEA Saclay, DSM/IRFU/SAP, Bâtiment 709, F-91191 Gif-sur-Yvette, Cedex, France

Abstract. The formation of the brightest cluster galaxies (BCG) is a challenge for galaxy formation theory. We performed high resolution cosmological hydrodynamical simulations with the AMR code RAMSES to study the properties of the BCG which forms at the center of a Virgo–like cluster. We compare the results of 2 galaxy formation scenarios, one in which only supernovae feedback is included, and one in which also AGN feedback is considered. Properties of the simulated BCG which are comparable with those of observed massive elliptical galaxies and BCGs cannot be obtained if AGN feedback is not considered. The stellar-to-halo mass ratio in simulations without AGN feedback appears too large when compared to observations, while it is compatible the observationally determined values when AGN feedback is included. The kinematical and structural properties of the BCG are extremely different in the two models. When we do not include AGN feedback, the BCG is quickly rotating, with high Sérsic index, a clear mass excess in the center and a very large stellar mass fraction. When AGN feedback is considered, the BCG is slowly rotating, with a significantly cored surface density profile and low stellar mass fraction.

Keywords. Galaxy Formation – Theory, Galaxy Clusters, Black Holes, Cosmology, Numerical Methods

1. Introduction

Galaxy formation models that not include AGN feedback are known to overpredict the masses and star formation rates of BCGs at redshift $z = 0$. AGN are expected to provide enough gas heating to prevent excessive star formation, therefore the inclusion of AGN feedback in theoretical models is extremely important when trying to reproduce the properties of BCGs. We use the AMR code RAMSES (Teyssier 2002) to perform two hydrodynamical cosmological simulations of a Virgo-like cluster of galaxies with a virial mass of $M_{\rm vir} = 10^{14}$ M$_\odot$ in the context of the ΛCDM cosmological scenario. We achieve a mass resolution of 8.2×10^6 M$_\odot$ and a spatial resolution of $\sim$500 pc, that allow us to resolve the formation and evolution of the brightest central galaxy in the cluster. The simulations include gas dynamics and standard galaxy formation recipes, including gas cooling, the effect of UV background, star formation and supernovae feedback. In one run, labeled as AGN-ON, we include AGN feedback adopting a modified version of the Booth & Schaye (2009) model. In the second simulation we do not include AGN feedback, therefore we labeled it as AGN-OFF. Comparison between the two runs and observational results allowed us to study the relevance of AGN feedback in shaping the

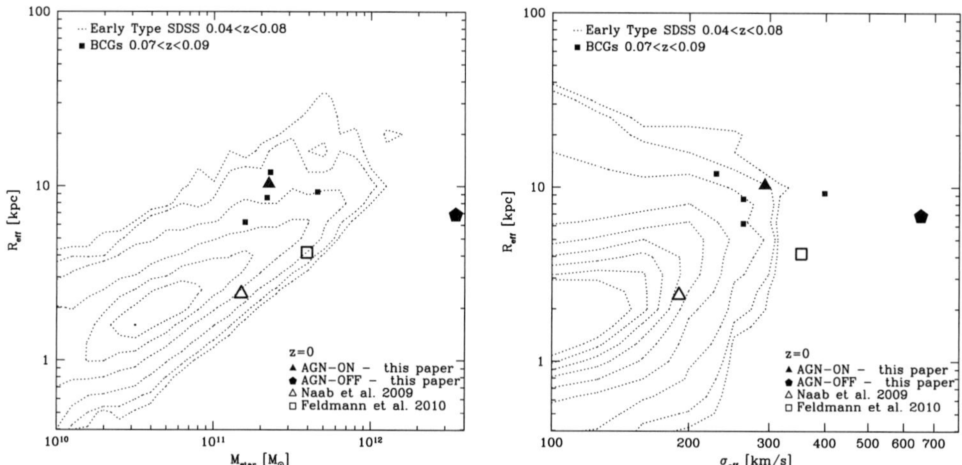

Figure 1. Mass-size (left) and velocity dispersion-size (right) relation of early-type galaxies at redshift $z = 0$ from SDSS data (van der Wel *et al.* 2008), compared to four early-type galaxies from different cosmological simulations. The black dotted lines are contours of the number of early-type galaxies per bin in the $0.04 < z < 0.08$ sample; going from outside-in we show contours for 5, 10, 30, 100, 200, 300, 400, 500, 600 galaxies per bin. Each bin has a size $\Delta \log(M_{\mathrm{star}}) = \Delta \log(R_{\mathrm{eff}}) = \Delta \log(\sigma_{\mathrm{eff}}) = 0.1$. The four BCGs analysed by Brough *et al.* (2011) are also shown as black filled squares. Plot taken from Martizzi *et al.* (2012).

properties of BCGs. Further details about our simulations can be found in Martizzi *et al.* (2012).

2. Results

We analyse the masses, effective radii and mean velocity dispersions within the effective radius of the BCGs in our models. A comparison is made with the simulations of Naab *et al.* (2009) and Feldmann *et al.* (2010) which reproduce some of the observed properties of field elliptical galaxies and central galaxies in groups without AGN feedback. We also compare our model to a sample of elliptical galaxies at $0.04 < z < 0.08$ (van der Wel *et al.* 2008) and to four BCGs at $0.07 < z < 0.09$ (Brough *et al.* 2011). Figure 1 shows the mass-size and velocity dispersion-size of early-type galaxies and BCGs at low redshift compared to the result of our two simulations and to the results of Naab *et al.* (2009) and Feldmann *et al.* (2010). Our results show that the inclusion of AGN feedback helps in bringing the stellar mass, velocity dispersion and size close to those of observed galaxies. In Figure 2 we show the stellar mass vs. halo mass relation predicted by abundance matching (Moster *et al.* 2010) and we compare it to our simulations. The AGN-OFF model produces a stellar mass that is ~ 10 times the expected value, whereas the AGN-ON model predicts a mass that well matches the abundance matching constraints.

We have also analysed the kinematic and structural properties of the simulated BCGs. When we do not include AGN feedback, the BCG is quickly rotating. When AGN feedback is considered, the BCG is slowly rotating. Figure 3 shows the stellar mass surface density profiles compared to Sérsic fits. The AGN-OFF BCG has a high Sérsic index and a mass excess with respect to the fit in the center, a product of a long sequence of wet mergers followed by intense star formation activity. The AGN-ON BCG has a significantly cored surface density profile (very shallow inner slope), a mass deficit with

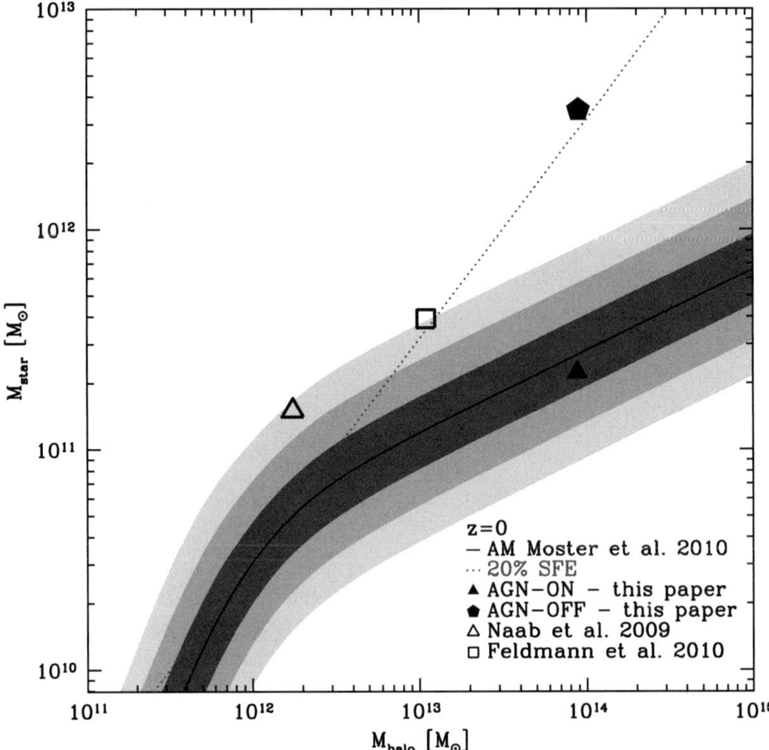

Figure 2. Comparison of the stellar-vs-halo mass relation in 4 early-type galaxies from different cosmological simulations (filled and empty black dots). The dotted line is the relation expected for a 20% star formation efficiency from the universal baryon fraction. The solid black line is the prediction from abundance matching (Moster *et al.* 2010). The grey shaded areas represent the 1σ, 2σ and 3σ scatter bars around the average relation. Plot taken from Martizzi *et al.* (2012).

respect to the Sérsic fit. In Martizzi *et al.* (2012b) we have shown that the formation of such a core is related to repeated, AGN driven impulsive gas motions which modify the gravitational potential in the central region of the cluster and generate irreversible modifications in the dark matter and stellar mass distribution. Further results concerning our BCG simulations and their interpretation are discussed in detail in Martizzi *et al.* (2012).

3. Conclusions and Future Work

The analysis of our two simulations and their comparison with observational results lead us to conclude that AGN feedback is an extremely relevant process which deeply influences the evolution of the most massive galaxies in clusters. This effect cannot be neglected when one is interested in modeling the BCGs. We are currently building a catalog of ~ 100 cosmological simulations of clusters of galaxies performed with RAMSES and including AGN feedback. The aim of this project is to study the properties of a large number of simulated clusters with different masses and histories, then compare the results to observations and realise constraints on theoretical models.

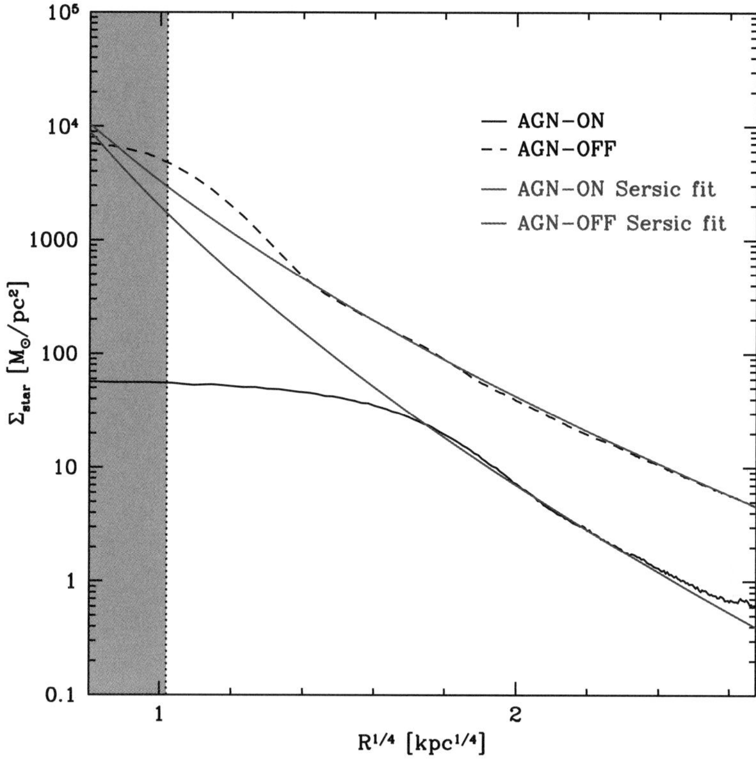

Figure 3. Stellar mass surface density profile of the BCG in our two models, compared with Sersic profiles (see text for details). The black solid lines are for the AGN-ON model and the black dashed lines are for the AGN-OFF model. In both panels the grey shaded area shows the unresolved region of our simulations. Plot taken from Martizzi *et al.* (2012).

References

Teyssier, R. 2002, *A&A*, 385, 337

Booth, C. M. & Schaye, J. 2009, *MNRAS*, 398, 53

Naab, T., Johansson, P. H., & Ostriker, J. P. 2009, *ApJ*, 699, L178

Feldmann, R., Carollo, C. M., Mayer, L., *et al.* 2010, *ApJ*, 709, 218

Martizzi, D., Teyssier, R., & Moore, B. 2012, *MNRAS*, 420, 2859

Moster, B. P., Somerville, R. S., Maulbetsch, C., *et al.* 2010, *ApJ*, 710, 903

Brough, S., Tran, K.-V., Sharp, R. G., von der Linden, A., & Couch, W. J. 2011, *MNRAS*, 414, L80

van der Wel, A., Holden, B. P., Zirm, A. W., *et al.* 2008, *ApJ*, 688, 48

Martizzi, D., Teyssier, R., Moore, B., & Wentz, T. 2012, *MNRAS*, 422, 3081

The intriguing life of massive galaxies
Proceedings IAU Symposium No. 295, 2012
D. Thomas, A. Pasquali & I. Ferreras, eds.

© International Astronomical Union 2013
doi:10.1017/S1743921313005383

Structure and dynamics of massive galaxies at z=0 in a fully cosmological simulation

E. Ricciardelli[1]**, J. Navarro-González**[1]**, V. Quilis**[1]**, and A. Vazdekis**[2,3]

[1] Departament d'Astronomia i Astrofisica, Universitat de Valencia, c/ Dr. Moliner 50,
E-46100 - Burjassot, Valencia, Spain, email: elena.ricciardelli@uv.es

[2] Instituto de Astrofísica de Canarias, Vía Lactea s/n, E-38200 La Laguna, Tenerife, Spain

[3] Departamento de Astrofísica, Universidad de La Laguna, E-38205, Tenerife, Spain

In this contribution we present the results of an Eulerian adaptive mesh refinement (AMR) hydrodynamical and N-body simulation in a ΛCDM cosmology. The simulation used was performed with the cosmological code MASCLET (Quilis *et al.* 2004). Galaxies have been identified in the simulation outputs by means of an adaptive friends of friends algorithm applied to the star particles. To give light to our virtual galaxies we have assigned a spectrum to each stellar particle using the MIUSCAT stellar population models (Vazdekis *et al.* 2012; Ricciardelli *et al.* 2012).

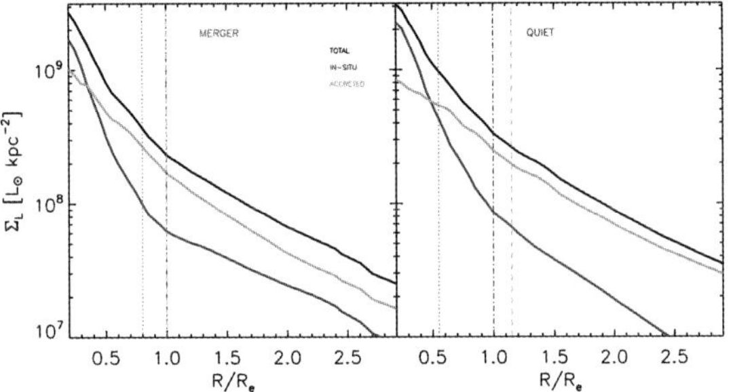

Figure 1. Median luminosity density profiles for three merger (*left-hand panel*) and three quiet (*right-hand panel*) galaxies. The contribution from stars formed in-situ and ex-situ are shown by the orange and purple lines, respectively. Vertical lines indicate $1R_e$ of the components: total (black, dot-dashed), in-situ (orange, dashed) and ex-situ (purple, dotted).

We focus our analysis on the most massive galaxies ($M_{star} > 10^{11} M_\odot$) in the simulation at redshift $z = 0$ and discuss their properties according to their morphological types, merging histories and dynamical properties (Navarro-González et al., in prep.). The most important factor in shaping the present-day structure of our simulated galaxies turns out to be the merging history. Indeed, galaxies having undergone an important merger event exhibit kinematical and metallicity gradients significantly different from those galaxies having experienced a more quiet life. We also study the accretion history of the galaxies in our sample, by classifying the stellar particles according to whether they formed in the main progenitor (in-situ) or formed outside it and were accreted later-on (by mergers or smooth accretion). As shown in Figure 1, the two populations of stars exhibit very different luminosity profiles. In the very central regions ($r < 0.5R_e$), the light distribution is dominated by the in-situ stars, whereas the outskirts are shaped by the accretion of stars formed outside the main body of the galaxy.

References

Quilis, V. 2004, *MNRAS*, 352, 1426
Ricciardelli, E., Vazdekis, A., Cenarro, A. J., & Falcón-Barroso, J. 2012, *MNRAS*, 424, 172
Vazdekis, A., Ricciardelli, E., Cenarro, A. J., *et al.* 2012, *MNRAS*, 424, 157

Future prospects and final discussion

The intriguing life of massive galaxies
Proceedings IAU Symposium No. 295, 2012
D. Thomas, A. Pasquali & I. Ferreras, eds.

© International Astronomical Union 2013
doi:10.1017/S1743921313005395

Future prospects in observational galaxy evolution: towards increased resolution.

Karl Glazebrook[1]

[1]Swinburne University of Technology, PO Box 218, Hawthorn, Vic 3122, Australia
email: `kglazebrook@swin.edu.au`

Abstract. Future prospects in observational galaxy evolution are reviewed from a personal perspective. New insights will especially come from high-redshift integral field kinematic data and similar low-redshift observations in very large and definitive surveys. We will start to systematically probe the mass structures of galaxies and their haloes via lensing from new imaging surveys and upcoming near-IR spectroscopic surveys will finally obtain large numbers of rest frame optical spectra at high-redshift routinely. ALMA will be an important new ingredient, spatially resolving the molecular gas fuelling the high star-formation rates seen in the early Universe.

Keywords. galaxies: evolution, galaxies: formation, galaxies: high-redshift, telescopes, instrumentation: miscellaneous

---

1. Introduction

I would like to thank the organisers for their kind invitation to review the future observational prospects in galaxy evolution, and in particular for massive galaxies, the theme of this Symposium. I am going to attempt to look forward about five years, this seems a sensible time frame on which to make predictions of what will be the most highly impactful observations.

If we review *the last five years* for comparison, it is quite startling to see the unexpected discoveries and developments that came about. Here are the ones that stick most in my mind (and references are intended to be illustrative not complete!):

(*a*) The dramatic size evolution found in elliptical galaxies — up to a factor of five since $z \sim 2$ (van Dokkum *et al.* 2008, Cimatti *et al.* 2008, Damjanov *et al.*2009).

(*b*) The existence of an evolving star-formation rate–stellar mass 'main sequence' for star-forming galaxies (Noeske *et al.* 2006).

(*c*) That most stellar mass growth in massive galaxies occurs via *in situ* star-formation and not via mass delivery in mergers (Conselice *et al.* 2012).

(*d*) That massive star-forming galaxies at $z \sim 2$ show a large fraction of rotating disks (Genzel *et al.* 2006).

(*e*) That the clumpy morphologies of high-redshift galaxies are likely due to giant star-formation complexes driven by the Jean's scale in turbulent high-velocity dispersion disks (Bournaud *et al.* 2009).

(*f*) That the universality of the Initial Mass Function (IMF) is now back in question (van Dokkum 2010, Hoversten & Glazebrook 2008).

(*g*) That the various physical properties of galaxies on the 'red sequence' or 'blue cloud' seem to be set solely by their stellar mass and to be independent of environment (e.g. Balogh *et al.* 2004, Baldry *et al.* 2006, Moucine, Baldry & Bamford 2007, Mocz *et al.* 2012, Peng *et al.*2010, Thomas *et al.* 2010), i.e. the only effect of environment seems to be in setting the numbers of red vs blue objects, perhaps via a threshold effect.

Given the recent history of unexpected developments in galaxy evolution this seems to make predicting the next five years fairly perilous! One thing that makes it slightly easier is that no major new telescopes will be commissioned during the period, indeed the new generation of Extremely Large Telescopes (ELTs) won't arrive until at least 2018. Other new large facilities such as the Large Synoptic Survey Telescope and the Square Kilometre Array are destined for the 2020's.

In this look forward I am going to focus on three major areas that I have picked on due to upcoming new capabilities: (i) galaxy structures and kinematics, (iii) high-redshift imaging and spectroscopic surveys and (iii) the imminent revolution in sub-mm astronomy from the Atacama Large Millmetre Array (ALMA).

2. Galaxy Structures and Kinematics

Integral Field Spectroscopy (IFS) has revolutionised the study of the kinematics of high-redshift star-forming galaxies and we now have about 100–200 high-quality observations of galaxies at $z \gtrsim 1$ from various surveys and nicely reviewed in S. Wuyt's talk at this Symposium. At these redshifts we see a picture where galaxy kinematic classes appear three-way split into (i) rotating objects with clearly disk-like velocity fields (ii) objects with kinematic structures but no uniform disk-like pattern (sometimes said to be 'mergers') and (iii) objects with no kinematic structure (sometimes referred to as 'dispersion dominated', Law $et\ al.$ 2007). The split here is around 20–40% in each class but this is sensitive to the particular survey and selection function and the fraction of disks seems to increase towards higher stellar masses (Förster-Schreiber $et\ al.$ 2009). Objects with disk kinematics seem to follow a Tully-Fisher relation in that they have the tightest scatter around a luminosity (or stellar mass) vs circular velocity line with a similar slope to, but a small offset from, the local Tully-Fisher relation (Puech $et\ al.$ 2008, Cresci $et\ al.$ 2008). A particular development at this symposium is a nice Tully-Fisher relation at $z \sim 1.2$ form the MASSIV survey (P. Amran talk), a redshift in which there was previously somewhat of a gap.

One key upcoming development is the advent of the KMOS IFS (Sharples $et\ al.$ 2004) which is to be commissioned on the Very Large Telescope at the end of 2012. This offers the first near-IR multiplexed IFS on a large telescope and IFS observations of up to 24 galaxies can be performed simultaneously. This will enable two important advances: first, and obviously, much larger high-redshift IFS kinematic samples will be obtainable allowing statistical trends to be studied. Secondly the large multiplex means it will be efficient to study much fainter galaxies with longer exposure times. Current IFS surveys are restricted to studying the more luminous (in emission lines) objects, typically around $\sim L^*$ in Hα at $z \sim 2$, thus being able to tackle even small numbers of sub-L* objects will allow selection biases to be studied.

One open question, in my mind, to be tackled by future surveys is the evolution of the galaxy merger rate. IFS surveys typically identify 20–30% of galaxies as mergers via kinematics at $0.5 < z < 2$ (Yang $et\ al.$ 2008, Lopez-Sanjuan $et\ al.$ 2012, Förster-Schreiber $et\ al.$ 2009) which is in stark contrast to the local value of $\sim 4\%$. Are the merger rates identified via kinematics consistent with those measured by close-pair counts (e..g Y. Peng, this Symposium)? Can we even objectively identify mergers in kinematic maps? Pioneering work in this latter topic was done using kinemetry by Shapiro $et\ al.$ (2008) but needs to be further developed, especially with respect to local calibration samples. In this Symposium P. Amran showed a new and different approach to quantitively identifying mergers. This is an excellent area for the future development of parametric and

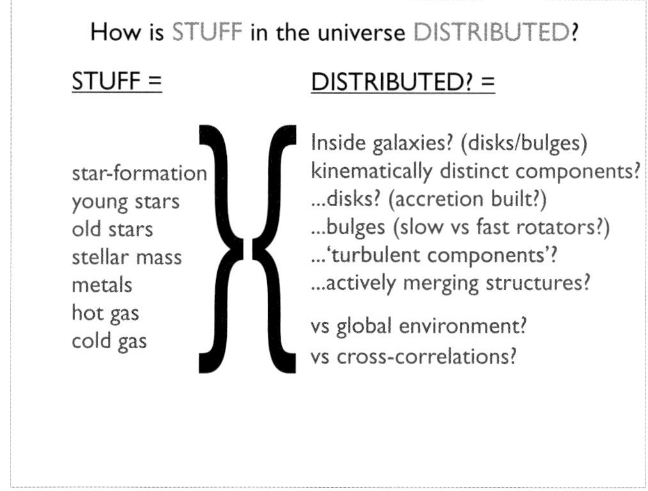

Figure 1. The meta-questions of IFS surveys. What is the mapping between the left and the right columns?

non-parametric statistics. A related question is can we go to the next step and measure mass ratios and merger timescales from IFS maps?

I believe the other key development will be the carrying out of large-scale local IFS surveys, a 'kinematic SDSS'. Current local IFS samples are of order several hundred galaxies, diversely selected and with heterogeneous data. This is analogous to the situation for imaging and 1D spectroscopic surveys before the 2dF Galaxy Redshift Survey and the Sloan Digital Sky Survey (SDSS). The next five years will see surveys of several thousand, perhaps tens of thousands of local galaxies done with multiplexed IFS instruments. Projects actively building instruments and planning observational campaigns in the near term are the SAMI consortium (Croom *et al.* 2012), who will use the Anglo-Australian Telescope, and the MaNGA team (P.I. Kevin Bundy) planning to use the SDSS telescope. These instruments typically deploy ~ 20 integral field units in a 2–3° field-of-view. This will allow the statistical study of the distribution of resolved kinematic structures in the local Universe and other meta-questions (Figure 1). In particular we will move away from scaling relations such as Tully-Fisher to the study of true kinematic distribution functions where space-density plays a key role in comparing with theoretical models. These surveys will also provide a cornerstone for quantitative comparison with high-redshift surveys, for example by providing a high-quality merger sample where mergers are identified by kinematics and photometry (e.g. tidal tails and other low surface brightness features that may not be visible at high-redshift). They can also be used to find rare local analogues of high-redshift galaxies: because they are nearby they can then be followed up in exquisite detail to see what makes the tick astrophysically. One example of this is the work of Green *et al.* (2010) where we identified candidate local turbulent disks with high star-formation rates. We are currently engaged with HST, Gemini IFS and other facilities to prove if they are indeed analogues and how the star-formation is driven.

We have also seen some nice work presented in this symposium on the kinematics and structures of red galaxies from high to low redshift. The so-called 'two-phase model' for the assembly of red galaxies (Forbes *et al.* 2011, Figure 2) is becoming popular where red

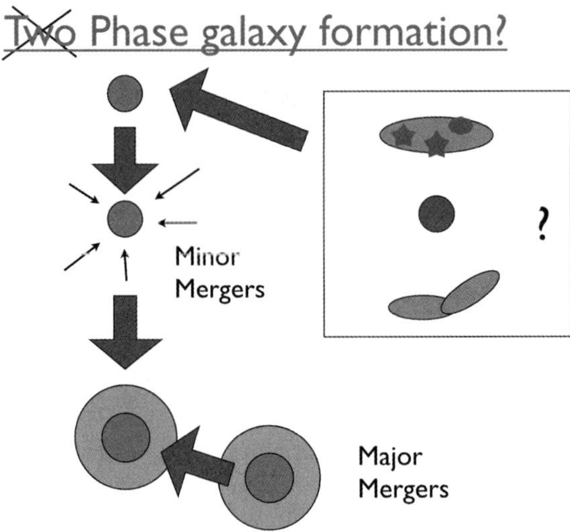

Figure 2. The 'Two-Phase Model' of galaxy formation? A red nugget at $z \sim 2$ grows a stellar halo and a considerable size increase via minor mergers. In some cases it may undergo major mergers to build a massive red galaxy. But what is Phase Zero? How does the red nugget get there in the first place from some blue predecessor? Possible mechanisms include fading of a clumpy disk, of a blue nugget or from a disk merger are illustrated. All would predict different spatial and kinematic morphologies for the red nugget.

galaxies start out as compact and very dense primordial 'red nuggets'† and then accrete a stellar halo via minor mergers as the core loses density. This allows a considerable amount of evolution of effective size per unit stellar mass increase and seems to be the emerging consensus explanation of size evolution in red galaxies. This does beg the question as to how the initial red nugget forms, is it via dissipative monolithic collapse and rapid starburst of a primordial gas cloud? Or the quenching or merging of high-redshift disks? Is this consistent with the axial ratios and Sersic indices being found at high redshift? (e.g. Damjanov, this symposium, Chevance *et al.* 2012.) We now have a limited number of velocity dispersion measurements, from absorption lines, of the most massive high-redshift ellipticals which seem to supper the minor-merger hypothesis (e.g. I. Trujillo's review in these proceedings). What we do not yet have is *resolved* kinematic measurements, for example are the red nuggets very rapidly rotating disks? Absorption line measurements are very difficult but future deep IFS observations such as those of KMOS can address this question. So will deep imaging using multi-conjugate adaptive optics (AO) which will deliver resolution 2–3× that of HST (McGregor *et al.* 2004).

At low redshift it remains to be seen if the two-phase model can reproduce the distribution of elliptical galaxies between slow and fast rotators which has now been measured in the field and in very dense environments (see talk by R. Davies). Does the real cosmological merger history deliver the right final angular momentum distribution? This is a challenge for theory as well as observers (e.g. Burkert *et al.* 2008). Surveys such as MaNGA and SAMI will deliver much better statistics but hydrodynamic simulations of massive galaxies embedded in large cosmological volumes remains supercomputer-intensive.

One final question that is perhaps unlikely to be answered in the next five years is the nature of the dispersion dominated compact *star-forming* galaxies that seem to constitute almost a third of the population. These are lower mass ($< 5 \times 10^{10} M_\odot$) so may not be related to the red nuggets even though they are a similar size ($\sim$1–2 kpc). Are they

---

† Confession: my own invented phrase, now seems increasingly apt!

purely dispersion dominated or do these conceal very compact disks that are unresolved even with AO IFS? This may require AO on ELTs to resolve, though spectroastrometry (Gnerucci *et al.* 2011) may allow information to be gleaned in the nearer term.

3. High-Redshift Imaging and Spectroscopic Surveys

In the last five years it has become routine for deep optical imaging surveys ($AB \sim 26$–27) to cover tens to hundreds of square degrees. At these depths galaxies are surveyed to $z \sim 6$. In the next five years even more gigapixels on sky will allow surveys such as the Dark Energy Survey (Flaugher 2005) and the Hype Suprime-Cam survey (Takada 2010) to cover thousands of square degrees at these depths. VISTA will similarly allow deep and wide near-IR surveys (McPherson *et al.* 2004). As outlined by D. Capozzi in these proceedings these imaging surveys will contribute to galaxy evolution studies via accurate measurements of photometric redshifts, luminosity functions, galaxy clustering, etc.

However at the risk of some controversy I predict that the most important applications to galaxy evolution from the new imaging surveys will come from the use of galaxy lensing enabled by such large areas. Weak lensing will enable the direct statistical measure of dark matter in galaxy and cluster haloes — some very nice work along these lines using the CFHT Legacy Survey was presented by M. Hudson in this symposium showing a good correlation between dark halo mass and stellar mass fraction in red and blue galaxies very suggestive of possible physical mechanisms. Strong lensing is also very powerful especially when combined with kinematic data (e.g. T. Treu talk in this symposium) as it allows mass structures and the IMF to be measured in the lensing galaxy. It is also very good for studying the lensed galaxy due to the large magnification of the light, making it brighter but also allowing smaller spatial scales to be resolved if the lens model can be inverted. The prospects of wider imaging surveys contributes to both weak lensing, via better statistics, and to strong lensing allowing more of these rare phenomena to be found.

In spectroscopy the instrument that I am personally most excited about is MOSFIRE, the near-IR multislit spectrograph commissioned on Keck in mid-2012. This cryogenic instrument operates from 0.9–2.4μm and allows slit spectroscopy of up to 46 targets simultaneously (McLean *et al.* 2011). In my view it offers the first combination of three key features required to make near-IR spectroscopy succeed for faint high-redshift targets: (i) sufficient spectral resolution ($R = 3300$) to well-resolve the airglow OH background out and 'get between the lines'. (ii) low scattered light and thermal background meaning it is truly dark between the sky lines; the measured interline background of MOSFIRE is very dark and comparable to the measurements of Maihaira *et al.*(1993). (iii) low readout noise and (iv) high instrument throughput 30–40%. Other similar instruments exist (such as F2 on Gemini) but do not offer the same spectral resolution for the one arcsec slit sizes required and have yet to be demonstrated on sky. The performance of MOSFIRE is shown by the detection of Hα in normal Lyman Break Galaxies at $z \sim 2$ in exposure times as short as 30 minutes! †

The key science area which will be tackled by MOSFIRE is the routine continuum spectroscopy of normal galaxies at high-redshift in large numbers in the rest-frame optical for detailed comparison with low redshift surveys such as SDSS. These spectra will measure spectroscopic redshifts, stellar populations, metallicities and velocity dispersions for homogenous samples. Without an instrument such as MOSFIRE this has been very

† See 'first light presentation' on http://irlab.astro.ucla.edu/mosfire/

difficult and most work in the last decade has relied on photometric redshifts. Even the very simplest product — redshift — should not be ignored as it allows clusters, environments and larger scale structures to be defined at high-redshift. These are the context of high-redshift galaxy evolution and current spectroscopic samples are highly biassed towards subsets of the population such as Lyman Break Galaxies. Photometric redshifts do not have the accuracy to measure such 3D environments though the most accurate ones, with medium band filters, do start to identify large scale structures and clusters (Spitler *et al.* 2012, Labbé talk this symposium) but require spectroscopy to confirm. The prospects for MOSFIRE surveys are excellent with high-quality very deep high-quality near-IR imaging data for selection already available from HST (the CANDELS survey, Grogin *et al.* 2011) and from the ground with medium bands. Because of this nexus we will now see a renaissance in high-redshift spectroscopy. It is interesting to note that this capability was in fact a key original science goal of 8m class telescopes and in the next five years we will finally see it delivered.

Towards the end of the five year forecast we may see the Subaru Prime Focus Spectrograph arrive (Ellis *et al.* 2012) offering a 50-fold increase in optical near-IR multiplex and field-of-view over current systems (though being non-cryogenic will operate at wavelengths $< 1.5\,\mu\mathrm{m}$). This will open the exciting prospect of using galaxies at $z >> 1$ for *cosmology* as well as galaxy evolution.

4. The Age of ALMA

As I write one very significant new telescope is being commissioned: ALMA (Hills & Beasley 2008). Virtually no ALMA results were presented at this symposium as very few people actually have any ALMA data.† So far no more than about 1000 hours of ALMA science time has been available to the community. However if we have a conference such as this in five years time I fully expect ALMA results to dominate the conference.

Why do I say this? Today high-redshift is dominated by optical and near-IR observations which are mainly sensitive to stars and hot ionised gas (e.g. from star-formation or AGN). However we need to consider the fuel as well as the fire. We know from current sub-mm observations that the molecular gas fractions of massive galaxies rises from a mere 5–10% at $z = 0$ to $\sim 50\%$ at $z \sim 2$ (Daddi *et al.* 2010, Tacconi *et al.* 2010). This probably accounts for the high prevalence of unstable, clumpy, turbulent disks (e.g. Genzel *et al.* 2008) and necessitates high inflow rates of cosmic material to sustain them (Dekel *et al.* 2009).

However current sub-mm telescopes barely resolve high-redshift galaxies with 0.5–1 arcsec beams and require many hours of integration per target. ALMA will improve this by factors of ten and enable kpc-resolution morphology and kinematics of molecular gas and dust in normal star-forming galaxies to be routinely made. We predict the clumpy disks to be gas rich and thick. Will we see thick cold molecular gas disks co-rotating and aligned with the young stars seen by the near-IR IFS observations? Will we see *supergiant molecular clouds* associated with the giant star-forming regions see in the UV? I predict we will!

A particularly important question for ALMA's spatial resolution is the nature of the star-formation law relating gas density to star-formation rate, a critical theoretical ingredient of galaxy formation simulations (the 'sub-grid physics'). Around 80% of the stars in the Universe formed at $z > 1$ but we have seen throughout this conference that galaxies in the the high-redshift Universe are very different to today. Will the star-formation law

† A show of hands at the symposium revealed at most 2–3 hands up in the audience.

be the same or quite different? The classical Kennicutt-Schmidt law (Kennicutt 1998) simply relates surface densities of gas and star-formation via a power law. Even locally there are many variations on this theme (a topic extensively discussed in Symposium 292 the previous week), for example there may be 'thresholds' or a volumetric relation may be more appropriate (Krumholz, McKee & Tomlinson 2009). At high-redshift Daddi *et al.* (2010) suggested there are in fact two relations — a 'sequence of starbursts' and a 'sequence of disks' but which may be unified by introducing a dynamical time in to the formulation. ALMA will bring a highly superior set of data to bear on this problem and I will predict some surprises!

Finally one interesting prediction that could perhaps be tested by ALMA is the existence of *dark* turbulent disks (Elmgereen & Burkert 2010). The prediction is that turbulence in gas disks starts initially in an accretion driven phase lasting for ~ 180 Myr before star-formation turns on. The gas would be cold and molecular — the visibility of such objects to ALMA has not yet been calculated, but would make for an interesting paper.

5. Final Words

Some firm predictions for the next five years:

(a) We will see a move back to real spectroscopic surveys at $2 < z < 5$.

(b) A 'Golden Age' of Integral Field Spectroscopy of large samples including definitive local surveys.

(c) We will probe the 'fuel for the fire' with ALMA.

(d) We will *still* be arguing about stellar population synthesis model ingredients (if this conference is anything to go by!).

Finally it is amusing to note that at this conference we saw Carlos Frenk (doyen of semi-analytic modelers) saying that 'galaxy formation is complicated' and Simon Lilly (the archetypal observer) saying 'galaxy formation is simple'! This appears to be a reversal of the theory-observer dichotomy of ten years ago to my memory, however I will dare to suggest that they are both in fact wrong! I think in the next 5–10 years we will see basic physical questions of star-formation and quenching (i.e. the formation of the red sequence) ironed out through better spatially-resolved observations as described above and there will be less need for 'recipes' in both camps. I speculate these observations will reveal new simplicities but also more complexity then the over-simplified picture that has arisen from large surveys with integrated spectra.

References

Baldry, I. K., Balogh, M. L., Bower, R. G., Glazebrook, K., Nichol, R. C., Bamford, S. P., & Budavari, T. 2006, *MNRAS*, 373, 469

Balogh, M. L., Baldry, I. K., Nichol, R., Miller, C., Bower, R., & Glazebrook, K. 2004, *ApJ*, 615, L101-L104

Bournaud F., Elmegreen B. G. 2009, *ApJ*, 694, L158

Burkert, A., Naab, T., Johansson, P. H., & Jesseit, R. 2008, *ApJ*, 685, 897

Chevance M., Weijmans A.-M., Damjanov I., Abraham R. G., Simard L., van den Bergh S., Caris E., Glazebrook K. 2012, *ApJ*, 754, L24

Cimatti, A., Cassata, P., Pozzetti, L., *et al.* 2008, *A&A*, 482, 21

Cresci, G., *et al.* 2009, *ApJ*, 697, 115

Conselice, C. J., *et al.* 2013, *MNRAS*, 430, 1051

Croom, S. M., Lawrence, J. S., Bland-Hawthorn, J., *et al.* 2012, *MNRAS*, 421, 872

Daddi, E., Elbaz, D., Walter, F., *et al.* 2010, *ApJ*, 714, L118

Damjanov, I., *et al.* 2009, *ApJ*, 695, 101

Dekel A., *et al.* 2009, *Nature*, 457, 451

Ellis, R., Takada, M., Aihara, H., *et al.* 2012, *Extragalactic Science and Cosmology with the Subaru Prime Focus Spectrograph (PFS)*, arXiv:1206.0737

Flaugher, B. 2005, *International Journal of Modern Physics A*, 20, 3121

Forbes, D. A., Spitler, L. R., Strader, J., *et al.* 2011, *MNRAS*, 413, 2943

Førster Schreiber N. M., *et al.* 2009, *ApJ*, 706, 1364

Genzel R., *et al.* 2006, *Nature*, 442, 786

Gnerucci *et al.* 2011, *A&A*, 533, 124

Green A. W., *et al.* 2010, *Nature*, 467, 684

Grogin, N. A., Kocevski, D. D., Faber, S. M., *et al.* 2011, *ApJS*, 197, 35

Hills, R. E. & Beasley, A. J. 2008, *SPIE*, 7012

Hoversten, E. A. & Glazebrook, K. 2008, *ApJ*, 675, 163

Kennicutt R. C., Jr. 1998, *ApJ*, 498, 541

Krumholz M. R., McKee C. F., Tumlinson J. 2009, *ApJ*, 699, 850

Law *et al.* 2007, *ApJ*, 669, 929

López-Sanjuan *et al.* 2012, *AJ* in press (2012) arXiv:1208.5020

Maihara *et al.* 1993, *PASP*, 105, 940

McGregor P., *et al.* 2004, *SPIE*, 5492, 1033

McLean I. S., *et al.* 2010, *SPIE*, 7735

McPherson, A. M., Born, A. J., Sutherland, W. J., & Emerson, J. P. 2004, *SPIE*, 5489, 638

Mocz, P., Green, A., Malacari, M., & Glazebrook, K. 2012, *MNRAS*, 425, 296

Mouhcine M., Baldry I. K., Bamford S. P. 2007, *MNRAS*, 382, 801

Noeske, K. G., *et al.* 2007, *ApJL*, 660, L43

Peng, Y.-J., Lilly, S. J., Kovač, K., *et al.* 2010, *ApJ*, 721, 193

Puech *et al.* 2008, *A&A*, 484, 173

Shapiro *et al.* 2008, *ApJ*, 682, 231

Sharples, R. M., Bender, R., Lehnert, M. D., *et al.* 2004, *SPIE*, 5492, 1179

Spitler L. R., *et al.* 2011, *ApJ*, 748, L21

Tacconi L. J., *et al.* 2010, *Nature*, 463, 781

Takada, M. 2010, *American Institute of Physics Conference Series*, 1279, 120

Thomas *et al.*(2010)]2010MNRAS.404.1775T Thomas, D., Maraston, C., Schawinski, K., Sarzi, M., & Silk, J. 2010, *MNRAS*, 404, 1775

van Dokkum P. G., *et al.* 2008, *ApJ*, 677, L5

van Dokkum, P. G. & Conroy, C. 2010, *Nature*, 468, 940

Yang *et al.* 2008, *A&A*, 477, 789

The intriguing life of massive galaxies
Proceedings IAU Symposium No. 295, 2012
D. Thomas, A. Pasquali & I. Ferreras, eds.

© International Astronomical Union 2013
doi:10.1017/S1743921313005401

The Intriguing Life of Massive Galaxies: Introducing the Final Discussion

Alvio Renzini[1]

[1] INAF – Osservatorio Astronomico di Padova
email: alvio.renzini@oapd.inaf.it

Abstract. This is a brief introduction to the closing discussion of the IAU Symposium 295, "The Intriguing Life of Massive Galaxies", that was held in Beijing from August 27 through 31, 2012. The discussion was focused on only four hot items, namely 1) the redshift evolution of the size of passively evolving galaxies, 2) the evolution with redshift of the specific star formation rate, 3) quenching of star formation in galaxies and dry merging, and 4) the IMF.

Keywords. Galaxy Evolution; Elliptical Galaxies; Star Formation; IMF

1. Introduction

I have been asked by the organizers to introduce the final discussion of this IAU Symposium 295 "The Intriguing Life of Massive Galaxies" thus trying to *provoke* a wide participation by the audience. I have chosen to focus the discussion on just a few among the many issues that have been addressed and debated at this exciting meeting. For each topic, after five minutes of introduction, some ten or more minutes of general discussion followed. All this was recorded but at this point it is not clear whether it will become accessible on line on the site of the symposium. Therefore, this short article is limited to a report of my introduction to each topic, in the wish of attracting readers towards the actual event, if the records will become available.

I wish to start by confessing that already quite some time ago I lost faith in the ability of theoretical models to tell us how real galaxies form and evolve, being them of the semianalytic or the hydro-simulation variety. The reason is that once the mess of baryonic physics is added to the clean elegance of dark matter N-body realizations most –if not all– predictive power is lost in a plethora of adjustable parameters meant to describe the many physical processes at work. Just to name a few: star formation, galactic winds, cold streams, supermassive black hole formation, nuclear activity and its feedback, chemical evolution, galaxy mergers, starbursts, disk instabilities, multiphase ISM, supernova feedback, ram pressure stripping, dust formation, radiative transfer and many more. For example, Benson & Bower (2010) list over two dozens adjustable parameter for their semi-analytic model. So, being an infidel, my view of theory would be biased and I preferred to drive the discussion towards some specific points that direct observations may help to clarify.

2. The Size Evolution of Passively Evolving Galaxies

Over a dozen talks have been dealing with the size of passively evolving galaxies (PEG) at the various redshifts. Sizes are straightforward to visualize and effective radii ($R_{\rm e}$) are relatively easy to measure, thus over one hundred papers have been dedicated in recent years to report on how high redshift PEGs are smaller and denser at given stellar mass compared to their local analogs, and what processes may account for their apparent growth by up to a factor of ~ 4 from $z \sim 2.5$ to $z \sim 0$. Indeed, at fixed stellar mass the average R_e of PEGs ($< R_e >$, as normalized to the local $R_e - M_*$ relation) steadily declines with increasing redshift, a trend that –with few exceptions (e.g., Valentinuzzi et al. 2010; Cassata et al. 2011; Newman et al. 2012)– has been often entirely ascribed to the physical growth of individual galaxies. Accretion of an envelope of small satellites (minor mergers) has been widely entertained as the leading mechanism to *puff up* the individual high-z compact PEGs until they reach their due, final dimension. Yet, this is only part of the story, and likely only a minor one.

The fact is that the comoving number density of massive PEGs is increasing by a large factor (~ 25) between $z = 2.5$ and 0, with most of this increase taking place between $z = 2.5$ and ~ 1. Hence, the run of the average size of PEGs primarily reflects the size with which at each redshifts they first appear as quenched, rather than the subsequent grow of each of them. Thus, the real challenge is to understand why galaxies which are quenched at later times are born bigger than those formed earlier, rather than (or not only) to understand to which extent individual PEGs secularly increase their size. Precursors to PEGs must be star forming galaxies likely spending most of their time on their *main sequence* growing inside-out. Hence, the later they are quenched the bigger they are, and presumably the bigger their passive remnant. Is just this the main story?

An intriguing aspect is that the number density of compact PEGs actually appears to *increase* with cosmic time, peaking at $z \sim 1$ (e.g., Cassata et al. 2011), right when $< R_e >$ is most rapidly increasing (!), and then starts to drop at lower redshifts. Thus, between $z \sim 2.5$ and ~ 1 the distribution of effective radii of newly formed PEGs is rapidly evolving, and does so in such a way that the birth rate of *normal size* PEGs is higher than a sill increasing birth rate of compact ones. It is only below $z \sim 1$ that the number density of compact PEGs starts to drop, and some growth of individual galaxies must take place. Thus, understanding the redshift evolution of $< R_e >$ requires to follow *simultaneously* the complex interplay between the growth of individual galaxies *and* the birth of new PEGs and their size distribution. A key issue that remains to be settled concerns the number density of compact PEGs in the local Universe: how does it compare with the number density at redshift 1 or 2?

3. The Specific Star Formation Rate

The specific star formation rate (sSFR) of *main sequence* galaxies as a function stellar mass and cosmic time, conveniently parameterized as $\mathrm{sSFR}(M_*, t) \propto M_*^{\beta} t^{-\gamma}$, plays a pivotal role in galaxy evolution. It controls the growth rate of galaxies, the evolution of their mass function, and can have a direct effect even on quenching of star formation itself. Despite its importance, the precise value of β is still uncertain, with values in the literature spanning a very wide range ($-0.4 \lesssim \beta \lesssim 0$), the result depending on how star forming galaxies are selected, and how SFR and M_* are measured.

There is clear evidence that the sSFR drops systematically with time since $z \sim 2$, by at least a factor of ~ 20, hence $\gamma \simeq 2.2 - 2.5$. This drop with time (increase with lookback time and redshift) runs almost parallel to the specific mass increase rate of merging dark matter halos, thus hinting for a direct link between star formation rate and accretion rate of the haloes. A simple interpretation of such quasi-parallelism is that the supply of fresh baryons (gas) feeding star formation in galaxies follows the dark matter halo accretion rate, as if baryons and dark matter were fairly bound to each other. However, beyond redshift ~ 2 the sSFR appears to flatten, then remaining nearly constant (i.e., $\gamma \simeq 0$) all the way to very high redshifts (e.g., Gonzalez et al. 2010). On the contrary, the specific increase rate of dark matter haloes keeps increasing.

This divergence from parallelism beyond $z \sim 2$ has caused some concern on the theoretical side (e.g., Weinmann, Neinstein & Dekel 2011), as it would apparently demand drastic modifications to widely adopted assumptions, as if baryons and dark matter were substantially decoupled at early cosmic times. Thus, measuring the actual value of γ for redshifts beyond ~ 2.5 has potentially strong implications for the relative behavior of baryons and dark matter. It may also be especially relevant for reionization, but may be relatively unimportant for the actual mass growth of galaxies, which most takes place at lower redshifts. In phenomenological models of galaxy evolution (e.g., Peng *et al.* 2010), compared to a $\gamma = 0$ case a $\gamma > 0$ beyond $z \sim 2.5$ would imply a different ratio of the final galaxy stellar mass to the mass of the initial seed, having otherwise rather irrelevant effect on the final outcome. But a precise measurement of the run of the sSFR with mass and time (i.e., of β and γ) remains a central issue for a proper understanding of galaxy formation and evolution. Contrary to previous results, it has been reported at this meeting that the sSFR may still be increasing somewhat even well beyond $z \sim 2$ (e.g., Stark *et al.* 2013). Thus, the issue remains unsettled and worth attracting further studies.

4. Quenching and Dry Merging

Quenching of star formation then turning to passive evolution is perhaps the most salient event in the life of a galaxy. Observations tell us that the fraction of quenched galaxies is a strong function of both stellar mass and local overdensity. Thus, two distinct physical processes must exists, dubbed *mass quenching* and *environment quenching*, that act independent of each other (Peng *et al.* 2010). What remains to be established is the physical nature of these two processes: what is environment quenching? and, what is mass quenching?

Is environment quenching just ram pressure stripping? or *strangulation*? or a combination thereof? Is mass quenching a process internal to galaxies themselves, such as e.g., AGN feedback? or is it an external process, also related to the environment? (e.g., galaxies are quenched when the mass of the host dark matter halo exceeds a threshold value). Does mass quenching work by ejecting gas *out of* galaxies or by preventing accretion of cold gas *into* galaxies? In other words, what is *mass* in mass quenching? Is it M_* or $M_{\rm h}$? Theoreticians appear to be equally divided, and probably only observations can answer the question.

Yet, quenching may not be the last event in the evolution of massive galaxies. Post-quenching (dry) merging can also take place, and in a popular cartoon such further mass increase may be as large as a factor ~ 10 (see Figure 1). This scenario was motivated by the perception that star-forming (SF) galaxies as massive as the most massive quenched galaxies would not exist, hence massive PEGs ($M_* \gtrsim 10^{11}\, M_\odot$) could not form by just quenching star formation in a *blue cloud* galaxy, then moved to the *red sequence*, i.e., there would be not enough massive blue galaxies in the distant Universe with masses comparable to those of the brightest red galaxies. Actually, massive enough SF galaxies do exist, but they are very rare. The reason why they are rare is precisely because they are growing in mass so rapidly that their life expectancy (as star forming) is very short: they soon are going to be *mass quenched*! Thus, the cartoon is missing the main point: most of the action, i.e., most of the *mass quenching*, takes place near the top end of the mass function of SF galaxies, for $M_* \gtrsim 10^{11}\, M_\odot$, within the open ellipse in Figure 1, a region left empty in the original plot.

Now, dry merging does certainly exist, but its role is not as prominent as sometimes envisaged. In high density regions the Schechter M_*^* of quenched galaxies is just ~ 0.1 dex higher than in low density regions, where merging is almost absent (Peng *et al.*

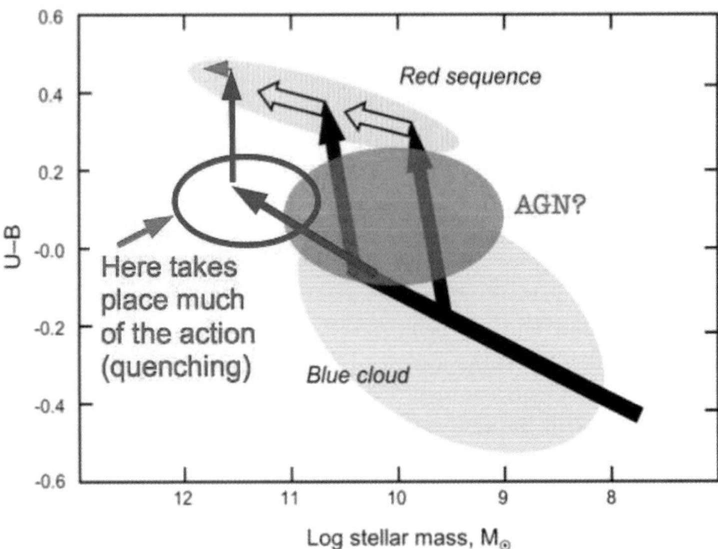

Figure 1. A modification of the original cartoon with an idealized sketch of the transition from star-forming "blue cloud" galaxies to quenched "red sequence" galaxies (black and grey scale, Faber *et al.* 2007). In color the actual path of galaxies being *mass quenched* is indicated (in blue), together with a modest further mass increase by dry merging (short red arrow).

2010). Thus, dry mergers make just an average $\sim 20\%$ mass increase, certainly not a factor ~ 10.

5. The IMF

Finally the IMF (usually described by a power law $dN \propto M^{-s} dM$). It is currently common practice to adopt a universal IMF for a broad range of astrophysical applications, such as e.g., to estimate stellar masses and star formation rates of galaxies from the local to the most distant Universe so far explored. Yet, it is perfectly legitimate to entertain the notion that the IMF may not be universal. Indeed, from time to time one appeals to different IMFs to ease perceived discrepancies between some theoretical models and observations (e.g., Davé 2008), or between the dynamical and stelllar population mass to light ratios of galaxies (e.g., Cappellari *et al.* 2012). Thus, sometimes one appeals to a top heavy IMF with a lot of massive stars boosting the luminosity and metal production rate, sometime to a bottom heavy IMF, with a lot of low mass stars, making just mass but little light and no metals. In particular, in a recent surge of papers observational evidence has been presented that would favor a very bottom heavy IMF in massive elliptical galaxies. This was based on the strength of the Na I $\lambda 8190$ Å doublet and the FeH Wing-Ford band at $\lambda 9900$ Å which both are strong in dwarfs and weak in giants (van Dokkum & Conroy, 2012; Ferreras *et al.* 2013; and references therein). These features appear to be stronger in such galaxies than in synthetic stellar population models which adopt a Milky Way (bottom light) IMF, as illustrated in Figure 2 (from van Dokkum & Conroy 2012).

Notice that the depth of the NaI feature is just a few percent of the (pseudo)continuum, and then a drastic variation of the IMF (with the low-mass slope changing from s = 1.35 to 4) leads to a variation of the central line depth from $\sim 96.2\%$ to just $\sim 95.2\%$. Formally, as shown in Figure 1, even steeper IMFs would be required for the ellipticals with the highest velocity dispersion. Clearly, the NaI feature is very insensitive to the

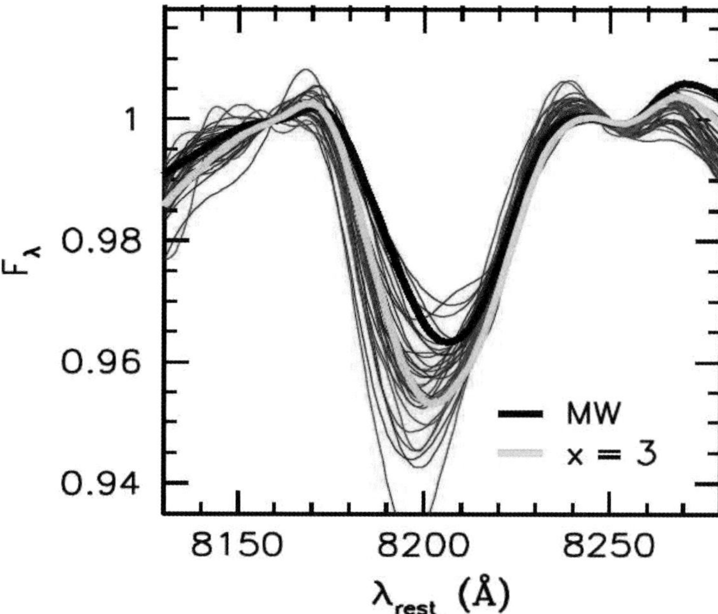

Figure 2. The strength of the NaI $\lambda8190$ Å doublet (here blended by velocity dispersion) in elliptical galaxies, color coded blue to red with increasing velocity dispersion (from van Dokkum & Conroy 2012). Two synthetic stellar population spectra are also shown for an age of 13 Gyr and Milky Way bottom light IMF ($s = 1.35$) and a bottom heavy IMF ($s = 4$ equivalent to a logarithmic slope $x = 3$ as $s = x + 1$).

IMF slope and, moreover, deriving such slope from the strength of this feature rests entirely on the reliability of the stellar population models used to draw synthetic spectra such as those shown in Figure 1. Thus, to trust the resulting IMF one has to trust the ability of synthetic models to reproduce the feature with exquisite accuracy. Therefore, a check of such models should be mandatory before taking into serious consideration systematic IMF variations as a function of the velocity dispersion of elliptical galaxies. This is especially true given that models are particularly uncertain (i.e., uncalibrated) for the super-solar metallicities typical of the most massive elliptical galaxies.

Any inference on the IMF from integrated light ultimately rests on synthetic stellar population models and therefore on their reliability. For example, Cappellari *et al.* (2012) have convincingly demonstrated that mass closely follows light in the core of elliptical galaxies, thus contributing to break the IMF-dark matter degeneracy. Having constrained as marginal the dynamical effect of dark matter, they find that in the core of elliptical galaxies the M/L ratio as inferred from dynamical modeling increases with M_* and central velocity dispersion σ more than current stellar population models predict. They then interpret this departure in terms of a systematic trend of the IMF with increasing galaxy mass and/or σ. The M/L ratio is certainly a more fundamental, bulk property of galaxies compared to a weak spectral feature. Yet, also in this case the inference on the IMF is correct only as long as stellar population models are correct. Worth noting is indeed that also metallicity increases with mass and/or σ and gets super solar at high mass and σ: right where both the heavier IMF is demanded, and where population models are less reliable. There is in fact another degeneracy yet to be broken, one between the IMF and stellar population models, whose use is unavoidable to infer anything about the IMF.

                    A. Renzini

References

Benson, A. J. & Bower, R. 2010, *MNRAS*, 405, 1573

Cappellari, M. *et al.* 2012, *Nature*, 484, 485

Cassata, P., Giavalisco, M., Guo, Y. *et al.* 2011, *ApJ*, 743, 96

Davé, R. 2008, *MNRAS*, 385, 147

Ferreras, I. *et al.* 2013, *MNRAS*, 429, 15

Gonzalez, V. *et al.* 2010, *ApJ*, 713, 115

Newman, A. B., Ellis, R. S., Bundy, K. *et al.* 2012, *ApJ*, 746, 162

Peng, Y. *et al.* 2010, *ApJ*, 721, 193

Stark, D. P. *et al.* 2013, *ApJ*, 763, 129

Valentinuzzi, T., Poggianti, B. M., Saglia, R. P. *et al.* 2010, *ApJ*, 721, L19

van Dokkum, P. & Conroy, C. 2012, *ApJ*, 760, 70

Weinmann, S. M., Neinstein, E., & Dekel, A. 2011, *MNRAS*, 417, 2737

Author Index